DEVELOPMENTS IN QUATERNARY SCIENCES 12
SERIES EDITOR: JAAP J.M. VAN DER MEER

ADVANCES IN QUATERNARY ENTOMOLOGY

Developments in Quaternary Sciences
Series editor: Jaap J.M. van der Meer

Volumes in this series

1. The Quaternary Period in the United States
   Edited by A.R Gillespie, S.C. Porter, B.F. Atwater
   0-444-51470-8 (hardbound); 0-444-51471-6 (paperback) – 2004

2. Quaternary Glaciations – Extent and Chronology
   Edited by J. Ehlers, P.L. Gibbard
   Part I: Europe ISBN 0-444-51462-7 (hardbound) – 2004
   Part II: North America ISBN 0-444-51592-5 (hardbound) – 2004
   Part III: South America, Asia, Australasia, Antarctica ISBN 0-444-51593-3 (hardbound) – 2004

3. Ice Age Southern Andes – A Chronicle of Paleoecological Events
   By C.J. Heusser
   0-444-51478-3 (hardbound) – 2003

4. Spitsbergen Push Moraines – Including a translation of K. Gripp: Glaciologische und geologische Ergebnisse der Hamburgischen Spitzbergen-Expedition 1927
   Edited by J.J.M. van der Meer
   0-444-51544-5 (hardbound) – 2004

5. Iceland – Modern Processes and Past Environments
   Edited by C. Caseldine, A. Russell, J. Hardardóttir, Ó. Knudsen
   0-444-50652-7 (hardbound) – 2005

6. Glaciotectonism
   By J.S. Aber, A. Ber
   978-0-444-52943-5 (hardbound) – 2007

7. The Climate of Past Interglacials
   Edited by F. Sirocko, M. Claussen, M.F. Sánchez Goñi, T. Litt
   978-0-444-52955-8 (hardbound) – 2007

8. Juneau Icefield Research Project (1949–1958) – A Retrospective
   By C.J. Heusser †
   978-0-444-52951-0 (hardbound) – 2007

9. Late Quaternary Climate Change and Human Adaptation in Arid China
   By David B. Madsen, Chen Fa-Hu, Gao Xing
   978-0-444-52962-6 (hardbound)-2007

10. Tropical and Sub-Tropical West Africa – Marine and Continental Changes During the Late Quaternary
    By P. Giresse
    978-0-444-52984-8 – 2008

11. The Late Cenozoic of Patagonia and Tierra del Fuego
    By J. Rabassa
    978-0-444-52954-1 – 2008

12. Advances in Quaternary Entomology
    By S.A. Elias
    978-0-444-53424-8 – 2010

For further information as well as other related products, please visit the Elsevier homepage (http://www.elsevier.com)

Developments in Quaternary Sciences, 12
Series Editor: Jaap J.M. van der Meer

ADVANCES IN QUATERNARY ENTOMOLOGY

by

S.A. Elias

Centre for Quaternary Research, Geography Department,
University of London, UK

ELSEVIER

Amsterdam • Boston • Heidelberg • London • New York • Oxford
Paris • San Diego • San Francisco • Singapore • Sydney • Tokyo

Elsevier
Radarweg 29, PO Box 211, 1000 AE Amsterdam, The Netherlands
Linacre House, Jordan Hill, Oxford OX2 8DP, UK

First edition 2010

British Library Cataloguing in Publication Data
A catalogue record for this book is available from the British Library

Library of Congress Cataloging-in-Publication Data
A catalog record for this book is available from the Library of Congress

ISBN: 978-0-444-53424-8 (this volume)
ISSN: 1571-0866 (Series)

For information on all Elsevier publications
visit our website at elsevierdirect.com

Printed and bound in Hungary

10 11 12 13 14   10 9 8 7 6 5 4 3 2 1

Contents

Acknowledgments

I owe a debt of gratitude to many people who helped with illustrations, editing, and encouragement during the last 2 years. I thank Russell Coope for writing a Foreword, and for his editorial comments. I thank Tirza van Daalen and Linda Versteeg-Buschman at Elsevier for their support in the development of this manuscript. Professor Jaap J.M. van der Meer edited the entire manuscript, providing many useful suggestions for its improvement. Dr. Nicholas Porch, Australian National University, and Dr. Maureen Marra, University of Waikato, New Zealand, reviewed the Australian and New Zealand sections of Chapter 13, respectively. Dr. Darren Gröcke, Geography Department, Durham University, kindly provided editorial comments on Chapter 15. Dr. Ian Barnes, School of Biological Sciences, Royal Holloway, University of London, kindly provided editorial comments on Chapter 16. My ancient DNA research with Dr. Barnes and my research on interglacial faunas from Alaska were supported by a grant from the Leverhulme Foundation, F/07 537/T.

Foreword

When Scott Elias sent me the text of this book my response was the traditional greeting of grandparents: ''My! but you *have* grown.'' This book shows how the subject has thrived in the last decade and the science of Quaternary entomology has now spread throughout almost the whole world. To the uninitiated, it should be pointed out that this science investigates insect remains from the latest geological period, the Quaternary, which spans the time from just over two million years ago right up to the present day. These remains are preserved still in their original chitin and are best thought of as subfossils. They represent the ancestors of the modern insect fauna.

It is interesting to note that Japanese and western scientists had been investigating subfossil insects for almost as long as one another, yet each were unaware of the others, activities: that is until Scott's first book was published. This second book bridges this gap still further and brings together a worldwide community of entomologists working under very different conditions, sometimes, where the current taxonomic knowledge is still in an embryonic state that makes it difficult to know whether a fossil species is actually extinct or merely awaiting discovery.

Several universals are now emerging from the systematic study of subfossil insects, universals that cannot be derived from studies of their present day representatives alone. Firstly, the long-term morphological constancy of many insect species, often over millions of generations, can now be shown to be a worldwide phenomenon. Furthermore their physiological stability, in so far as it determines their environmental preferences, can also be inferred with increasing confidence by the regular occurrences of fossil assemblages of species that make ecological sense in themselves and also agree with inferences drawn from other palaeo-environmental indicators, both biological and physical.

Since the Quaternary period encompasses the time when intense climatic oscillation of the glacial/interglacial cycles that become more and more numerous and complex as our knowledge increases, species stability may seem counterintuitive in the face of the general belief that environmental change stimulates evolutionary change in which species adapt in a Darwinian sense to the new conditions. The solution to this enigma may lie in the way in which insects have changed their geographical ranges on an enormous scale in response to past climatic changes. In many cases, the fossil record shows that range changes of species involving thousands of kilometers have taken place within the latest glacial/interglacial cycle alone. Scott's book draws together examples from around the world that show that this geographical dynamism is also a universal and not just confined to the small areas in the Northern Hemisphere that have been most intensively investigated. Vast range changes have taken place again and again as successive ice ages have come and gone with their global sequences of environmental changes in train. Without the subfossil record there would be no way of recognizing this biogeographical dynamism, still less of understanding its evolutionary significance.

The Quaternary fossil record shows that evolutionary stasis amongst many insect groups can better be understood as a *consequence* of environmental instability. Thus, if species respond to geographical changes in the locations of acceptable environments by tracking them across the continents then, when seen from the insects, viewpoint, environments stay effectively constant (they can have no notion that the geography has shifted). Their forced marches ensure for generation after generation the environment in which they lived remained effectively the same and the intricacies of their adaptations reenforced over millions of reproductive events. This book shows that species stability amongst insects is worldwide and that large-scale range shifts take place in the past wherever the geography allows thus reflecting global environmental instability. Significant environmental changes were not confined to the glaciated latitudes where such changes are most obvious, but were a worldwide phenomenon.

For those species of insect that have left subfossil evidence, examples of global extinctions during the last million years are extremely uncommon, perhaps *because of* their efficiency in solving the problems of climatic change by frequent adjustments in their geographical ranges as the new conditions dictated.

But what of the insects on oceanic islands? On remote islands, there can be no escape when environments change, simply because there is no elsewhere else to go. The recent discovery of Holocene fossil insects on oceanic islands presents a fascinating opportunity to study the difference of their response to environmental changes compared with that on the continental masses. Work is now beginning on the subfossil insect faunas of island archipelagos such as the Galapagos and Hawaii where Nick Porch has shown that the impact wrought by the human invasion of these vulnerable ecosystems has meant that many insect species have followed their well-known vertebrate contemporaries into oblivion. If more ancient deposits can be found on these islands, what evolutionary treasures may also be uncovered, albeit rather atypically fast compared with their continental cousins?

From a geological perspective, Quaternary insects are now revealed as sensitive indicators of past ecologies and climates, responding to rapid changes with extraordinary alacrity. After episodes of sudden and intense climatic shifts, many orders of insect respond more rapidly than any other animals or plants. Under these conditions, ecosystems can develop that have temporarily no analogue at the present day; a situation that might suggest to the unwary that the environment had similarly no present day equivalent. Genuine cases of geographically nonanalogue insect faunas do occur occasionally in the Quaternary fossil record and these require sophisticated ecological analysis.

Scott hints at other exciting avenues of research that are only now beginning to be explored. Notable amongst these are various biochemical techniques that may help in establishing genetic relationships between the fossils and their present day representatives. Isotopic analysis might also provide environmental clues with which to cross-check conclusions that have been based on morphological criteria. In the area of radiocarbon dating, the chemical dismantling of insect fossil cuticle is beginning to establish why dates obtained from beetle exoskeletons sometimes differ significantly from those obtained from plant remains out of the same deposit.

Quaternary entomological studies are proving to be essential to any student of insect evolution, biogeography, endemism, past climates, and environments. Far from being ''19th century'' science as the naïve would have us believe, this subject has direct relevance to our understanding and conservation of the largest group of organisms on this earth. Human activities that are now exploiting the resources of this planet on an enormous scale are imposing extraordinary stresses on our ecosystems. Their survival, and our own, may now depend on our understanding of the ways in which these have responded to similar stresses in the past.

G. Russell Coope, Foss, Scotland

Glossary of Terms

Aedeagus: male reproductive organ of insects.

Allerød pollen zone: an interval of climatic amelioration during the late glacial, from 12,000–11,000 yr BP, which favored the growth of birch, pine, and willow in Europe.

Altithermal: an interval of maximum warming in the Holocene, which is thought to have occurred from 9000-5000 yr BP.

Amino acid racemization: a method of dating fossils by measuring the extent to which the lefthand configuration of certain amino acids has racemized to the righthand configuration. This process is used as a measure of time since the organism died.

Anoxic sediment: sediment depleted of oxygen, in which decomposition is slowed or halted.

Anthropogenic: changes caused by humans.

Archeoentomology: the study of insect fossils from archeological sites.

Bedding plane: a planar bedding surface that visibly separates each successive layer of strata in a section.

Beringia: North Eastern Siberia, Alaska, the Yukon Territory west of the Mackenzie River, and the continental shelf region between Siberia and Alaska. This region was largely unglaciated during the late Pleistocene, and formed a high-latitude refugium for arctic biota.

Biogeography: the branch of biology that deals with the geographic distribution of plant and animals.

Biomass: the total quantity of living organisms of one or more species per unit of space at a given time, or of all the species in a community.

Bog: an undrained or poorly drained area with vegetation of sedges and mosses (especially Sphagnum), often accumulating in layers of peat.

Bølling pollen zone: an interval of lateglacial time, from 13,000–12,000 ^{14}C yr BP, during which there was climatic amelioration, favoring birch and park-tundra vegetation in Europe.

Bootstrapping: in statistics, the practice of estimating properties of an estimator (such as its variance) by measuring those properties when sampling from an approximating distribution. This is done by constructing a number of resamples of the observed dataset, each of which is obtained by random sampling with replacement from the original dataset.

Brachypterous: with short or abbreviated wings; unable to fly.

Bronze Age: in archeology, a cultural level (principally in Europe) from about 5500 to 3000 ^{14}C yr BP, characterized by the development of bronze technology.

Bucket sieve: a bucket without a bottom, in which a sieve screen has been fixed near the base, in order to screen large quantities of sediment.

Carina: an elevated ridge or keel.

Caste: in entomology, the various forms or kinds of adult individuals among social insects, such as workers, soldiers, and queens.

Cephalothorax: the united head and thorax of arachnids (spiders, mites, scorpions, etc.).

Clade: a group of biological taxa or species that share features inherited from a common ancestor.

Cladistic analysis: a taxonimic system based on quantitative analysis of comparative data and used to construct cladograms showing the phylogenetic relations and evolutionary history of groups of organisms.

Cluster analysis: a statistical method of grouping variables according to magnitudes and interrelationships between correlation coefficients.

Clypeus: the part of the head of an insect, anterior to the frons, to which a upper lip (labrum) is attached.

Cold-hardiness: the capacity of an organism to tolerate cold temperatures.

Coprolite: fossilized excrement.

Cordilleran Ice Sheet: an ice sheet that covered most of western Canada and the northwestern United States during the Wisconsin glaciation.

Cytochrome oxidase: an iron-containing enzyme found in the mitochrondria of animal and plant cells that is very important in cellular respiration.

Depigmentation: loss of pigment in the cuticle, often found in subterranean or cave-dwelling insects.

Dichotomous key: a key to the identification of a taxonomic group, in which the key characters are split into couplets.

DNA nucleotides: the four nucleotide molecules, adenine, thymine, guanine and cytosine, whose sequence forms the genetic code in strands of DNA.

Dorso-ventral compression: flattening of the body so that the dorsal and ventral surfaces are broadened and lateral regions are reduced.

Eemian interglaciation: in the European stratigraphic scheme, the last interglaciation.

Elytron: the anterior leathery or chitinous wings of beetles, which cover the hind wings and often the abdomen.

Endemic species: a species which is thought to have originated and remains only in a narrowly defined region.

Eukaryote: Single-celled or multicellular organisms whose cells contain a distinct membrane-bound nucleus.

Eutrophic: bodies of water which are rich in mineral nutrients and organic materials and therefore productive. Oxygen may be seasonally deficient.

Exoskeleton: the external skeleton of insects.

Exposure: in geology, a place where sediments or rock outcrops are exposed (either natural or man-made), such as a stream bank, a lake shore, a cliff, or an irrigation ditch.

Extirpation: the extinction of a species from a given region.

Extra-cellular freezing: freezing of interstitial liquids in the body of an organism that occurs outside the cells.

Fen: a tract of low, marshy ground containing alkaline peat, rich in mineral salts, situated in the upper parts of old estuaries and surrounding freshwater lakes.

Fluvial sediment: sediment laid down in running waters.

Forensic entomology: the study of insects as applied to the reconstruction of crimes, especially murder.

Frons: the frontal sclerite of the head capsule of insects, posterior to the clypeus.

Frontoclypeus: on the head capsules of caddisfly larvae, the fused frons and clypeus sclerites.

GenBank: a DNA sequence database that is allows open access via the internet; an annotated collection of all publicly available nucleotide sequences and their protein translations.

Genitalia: reproductive organs.

Genotype: the genetic constitution, or total sum of the genes, of an organism.

Geochronology: the reconstruction of the timing of events in geologic time, achieved through both absolute and relative dating methods.

Geomorphology: the study of the origin, nature and development of landforms.

Glaciation: the covering of large land masses by glaciers or ice sheets.

Gulf Stream: an oceanic current, originating in the Gulf of Mexico and transporting warm water to northwest Europe.

Halobionts: organisms that live in saline habitats.

Hemelytra: the anterior wing of bugs (Hemiptera and Homoptera); the basal portion is thickened, the apical portion is membranous.

Holocene: an epoch of the Quaternary Period, spanning the interval after the last glaciation (10,000 ^{14}C yr BP to recent).

Horizon: in stratigraphy, an interface indicative of a particular position in a sequence, often a distinctive, thin bed.

Instars: the succession of larval stages of insects.

Interglacial: a lengthy interval separating two glacial epochs in which climatic conditions were as warm as or warmer than modern.

Interstadial: a warm substage of a glaciation, marked by temporary retreat of ice margins.

Iron Age: in archeology, a cultural level (principally in Europe) beginning about 3000 yr BP, characterized by the development of iron technology.

Isodiametric mesh: meshes of equal diameter; on insect cuticle, these are often hexagonal in shape.

Isotope: an isotope is a variety of a chemical element that may contain different numbers of neutrons. All the isotopes of an element have the same chemical characteristics, but vary in their atomic mass.

Kettlehole: a steep-sided basin, commonly without surface drainage, in glacial drift deposits (especially outwash), formed by the melting of a large, detached block of stagnant ice.

Lacustrine deposit: sediments accumulated on the bottom of a lake.

Larval exuviae: the molted skins of larvae or nymphs at metamorphosis.

Lateglacial: the interval of time during the waning of the last glaciation, immediately preceding the Holocene.

Laurentide ice sheet: an ice sheet covering most of eastern and central Canada and the northeastern and north-central United States during the Wisconsin glaciation.

Lignite: a brownish-black coal, intermediate between peat and subbituminous coal.

Littoral zone: the zone in a lake or pond extending from the shore through the depth at which plants are rooted.

Macroclimate: the climate of an entire region.

Macropterous: possessing large wings.

Macula: a colored mark, larger than a spot.

Magdalenian: in European archeology, a cultural period in the upper Paleolithic, at the end of the last glaciation.

Mammalian megafauna: large mammal species, with adults weighing more than 44 kg.

Marine isotope stages: Chronologic stages in geologic time, based upon major shifts in oxygen isotope ratios in marine invertebrate shells (see Shakleton and Opdyke, 1973).

Marine transgression: the spread of seas over land areas; the spread of marine deposits over large regions previously above sea level.

Marker horizon: a layer forming a sharp boundary in a stratigraphic sequence.

Mesolithic: In archaeology, the middle division of the stone age, characterized by hunting and gathering cultures of the early Holocene.

Microclimate: the climate of a small habitat or locality.

Micropaleontological slide card: a card, 2.5 × 7.5 cm, with a rectangular cavity in which micropaleontological specimens are placed. The card is covered with a glass slide, held by an aluminum holder.

Microsculpture: in entomology, microscopic sculpture, including striations, punctures, and meshes, on the surface of sclerites.

Milankovitch cycles: long-term climatic changes, which, Milutin Milankovitch (1941) theorized are caused by changes in the Earth's orbit around the sun (including eccentricity, tilt of rotation, and longitude of perihelion).

Mitochrondrial DNA: DNA located in the mitochondrion, the organelle in a cell that produces chemical energy in the form of adenosine triphosphate (ATP) and regulates cellular metabolism. Although most of a cell's DNA is contained in the nucleus, the mitochondrion contains its own independent genome.

Multiple regression analyses: a statistical method in which regresses one variable on a series of other variable, taking various combinations of these to obtain a minimum of unexplained variance.

Neolithic: in archeology, the youngest division of the Stone Age, characterized by the development of agriculture and animal domestication.

Nicotinamide adenine dinucleotide (NADH): a small organic non-protein molecule that carries chemical groups between enzymes. NADH has several essential oxidation and reduction roles in cellular metabolism.

Oligocene: an epoch of the early Tertiary period, spanning the interval from about 33 million to 22 million years ago.

Orbital forcing: climatic change brought about by changes in the Earth's orbit about the Sun (see Milankovitch cycles).

Organic deposit: a sedimentary deposit rich in organic materials, such as peat or organic detritus concentrated in silts and sands.

Oxidized sediment: sediments deposited in an aerobic environment, fostering decomposition of organic matter.

Oxygen isotope ratio: the ratio of oxygen-18 to oxygen-16 isotopes in oxygen-bearing geologic materials (such as carbonate shells of marine organisms), used as a measure of past temperatures.

Packrat midden: well-preserved fragments of plants and animals accumulated locally by packrats (*Neotoma* spp.) and often encased in crystallized urine.

Palearctic region: a biogeographic realm, including Europe, North Africa, and northern and central Asia.

Paleoecology: the study of the relationships between ancient organisms and their environments.

Paleoentomology: the study of fossil insects and allied arthropod groups.

Paleolimnology: the study of conditions and processes in ancient lakes.

Paleosol: a buried soil horizon.

Palynology: the study of fossil and modern pollen.

PCR primer: a strand of nucleic acid that serves as the starting point for DNA replication in the polymerase chain reaction (PCR). Primers are short, chemically synthesized nucleic acid polymers with a length of about 20 bases. These are hybridized to target the DNA of a specific taxon, which is then copied by the polymerase in the chain reaction.

Periglacial environments: environments at the immediate margins of glaciers and ice sheets, greatly influenced by the cold temperature of the ice.

Permafrost: permanently frozen ground, found in arctic, subarctic, and alpine regions.

Phytophagous: plant-eating.

Piston core: a coring device which employs a piston inside a cylinder which reduces friction by creating suction.

Plant macrofossil: macroscopic remains of ancient plant fragments, including roots, stems, leaves, and fruits.

Pleistocene: an epoch of the Quaternary Period, spanning the interval from 2.6 million years ago to 10,000 years ago. The Pleistocene is characterized by a series of major glaciations.

Pliocene: an epoch of the late Tertiary period, spanning the interval from about five million to 2.6 million years ago.

Poikilotherms: an animal which lacks the capacity to control its body temperature.

Polar front: the southern boundary of the cold, polar water mass in the North Atlantic Ocean (see Ruddiman and McIntyre, 1981).

Polder: a tract of lowland reclaimed from a body of water.

Polypeptide: strings of short polymer molecules (peptides) made of linked amino acids. Peptides combine in long chains to form proteins.

Post-mortem: in Quaternary entomology, changes in the appearance of insect exoskeletons that occur after death.

Preboreal pollen zone: a term used primarily in Europe for the interval from 10,000–9000 ^{14}C yr BP when pollen evidence indicates that climate was somewhat cooler and wetter than the subsequent Boreal zone.

Principal components analysis: a multivariate statistical technique for calculating variance along principal axes of trends in data.

Pro-glacial lakes: lakes formed at the margins of ice sheet or glacial lobes.

Pronotum: the dorsal thoracic shield of beetles and some other insects.

Proxy data: in Quaternary studies, data from fossil organisms, sediments, ice cores, etc., used to reconstruct past environments; proxy data serves as a substitute for direct measurements of such phenomena as past temperatures, precipitation, and sea level.

Pseudoscorpion: an order of small arachnids, superficially resembling scorpions but lacking stingers.

Punctulae: in entomology, small punctures on the surface of insect exoskeletons.

Quaternary: the second period of the Cenozoic era, following the Tertiary and spanning the interval from about 2.6 million years ago to present.

Regression equations: statistical methods of estimating the relationship of one variable with another by expressing one variable in terms of a linear (or more complex) function of the other.

Restriction sites: in DNA molecules, these are specific sequences of nucleotides that are recognized by restriction enzymes. These enzymes exist in bacteria, and act as defense mechanisms to attack invading viruses. Inside the bacterium, restriction enzymes selectively cut up foreign DNA in a process called restriction.

Riparian: the land bordering a stream, lake, or tidewater.

Rugose: wrinkled.

Sangamon: in North America, the term used for the last interglacial.

Sclerites: in entomology, individual plates of an insect exoskeleton, separated from other plates by sutures.

Seasonality: the degree to which annual climatic variability is expressed in seasons.

Seta: in entomology, a hair-like appendage developed as an extension of the epidermal layer.

Setaceous puncture: a puncture on the surface of insect exoskeleton, containing a seta.

Solifluction: the slow, viscous, downslope movement of waterlogged soil in regions underlain by frozen ground.

Speciation: the processes in evolution by which new species are formed.

Stable isotope: a chemical isotope that is not radioactive (i.e., it has not been observed to decay).

Stadial: a substage of a glaciation, marked by glacial advances.

Steppe: an extensive region of dry grassland, most often used in reference to the grasslands of southwestern Asia and southeastern Europe.

Steppe-tundra: a mixture of steppe and tundra vegetation, developed in cold, dry conditions in Beringia during Pleistocene stadials.

Stratigraphy: the arrangement of strata as to geographic position and chronologic sequence.

Striation: in entomology, a fine, impressed line on the surface of the exoskeleton.

Subalpine zone: the uppermost forest zone in mountains.

Super-cooling: the cooling of a liquid to below that liquid's freezing point, without the formation of ice.

Synanthropic: organisms that live in close association with humans, and are more-or-less dependent on the environment created by human habitations.

Taphonomy: a branch of paleontology, dealing with the manner of burial of plant and animal remains.

Tephrochronology: the collection, preparation, petrographic description, and dating of tephra, or volcanic ash.

Tertiary: The first period of the Cenozoic era, spanning the interval 65 million to 2.7 million years ago.

Thermoluminescence dating: a method of dating loess and materials that have once been heated, by measuring the light emitted when the sample is heated in the laboratory. Some of the energy produced by radioactive decay is stored as trapped electrons, which are released as light when heated.

Thermophilous: warmth-loving; animals that require warm environments to complete their life cycle.

Trackway: cut tree branches and other woody vegetation laid down across moors in pre-historic times, in an attempt to create a raised surface for transportation.

Treeline: the upper limit of trees in mountainous regions, or the latitudinal limit of trees in the subarctic or subantarctic regions.

Trophic condition of water: the level of nutrients in a body of water, ranging from oligotrophic (nutrient-poor) to eutrophic (nutrient-rich).

Unconsolidated sediments: sediment with particles not cemented together or turned to stone.

Upper Paleolithic: the youngest division of the Paleolithic, or Old Stone Age, characterized by the appearance of modern humans, *Homo sapiens sapiens*.

Weichselian: the last glaciation in northern Europe.

Wisconsin: the last glaciation in North America.

Würmian Ice Sheet: the ice sheet that covered much of the Alps during the last glaciation.

Xeric-adapted: plants or animals adapted to dry conditions.

1

The History of Quaternary Insect Studies

> The great resemblance of the insects to those now living, in most cases amounting to identity, shows that it takes a long time to effect a change in the Coleoptera.
> – Fordyce Grinnell (1908)

1. Introduction

The opening chapter of the predecessor of this book, *Quaternary Insects and Their Environments* (Elias, 1994) started out by declaring that ''Quaternary entomology is achieving considerable success in unraveling the history of changing environments and biotic responses to waxing and waning continental ice sheets.'' That book attempted to summarize the first 40 years of the discipline. Since then, an additional 16 years of research has taken place. My 1994 statement, that ''the study of Quaternary insect fossils has expanded rapidly during the last five decades,'' remains true today, but the discipline has since taken off in several new directions and in several new regions of the world. This is a technical book, but I have tried to make it accessible to a wider audience by avoiding jargon wherever possible. The use of some technical terms is unavoidable, however, so when these terms are used, they appear in **boldface**. All boldfaced terms appear in the glossary, starting on page xi.

Unlike most other paleontological endeavors, **Quaternary** insect studies deal with the fossil remains of species which still exist. Thus, the process of identification of Quaternary insects primarily involves matching fossils to modern species. This, in turn, provides unusually accurate information on past environments, because the living populations of the insect species found in Quaternary assemblages provide the necessary information on environmental tolerances, behavior, and distribution pattern. This aspect of the work has provided a treasure trove of information on the shifting environments of the Quaternary.

In contrast, the development of Quaternary historical information, based on fossils, provides new insights into the modern fauna. For the first time, we have data to begin answering the important questions about species longevity, centers of origin, and the stability of insect communities. Previously, these questions have been broached by theoreticians, using only modern data. A body of fossil data is now in hand which is providing hard facts, some of which conflict sharply with previous theories. The principal aim of this book is to provide an up-to-date synthesis of the fossil insect evidence, so that neontologists, paleontologists, and others may understand its importance, and employ it in their own fields of study.

In 1994, I wrote that ''communication between paleontology and modern biology has been limited.'' Unfortunately, even in this age of wide electronic dissemination of journal articles and Web searching, my statement remains true. We live in an age of increasing specialization within scientific disciplines; few people are capable of finding and digesting all the necessary information in their own fields, much less the interdisciplinary aspects of topics outside their immediate interests. Therefore, this book carries forward the task of highlighting some of the important, if dimly perceived, connections between Quaternary and modern entomology.

I shall focus on the Quaternary period, with the exception of Chapters 9 and 10, in which information on some Late **Tertiary** insects receives attention. Most insect fossil studies to date have dealt with Late Quaternary faunal assemblages, because younger deposits are more plentiful, and most of the insect fossils they contain are in a better state of preservation than those found in deposits of Early Quaternary age. In this chapter, I start with an overview of the development of Quaternary entomology. Chapters 2–6 deal with methods, and the value of insects in **paleoecology**, paleoclimatology, and zoogeography. Chapter 7 discusses the usefulness of insect fossils in environmental reconstructions from archeological sites. Chapters 8–11 provide summaries of studies from Europe, Asia, and the Americas. Chapter 12 deals with the Japanese fossil record, and Chapter13 discusses the fossil records of Australia and New Zealand. Chapter 14 discusses the use of **stable isotopes** extracted from fossil beetle chitin as an independent method of paleoclimate reconstruction. This is a developing line of research that began in the 1980s, but it has received a new impetus with recent technological advancements that allow the rapid analysis of increasingly smaller samples. Chapter 15 explores another new scientific endeavor, the analysis of ancient DNA from insect fossils. Chapter 16 offers conclusions and a prospectus for the future of the discipline.

While the field of Quaternary entomology has dealt with insects from many orders, most studies have focused on the remains of the hard-bodied (highly sclerotized) groups, especially beetles (order Coleoptera). Beetles, therefore, take center stage in most of the faunal assemblages discussed in this book, although other insects, especially midges (principally the family Chironomidae in the order Diptera) are playing an increasingly important part in paleoenvironmental reconstructions, as explained in Chapter 3. Fossil chironomid analysis has

DEVELOPMENTS IN QUATERNARY SCIENCES
VOLUME 12 ISSN 1571-0866

expanded greatly in recent decades. According to *The Chironomid Home Page* (http://insects.ummz.lsa.umich.edu/~ethanbr/chiro/), there are now more than 500 researchers in this field (although not all of them study fossil chironomids).

A small but growing cadre of scientists is applying Quaternary insect studies to new regions and various types of **organic deposits**. Most of the researchers in Europe and North America were trained by Russell Coope at the University of Birmingham, or by his former students. In the last 20 years, however, new centers of research have sprung up independently, in Russia, Japan, Australia, and New Zealand (although researchers in the latter two countries have received training from European and North American scientists). The recent success of Quaternary entomology was preceded by almost a century of false starts and misconceptions. It is necessary to step back and examine this history, if only to gain a better appreciation of recent advances.

2. The Beginnings of Quaternary Entomology

The intensive, systematic studies summarized in this book were preceded by a series of intermittent, isolated studies in Europe and North America. These efforts, while well intentioned, were not very useful to the developing science of paleoecology, because for the most part they were based on a false premise. The early investigators assumed that all fossil specimens, even from Late Quaternary deposits, represented extinct species. As we will see, this idea was pervasive not only in the 19th century, but also well into the 20th. Armed with this notion of **Pleistocene** extinctions, a few early workers began describing Quaternary insect fossils, applying

Table 1. Beetle species described as extinct and subsequent extant identifications.

"Extinct" taxon	Locality	Reference(s)	Extant species	Reference(s)
Carabidae				
Elaphrus clairvillei lynni Pierce	Lynn Creek, British Columbia	Pierce (1948)	*E. clairvillei* Kby.	Goulet (1983)
Elaphrus irregularis Scudder	Scarborough, Ontario	Scudder (1890)	*E. parviceps* VD or *E. americanus Dej.*	Goulet (1983)
Elaphrus ruscarius foveatus Pierce	McKittrick, California	Pierce (1948)	*E. finitimus* Csy.	Goulet (1983)
Patrobus gelatus Scudder	Scarborough, Ontario	Scudder (1890)	*P.* cf. *stygicus* Chd.	Darlington (1938)
Patrobus decessus Scudder	Scarborough and Toronto	Scudder (1900)	*P.* cf. *stygicus* Chd.	Darlington (1938)
Patrobus frigidus Scudder	Toronto, Ontario	Scudder (1900)	*P.* cf. *stygicus* Chd.	Darlington (1938)
Hydrophilidae				
Helophorus pleistocenicus Lomn.	Borislav, Ukraine	Lomnicki (1894)	*H. sibiricus* Angus	Angus (1973)
Helophorus dzieduszckii Lomn.	Borislav, Ukraine	Lomnicki (1894)	*H. aquaticus* L.	Angus (1973)
Helophorus kuwerti Lomn.	Borislav, Ukraine	Lomnicki (1894)	*H. oblongus* LeC.	Angus (1973)
Silphidae				
Nicrophorus guttula labreae Pierce	Rancho La Brea, California	Pierce (1949)	*N. marginatus* Fab.	Miller and Peck (1979)
Nicrophorus obtusiscutellum Pierce	Rancho La Brea, California	Pierce (1949)	*N. marginatus* Fab.	Miller and Peck (1979)
Nicrophorus mckittricki Pierce	McKittrick, California	Pierce (1949)	*N. marginatus* Fab.	Miller and Peck (1979)
Nicrophorus investigator alpha Pierce	Rancho La Brea, California	Pierce (1949)	*N. nigrita* Mannh.	Miller and Peck (1979)
Tenebrionidae				
Apsena labreae Pierce	Rancho La Brea, California	Pierce (1954a)	*A. laticornis* Csy.	Doyen and Miller (1980)
Coniontis blissi Pierce	Rancho La Brea, California	Pierce (1954b)	*C. abdominalis* LeC.	Doyen and Miller (1980)
Coniontis tristis alpha Pierce	Rancho La Brea, California	Pierce (1954b)	*C. abdominalis* LeC.	Doyen and Miller (1980)
Eleodes elongatus Grinnell	Rancho La Brea, California	Grinnell (1908)	*E. grandicollis* LeC.	Doyen and Miller (1980)
Eleodes behri Grinnell	Rancho La Brea, California	Grinnell (1908)	*E. osculans* (LeC.)	Doyen and Miller (1980)
Scarabaeidae				
Canthon simplex antiquus Pierce	Rancho La Brea, California	Pierce (1946)	*C. simplex* LeC.	Miller *et al.* (1981)

Fig. 1. Modern (left) and fossil (right) specimens of Helophorus obliquus, *misidentified by Lomnicki as* H. kuwerti *(photos by Robert Angus, used by permission).*

such evocative names as *Helophorus pleistocenicus* (Lomnicki, 1894), *Olophrum interglacialie* (Mjöberg, 1904), *Platynus exterminatus* and *Lathrobium antiquatum* (Scudder, 1900). Unfortunately, much of the fossil material described in these early papers has been lost, but many of the remaining specimens have subsequently been reidentified as belonging to extant species. A sampling of these reinterpreted fossils appears in Table 1.

Revisions of misidentified material by modern workers merit our attention, because they show the extent and nature of errors made by early workers. The first is Angus' (1973) study of *Helophorus* (Hydrophilidae) fossils described by M. Lomnicki from Pleistocene deposits in the Ukraine. Lomnicki placed the *Helophorus* fossils from the Borislav site in five species, all described as new. Angus examined these fossils in Lomnicki's collection at Lvov (Fig. 1), and because of their excellent state of preservation, was able to perform a detailed study, including examination of male **genitalia** of several specimens. All the fossils were found to represent four extant species (Table 1).

In 1877, Samuel H. Scudder (Fig. 2) published his first paper about insect fossils from the Late Quaternary deposits at Scarborough, Ontario. During the next 20 years, Scudder described hundreds of beetle **sclerites** from the Scarborough Bluffs as extinct Pleistocene species. Scudder's work on the Late Pleistocene insects from the Scarborough bluffs (Fig. 3) and elsewhere was an important, albeit flawed, first step for North American studies.

Scudder enjoyed dual careers in paleontology and entomology. He served many years as paleontologist to the U.S. Geological Survey, but also wrote lengthy monographs on the modern Lepidoptera of eastern North America. Most of his work on insect fossils dealt with

Fig. 2. Samuel H. Scudder (1837–1911).

Tertiary specimens. All told, he described 1,144 insect species based on fossil material. Among these are 54 species of Pleistocene beetles. All but two were named as extinct species. Scudder had a reputation as an ''excessive splitter'' of genera (Cockerell, 1911), and his contemporaries observed that he described too many species based on specimens which were preserved inadequately for satisfactory classification. Nevertheless, his contribution to **paleoentomology** was likened to the vertebrate paleontological contributions of ''Leidy, Cope and Marsh combined'' (Cockerell, 1911).

One gains a better understanding of Scudder's difficulties in dealing with isolated fossil sclerites by

Fig. 3. Scarborough Bluffs, Toronto, Ontario (photo by the author).

noting some of the names he gave to them. For instance, the ground beetles he described include *Bembidion fragmentum*, *B. expletum*, and *B. damnosum*; *Pterostichus destitutus*, *P. depletus*, and *P. destructus*; *Platynus dissipatus* and *P. dilapidatus* (Fig. 4).

Unfortunately, much of Scudder's fossil material has been lost. But the specimens that have been reexamined by modern taxonomists (Table 1) have been combined with extant species (Darlington, 1938; Goulet, 1983). Moreover, recent studies of insect fossils from the Scarborough Bluffs have yielded only specimens attributable to modern species (Morgan, 1972, 1975; Morgan and Morgan, 1976; Williams *et al.*, 1981).

Carl Lindroth (Fig. 5) played an important role in establishing the modern aspects of the discipline of Quaternary entomology by his revision of previous Swedish studies of **interglacial** insects from Frösön, Härnön, Hälsingland, Pilgrimstad, and Angermanland (Lindroth, 1948). In particular, Eric Mjöberg named a number of fossil beetle sclerites from the Härnö site as representing extinct species (Mjöberg, 1904, 1905, 1915, 1916). Lindroth (1948) revised the Härnö material, and found that ''Mjöberg's opinion was extremely weakly founded.'' Some of Mjöberg's extinct species represented extant taxa, and others had been placed in the wrong genera. Thus, Lindroth concluded that all of Mjöberg's species names from the Härnö sites were junior synonyms of extant species. Mjöberg had attempted to identify various sclerites, such as staphylinid elytra, to the species level (an impossible task for most specimens). He was also faced with many **postmortem** changes, which led to more difficulties. Lindroth summarized Mjöberg's problems as follows:

> It is not difficult to understand how the Härnö material could make so strange an impression on Mjöberg that he suggested it to be the remains of a partly extinct fauna. He had met with the manifestation of a fact, the

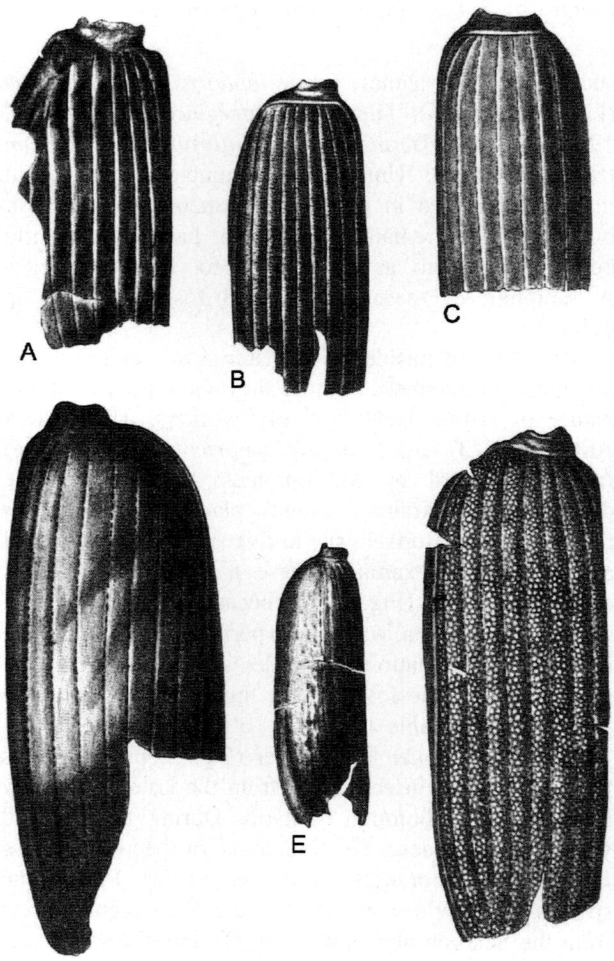

Fig. 4. Fossil beetles identified by Scudder from the Scarborough Formation, illustrated by Henry Blake in Scudder (1900). (A) Bembidion expletum, *(B)* Badister antecursor, *(C)* Pterostichus depletus, *(D)* Patrobus decessus, *(E)* Bembidion damnosum, *and (F)* Patrobus frigidus.

Fig. 5. Carl H. Lindroth (1905–1979). Photograph courtesy of Terry Erwin, U.S. National Museum of Natural History.

difficulty of which could be expressed in the form of the following question: To what extent do fossils and subfossils change by post-mortal processes?

By the time Lindroth was preparing his revision, the question of postmortem changes in fossil insect sclerites had been the subject of some debate in Scandinavia (Henriksen, 1933). Lindroth noted that the normal **punctulae** on the **exoskeletons** of ground beetles (Carabidae), rove beetles (Staphylinidae), and pill beetles (Byrrhidae) were deepened in the interglacial fossil specimens. He tried to reproduce these changes by treating modern specimens with several acids and bases, but could not achieve the same effects. He concluded that the alterations were due principally to postmortem changes.

Lindroth made several other important advances in this paper, including the observation that in dealing with fossil material, the most useful taxonomic characters are in the cuticular **microsculpture** (microscopic lines and meshes formed on the surface of the exoskeleton). This character system has been employed ever since (Coope, 1970, 1986). Lindroth identified 42 species of Hemiptera and Coleoptera from the interglacial assemblages, and made paleoenvironmental reconstructions based on the species' modern distributions and ecological requirements.

Strobel and Pigorni (1864) published one of the first studies to attempt a paleoenvironmental reconstruction based on fossil insect species. Their work dealt with an archeological site in Italy. Other early European studies were mostly paleontological in focus. These include work in France (Fliche, 1875, 1876), Germany (Flach, 1884; Kolbe, 1894; Schaff, 1892), Switzerland (Heer, 1865), Denmark (Wesenberg-Lund, 1896), and Finland (Andersson, 1898). British studies began with Bolton (1862) and Wollaston (1863). Shotton *et al.* (1962) reexamined Wollaston's specimens, and found them to represent extant taxa. The practice of naming extinct species of beetles from Pleistocene assemblages was continued through the first half of the 20th century. Coope (1968a) discussed this problem in his reinterpretation of Lesne's unpublished (ca. 1920) Pleistocene ''arctic'' fauna from Barnwell Station, Cambridge.

Nineteenth century studies in North America have been summarized by Ashworth (1979). They include work in the eastern United States by Horn (1876), other Canadian and eastern United States sites studied by Scudder (in Ami, 1894; Scudder, 1898), and in Illinois by Wickham (1917). All of these papers primarily describe fossils as extinct Pleistocene species. However, in 1919, Wickham published a paper on fossil insects from Vero Beach, Florida, and stated that the fossils probably represented extant species, an hypothesis later supported by Young (1959).

The final chapter in this history of first attempts and foibles comes from the famous asphalt deposits of southern California. These studies have been summarized by Miller (1983). Fordyce Grinnell began fossil insect work at the Rosemary site, near Los Angeles, in 1908. He was perhaps the first worker to abandon the assumption that all Pleistocene fossils must represent extinct forms, and made matches of the fossil material with many modern species (Ashworth, 1979). Pierce began publishing his extensive work on the Rancho La Brea and McKittrick sites in 1944, and continued for two decades (Pierce, 1946, 1948, 1949, 1954a,b, 1957). While Pierce appreciated the significance of Pleistocene insects as paleoenvironmental indicators, his taxonomic work is fraught with errors. He continued the practice of naming species and subspecies of ''extinct'' beetles. Most of his fossil material has been reexamined recently, and nearly all of the specimens have been matched with extant species (Miller and Peck, 1979; Doyen and Miller, 1980; Miller *et al.*, 1981; Goulet, 1983; Miller, 1983; Wilson, 1986). However, Miller *et al.* (1981) could not find modern species to match two scarab taxa described by Pierce: *Onthophagus everestae* and *Copris pristinus*. The species most closely related to these in modern collections live in Texas and Mexico, and feed on mammal dung. Miller (1983) has speculated that the extinction of many mammalian species at the end of the Pleistocene would have caused a reduction in dung availability for these beetles, leading to their extinction. Increasing aridity also may have played a role in the beetles' demise. However, Miller noted that these species may yet be found living today in inadequately studied regions of Mexico.

This last possibility was demonstrated by Russell Coope (Fig. 6), in his search for a modern counterpart to a fossil dung beetle species from the British Pleistocene. Specimens of this *Aphodius* were common enough in deposits from the middle of the last **glaciation**, but for years Coope failed to find a modern species that matched it. Finally, he searched through uncatalogued beetle collections from the 1924 British Everest Expedition, housed in the British Museum, and found several unnamed specimens that were a perfect match for the fossils (Coope, 1973). The modern specimens were determined to be *Aphodius holdereri*, a species known today only from the Tibetan Plateau (Fig. 7). The correspondence of fossils to modern specimens was confirmed through the comparison of male genitalia (Fig. 8). Male genitalia are considered one of the most diagnostic elements of the sclerotized body parts of beetles.

3. Coope's Pioneering Work in Britain

In 1955, Russell Coope began studying Quaternary insect fossils at the Upton Warren site, near Birmingham. When he began work on this site, he had little knowledge of paleoentomology. In fact, Coope's expertise was in coral fossils from pre-Quaternary marine beds in Britain. He first visited the Upton Warren site on behalf of the geological museum in his department, to collect Pleistocene mammal bones protruding from the **exposure**. However, an examination of the organic sediments surrounding the bones revealed numerous, shiny fragments of beetles. These fossils were initially in an excellent state of preservation, and Coope decided to try to identify them. In fact, he was ''blissfully unaware'' of the difficulties he would encounter in dealing with taxonomists who had been taught that all Pleistocene insect fossils must represent extinct species (Coope, 1965). By patient comparisons with modern specimens in the Natural History collections of the Birmingham Museum, Coope matched most of the Upton Warren material with modern species, and gradually became convinced that all of the fossil material could eventually be identified to species included in the modern fauna. His refusal to accept conventional thinking (i.e., the dogma of Pleistocene extinction) and his keen interest in the fossils led him to persevere in their study until he managed to win over many of his paleontological and modern entomological colleagues. This first endeavor, and his first (1959) paper on the subject (concerning the Chelford site) were greatly aided by the support of Coope's departmental head in the Geology Department at Birmingham, Professor Fred Shotton, who provided ''constant encouragement … boundless enthusiasm, and … practical assistance during the course of the work.'' The Upton Warren paper was published a few years later (Coope *et al.*, 1961).

Fig. 6. Russell Coope (photo courtesy of the Geography Department, Royal Holloway, University of London).

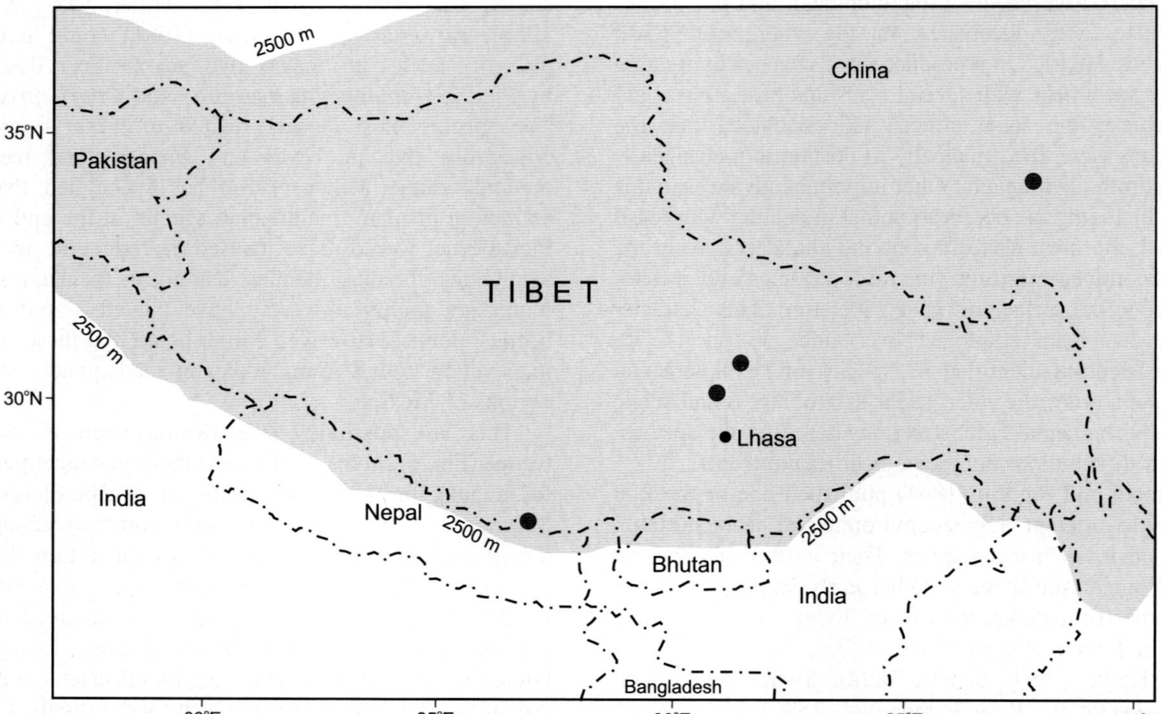

Fig. 7. Modern collecting localities for Aphodius holdereri *(data from Coope (1973) and Morgan (1997)). Shaded area represents the approximate boundaries of the Tibetan Plateau (regions above 2,500 m elevation).*

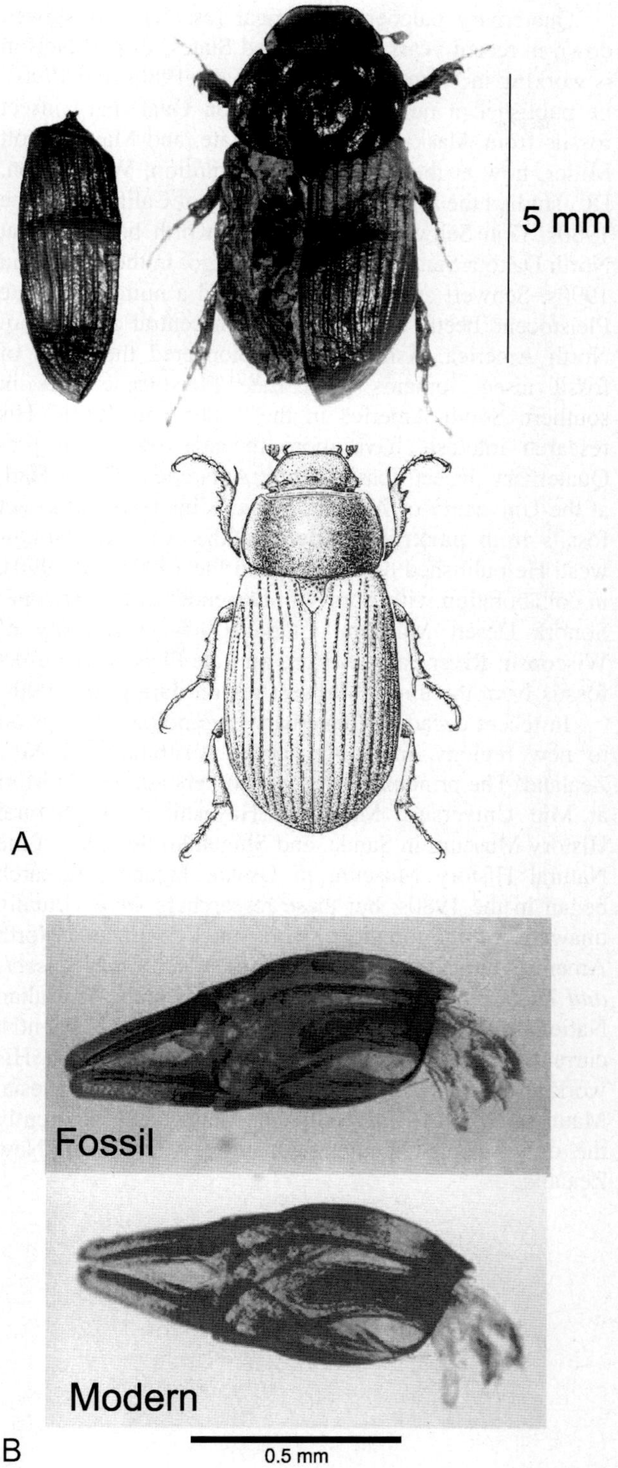

Fig. 8. (A) Fossil and modern specimens of A. holdereri. *Fossil specimen from Late Pleistocene assemblage from the Ukraine; modern specimen drawing from Tibet. (B) Fossil and modern aedeagi of* A. holdereri. *Fossil specimen from England; modern specimen from Tibet (photos courtesy of Robert Angus, Royal Holloway, University of London).*

Coope is a dedicated naturalist with wide-ranging interests and a driving curiosity. His determination, often in the face of a great deal of opposition from entomologists and Quaternary scientists alike, brought the field of Quaternary entomology out of obscurity and into the limelight. Since these seminal works, Coope has gone on to study assemblages throughout the British Isles and Western Europe, publishing more than 200 papers on Quaternary insects. Coope's contribution to the field is outstanding. He has established most of the important principles in the field, and has trained dozens of students, several of whom have established their own laboratories. Moreover, he is a gifted public speaker, and his great enthusiasm for Quaternary entomology has captured the imagination of audiences at geological and entomological meetings for more than three decades.

4. The Development of Quaternary Entomology Outside Britain

Coope and his colleagues Peter Osborne and Fred Shotton at the University of Birmingham worked more-or-less in isolation through the 1950s and early 1960s. However, they trained students, and by the 1970s, some of these students established their own laboratories and research programs in various regions. As of this writing, 40 scientists in 13 countries are studying the fossil remains of Quaternary or Late Tertiary beetles. Great Britain, the cradle of Quaternary insect studies, still has the greatest concentration of workers per square kilometer. Fred Shotton died in 1990. Russell Coope and Peter Osborne retired from the University of Birmingham, but Coope remains active in the field. Paul Buckland has lately been at the University of Bournemouth, Harry Kenward is in York University, Mark Robinson is at Oxford University, John Sadler and David Smith are at the University of Birmingham, Nicki Whitehouse is at Queens University, Belfast, Northern Ireland, Eva Panagiotakopulu is at University of Edinburgh, and I am at Royal Holloway, University of London. The majority of British workers are principally concerned with beetle remains from archeological sites. My own work continues to include Quaternary and Late **Pliocene** insects from Alaska and the Yukon Territory, as well as research on British insect fossils.

Continental Europe remains underrepresented in Quaternary entomology, with few workers. Philippe Ponel studies Late Pleistocene and **Holocene** insect assemblages in his laboratory in Marseille, France. J.H. Yvinec is in Compiègne, France, where he studies insects from archeological sites. In Amsterdam, the Netherlands, Bas Van Geel studies insects from archeological sites. Alexander Klink, in Wageningen, studies aquatic insect fossils, as does Wolfgang Hofmann, in Plon, Germany. In Scandinavia, Jens Böcher (Copenhagen) is the only Danish researcher. He has studied fossil insects from Greenland. In Sweden, Geoffrey Lemdahl (Kalmar) studies Late Pleistocene and Holocene insect assemblages from a variety of European sites, Phil Buckland (Umeå) has recently completed a PhD project that focused on the development of the BUGS fossil beetle software package, and Magnus Hellqvist (Falon) and Gunnar Gustavson (Varberg) have worked on fossil insect assemblages from archeological sites, mainly in Scandinavia. Alessandro Minelli (Padova, Italy) has studied Late Pleistocene beetle assemblages from Italy.

Russian scientists in this field have worked in relative isolation from their western colleagues until recently. Nevertheless, they have developed extensive research programs. Most studies have focused on the arctic and subarctic fossil faunas of Siberia. Vertebrate paleontologist Andrei Sher played an important role in getting the Siberian studies underway. He was able to obtain permission, at a time when such permission was seldom given, to send samples from the Kolyma region to John Matthews in Canada. Matthews' preliminary report stimulated further research by the Russian scientist, Sergei Kiselyov, who went on to publish a monograph on the fossils from Quaternary deposits of the Kolyma lowland, and other Siberian regions. Since the 1970s, additional studies have been performed by Arkady Alfimov and Daniel Berman in Magadan (Russian Far East), by Evgenij Zinovjev in Ekaterinburg and by Svetlana Kuzmina and Dmytry Krivolutsky, in Moscow. Sadly, three Russian paleoentomologists have died in recent years. Vladimir Nazarov (Mink, Belorussia) died in 1996, Sergei Kiselyov (Moscow) died in 2005, and Andrei Sher (Moscow) died in 2008. Fedor Bidashko is now working on fossil insect assemblages in Kazakhstan.

North American studies of the modern era began in 1968 with John Matthews's graduate studies of Late Quaternary insects from the Alaskan interior. Matthews worked throughout his career at the Geological Survey of Canada at Ottawa, and has recently retired. His former technician, Alice Telka, continues as an independent researcher, focusing mainly on Pleistocene assemblages from the Yukon. Other Canadian researchers include two people at University of Laval, Quebec: Allison Bain, who studies insects from archeological sites, and Claude Lavoie, who studies Holocene insect fossil assemblages. Randy Miller occasionally continues to work on Quaternary insects at the New Brunswick Museum in St. John, and his mentor, Alan Morgan, has been researching Pleistocene insect fossils from a wide variety of sites during the last 30 years.

Quaternary paleoentomological research has slowed down in recent years in the United States. Robert Nelson is working in Waterville, Maine. In the 1980s and 1990s, he published a number of papers on Quaternary insect fossils from Alaska, Washington state, and Maine. Scott Miller, now at the Smithsonian Institution, Washington, DC, studied the asphalt deposit insects of California in the 1980s. Don Schwert and Allan Ashworth both work at North Dakota State University, in Fargo. In the 1980s and 1990s, Schwert and Ashworth studied a number of Late Pleistocene beetle assemblages from central and eastern North America. Ashworth also pioneered the study of fossil insect remains from Late Pleistocene sites in southern South America in the 1980s and 1990s. His research interests have more recently shifted to pre-Quaternary insect fossils from Antarctica. Gene Hall, at the University of Arizona, remains interested in insect fossils from **packrat middens** in the American Southwest. He published on this topic in the 1980s and 1990s, in collaboration with Tom Van Devender of the Arizona-Sonora Desert Museum. Clarke Garry, University of Wisconsin-River Falls, worked on Late Pleistocene insect fossils from the upper Midwest region during the 1990s.

In recent decades, Quaternary entomology has spread to new regions, including Japan, Australia, and New Zealand. The principal Japanese workers are Yuichi Mori at Mie University, Masakazu Hayashi at the Natural History Museum in Sanda, and Shigehiko Shiyake, at the Natural History Museum in Osaka. Japanese research began in the 1980s, but these researchers were virtually unaware of the previous work in Europe and North America until the publication of *Quaternary Insects and Their Environments* in 1994. Nick Porch, Australian National University, Canberra, is the only scientist currently studying Quaternary insects from Australia. His work also extends to sites in New Guinea and Polynesia. Maureen Marra, University of Waikato, is currently the only scientist studying Quaternary insects in New Zealand.

2

Methods

… the main problem when dealing with Quaternary insect fossils is embarrassment of riches.

– Russell Coope (1986)

In this chapter, I discuss the methods used to excavate, extract, identify, and interpret Quaternary insect assemblages, particularly fossil beetle assemblages. The methods used in the study of aquatic insect fossils, such as midges (Chironomidae) and their allies, are discussed in the recent literature by Hofmann (1986) and Walker (1995, 2001, 2007).

1. Fossil Preservation

The remains of robust insects, especially beetles and ants, caddis fly larvae, bugs, and various genera within other orders, are abundant in **unconsolidated** organic deposits of Quaternary age. In the arctic, this type of preservation extends back several million years in permanently frozen sediments of Late Tertiary age (Matthews, 1977a). The remains found in these sediments are actual exoskeletons, not mineral replacements or impressions in fine-grained sediments. Insect tracks, or *lebensspuren*, are also found occasionally in Holocene muds (Ratcliffe and Fagerstrom, 1980). Tunnels dug by beetles are often seen in sediments, but these are rarely, if ever, diagnostic to the species that made them. The remarkable preservation of insect exoskeletons in Quaternary sediments is due, in large part, to their composition. The sclerites that comprise the body walls of insects are composed of two principal layers: a thin, more-or-less waxy, outer epicuticle that serves to prevent water loss and a thicker procuticle just beneath (Fig. 1). The procuticle is composed of an outer layer (the exocuticle) and an inner layer (endocuticle). The exocuticle contains a tough, durable substance called sclerotin. This forms during the initial hardening of the exoskeleton of the emerging adult, when cuticular proteins are exposed to secreted quinones that cause hardening by the formation of cross-links between protein molecules. One reason that beetles are the dominant group in most Quaternary fossil assemblages is that many beetle exoskeletons are very heavily sclerotized – a fact readily apparent to anyone who has attempted to put an insect pin through modern specimens of weevils (Curculionidae), darkling beetles (Tenebrionidae), and ground beetles (Carabidae).

The procuticle is made chiefly of chitin, a nitrogenous polysaccharide $(C_8H_{13}NO_5)_n$. Chitin is a very resistant substance that is insoluble in water, alcohol, dilute acids, and bases (Borror *et al.*, 1981). Furthermore, it is not broken down by the digestive enzymes of mammals. This is noticeable in mammal scat, in which many insect parts remain intact (see Elias and Halfpenny, 1991; Coope, 2007c).

2. Types of Sediments Containing Insect Fossils

Insect exoskeletons are found chiefly in **anoxic sediments** that contain abundant organic detritus. Insects decompose rapidly in heavily **oxidized sediments**, leaving either thin, partially preserved sclerites, or no trace. Water-lain sediments are generally the best source of insect fossils, because water acts to concentrate the insects. In contrast, ancient soils (**paleosols**) contain few fossils, because insects that die on a soil surface are very widely dispersed across a landscape. Chemical processes in soil formation also cause degradation of insect fossils.

Lakes, ponds, and **kettleholes** serve as reservoirs that collect insects, and sediments that accumulate in these waters act rapidly to cover their remains, preventing oxidation. The best lake sediments for study are those that contain abundant organic detritus (**plant macrofossils**, insect exoskeletons, and other organic debris). This type of sediment usually accumulates in the shallow waters of the **littoral zone**. Sediments rich in organic detritus are found where a small stream enters a lake, because the flotsam carried by the stream tends to settle out rapidly where the stream enters the standing water (Fig. 2) (see Elias, 1985, p. 31).

Fluvial sediments often accumulate detritus, especially in secondary channel bends, backflows, and pools between riffles. Oxbow lakes, formed when a bend in a large stream becomes cutoff from the main channel, accumulate a sequence of organic sediments representing the transition from fluvial to **lacustrine** environments, often finishing with infilling by **bog** vegetation.

The aim of sampling these sediments is to obtain a sufficient quantity to yield at least a liter of organic detritus. Generally, the quantity of sediment obtainable by piston coring of lake and pond sediments is insufficient to yield adequate amounts of detritus. While some success has been achieved by taking multiple, large diameter cores, this is not often practical, and may be hampered by difficulties in correlating the **horizons** between cores, unless numerous **marker horizons** are present (see Hoganson and Ashworth, 1992, pp. 102–104). The best results are obtained by sampling exposures of sediments, either natural or man-made. Natural exposures include cut

DEVELOPMENTS IN QUATERNARY SCIENCES
VOLUME 12 ISSN 1571-0866

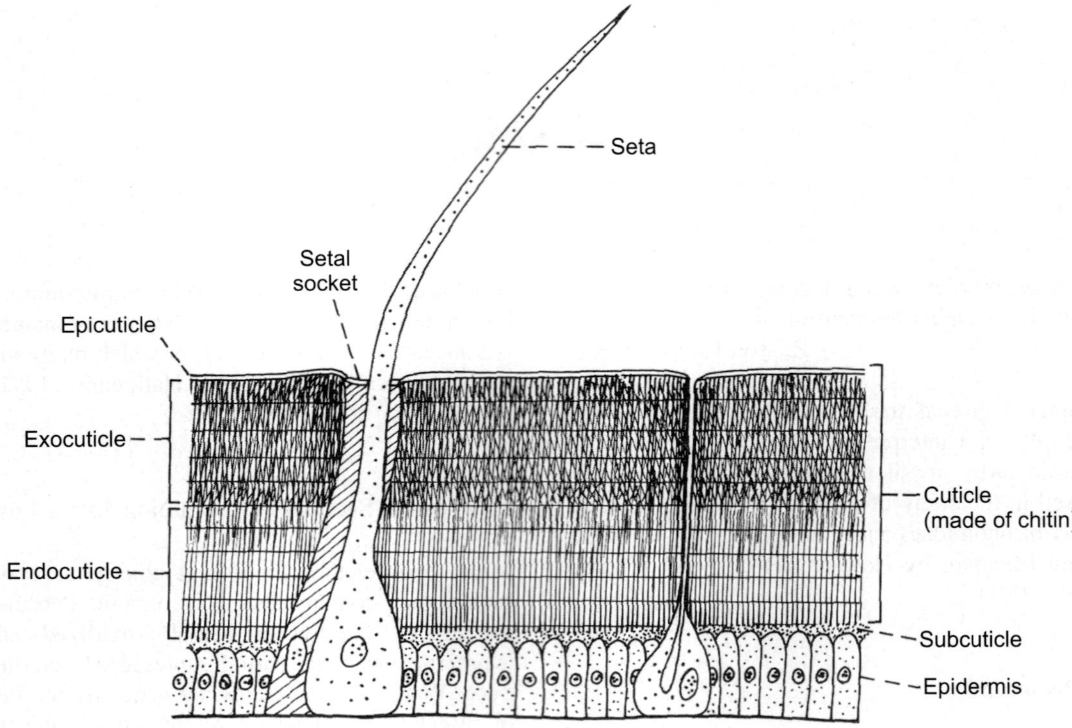

Fig. 1. Cross section of insect cuticle (after Borror et al.*, 1981). cut, cuticle; end, endocuticle; ep, epidermis; epi, epicuticle; exo, exocuticle; se, seta; ss, setal socket.*

Fig. 2. Deltaic deposit at Lake Isabelle, Colorado (photo by Susan K. Short, used with permission).

banks of streams, and ocean and lake bluffs (Figs. 3 and 4). Man-made exposures include gravel and clay pits (Fig. 5), irrigation ditches, trenches, and building sites. At times, even man-made catastrophes are put to good use. For instance, organic sediments deposited along the banks of the Roaring River, Colorado, were exposed for the first time in 1982 when an earthen dam failed several kilometers upstream, bringing a torrent of water that scoured the river channel (Elias *et al.*, 1986). Similarly, lake sediments were exposed at Lake Emma, Colorado (Fig. 6), when a mine roof beneath the lake collapsed in 1978, catastrophically draining the lake (Elias *et al.*, 1991).

Organic-rich sediments in many of these exposures are made up of detritus dispersed in sand, silt, and clay. For insect fossil analysis, the best results are obtained from organic detritus in silts. Not only does the silt readily disaggregate and separate from the organics in water screening, its limited compressibility ensures that the fossils are preserved in their original shape (Coope, 1986).

Bogs and fens are rich sources of insect fossils. Mosses and sedges may accumulate so rapidly in shallow water that they do not decompose, but rather build into layers of peat that may also trap insect exoskeletons (Figs. 7 and 8). Buckland (1976a) noted that **fen** peats tend to be richer in insect remains than acid peats (such as *Sphagnum* peats), because fen peats offer a greater diversity of insect habitats. Some peat bogs in northern countries may cover thousands of square kilometers. The centers of large bogs tend to be rather sterile in terms of insects. The best samples for insect fossils are obtained from the edges of bogs. As with lake sediments, peat samples are taken most readily from exposures.

Peats overlain by glacial ice or by large volumes of other sediments may become compressed or felted, making insect fossil extraction difficult. Coope (1961) and his colleagues at Birmingham first dealt with felted peats by splitting them along **bedding planes**, and

Fig. 3. Organic horizon in coastal bluff exposure, Kvichak Peninsula, Alaska (photo by author).

Fig. 4. Peat profile exposed in bluff at Ennadai Lake, Keewatin, Northwest Territories, Canada (photo courtesy of Harvey Nichols).

Fig. 5. Late Pleistocene (Scarborough Formation) organic sediments exposed in Don Valley Brick Pit, Toronto, Ontario (photo by author).

examining those surfaces for fossils. This technique had the advantage of yielding some nearly intact specimens, found together on the bedding planes. However, the procedure tended to overemphasize large, brightly colored specimens and to miss small, dull specimens. The latter generally represents the majority of species in any sample. In spite of this problem, Japanese workers prefer to split open peat layers to find fossil beetles from

Fig. 6. (Top) Lake Emma, Colorado, in 1921 (San Juan County Historical Society photograph). (Bottom) Lake Emma basin, after catastrophic drainage of the lake into mine in 1979 (photo by Paul Carrara, used with permission).

Pleistocene deposits (Fig. 9). They leave the fossil sclerites *in situ* on the surface of these layers, in order to describe them (Hayashi, 2007).

Peats and layers of organic detritus that are deposited in **permafrost** yield some of the best-preserved insect specimens yet studied. The frozen state of the sediments greatly retards bacterial decomposition, and the frozen matrix of sediments lock the specimens in place, preventing reworking, except in cases where frost activity displaces whole layers of sediments (Fig. 10). Organic layers have been preserved intact for several million years in the high arctic, because of permafrost. Along major rivers in the Yukon, Alaska, and Siberia, there are exposures of sediments spanning more than a million years in some cases (e.g., the Palisades site, described by Matheus *et al.*, 2003). High river bluffs often contain alternating beds of glacial-age silts, interspersed with organic deposits, typically called ''forest beds,'' that contain extremely well-preserved plant macrofossils, pollen, and insect remains that were deposited during interglacial periods (Fig. 11).

In the last 10 years, Quaternary insect fossils have been discovered in packrat (*Neotoma* spp.) middens in the American southwest. Packrats accumulate objects in their nests, including edible plants, cactus spines, vertebrate remains, insect remains, small pebbles, and feces cemented into black tarry masses by packrat urine (Fig. 12). Packrats are known to bring objects to their den site for a variety of reasons including food, curiosity, and protection. As packrat urine dries, it hardens, cementing layers of midden material which dry in rock shelters, preserving a paleoecological record for thousands of years (Fig. 13). Packrat middens are different from water-lain deposits; physical factors (current speed, size of catchment basin, facies, etc.) generally determine the composition of water-lain deposits, whereas the packrat is the most important agent of accumulation of organic material preserved in middens. Modern studies have shown that the majority of insects associated with packrat nests in rock shelters and

Fig. 7. Sedge peat exposure at La Poudre Pass, Rocky Mountain National Park, Colorado (photo by author).

Fig. 8. Sphagnum *peat exposure on the Foraker River, Denali National Park, Alaska (photo by the author).*

Fig. 9. Fossil elytron of Elaphrus japonicus, *exposed on a layer of peat (photo by M. Hayashi, Encyclopedia of Quaternary Science, copyright Elsevier, 2007).*

caves represent species which generally live outside such shelters, but come in to share the microenvironment created by the packrat. These insects, called facultative inquilines, either prey on other insects in and around the rat's nest or scavenge plant materials there (Elias, 1990a).

3. Sampling Procedures

The most informative paleoenvironmental reconstruction of a given site is not based on isolated study of any one type of fossil. It is rather the combined effort of a team of investigators collaborating in the joint study of a broad spectrum of fossils including not just insects, but also pollen, plant macrofossils, diatoms, ostracodes, mollusks, and vertebrates. **Stratigraphy**, **geomorphology**, and **geochronology** also play a vital role (see the study of Lobsigensee, Switzerland, described in Ammann *et al.*, 1983). Obviously, this type of collaborative effort requires much advance planning before fieldwork is undertaken. For fossil insect work, as for these other disciplines, it is important to locate and sample exposures in such a way as to allow the collection and compilation of other types of data, such as seeds, pollen, material for dating, etc. (Coope *et al.*, 1961; Coope, 1986).

Before sampling an exposure for insect fossils, the investigator must cut into the exposed face, to remove material that has been exposed to repeated wetting and drying, and to eliminate contamination by modern insects that burrow into exposed banks. This procedure will also help to ensure that potential radiocarbon samples will not be contaminated by modern carbon from rootlets, mosses, and other recent organic material (Morgan, 1988).

Once the exposure has been cleaned, and a baseline stratum is established (an easily observable feature such as a distinct sand lens, a lens of volcanic ash, etc.), horizons to be sampled for insects are generally measured in 10 cm increments. It is useful to photograph and sketch the exposure before sampling; many workers now take digital photographs of exposures, to accompany their field notes. A video camcorder is another useful tool in the field, to more fully document sites.

Some discrete, organic-rich horizons in lake, pond, and fluvial sediments yield more insect fossils than bulk sediments with little organic content. It is preferable to sample these units in intervals of not greater than 5 cm thickness, so that the insect assemblages from a single sample represent a discrete time interval. However, in some cases, organic-rich lenses in an exposure may be too few to document a complete sequence of changes in insect faunas through time.

Then, less organic-rich sediments have to be sampled, necessitating sieving large quantities of sediment to obtain sufficient organic detritus to study. This is best done in the field with a **bucket sieve**, with a 300 μm screen (Fig. 14). Sampling laterally for several meters along an exposure may yield tens or hundreds of kilograms of silts and fine sands that contain 3–5% organic detritus. It is possible to rapidly wash large volumes of fine-grained sediments

Fig. 10. Frost heaved organic sediments on the Beaufort Sea coast, Barter Island, Alaska. Note that the bottom layer of sediment has been contorted so that it stands at the top of the section (photo by the author).

(except clays) through the sieve screen near the base of the bucket, accumulating a quantity of organic detritus on the screen. This process requires a water source in close proximity to the exposure (not to mention a strong back). Ponds and rivers in which samples are sieved often contain entrained insect fragments, either modern or fossil. These fragments may get into the sieved residue if water is allowed to wash in over the top of the bucket sieve. The volume of sediments required to provide sufficient organic detritus varies from site to site and from horizon to horizon. For instance, to obtain 1 kg of organic detritus, Nelson (1982) removed about 1,000 kg of sediments from one horizon in an exposure on the Ikpikpuk River on the North Slope of Alaska. In contrast, only 20 kg of lake sediments from Lobsigensee, Switzerland, provided the requisite liter of organic detritus (Elias and Wilkinson, 1983).

Russian workers have developed a procedure similar to bucket sieving, but in their method, they use a wooden frame with a screen fixed to the bottom (Fig. 15). This tool may be collapsed for ease of transport to and from the field. As in the bucket-sieving technique, lake or river water is used to perform the wet screening.

Frozen organic-rich sediments can be very challenging to extract from the permafrost. If the sampling can be timed to occur towards the end of the summer, then the surfaces of exposed sediments along rivers and coastal bluffs will have thawed to the maximum extent, and this thawed layer can be sampled for analysis. This is particularly useful when sampling along the shores of high-latitude rivers. These rivers generally freeze over in winter, and the breakup of river ice acts as a powerful erosional agent, scouring the banks of the river, and exposing new layers of sediments to the air, often for the first time in many thousands of years. Such was the case when my research team sampled the scoured banks of the Foraker River in Denali National Park, Alaska. The spring breakup of river ice exposed an amphitheater of frozen sediments (Fig. 16A). By patiently scraping away at the thawing surface of organic lenses that were adjacent to a large body of ground ice (Fig. 16B), we were able to extract sufficient amounts of organic detritus to yield some

Fig. 11. High bluffs along the Toklat River, Denali National Park, Alaska. Exposures such as these frequently contain alternating sequences of glacial silts and sands, interspersed with organic deposits laid down during interglacial periods (photo by the author).

Fig. 12. Late Pleistocene packrat midden, Emery Falls site (photo by T. R. Van Devender, used with permission).

Fig. 13. Maravillas Canyon Cave, Texas. Note packrat midden fragments on floor of cave – the remains of a midden that had previously been sampled (photo by the author).

insect fossil assemblages (Elias *et al.*, 1996a). Sometimes, these newly thawed deposits slump, covering the original sedimentary sequence with a debris apron (Fig. 17). In such cases, a great deal of additional work is required, so that the underlying sediments are once again exposed.

Russian workers typically take very large samples (up to 150 kg of sediment) from sites in the permafrost zone of northeast Siberia. This is accomplished by chopping out blocks of frozen sediment with an axe (Fig. 18). Once the sediments are sampled, they are allowed to thaw, and then processed on site, using river or lake water for wet screening. In northern Alaska, I have cut out blocks of frozen peat with a chain saw (Fig. 19). Needless to say, this kind of work is muddy, arduous, and somewhat dangerous. The paleontological rewards can be great, however, because of the excellent state of preservation of the fossils obtained from these permanently frozen sediments.

The chronicles of permafrost sampling would not be complete without a discussion of the samples from Devil Mountain region of the northern Seward Peninsula, Alaska. About 21,000 years ago, there was a massive volcanic eruption in this region that blanketed more than 1,200 km^2 of land surface in at least 1 m of tephra (Goetcheus and Birks, 2001). In places where the tephra was thicker than the summer-thawed active surface layer, it preserved the plant and animal remains present on the surface by *in situ* freezing. The tephra layer was subsequently buried by loess deposits, and was not discovered until the 1960s. The tephra burial of the landscape surface created a geologic Pompeii, in which the flora and insect fauna that were living in this region

Fig. 14. Wet screening of sediments in the field, using a bucket sieve with a 0.30 mm diameter mesh (photo by the author).

Fig. 15. (A) Wet screening of sediments in the field, using a large handmade box with 0.50 mm mesh sieve. (B) In Siberian research, the screened residues are dried in the field before transportation to the laboratory. This greatly reduces the weight of the samples. In North America, field-screened samples are kept moist. While this increases the weight of samples to be shipped back to the laboratory, it ensures that specimens do not crack and split as they dry (photos by Svetlana Kuzmina, University of Alberta, used with permission).

during the last-glacial maximum (LGM) were preserved *in situ* on the surface buried by the ash. All of these sediments have been permanently frozen since that time, so the fossils are beautifully preserved (Fig. 20). Beetle remains have been preserved at several sites (Kuzmina *et al.*, 2008). Many of these are more or less articulated exoskeletons, presumably representing individuals that died on the day the volcanic ash smothered the landscape.

Once the organic detritus samples are collected (either through sampling of organic-rich horizons or through the concentration of organics by field sieving), each sample should be placed, still damp, in heavy gauge plastic bags and sealed shut. If the sample is allowed to dry, the insect sclerites contained in the sample tend to become curled and split into small fragments. Each bag must be clearly labeled.

Peat exposures may be sampled in blocks from a cleaned face (Fig. 21). The blocks should be at least 10–15 cm across and 10–15 cm deep, and taken in vertical increments of 5 cm each. However, in order to expedite the fieldwork, larger blocks may be taken (representing depth intervals of 10, 15, or 20 cm), and then split into 5 cm depths at a later date in the laboratory.

The peat blocks are wrapped in heavy gauge aluminum foil, labeled to show the orientation of the block (top, bottom, front, etc.) and the depth interval. The wrapped block is then sealed in a plastic bag that is also clearly labeled. If time allows, a cleaned peat face at an

Fig. 16. (A) Ice body and frozen sediments exposed along the Foraker River, Denali National Park, Alaska. Scouring by river ice creates new exposures along arctic and subarctic rivers every summer (photo by the author). (B) Sampling of recently thawed surficial layers of organic-rich sediments at the Foraker River site (photo by the author).

Fig. 17. Sediments exposed along the banks of the Noatak River in northwest Alaska are covered by an apron of debris that has slumped over the exposure (photo by the author).

exposure may be sampled in 5 cm increments, and the samples placed in heavy gauge plastic bags for transport to the laboratory (Fig. 22).

In most cases, the study of a fossil site is literally a once in a lifetime opportunity, either because of the high cost of travel or the ephemeral nature of the sampling site itself. Ditches tend to be filled in; stream-cut exposures slump or disappear after the stream changes course; frozen sediments thaw, causing exposures to be ever changing; and building sites are covered with concrete. In light of this, redundancy in sampling is very important. An archive of every sample studied should be saved, whenever possible, for future study. When a sample is sieved in the field, it is also advisable to collect an unsieved sample for laboratory preparation, as a check against contaminants in the sieved sample. Redundancy of sampling also ensures that even if the samples become lost by baggage handlers or mail clerks, or are left in cold storage for a decade, the investigator will be assured of a useful set of samples.

It is worth noting here that not all valuable insect fossil samples have come from researchers exploring well-developed exposures. For instance, I have studied insect fossils from terrestrial peats from near the top of marine sediment cores, taken onboard a ship in the shallow waters of the Chukchi and Bering seas, off the northwest coast of Alaska. The geologists who took these cores were more interested in the marine sediments they contained, but the terrestrial peats near the top of the cores contained insects that once lived on the exposed continental shelves of the Bering Land Bridge (Elias *et al.*, 1992a,b, 1997). I have also retrieved useful insect assemblages from pond sediment scraped from the brain case of a Columbian mammoth in the Wasatch Mountains of Utah (Fig. 23) (Elias, 1990b), and from dung taken from the frozen carcass of a woolly mammoth in Alaska (Elias, 1992a), and Columbian mammoth dung from the floor of a dry cave in Utah that was apparently used as a megafaunal toilet (Elias *et al.*, 1992b). Some very valuable information may come from insects extracted from samples taken from unorthodox settings.

The collection of insect fossil samples from archeological sites is somewhat specialized (Buckland, 1976a). It is vital that proper sampling methods be used, and that the fossil insect investigation is coordinated with the archeological project as a whole. Sampling intervals from archeological sites may need to be smaller than those in natural deposits, because the paleoentomologist is often called upon to develop a paleoenvironmental reconstruction of a discrete interval (e.g., a single human occupation horizon in a sequence). As in strictly paleontological studies, it is best for researchers to take their own samples, in order to develop a clear understanding of the site and its peculiarities.

Fig. 18. A Russian research student chopping samples of frozen sediments at the Mamontovy Khayata (MKh) fossil site on the Lena Delta, Bykovsky Peninsula, northeastern Siberia (photo by the late Andrei Sher).

Fossil insect samples have been taken from a wide variety of archeological contexts. These fall into two broad categories: **anthropogenic** deposits and natural deposits. The former consists of organic-rich detritus sampled from localities such as houses, barns, graves, granaries, latrines, moats, and other structures made by people. The latter consists of organic deposits from natural settings, such as lakes, ponds, or bogs that were in close proximity to a site of human occupation. The purposes of sampling from these two classes of ancient environments are generally quite different. The aim of sampling from anthropogenic deposits is to help reconstruct the lifestyle of the people living there at that time. The aim of sampling from natural deposits associated with human habitation is to reconstruct the natural environments associated with the particular time of human occupation. However, the two aims are not mutually exclusive, and sometimes they overlap. For instance, in recent years, I have been involved in paleoenvironmental reconstructions from the lower Thames Valley, England (Elias *et al.*, 2009). These reconstructions focus on Bronze-Age bogs and ponds where people built wooden **trackways**. By studying the insect fossil faunas associated with the time intervals in which these trackways were built, our research team has attempted to interpret the reasons behind their construction. Were they built during particularly wet intervals, to facilitate the transportation of people and livestock from one patch of high ground to another, or were they built to exploit wetland habitats during times of relatively dry climate? The discovery of insects associated with cereal crops in these fossil assemblages gives clear indication of yet another human activity on a nearby landscape.

Sampling for insect remains in anthropogenic deposits often involves taking discrete samples of organic detritus from beneath ancient floorboards (Fig. 24), the bottoms of cesspits and latrines (Fig. 25), and the sediments that accumulated in moats.

4. Extraction and Concentration of Insect Fossils from Sediments

The sequence of steps in extraction and mounting of insect fossils is summarized in Fig. 26. Fortunately, fossil insect extraction is relatively safe, cheap, and easy. No

Fig. 19. Sampling blocks of frozen peat with a chain saw. The peat was exposed in a thermokarst gully near Barrow, Alaska (photo by the author).

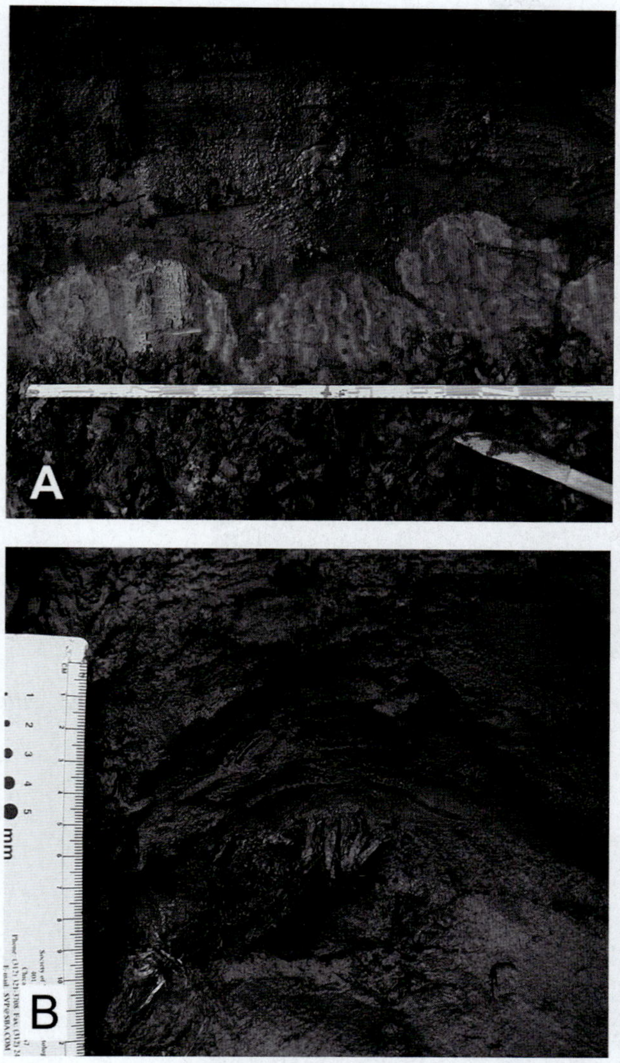

Fig. 20. (A) Exposure of frozen sediments on the shore of Tempest Lake, Seward Peninsula, Alaska. Note the dark tephra layer draped over the lighter, hummocky surface beneath (the ancient land surface) (photo by Svetlana Kuzmina, used with permission). (B) Fossil sedges preserved in situ on the ancient land surface buried by the tephra and preserved in permafrost (photo by Svetlana Kuzmina, used with permission).

Fig. 21. The peat bank at La Poudre Pass, Colorado, after blocks of peat have been taken for fossil insect study. These blocks were in depth increments of 15–25 cm. They were later split into 5-cm intervals for fossil insect analysis (photo by the author).

costly chemicals or equipment are required, and less time is needed than for the preparation of pollen or diatom samples, for instance. The only lengthy process that may be involved is the pretreatment of samples to disaggregate the organic detritus from inorganic matrix (such as calcareous sediments and clays), or to soften and then disperse felted peats or **lignite**. Such procedures may last hours or days, and may have to be repeated several times in order to be effective.

Processing procedures are largely a matter of personal preference and availability of equipment, and some workers use a slightly different procedure than the one discussed below (Fig. 26). Once disaggregated organic detritus is obtained, the next step is to wet screen the sample in a 300 μm sieve. This process removes fine particles, such as silt, that may fill the concavities of rounded insect sclerites, such as head capsules and the elytra of some weevils. If the silt is not removed, the affected sclerites may not rise to the top in the subsequent kerosene flotation procedure. Some fossil beetle workers prefer a slightly finer screen size (200–250 μm), and the study of smaller organisms requires even finer sieves.

Some Russian workers prefer to air-dry the wet-screened residues, and pick through them entirely. In their view, this allows them to find more of the heavy-bodied beetles with rounded heads, pronota, and elytra than they would recover by the kerosene flotation method. Their argument is that such specimens are often filled with tightly packed silt that does not wash out during wet screening of the samples. They admit, however, that by using this method, they may be overlooking the remains of some of the smaller bodied insects in their samples (Kuzmina *et al.*, in preparation). In fact, an examination of the fossil insect literature from study sites in Siberia does show a surprising lack of such small-bodied groups such as rove beetles (Staphylinidae). It remains unclear whether the lack of such groups in these faunal lists is due to the scarcity of this group in regional **steppe–tundra** communities of the Pleistocene, or due to sampling bias.

Fossil midge larvae samples (Diptera: Chironomidae) are sieved on a succession of 200 and 100 μm screens (Hofmann, 1986), while fossil Cladocera (water fleas) are caught on 63 μm screens (Frey, 1986). If collaborations

Fig. 22. Peat being sampled in Ireland (photo by Kate Denton, Royal Holloway, University of London, used with permission).

Fig. 23. A mammoth skull being jacketed in plaster before removal from the Huntington Canyon site, Utah. The mud which had filled the brain case of this specimen yielded sufficient numbers of insect specimens to provide a paleoenvironmental reconstruction (photo by David Madsen, used with permission).

are to be made with researchers studying these animals, either duplicate samples should be taken from the same horizon or the <300 μm fractions of samples should be retained for study by other specialists.

Once the detritus has been screened, the residual material is placed, still damp, in a large bowl with a spout, or a rectangular dishpan, and processed by kerosene flotation to isolate and concentrate insect fossils. This procedure should always be done in a room with good ventilation, or under a fume hood. Kerosene or other lightweight oil is added to cover the sample, and gently worked into the sample by hand for several minutes. The oil adheres to the insect sclerites but not to plant detritus. The remaining kerosene is decanted from the bowl and filtered into the stock bottle through a fine screen (200–300 μm) over a funnel. Since oil and water do not readily mix, the kerosene comes away cleanly from the mixture in the bowl, and may be used repeatedly. Cold water is vigorously added to the oily detritus in the bowl, with the aid of a hose to reach the bottom of the sample and stir it thoroughly. In most samples, nearly all of the insect sclerites will rise to the top, and float at the

Fig. 24. (A) Insect remains from beneath a medieval floor at the Bishophill I site, York, England (photo by Paul Buckland, used with permission). (B) Excavation of an ancient farmhouse at Storaborg, Iceland (photo by Paul Buckland, used with permission).

Fig. 25. Excavation of a historic latrine from the Intendant's Palace site, Quebec City (photos by Alison Bain, University of Quebec, used with permission).

oil–water interface. Within 15–60 min, most plant residue sinks to the bottom of the bowl, and the concentrated insects may be decanted onto a screen, to be washed gently in detergent and dehydrated in 95% ethanol before microscopic sorting.

In some peaty samples, notably those with abundant *Sphagnum* mosses, the plant macrofossils contain trapped air bubbles which prevent them from sinking. Other plant fossils, such as the shiny seeds of some plants, float quite readily. Some success in sinking this non-insect material in water may be achieved by prolonged soaking of the samples in water (for periods of a week or more). If this technique fails, or if both plant and insect fragments sink to the bottom in the flotation procedure, the entire sample must be sorted under the microscope. It is also necessary to check both sink and float fractions for fossils. Sometimes, for reasons not yet understood, the kerosene flotation fraction has a very low yield, even though the sample may be rich in insect fossils. One only learns this by checking both fractions. It should also be noted that the kerosene flotation method yields a biased sample, rich in some insect body parts and poor in others. Not all insect parts are equally buoyant, even when coated with oil. As discussed above, Russian workers avoid the kerosene flotation method because it misses out some sclerites of heavy-bodied beetles and convex sclerites filled with silt that are too dense to float in water.

One variation on the above procedure, used mostly with alluvial debris, lacustrine sediments, and peats, is to place the sample in water and while waiting for it to become disaggregated, decant all the material that floats on water through nested 5/20/80 mesh sieves. This removes most large wood pieces, many of the plants that float in water (and which would also float in kerosene), and provides a concentrated sample of fossils such as oribatid mites and chironomid heads that would float in the kerosene flotation process, but which would be diluted by the abundance of other fossils.

Small samples should be sorted completely, rather than processed with kerosene. Also, samples that may need to be submitted for radiocarbon dating should not be exposed to kerosene. Accelerator mass spectrometer (AMS) radiocarbon dating techniques have been refined to the point that it is now possible to obtain a ^{14}C age from just a few fossil insect specimens, such as 3–4 head capsules or 5–6 elytra. As discussed in Chapters 7 and 14, AMS radiocarbon dating has helped resolve problems associated with reworked faunal assemblages of mixed ages, but it is not completely straightforward.

Kerosene flotation samples may be safely stored after washing. If a fungicidal detergent is used to wash the

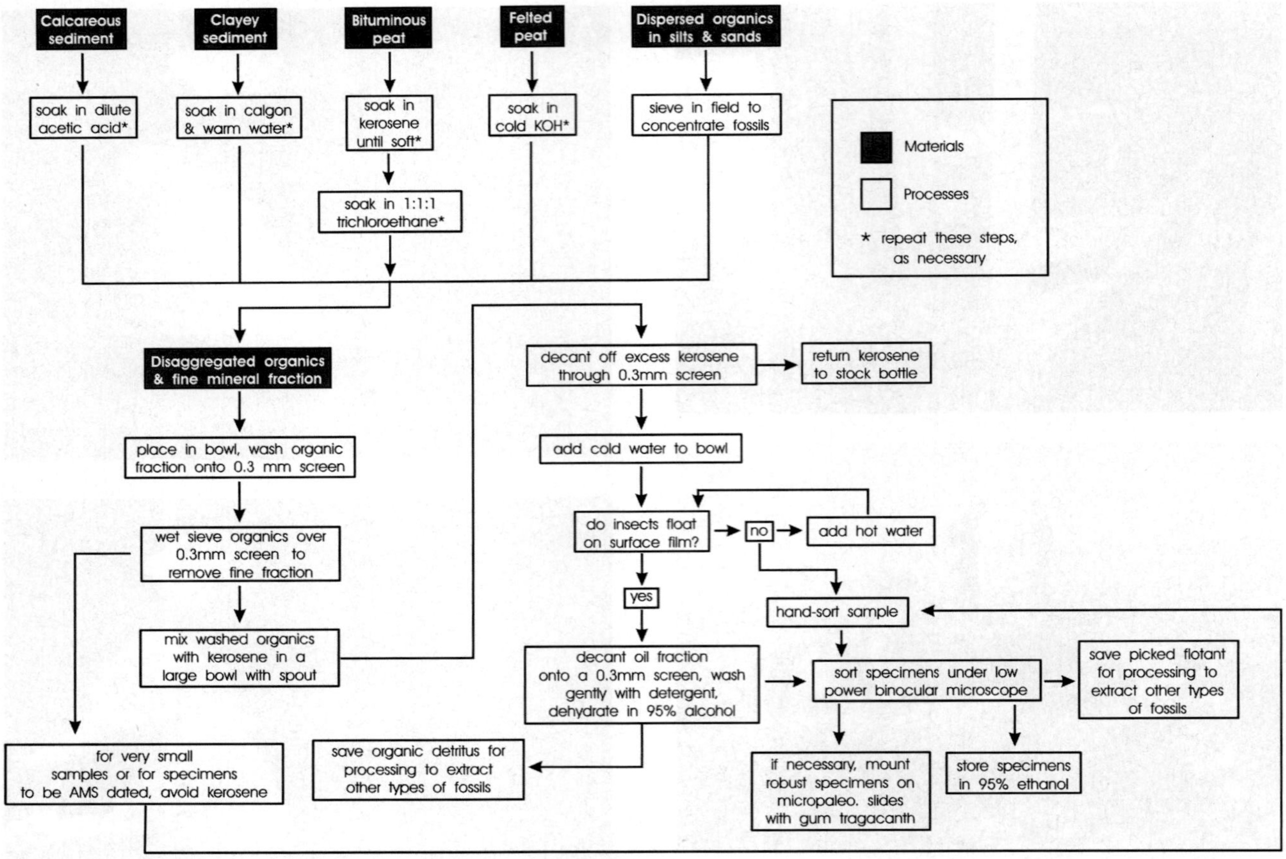

Fig. 26. Summary of methods for extraction of insect fossils from various types of sediments. Shaded boxes represent sediment types; white boxes represent processes. Note the treatment of calcareous, clayey, and bituminous sediments and felted peats (after Morgan, 1988).

sample, it may be stored for long periods in water without suffering fungal growth. This allows the sample to be picked in water, rather than alcohol. However, picked specimens should be stored on a long-term basis in dilute (30%) alcohol, since it retards fungal and bacterial attack. Specimen sorting is done under low power (10X) binocular microscope in alcohol or water. Vials of specimens in alcohol tend to dry out unless impermeable stoppers or caps are used. Only ethyl alcohol (ethanol) should be used for sorting and storing insects. Methanol releases poisonous fumes, and isopropyl alcohol forms a cloudy solution when mixed with water. Some workers add a small amount of glycerin to the vials of ethanol, to retard evaporation.

Robust beetle, bug, and ant fossils may be mounted with gum tragacanth (a water-soluble glue) onto **micropaleontological slide cards** with cover slip and aluminum holder. However, many specimens shrivel and break during the drying of the glue on the card, so it is best to leave most, if not all, specimens in alcohol. The micropaleontology slides have a rectangular cavity, 3 mm deep. The specimens are mounted on the floor of this cavity. This method works well for smaller specimens (up to a few millimeters in length), but large specimens, such as those found in many packrat midden samples, will not fit under the cover slip and must be stored in vials of alcohol, or on cards with deeper cavities.

Fossil caddis fly larval sclerites are examined with transmitted light. These must therefore by mounted under a cover slip on glass slides, using a mounting medium. Oribatid mite and chironomid fossils are also usually cleared and mounted on glass slides for viewing with a transmitted light microscope (Erickson, 1988; Walker,

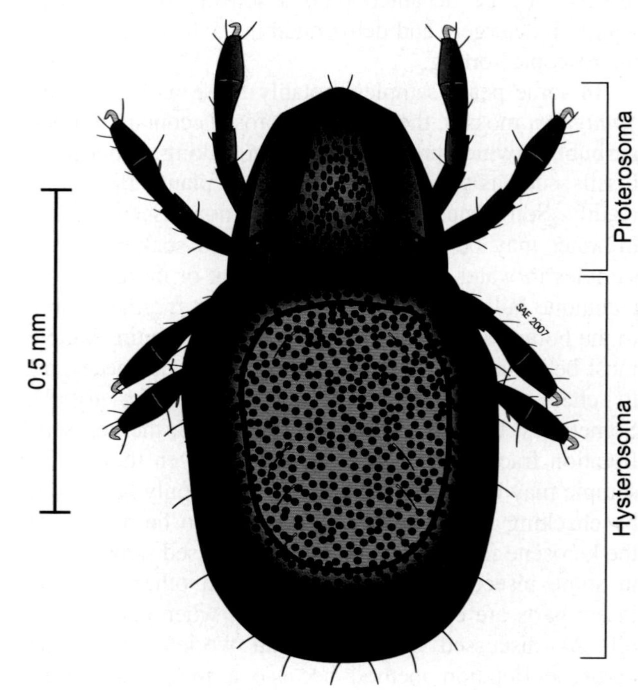

Fig. 27. Generalized drawing of an oribatid mite.

2001). Methods used in the study of caddis fly fossils are summarized by Williams (1988).

Many exoskeletons of oribatid mites (Fig. 27) are found nearly intact in Quaternary peats and lake sediments, especially from sediments deposited under cold climatic regimes. The extraction and preparation of oribatid mites for identification involves a different set of laboratory procedures, as discussed by Erickson and Platt (2007). However, standard 300 μm mesh sieving and kerosene flotation often yields substantial numbers of oribatids.

Special procedures are required to extract and identify fossil chironomid remains from lake sediments (Walker, 2001). The principal body part preserved is the head capsule. Fortunately, the head capsules of midges contain many diagnostic features. These head capsules may be very abundant in lake sediments. Walker (1987) reported that sediments from lakes in forested watersheds generally yield 50–100 fossil head capsules per milliliter.

The methods used in the sampling and extraction of fossil beetles have evolved during recent decades, but the most part remain much as I described them in 1994. It is interesting to note that the Japanese and Russian variants developed in isolation from the methods common to Europe and the Americas. The Japanese were unaware of Quaternary entomological research going on elsewhere in the world until they read my 1994 book (Hayashi, written communication, 1995). The Russians were more aware of Western European and North American research, but had little access to books and journals from these regions until after the dissolution of the Soviet Union in 1991.

3

Important Fossil Insect Groups and Their Identification

The most useful characters of subfossil remains of Coleoptera, especially when consisting merely of single or fragmentary elytra, lie in the microsculpture....

– Carl H. Lindroth (1948)

1. Use of Taxonomic Literature

The identification of most insect fossil sclerites from Quaternary deposits is a painstaking task, made more difficult by a general lack of suitable identification keys. Most **dichotomous keys** written for the identification of modern insects require entire specimens, or even a series of specimens of both sexes. Such fossil material is rarely available.

Fortunately, there are some exceptions to this rule. Some monographs on beetle families and genera include numerous illustrations, including useful photographs and line drawings, either of complete specimens or of prominent sclerites (head capsules, pronota, and elytra) that are also found in the fossil record. Carl Lindroth's (1961, 1963, 1966, 1968, 1969) series on the ground beetles of Canada and Alaska, and his revision (Lindroth, 1985, 1986) of the Fennoscandian ground beetles are good examples of systematic publications that are very useful to paleoentomologists. Other beetle taxonomists who have provided valuable illustrations in their generic revisions include Milton Campbell, Henri Goulet, and Aleš Smetana at the Biosystematics Research Centre, Agriculture Canada, Ottawa, who included descriptions and scanning electron micrographs (SEMs) of major sclerites frequently found as Quaternary fossils. Smetana's (1985) revision of North American *Helophorus* (Hydrophilidae) species includes SEM plates and detailed descriptions of the **pronotum** of each species. Goulet's (1983) revision of Holarctic Elaphrini (Carabidae) includes many SEM photos of head capsules, pronota, and elytra. Likewise, Campbell's revisions of the North American rove beetle genera *Acidota* and *Olophrum* (Campbell, 1982, 1983) include SEM photos and taxonomic discussions of the features observable on head capsules, pronota, and in the case of *Acidota*, elytra.

It is also possible to gain some familiarity with regional faunas by looking through monographs with abundant, accurate illustrations of regional faunas. Browsing through a series of good photographs or line drawings may provide clues useful in the identification of unknown specimen. The old saying, "one picture is worth a thousand words," illustrates this principle. Matching a fossil specimen and an illustration (either exact or nearly so) can be an important first step in its identification; this greatly reduces the amount of time involved in comparing fossil and modern specimens in a museum collection.

2. Use of Images on Web Sites

Since the publishing of my 1994 book on this topic, dozens of good entomological sites have been published on the World Wide Web. For instance, the BioImages Web site (http://www.bioimages.org.uk/Index.htm) offers a large selection of pictures of insects (mostly British), including more than 2,400 photos of beetles. The Coleopterist Web site has published photographs of 626 species of British beetles at http://www.coleopterist.org.uk/photos-list.htm. Russian beetle images are available from several Web sites. The Zoological Institute of the Russian Academy of Sciences, St. Petersburg has published a beautiful Atlas of the Beetles of Russia online at http://www.zin.ru/animalia/Coleoptera/eng/atlas.htm. This atlas provides color photos and line drawings of Russian beetles in all the major families, and many of the minor families.

A Russian Coleopterist, Konstantin A. Grebennikov, has published photographs of about 300 rove beetles, 100 ground beetles, and 50 Scarabaeidae (dung beetles and chafers), mostly from the lower Volga region (http://rove.front.ru/digital/).

The *Arbeitsgemeinschaft Rheinischer Koleopterologen* (Working group for Rheinland Coleoptera) published over 800 beetle photographs on their Web site (http://www.koleopterologie.de/gallery/index.html). This site is particularly useful, because the authors have published multiple photos of many of the species. North American beetle images have been published quite extensively on the Web. The Cedar Creek Long-Term Ecological Research site in Minnesota has published more than 400 photographs of the beetle species found in their study region (http://www.cedarcreek.umn.edu/insects/albumframes/orderframe.html). The Santa Barbara Museum has published photographs of 156 species of California beetles on their Web site (http://www.sbnature.org/collections/invert/entom/gallery1.php).

Another Web site that features multiple images of species is the site published by the Museum of Comparative Zoology at Harvard University (http://insects.oeb.harvard.edu/MCZ/index.htm). This may be the most ambitious effort to publish beetle photographs that has ever been attempted. The curators in this museum

DEVELOPMENTS IN QUATERNARY SCIENCES
VOLUME 12 ISSN 1571-0866

have published four to six views of more than 16,700 type specimens of insects in their collection, which contains one of the most extensive type collections for North American beetles. The photographs of more than 13,000 species of beetles are available online at this Web site.

3. Comparison with Museum Specimens

While photographs or drawings can help narrow the field when one is attempting to identify fossil beetle sclerites, there is no substitute for spending time in a good Natural History Museum, studying series of modern specimens of the species in question. Only by doing so will the investigator learn the range of variation of characters within each species of a given group, whether it be in body size, the color of the important body parts, or the surface sculpturing of the exoskeleton.

Most Quaternary insects are identified through direct comparison with modern, identified material. It is perhaps no coincidence that the first successful attempts at identifying Quaternary insect species took place in England. The British fauna is relatively small (only a few thousand species), and a synoptic collection of British beetles can be placed on a tabletop. Coope had such a collection when he was working at the University of Birmingham. Of course, having a collection of beetles is very different from having a good knowledge of the species in that collection. It is necessary to develop familiarity with the insect fauna that lives in a study region before attempting to identify insect fossils from that region. Furthermore, the piecing together of a paleoecological scenario based on insect assemblages from a study region must be based on a sound knowledge of the ecological requirements and interactions of the species found in the fossil assemblage. While some information may be gleaned from the literature, there is no substitute for prolonged study of modern material and hands-on experience, gained by years of observing, collecting, and identifying modern beetles from a given region.

4. Relative Size of Regional Faunas

If Coope had started working elsewhere, he might not have achieved such early success. The beetle fauna of North America comprises more than 30,000 species in 98 families (White, 1983). The fauna of Eurasia is at least as large. It is difficult to comprehend the superlative diversity of insects, except perhaps by comparison with other groups of organisms. In contrast, Anderson (1984) estimated that all the mammals of the Quaternary (both extinct and extant) comprise about 500 species. Wilson and Reeder (1993) came up with a much higher estimate of 4,629 extant species in the world, not including extinct Pleistocene taxa. Nevertheless, a mammalogist might still conceivably be competent to know the mammalian fauna of a continent. In contrast, no single entomologist can be expected to retain a comprehensive knowledge of the beetle fauna of a whole continent. Therefore, paleoentomologists seek the assistance of specialists in various families. Most of these taxonomists are housed at national museums of Natural History. Other specialists are scattered in colleges and universities, in government agricultural and forestry offices, and elsewhere. The budding Quaternary entomologist needs to develop contacts with these people, in order to find out who knows how to identify which groups, and who is willing to look at fossil material. In the end, however, it is the fossil worker's responsibility to verify any identifications made by taxonomic specialists.

The fauna of any single region is necessarily much smaller than the fauna of a whole continent, especially if the study region is in the higher latitudes, where species diversity is greatly diminished. I have found that after a few years of working on the fossil fauna of a given region, it is possible to develop enough familiarity with the taxa to be able to identify most of the fossil material without the help of specialists working on the modern fauna.

Fortunately, the exoskeletons of beetles, ants, caddisfly larvae, and some other insects and arachnids, exhibit a multitude of features useful in the separation of fossils to orders, families, genera, and species. The general success rate for the identification of fossils of major body parts (head capsules, thoraces, and elytra) to the species level is about 50%. That is, about half of the major sclerites in a given fossil assemblage will eventually be identified to species. This success rate varies from region to region. British workers have had far greater success with some samples. For instance, Coope and Angus (1975) identified 252 species, or nearly 90%, of a rich beetle fauna comprising 282 taxa from the Isleworth site in England. Of course, many insect body parts, such as leg segments, antennal segments, and abdominal sclerites, cannot be identified even to the family level. The investigator quickly learns which types of specimens are worthwhile studying in detail, and which are not.

5. Useful Characters for Fossil Beetle Identification

One reason that beetles preserve so well in sediments is that many species' exoskeletons form a hard, armor-like case over their bodies. Therefore, the major sclerites, especially on the dorsal side, are not just flimsy body coverings, but large, heavily sclerotized plates. The main exoskeletal parts that preserve in sediments are the head capsule, the pronotum (the dorsal thoracic shield), and the wing covers, or elytra (Fig. 1). Many of these plates are ornamented with a number of features that are preserved for as long as the sclerites themselves are preserved. These characters include striae (either isolated or in rows), **carinae** (Fig. 2B), grooves, tubercles, punctures, **setae** (or points of attachment of setae, if these have broken-off), scales (Fig. 2A), and rows of teeth. Each of these features represents character states that may exhibit a wide range of variation from one taxon to another. For instance, punctures range from deep to shallow and broad to narrow; they may contain a seta or not; they occur in dense patches, in rows, in small groups, or are widely dispersed; the shape of punctures varies from round to oblong to quadrate, and so on.

Microsculpture, some of which is visible only at high magnification (150–200X), is part of the exoskeleton of

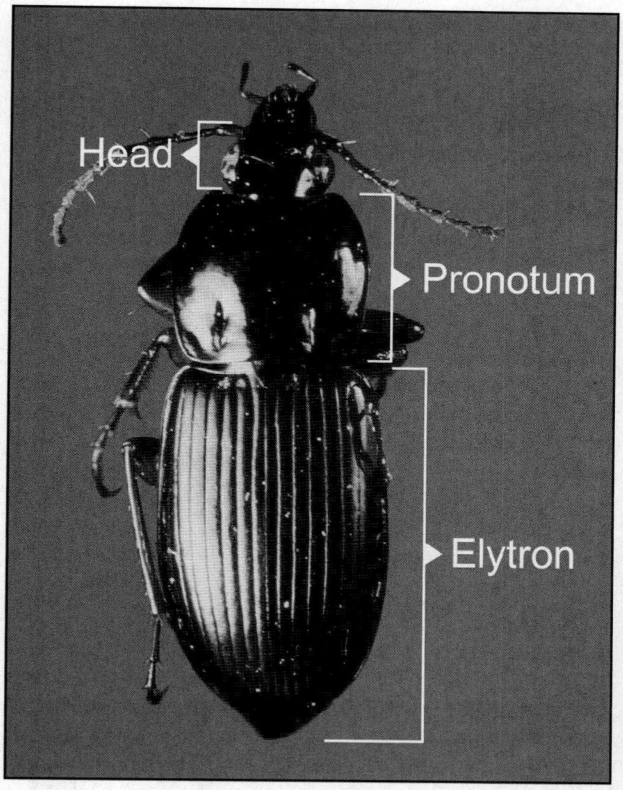

Fig. 1. Photograph of Pterostichus leconteianus, *showing the principle sclerites preserved in Quaternary fossil assemblages that are useful for identification (photo by the author).*

most insects. As Lindroth (1948) observed, microsculpture is very useful in identifying fossil beetles, because the microlines do not degrade or alter with time, and it is remarkably constant within species. Most microsculptural details are best viewed using a light source from a very oblique angle. Unfortunately, bright light reflecting on the shiny surface of a specimen makes viewing of microsculpture difficult. This problem is overcome by setting a small screen of opaque Mylar film between the light source and the specimen, as described by Campbell (1979). The screen effectively diffuses the light. Another approach is to use a surface-illuminating (petrographic) microscope. These are used by geologists to study surfaces of opaque minerals. They can easily reach a magnification of 400X, and they have a polarized light source that eliminates glare.

A few beetles are devoid of microsculpture, giving them a shiny appearance. Others have patches of microsculpture only at the margins of the pronotum and apex of elytra. Many species are completely covered with very dense patterns of reticulate lines, often in diagnostic shapes. These range from **isodiametric meshes** (Fig. 3A) to longitudinally or transversely elongated meshes (Fig. 3D). In some species, the meshes are broken into transverse rows of very fine lines (Fig. 3C). These serve as diffraction gratings that cause iridescence. Other beetles have a very dull, mottled appearance because of dense, granular microsculpture and ornamentation (Fig. 3B). Sometimes the surface microsculptural pattern is broken up by dense patches of **setaceous punctures** (Fig. 3F).

Microsculpture is thought to serve a variety of functions in insects, as discussed in Crowson (1981, pp. 300–302). Iridescence is thought by some authors to provide the beetle with some measure of protection from solar radiation, since it is most common in species living in low latitudes (Lindroth, 1974). Some microsculpture serves a cryptic function. One species of ground beetle, *Agonum bembidioides*, has alternating patches of short microsculptural lines at oblique angles to each other. This produces alternating patterns of silver and black spots. As Lindroth (1966) noted, this peculiar attire is extremely efficient camouflage for the beetle, greatly resembling opalescent pieces of charcoal in the newly burnt forests it inhabits.

Metallic coloration is fairly common in beetles, especially in some families (e.g., Cicindelidae, Carabidae, Scarabaeidae, Buprestidae, Meloidae, and Chrysomelidae). These are called structural colors, because they result from light interference from thin films of the cuticle. These colors persist even when the outer waxy layers of epicuticle are removed; they are probably derived from the outer part of the endocuticle (Crowson, 1981, p. 303). Some metallic-colored fossil specimens become darkened through time, but the metallic sheen reappears when the fossils are wetted. In such specimens, the colors may shift in hue. For instance, if a beetle was originally a metallic green its fossil sclerites may show a blue-green color when wetted (Coope, 1959). Bright coloration probably serves as a defense mechanism in insects. If a metallic or iridescent beetle runs from the shade into bright sunlight, it flashes a bright, startling color that may allow the beetle to elude a predator.

Another form of coloration is due to pigments. Pigment-based colors may or may not be preserved in insect fossils. Sclerites that were originally yellow, orange, and red tend to thin and become frail through time, whereas dark brown or black coloration due to tanning of chitin may persist (Coope, 1959). Hence, the black spots on a ladybird beetle (Coccinellidae) may preserve longer than the red or orange background, which disappears, leaving only the spots, like the smile on the Cheshire cat. Conversely, pale stripes or spots chemically erode, leaving only the surrounding dark field (Fig. 4F). However, pigmented ornamentation such as spots, stripes, and **maculae** persist well enough in most fossils to serve as identification aides. In fact, coloration patterns are sometimes diagnostic in identifying species of ground beetles, predaceous diving beetles (Dytiscidae), ladybird beetles (Coccinellidae), leaf beetles (Chrysomelidae), and others (Fig. 4). Often, pigmented color patterns accompany patterns in surface sculpture of beetle elytra or pronota. For instance, in the metallic wood boring beetle (Buprestidae) genus *Acmaeodera*, patches of yellow pigment are slightly elevated above the surrounding dark surfaces, giving the appearance that yellow paint was spilled on the surface of the elytra (Fig. 4A,B). In many leaf beetles, color patterns are accompanied by patches of impunctate, slightly elevated maculae, as in the genus *Pachybrachis* (Fig. 4C,D).

Generally, the ventral abdominal sclerites become separated from each other and isolated in sediments. However, whole abdomens of beetles are occasionally

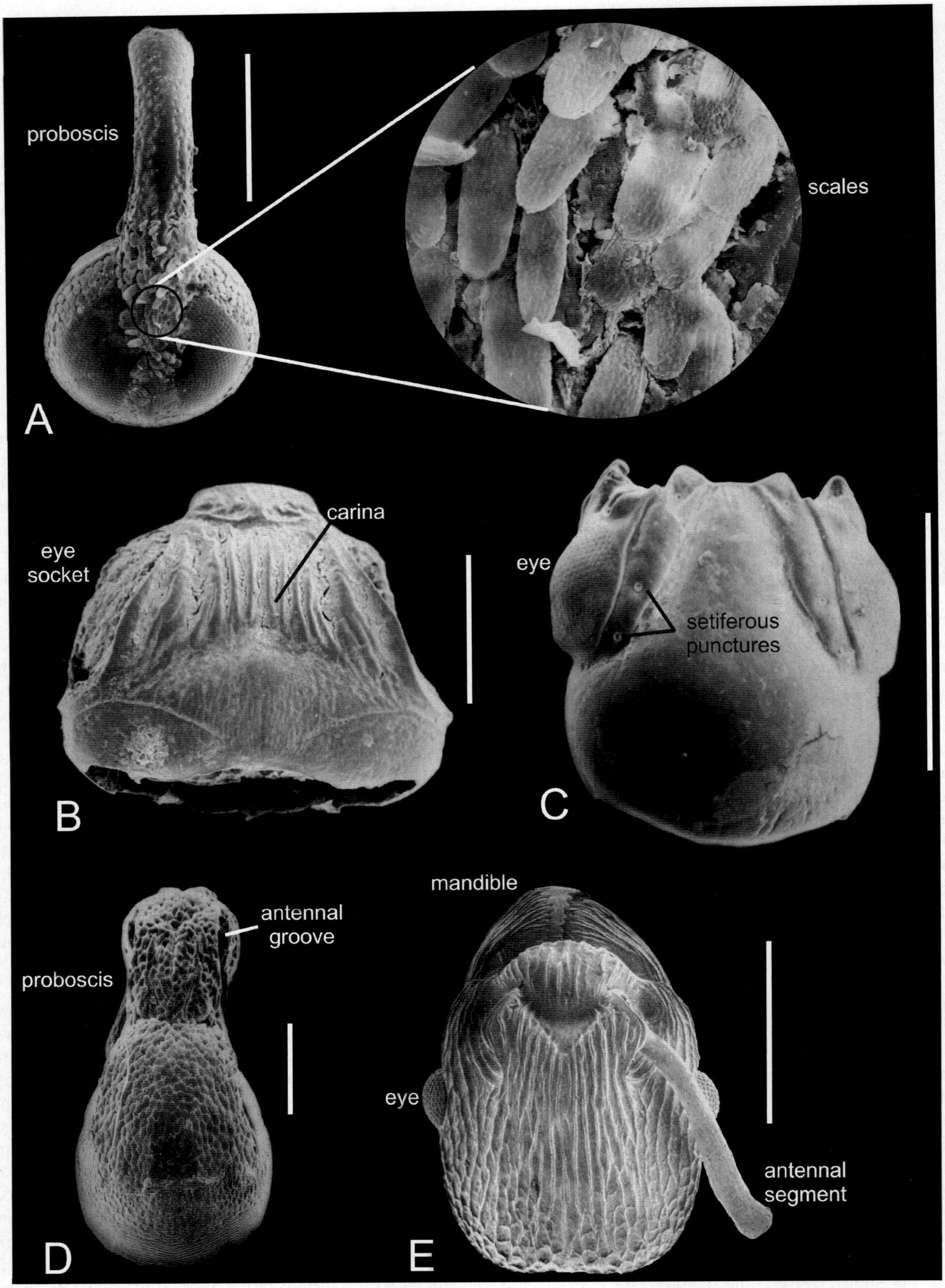

Fig. 2. SEMs of fossil insect head capsules. (A) The weevil Cylindrocopturus armatus, *showing preservation of scales (inset); (B) the ground beetle* Notiophilus borealis, *with longitudinal carinae on the frons; (C) the ground beetle* Bembidion, *showing setiferous punctures adjacent to the eyes; (D) the weevil,* Sapotes longipillis, *showing relatively short rostrum; (E) the ant* Myrmica alaskensis, *showing preserved eye, antennal scrobe (first segment of the antenna) and mandibles (photos by the author). Note the smooth surface of (C), the heavily impressed microsculpture of (D), and the reticulate sculpturing of (E). Each scale bar = 0.5 mm (all photos by the author).*

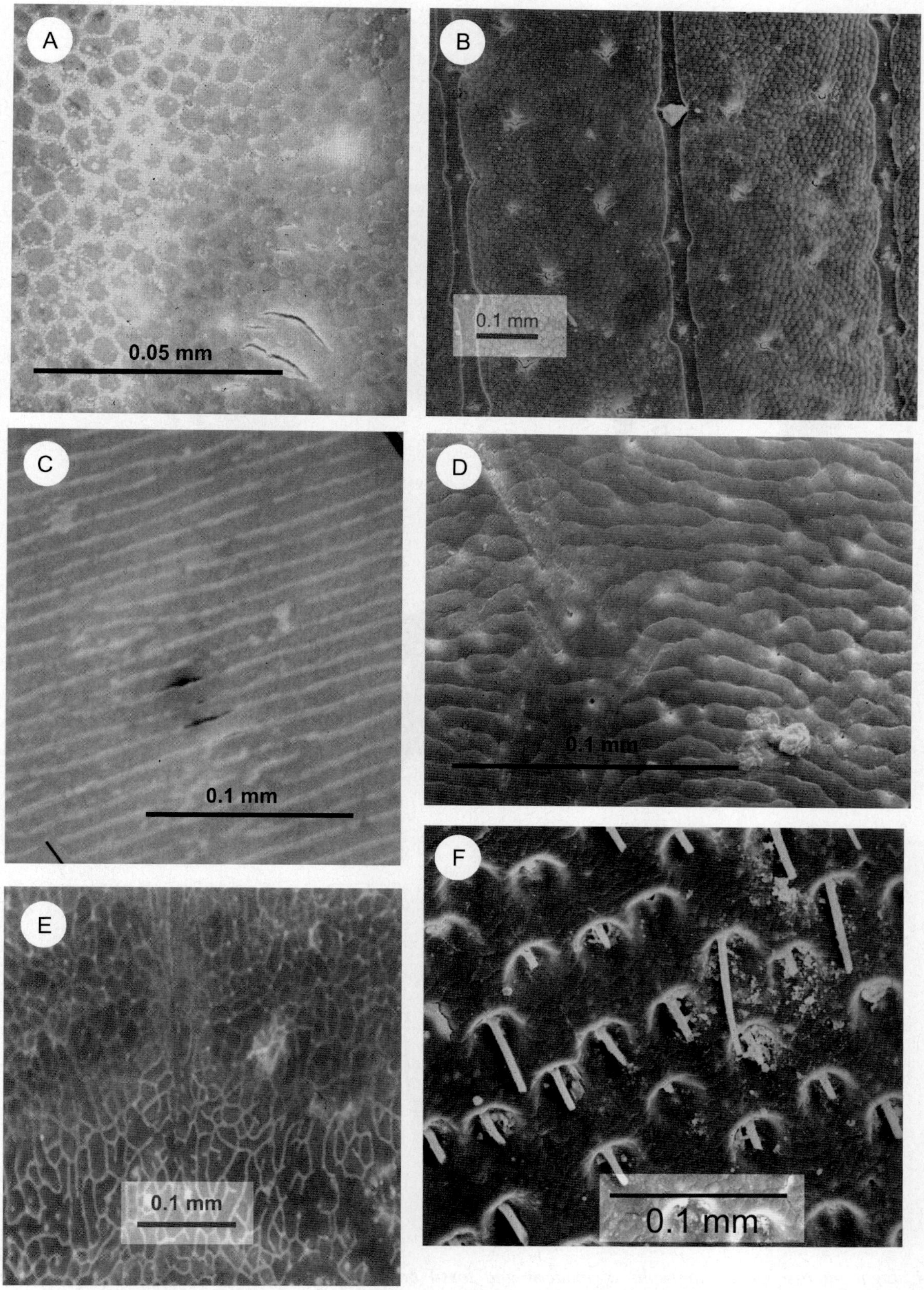

Fig. 3. SEMs of fossil beetles, showing microsculptural patterns. (A) Isodiametric mesh on the pronotum of the ground beetle Bembidion rapidum*; (B) dense, granular mesh on the elytron of the dung beetle* Onthophagus*; (C) fine transverse microsculptural lines on the elytron of the ground beetle* Selenophorus gagatinus *that cause iridescence; (D) horizontally stretched meshes on the pronotum of the rove beetle,* Tachinus brevipennis*; (E) alutaceous (leatherlike) microsculpture on the pronotum of the predaceous diving beetle,* Agabus arcticus*; (F) surface sculpture broken by numerous setaceous punctures on the pronotum of the carrion beetle* Thanatophilus truncatus *(all photos by the author).*

Fig. 4. Light microscope photographs of modern and fossil beetle specimens, showing pigmented color patterns. (A) Acmaeodera jamaicensis *(Buprestidae) modern specimen; (B)* Acmaeodera cuneata *fossil elytron from West Texas packrat midden; (C)* Pachybrachis mitis *(Chrysomelidae) fossil elytron from West Texas packrat midden; (D)* Pachybrachis hieroglyphicus *modern specimen; (E)* Hippodamia convergens *modern specimen (Coccinellidae); (F)* Hippodamia convergens *fossil pronotum from Lake Isabelle, Colorado; (G)* Hyperaspis trifurcata *modern specimen (Coccinellidae); (H)* Hyperaspis trifurcata *fossil elytron from West Texas packrat midden; (I)* Hyperaspis erythrocephala *modern specimen (photo A courtesy of the Natural History Museum, London; photo D courtesy of A. N. Posedko, Zoological Institute, Russian Academy of Sciences; photo E courtesy of Department of Entomology, Iowa State University; photo G courtesy of Joyce Gross, University of California, Berkeley; photo I courtesy of K. V. Makarov, Zoological Institute, Russian Academy of Sciences, St. Petersburg; and fossil photos by the author.*

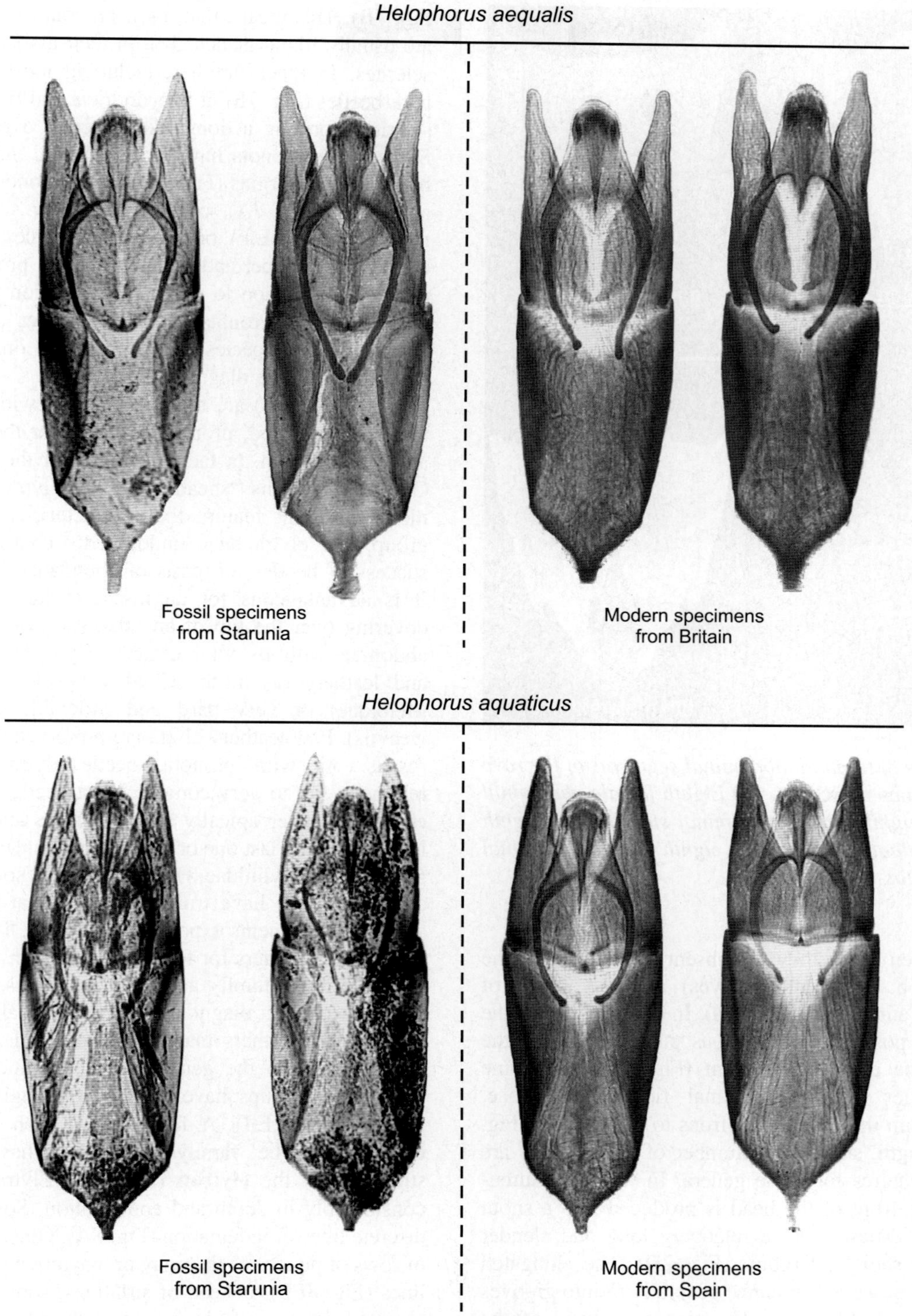

Fig. 5. Light microscope photographs of fossil and modern aedeagi of Helophorus aequalis *(above) and* Helophorus aquaticus *(below). The fossil specimen from the Starunia site in the Ukraine (see Chapter 1) was originally identified as* Helophorus dzieduszckii *by Lomnicki (1894) (photos by Robert Angus, Royal Holloway, University of London, used with permission).*

preserved intact. Inside the abdominal exoskeleton of male beetles is the **aedeagus**, or male genitalia (Fig. 5). This is also partially sclerotized, and so preserves well in many organic sediments. In addition, the terminal abdominal sclerites, especially in some rove beetles such as the genus *Tachinus*, are produced into a series of elongated lobes (Fig. 6). The shapes of these sclerites are the principal diagnostic features used to identify both modern and fossil specimens.

The head capsules of many families of beetles contain useful features, including the shape, position, and size of eye sockets, the position and shape of mouthparts (many mandibles are found separately from head capsules, but their position and points of attachment on the head can be

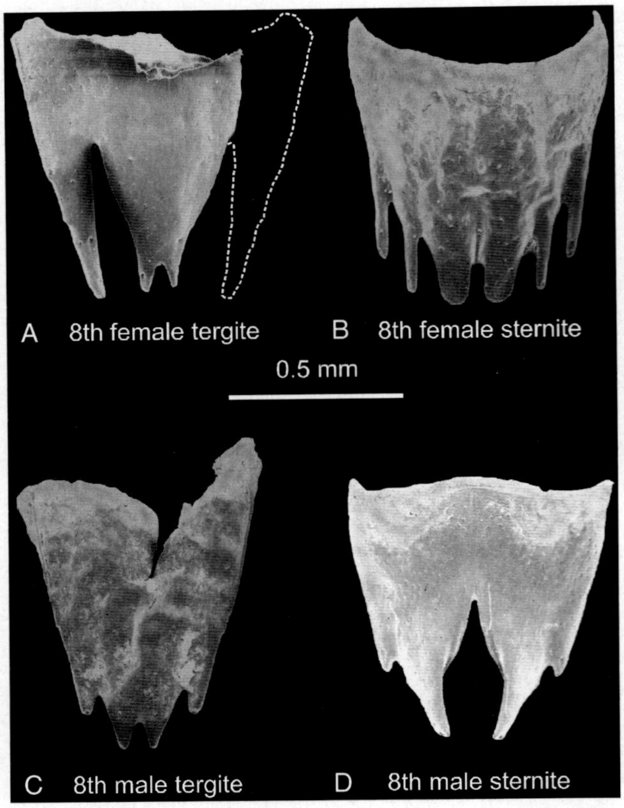

Fig. 6. SEMs of terminal abdominal segments of the rove beetle, Tachinus nearcticus. *(A) Eighth female abdominal tergite; (B) eighth female abdominal sternite; (C) eighth male abdominal tergite; (D) eighth male abdominal sternite (photos by the author).*

discerned even when they are absent), the shape of the frontal region (between the eyes) and the shape of the **clypeus** (anterior to the **frons**). In ground beetles, the number and position of setigerous punctures inside the margin of the eyes is important (Fig. 2C), and some ground beetles have longitudinal ridges or carinae, extending from the base of the frons to the clypeus (Fig. 2B). The length, shape, and number of these ridges are diagnostic features for some genera. In weevils (Curculionidae), the front of the head is produced into a snout or proboscis. These can be relatively long and slender (Fig. 2A) or short and robust (Fig. 2D). The elongated first segment of weevil antennae generally fit into grooves along the snout (Fig. 2D). The size and shape of the snout, and the shape and position of antennal grooves are important taxonomic features. The head capsules of ants (Hymenoptera: Formicidae) preserve relatively well, because many are heavily reinforced with chitin. The genus *Myrmica*, commonly called the black ant, is characterized by having a heavy reticulate sculpturing on its head capsule (Fig. 2E). The specimen pictured in Fig. 2E is exceptionally well preserved, with the eye facets, mandibles, and first antennal segment still in place.

The dorsal prothoracic shield, or pronotum, is often useful in beetle identification. In some families, such as most ground beetles and predaceous diving beetles (Dytiscidae), the pronotum is often a broad, flat plate that is completely divided from the ventral part of the prothorax Fig. 7B). This means that fossil pronota of these families are usually found detached from their associated thoracic sclerites. In other families, including most weevils and bark beetles (Fig. 7F), it is cylindrical and fixed to ventral sclerites, forming a dome-like covering over the thorax. Some beetle pronota have straight lateral margins, others have sinuate margins (Fig. 7D). Some pronota are smooth and shiny (Fig. 7A), others are heavily sculptured and **rugose** (Fig. 7C,E,F), or covered with scales or setae. The overall size, shape, and texture of many pronota provide enough information to enable identification to the family or genus level. Pronota of ground beetles offer the best characters for species determinations; only the male genitalia are more diagnostic.

Elytra (Fig. 8) are the modified forewings of beetles that are reinforced, protective covers for the hind wings and the abdomen. In fact, the name for the beetle order, Coleoptera, means ''sheath winged.'' Elytra are the single most important feature that characterizes beetles as a group, and elytra have undoubtedly contributed to the success of beetles, in terms of abundance and diversity. It is advantageous for an insect to have a protective covering over the top of an otherwise somewhat fragile abdomen with its vital organs. Elytra may be pliable and leathery (as in the Cantharidae, Lampyridae, and Meloidae) or very hard and inflexible (as in many weevils). Few leathery elytra are preserved as Quaternary fossils. As with pronota, beetle elytra range from relatively flat to very convex. Most beetle families have elytra that taper apically and cover the abdomen (or at least all but the last one or two segments). However, some families (Staphylinidae, Hydroscaphidae, some Silphidae, and Histeridae) have truncated elytra that leave several abdominal segments exposed (Fig. 8C,D). The usefulness of elytral characters for fossil identification varies greatly from family to family and genus to genus. Some elytra contain sufficient diagnostic characters to allow a species identification; others may not even be identifiable with any certainty to the generic level. Ground beetles and some other groups have elytra with longitudinal **striations** (Fig. 8A,E,I). A few beetles, such as the genus *Choleva* in the family Catopidae, have transverse striations on the **elytron** (Fig. 8B). Elytral striae vary considerably in depth and composition. Some consist of discrete lines of indentation (Fig. 8A). Other striae consist of rows of punctures that may or may not be connected by lines (Fig. 8E,I). Instead of striations, some beetles have elevated ridges, or carinae running their length (Fig. 8H). Many beetles have no striations on their elytra. Instead, they may be essentially smooth (Fig. 8D), or they may be covered with scattered punctures that do not form discernable rows (Fig. 8G). Finally, some beetles have ''confused'' striae that start and stop, curve rather than running straight to the apex of the elytron, or are interrupted by bare patches. The leaf beetle *Pachybrachis mitis* is a good example of this (Fig. 8F). It has striae with flat intervals near the base of the elytron, becoming convex intervals toward the sides and apex. Rows of punctures start and stop, and compete with patches of other punctures on the surface. The convexity of intervals between striations is another variable in this set of taxonomic characters. Some intervals between

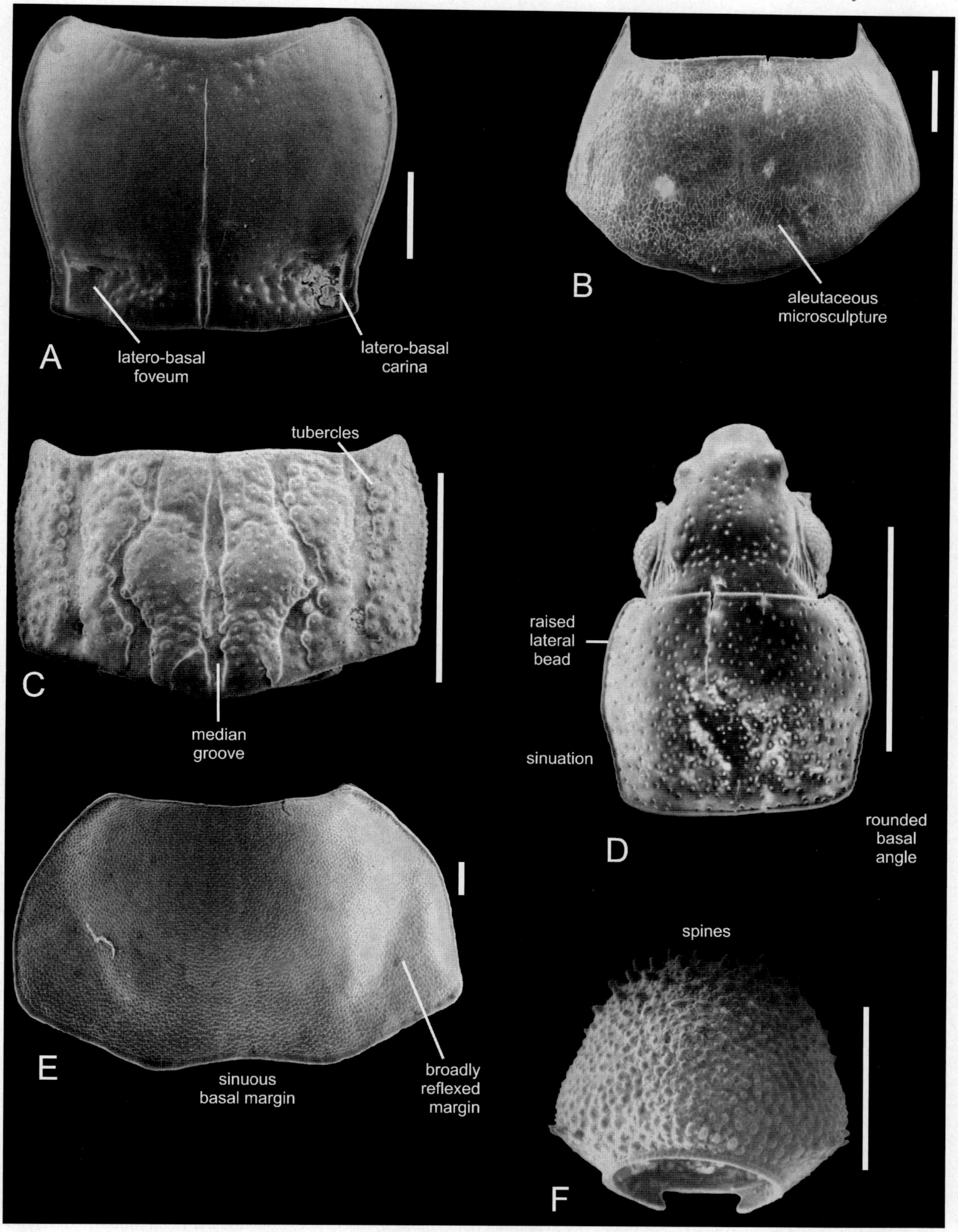

Fig. 7. SEMs of beetle pronota. (A) Patrobus foveocollis *(Carabidae); (B)* Agabus arcticus *(Dytiscidae); (C)* Helophorus auricollis *(Hydrophilidae); (D)* Pycnoglypta *near* P. lurica *(Staphylinidae); (E)* Thanatophilus truncatus *(Silphidae); (F)* Stephanopachys sobrinus *(Bostrichidae) (photos by the author). Scale bar = 0.5 mm.*

Fig. 8. SEMs of beetle elytra. (A) Selenophorus gagatinus *(Carabidae); (B)* Choleva americana *(Catopidae); (C)* Hypocaccus estriatus *(Histeridae); (D)* Tachyporus dimorphus *(Staphylinidae); (E)* Aspidoglossa subangulata *(Carabidae); (F)* Pachybrachis mitis *(Chrysomelidae); (G)* Stephanopachys sobrinus *(Bostrichidae); (H)* Rhagodera costata *(Zopheridae); (I)* Apion alaskanum *(Curculionidae) (photos by the author). Scale bar = 0.5 mm.*

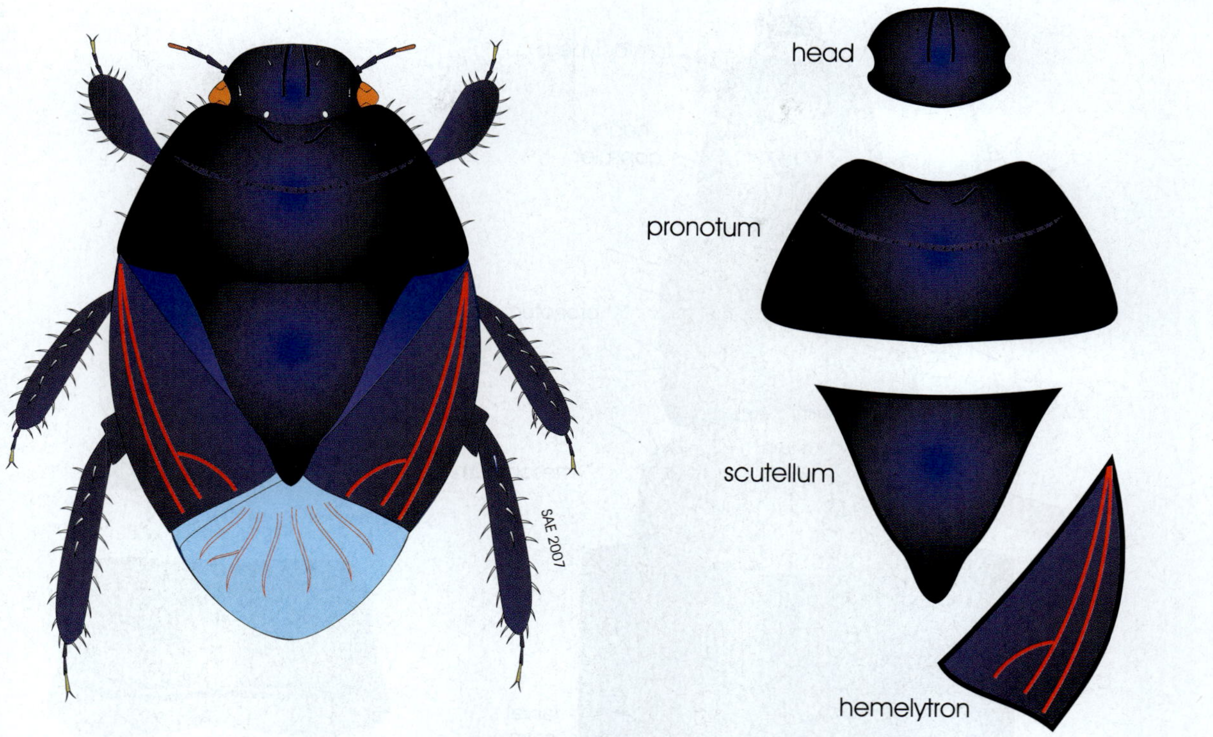

Fig. 9. Generalized drawing of a burrower bug (Cydnidae), showing sclerites frequently preserved as Quaternary fossils in desert regions (after Froeschner, 1960).

striae are essentially flat, as if the striae were lines etched in a rock face (Fig. 8A). Others have more-or-less convex intervals between striations (Fig. 8I), as if the lines were drawn across the surface of bread dough. These features vary between species, but not *within* species, so they are helpful in species diagnosis.

Different depositional environments (i.e., fluvial vs. lacustrine deposits) affect the numbers and types of insect body parts in assemblages. The various skeletal elements often have different hydrodynamic properties so that if the fossils are transported to any degree, say by fluvial action, they can become dissociated and sorted so that elytra may be concentrated in some sediments whilst the heads and pronota may be deposited elsewhere (Morlan and Matthews, 1983). Differences in buoyancy may also influence the representation of the various parts of an insect during kerosene floatation. Sediment retained in head capsules or thoraces causes them not to float as well as body parts that are free of sediment. Furthermore in some species the various parts of the exoskeleton have very different robustness (e.g., *Helophorus*), so that heads and pronota survive whereas the filamentous elytra may not.

6. Identification of Other Groups of Fossil Insects

Several other insect orders are frequently encountered in Quaternary sediments. Many ants (Formicidae) have heavily sclerotized head capsules that contain sufficient diagnostic characters to allow specific identifications (Fig. 2E). Ant mandibles are heavily sclerotized, and are often found in fossil assemblages. In some cases, it is not possible to identify ant remains to the species level, but even generic ant identifications may be of use in paleoenvironmental reconstructions. For instance, the mandible of a carpenter ant (genus *Camponotus*) offers evidence of the availability of wood at the fossil site at the time of deposition, since carpenter ants make their nests in rotting logs. However, identification of ant fossils is complicated by the fact that they form **castes** (queens, males, and workers), the head capsules of which are different. Most ants are workers, but the head capsules of queens have been found in some assemblages (e.g., Francoeur and Elias, 1985; Mackay and Elias, 1992). Ants, like many beetles, are general predators and scavengers not tied to specific prey or host plants; their fossils supply valuable information on past climatic conditions. For instance, in northern studies, ants provide evidence of conditions suitable for the growth of trees, since no ants are known to inhabit regions well-beyond arctic **treeline** (Gregg, 1972; Francoeur, 1983). Their fossils may occur in samples representing subarctic conditions.

Head capsules, thoraces, and the sclerotized portion of the hemelytra of bugs (Hemiptera and Homoptera) are also common in Quaternary sediments. Fossils of stink-bugs (Pentatomidae), seed bugs (Lygaeidae), and leaf-hoppers (Cicadellidae) are fairly common in terrestrial sediments, and water striders (Gerridae), shore bugs (Saldidae), water boatmen (Corixidae), and backswimmers (Notonectidae) are found in fluvial and lacustrine sediments. Recently, packrat midden assemblages from the Chihuahuan Desert have yielded head capsules and thoraces of Cydnidae (burrower bugs) (Fig. 9), and also head capsules of some Reduviidae (assassin bugs) that parasitize packrats (Elias and Van Devender, 1990, 1992). Coope *et al.* (1961) and Coope and Sands

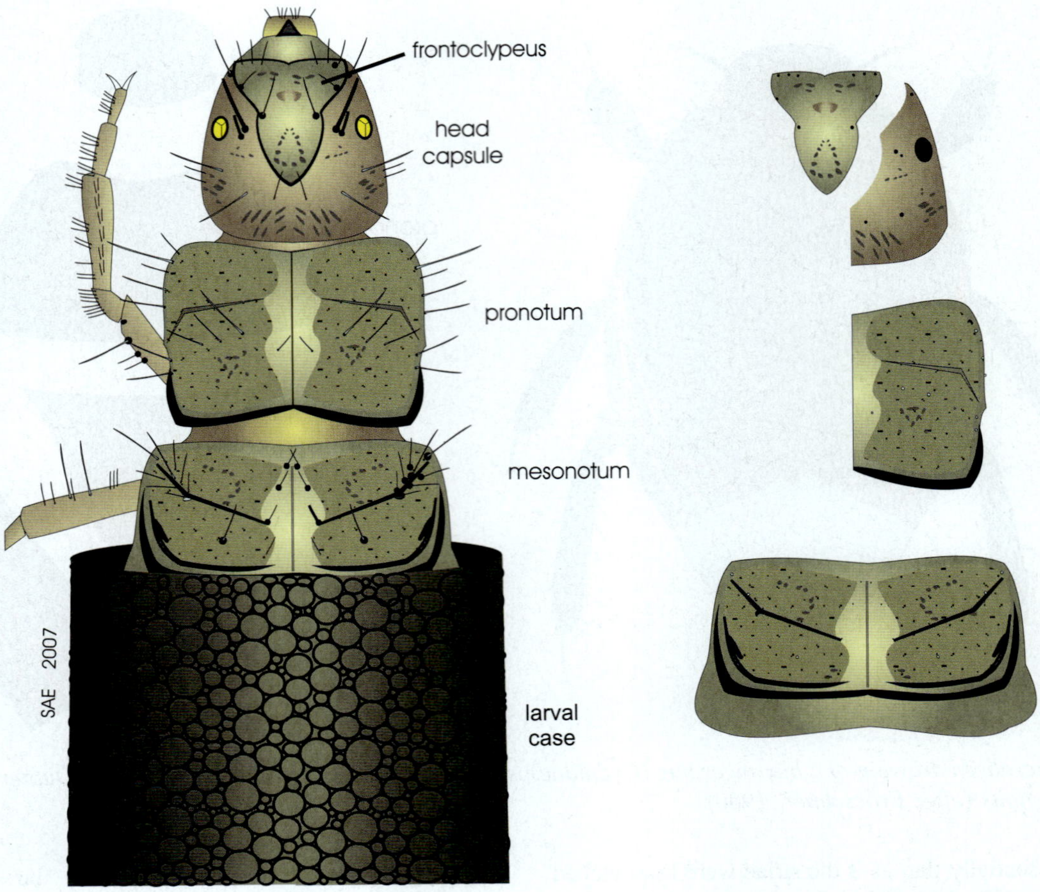

Fig. 10. Sketch of the head and thorax of a caddis fly larva (after Wiggins, 1977), showing sclerites frequently preserved as Quaternary fossils.

(1966) specifically identified both Hemiptera and Homoptera specimens. In the latter paper, the authors identified a non-British species of saldid, *Chiloxanthus stellatus* Curtis, that now ranges from northern Fennoscandia eastwards to Alaska and northern Canada.

The larvae of caddisflies (Trichoptera) are aquatic, and sclerites from the head capsule and thorax of caddis larvae are abundant in some lacustrine sediments (Fig. 10). The frons and clypeus of caddisflies are fused into a single sclerite, called the **frontoclypeus**. Besides the overall shape of the frontoclypeus, it also contains a number of other diagnostic features, including the position of setigerous punctures around its margins, surface sculpturing, and size, shape, and coloration of muscle scars that are revealed in transmitted light (Fig. 11). The pronotum and mesonotum are also useful for identification (Williams, 1988). The cases built by caddisfly larvae are occasionally found in ancient lake sediments. The size and shape of cases, as well as the materials used in their construction, are often diagnostic to the family or genus level. Though more rare, fossils of adult caddisflies have also been found.

Caddis larvae provide valuable information on the waters they inhabit, as many species have narrow thermal tolerances, as well as being sensitive to the trophic status and pH of the water (Wiggins, 1977). Some species require specific substrates and build larval cases from particular substances (sand grains of a certain size, reeds, and even snail shells). A recent article by Howard *et al.* (2009) discusses the use of caddis larval fossil data, in combination with aquatic beetle fossil data, to reconstruct the flow regimes of ancient rivers in England. By examining the fluvial environments associated with the aquatic insect faunas preserved in fluvial sediments, the authors were able to differentiate a series of flow regimes for the ancient River Trent for the Mid- to Late Holocene. The research group has used the Lotic invertebrate Index for Flow Evaluation, or LIFE method developed by Extence *et al.* (1999). This method assigns a flow-velocity score to each species of aquatic insect found in an assemblage. The scores range from I (semipermanent aquatic habitat) to VI (fast-flowing stream). This method was first applied to fossil assemblages by Greenwood *et al.* (2006). By sampling sediment monoliths across the river channel, Howard *et al.* (2009) were able to test the cross-channel replicability of the method. Their results confirmed that the method is successful and replicable across a river channel. It is, therefore, deemed to be a powerful paleolimnological tool.

The study of fossil midge larvae (Diptera: Chironomidae) also plays an important part in **paleolimnology**. The larvae are aquatic, and, like caddis larvae, the species composition of midges in lake sediments can provide a great deal of information on water quality and substrates (Walker, 2007). The head capsules of midge larvae (Fig. 12) are the most frequently encountered fossils.

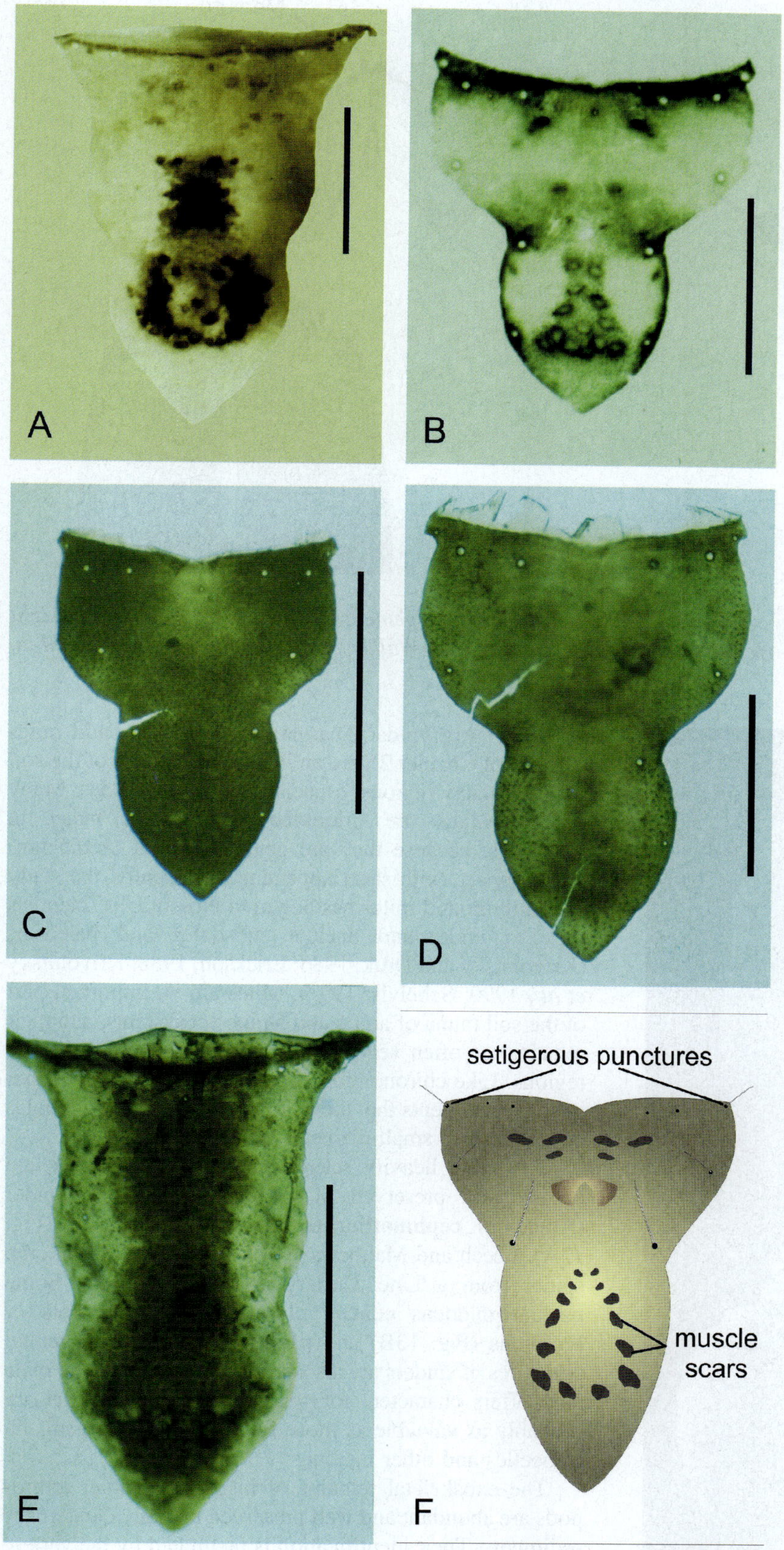

Fig. 11. Light microscope photographs of fossil caddis fly larval fossils. (A) Frontoclypeus of Agrypnetes *(Phryganeidae); (B) frontoclypeus of* Asynarchus *(Limnephilidae); (C) frontoclypeus of* Grensia *(Limnephilidae); (D) frontoclypeus of* Limnephilus*; (E) frontoclypeus of* Agrypnia *(Phryganeidae); (F) drawing of caddis fly larval frontoclypeus, showing principal diagnostic features (photos by the author).*

Phantom midge larvae (Diptera: Chaoboridae) have also been used as paleolimnological indicators (Uutala, 1990).

Fossil midge samples should be taken from the center of a lake, rather than from lake margins, where fossil beetle samples are taken (Hofmann, 1986). Midge head capsules occur in such great numbers in many lake sediments that even a small-diameter **piston core** of the type often used by palynologists provides samples with sufficient numbers of specimens for a detailed analysis. This is why the study of chironomid fossils is often performed in concert with pollen analysis, such as the study of paleoenvironments at Marion Lake, British Columbia, by Walker and Mathewes (1987).

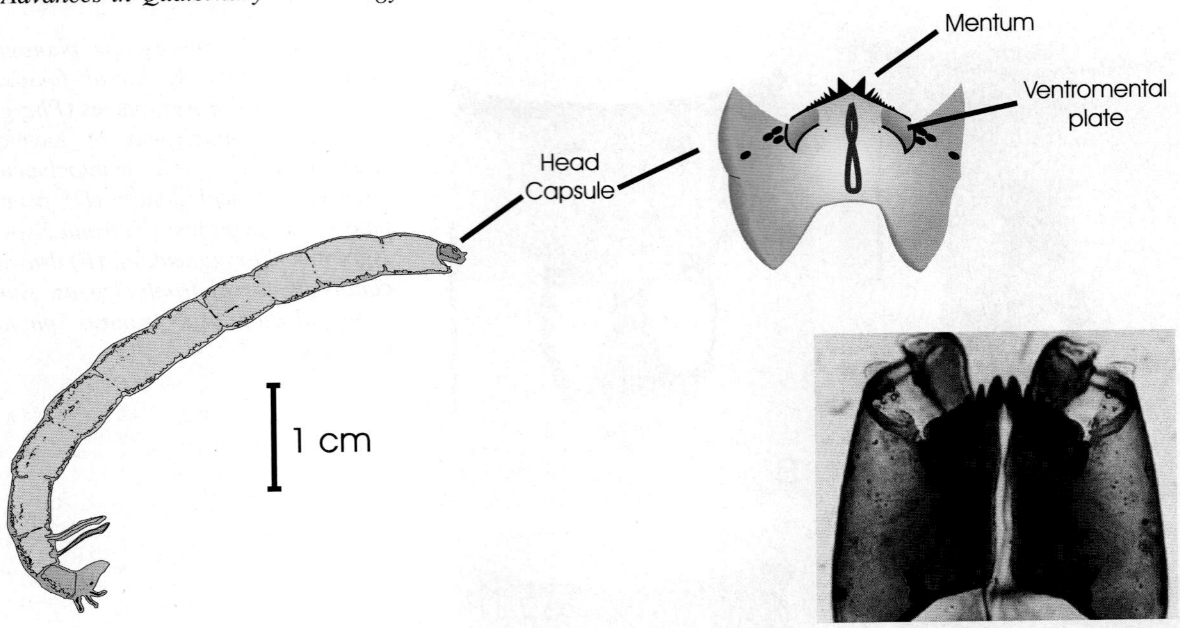

Fig. 12. Drawing of a midge larva (Diptera: Chironomidae), showing an enlargement of the head capsule. Inset: Light microscope photograph of head capsule of Heterotrissocladius (photo by Ian Walker, University of British Columbia, used by permission).

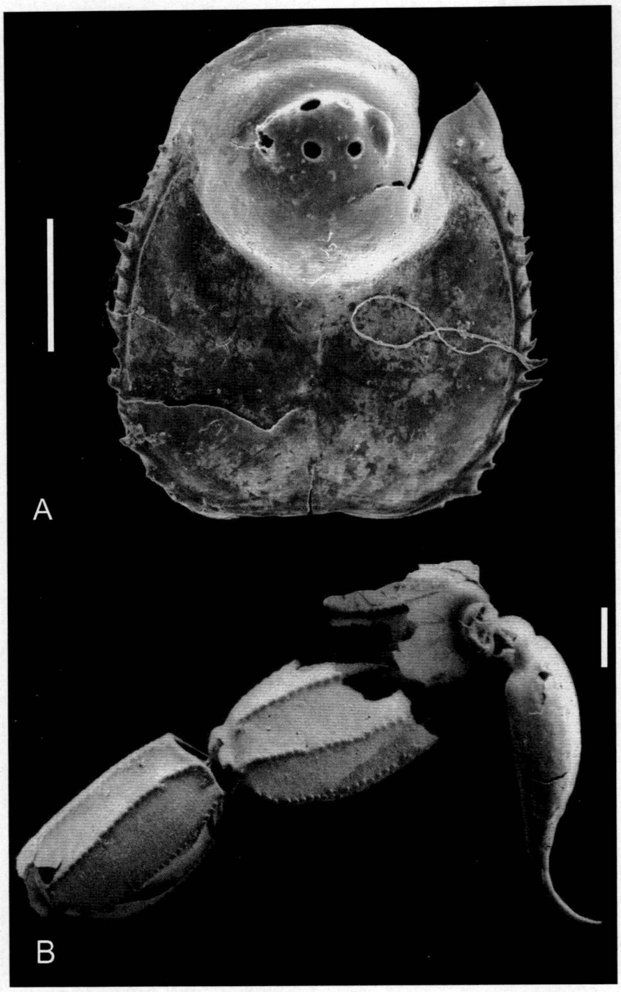

Fig. 13. SEMs of fossil arachnids. (A) Cephalothorax of the spider, Erigone dentiger; *(B) tail segments of a scorpion, cf.* Vaejovis *(photos by the author). Scale bar = 0.5 mm.*

Mites (Arachnida: Acarina), especially oribatid mites (Fig. 27 of Chapter 2), are an important element of the soil fauna of many regions, especially at high latitudes. Many oribatid mites are preserved more-or-less intact in sediments, because they are generally small (<0.5 mm) and compact, with short appendages. Recently, the study of fossil oribatid mites has begun to blossom, as scientists seek to understand ancient substrates and paleosols (Krivolutsky and Druk, 1986; Erickson, 1988; Krivolutsky *et al.*, 1990; Schelvis, 1997). Mites are an important part of the soil fauna of arctic and alpine ecosystems, and their remains are often very abundant in sediments from those regions. Like chironomids, oribatid mites are so abundant in some sediments that they can be obtained from samples collected from small-diameter piston cores.

The more heavily sclerotized parts of arachnids are occasionally preserved in Quaternary deposits. Spider (Araneida) **cephalothoraces** are not uncommon (Fig. 13A). Leech and Matthews (1971) described a fossil crab spider from a Late Tertiary deposit in Alaska. Some packrat middens contain claws and tail segments of scorpions (Fig. 13B) and **pseudoscorpions**. The genital structures of spiders are also occasionally found. The male palp offers characters for specific identification that are probably as valuable as those found in the male genitalia of beetles and other insects.

The exoskeletal remains of insects and other arthropods are abundant and well preserved in many Quaternary sediments. Their identification is facilitated by the variety of diagnostic features preserved on the fossil sclerites. The major obstacle in most fossil insect identifications is the investigator's lack of familiarity with the taxon in question. Nevertheless, as John Matthews (written communication, 1992) has said, there is no mystique in the identification of insect fossils. All it takes is patience and a good eye for detail.

4

The Value of Insects in Paleoecology

> In number of described species beetles represent the largest group of organisms ... their role in the operation of ecosystems, particularly on land, should never be underestimated.
>
> – M. Ghilarov in Crowson (1981)

1. Abundance and Diversity of Insect Assemblages

Insects are arguably the most successful group of organisms ever to have inhabited planet Earth. The current crop of insect orders has been in existence since long before the age of dinosaurs, and they represent roughly three-quarters of all animal species known at present. Insect abundance and diversity are important elements in their fossil record, because the fossil assemblages are likewise often abundant and diverse, providing an unusually rich record of past life.

This chapter focuses on beetles, because they are the most important insect order in the Quaternary fossil record. They are also the most diverse order, with more than 300,000 known species (White, 1983); about 1,500 new species are described each year (Arnett, 1973). Beetles account for 20% of all known species of organisms, which is more than all flowering plants combined. Little wonder then that the great defender of evolutionary theory, J. B. S. Haldane, said ''If one could conclude as to the nature of the Creator from a study of his creation it would appear that God has a special fondness for stars and beetles.''

The diversity of beetles is difficult to comprehend. Crowson (1981) lists 168 families of beetles. This taxonomic variety corresponds to an equally diverse ecological complexity. Beetles occupy almost every conceivable ecological niche and type of habitat on land and in freshwater. This makes them an important group in the fossil record, because their remains serve as **proxy data** for a wide variety of habitats and environmental conditions. They are found from arctic polar deserts (Danks, 1981) to the subantarctic islands (Crowson, 1981), and at elevations as high as 5,600 m in the Himalaya (Mani, 1968).

Beetles live in a bewildering variety of specialized habitats, thus their fossil remains provide a richness of paleoecological detail. Dead trees and associated fungi nurture hundreds of different species of beetles. Fossils of these species often survive to tell of ancient old-growth forests, even if the trees themselves are not preserved. Besides living in lakes, ponds, and streams of all sizes, some aquatic beetles also inhabit hot springs and brackish water.

Some ground beetles and rove beetles live in mammal and bird nests, where they prey on fleas, mites, and other parasites. The larvae of carrion beetles (Silphidae) feed on carrion, but adults also prey on other insects (especially maggots) found on carrion (Anderson and Peck, 1985). Again, the fossil remains of these insects tell us that their avian and mammalian hosts were present at the time and place in question, whether or not the hosts were preserved as fossils.

Predatory beetles are an important component in most fossil assemblages, and they have much to tell us about past environments. Predatory beetles are found in several families, including tiger beetles (Cicindelidae), ground beetles (Carabidae), predaceous diving beetles (Dytiscidae), rove beetles (Staphylinidae), hister beetles (Histeridae), ladybird beetles (Coccinellidae), and checkered beetles (Cleridae). Most ground beetles prey on a wide variety of arthropods and other invertebrates, but some genera specialize on one type of prey. Species in the genus *Calosoma* are caterpillar hunters, as are some species of *Calleida*. *Cychrus* feed on snails. *Dyschirius* is a **riparian** genus that burrows in soil, where it preys on a soil burrowing rove beetle, *Bledius*, and the variegated mud-loving beetles (Heteroceridae) which burrow in moist stream banks. The genus *Lebia* climb plants to prey on other insects. Some species prey exclusively on certain leaf beetles (Chrysomelidae), and some *Lebia* species mimic the appearance of their prey species (Lindroth, 1969). Ladybird beetles feed on plant lice and scale insects so effectively that they are used as biological control agents for these plant pests. Many checkered beetles live under bark, where they attack bark beetles (Scolytidae).

While predaceous and scavenging beetles show considerable diversity and degree of specialization, the abundance and diversity of **phytophagous** beetles is even greater (Fig. 1). Plant-feeding beetles are an important element in fossil reconstructions, because they provide information on past plant communities, including such vegetational aspects as the composition of plant communities (both aquatic and terrestrial), and the health and age class structure of tree stands. The majority of phytophagous beetles are found on the leaves and flowers of plants (7,950 North American species in 25 families), and a considerable number live under the bark of trees, or in rotting wood (4,025 North American species in 24 families). Some phytophages are generalists, but many are associated with a few or only one host plant species. The fossils of these beetles provide data on ancient plant communities, even when the plants' pollen signature is sparse or lacking (Elias, 1982a).

DEVELOPMENTS IN QUATERNARY SCIENCES
VOLUME 12 ISSN 1571-0866

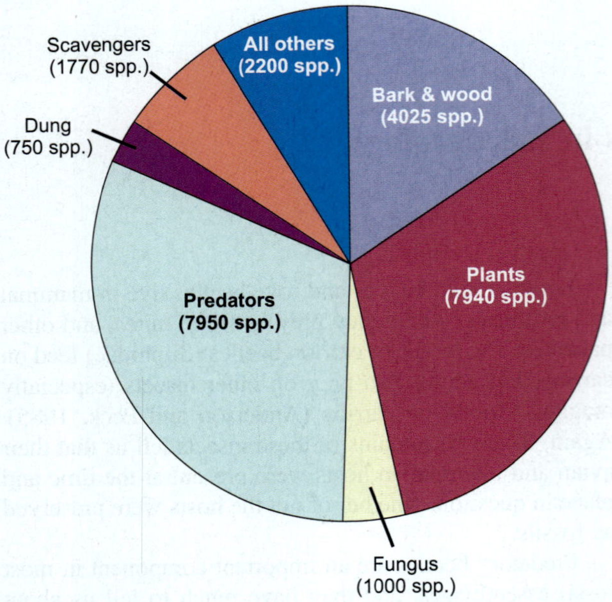

Fig. 1. Percent composition of North American beetle fauna by ecological group. Bark and wood feeders are mostly in the families Scolytidae, Curculionidae, Buprestidae, and Cerambycidae; plant feeders occur in many families, but especially Chrysomelidae and Curculionidae; fungus feeders are part of the Mycetophagidae, Ciidae, Erotylidae, and others; predators include members of the Cicindelidae, Carabidae, Dytiscidae, and Staphylinidae, as well as some other families; dung feeders are mostly in the family Scarabaeidae; scavengers include members of the Dermestidae and many other families (data derived from White, 1983).

The numerous small families of beetles serve to round out the paleoenvironmental scenario derived from fossil assemblages. There are families that specialize in feeding on mosses, fungi, slime mold, sap, algae, and pollen (a total of about 1,300 North American species in 20 families). Dung beetles (a part of the family Scarabaeidae) can also be generalists, or feed on the dung of only a few or just one species of vertebrate. The presence of a host-specific dung beetle in a fossil assemblage shows that the animal that produced the dung was also a part of the ancient community, whether or not any fossil remains of the dung maker are preserved.

Soil-dwelling insects in fossil records reveal types of past substrates and degree of soil development. Many types of beetles spend at least one part of their life cycle in the soil. Some beetle groups are adapted to particular soil types. For instance, many tiger beetles (Cicindelidae) are found only on sandy soils. Ground beetles that prey exclusively on snails frequent calcareous soils. Some beetles burrow deeply into soil cracks; others are adapted for a completely subterranean existence in caves. Many of the latter are blind, lacking in pigment, and **dorsoventrally compressed** (a character that allows them to penetrate narrow cracks more easily).

As we have seen, insect fossil records are not limited strictly to water-lain sediments in the middle to high latitudes. Packrat middens and cave deposits in desert regions are now yielding important information based on fossil insect assemblages (Elias, 1990a). The dried dung of Pleistocene mammals has been preserved in many caves in arid regions as well. Waage (1976) reported on fly larval and pupal remains (family Sciaridae) from dried dung in Gypsum Cave, Nevada. The dung was apparently that of the Shasta ground sloth, *Nothrotheriops shastensis*.

Desert-dwelling beetles have some remarkable adaptations to life in hot, dry environments. Darkling beetles (Tenebrionidae) have done particularly well in the world's deserts. This family dominates the beetle faunas of many desert regions, and is one of the few groups of arthropods that can remain active throughout hot days. The exceptionally thick, impermeable exoskeleton of some darkling beetles enables them to reduce moisture loss. One North American species covers its body with a thick secretion of wax during times of extreme aridity; the wax exhibits distinct color phases, ranging from light blue under conditions of low humidity to jet black under high humidity (Fig. 2A). The wax lowers water loss from cuticle, and the pale blue color exhibited during arid conditions lowers the body temperature by decreasing absorption of solar energy (Hadley, 1979). Another darkling beetle adaptation for water conservation concerns fused elytra. In extreme deserts, such as the Namib, up to 98% of darkling beetle species are flightless (Koch, 1962). Their elytra are fused together to form an impregnable, watertight shield over the abdomen (Fiori, 1977). Desert shrub-feeding weevils in the genus *Ophryastes* share some of the same adaptations discussed above for darkling beetles. Desert-dwelling *Ophryastes* beetles are covered in white scales, which reflect solar radiation (Fig. 2B,C). These weevils have very heavily sclerotized exoskeletons, and their elytra are frequently fused.

At the other end of the temperature spectrum, the arctic and alpine regions serve as an excellent repository for insect fossils. The adaptations of cold-hardy beetles are just as impressive as those of their desert-dwelling counterparts. **Cold hardiness** is brought about by a variety of metabolic responses to the changing seasons, as discussed in Danks (1978). The liquids in some beetles are able to **supercool** to temperatures as low as −40°C, preventing ice formation through the masking or removal of particles that could serve as nuclei for ice crystals. Other beetles are freeze tolerant, that is, they are able to survive **extracellular freezing** of their bodies. These animals secrete nucleators (proteins and **polypeptides**) that actually promote the growth of ice crystals at temperatures above 0°C. This in turn prevents the rapid formation of ice crystals in very cold supercooled liquids. Such ice formation may rupture cells, whereas the controlled freezing of extracellular liquids at temperatures above the normal freezing point of water prevents this kind of damage.

2. Species Constancy in the Quaternary

Ernst Mayr (1970) made the following comment on **speciation**:

> There is perhaps no other aspect of speciation about which we know as little as rate. Indeed, we shall

Fig. 2. Photographs of desert-dwelling beetles, illustrating adaptations for arid environments. A, Cryptoglossa verrucosa *(Tenebrionidae) – note the blue color, derived from a layer of wax filaments exuded from the exoskeleton; B,* Ophryastes argentatus *(Curculionidae); C,* Embaphion muricatum *(Tenebrionidae). Photo 4.2 (A), courtesy of Hartmut Wisch; photo 4.2 (B), courtesy of Kit Hubert, Beatty, Nevada; photo 4.2 (C), courtesy of Kojun Kanda, University of Arizona.*

probably never have very accurate information in this phenomenon. The splitting of one species into two is a short-time event that, as such, it is not preserved in the fossil record. For information we rely entirely on inference.

The fossil record that Mayr considered in this statement was essentially the pre-Quaternary fossil record, laid down in bedrock, full of temporal and spatial gaps, and generally lacking in continuity. Nevertheless, it is important to keep in mind that the Coleoptera fossil record goes back to the Permian (about 270 million years ago (mya)), and that most of the modern beetle families evolved in the Mesozoic (Arnol'di *et al.*, 1992). Some of the Mesozoic fossil beetles are quite well-preserved impressions, providing abundant evidence of taxonomically important characters. The Quaternary insect fossil record, albeit far from perfect, offers orders of magnitude more fossil material, far greater diversity of species, and far more precise chronological control (by radiocarbon dating and other radiometric methods) for many beetle families.

As discussed in Chapter 1, the first Quaternary entomologists described fossil beetles as extinct species, assuming that rapid insect speciation accompanied the extreme climatic fluctuations and the waxing and waning of continental ice sheets of the Pleistocene. In fact, the fossil beetle record of the Middle and Late Pleistocene provides no evidence of speciation (Coope, 1970; Matthews, 1976a,b). Unlike the Pleistocene **mammalian megafauna**, there is no evidence for any significant beetle extinction events in the Quaternary. But it appears that mammals may be the ''odd man out'' in their pace of evolution during the Quaternary. A study of speciation rates in birds (Klicka and Zink, 1997) indicated that most modern species diverged from their ancestors in the Late Pliocene, about 2.45 mya. Their mitochondrial DNA (mtDNA) evidence contradicted a popular hypothesis that North American songbirds evolved in response to Late Pleistocene glaciations. Stanley's (1985) paper on rates of evolution suggested that mammals have the shortest species duration of any major group of organisms (about 1–2 million years), while groups such as marine foraminifera and diatoms persist for more than 30 million years. Hewitt (2000) estimated that speciation rates vary greatly among mammals, ranging from 3–6 mya for European hedgehogs to 350,000–850,000 yr for some bears.

Some beetle groups have fairly well-documented fossil records extending back into the Mesozoic Era. For instance, dung beetles and chafers (family Scarabaeidae) have a fossil record extending back to 152 mya (Krell, 2006). Important tribes within this family have documented fossil records extending back to the Cretaceous (including Aegialiinae, Scarabaeinae, Sericinae, and Melolonthinae), while other groups apparently evolved in the Paleocene (Aphodiinae) or the Eocene (Cetoniinae) (Krell, 2006). It appears that dung feeding among certain groups of Scarabaeidae developed as early as the Cretaceous, presumably focusing on the grassy dung of herbivorous dinosaurs. The oldest extant species of scarabs, based on identifiable remains in fossil records, include *Aphodius rufipes*, *Oryctes nasicornis*, and *Lucanus servus*, all identified from fossils in Pliocene lake sediments at Willershausen, Germany, dated to ca. 3 mya (Schweigert, 2003).

Is there a characteristic rate of species evolution in insects? Stanley (1979) argued for this concept, but Wilson (1983) argues that there is no characteristic evolutionary rate in insects. A comparison of the numbers of extant species in the various important beetle families with their time of first appearance in the fossil record (Table 1) adds weight to Wilson's argument. While it is true that two of the largest groups (rove beetles and their allies – the Staphylinoidea, and ground beetles – Carabidae) are some of the oldest known groups, there are also some families that have been in existence for more than 100 million years, yet have only a few hundred extant species. Also, two of the largest families, Cerambycidae and Chrysomelidae, apparently

Table 1. Age of first appearance of beetle families, compared with number of known extant species in families.

Family or superfamily	First appearance (million yr BP)*	Number of extant species**
Staphylinoidea (Staphylinidae and allies)	170	35,100
Carabidae	150	25,000
Histeroidea (Histeridae and Sphaeritidae)	150	2,500
Scarabaeoidea (Scarabaeidae, Lucanidae, Trogidae)	150	18,800
Elateroidea (Elateridae and allies)	145	8,200
Hydrophilidae	145	2,400
Curculionidae	140	60,000
Dryopoidea (Dryopidae and allies)	125	1,053
Buprestoidea (Buprestidae and allies)	125	1,150
Dytiscidae	120	4,000
Dascilloidea (Dascillidae and Rhipiceridae)	115	495
Clavicornia (Cucujidae, Nitidulidae, Rhizophagidae)	115	16,000
Tenebrionidea (Tenebrionidae and allies)	115	25,100
Cerambycidae	115	20,000
Chrysomelidae	115	35,000
Nemonychidae	115	18
Gyrinidae	115	700
Cleroidea (Cleridae and allies)	70	3,400
Bostrichoidea (Bostrichidae and allies)	55	434
Dermestoidea (Dermestidae and allies)	55	819

*Data from Crowson (1975) and Arnol'di *et al.* (1992)
**Data from Britton (1970)

arose upwards of 50 million years after other important families. If a uniform rate of radiation were in action, then group longevity would equate with group size, and it clearly does not.

Some of the best evidence for beetle species longevity comes from Late Tertiary and Early Quaternary assemblages in the Arctic. Elias *et al.* (2006) summarized the available fossil insect data from sites in northern Siberia, arctic North America, and Greenland. They cataloged a total of 251 beetle species found in Late Tertiary and Early Quaternary fossil assemblages in these regions. Of these taxa, only 13 are known to have become extinct. This number represents 5% of the species. Some of the species identifications in the faunal lists were only tentatively matched with modern species. Some of these may also represent extinct species, but in contrast, some of the species thought to be extinct may eventually be discovered living in some other region of the modern world, albeit under a different name. An excellent example of this phenomenon is illustrated by the case of the fossil rove beetle, *Micropeplus hoogendorni*.

Matthews (1970) described two extinct species of the rove beetle genus *Micropeplus* from Pliocene deposits at the Lava Camp Mine, Alaska. *M. hoogendorni* (Fig. 3) and *M. hopkinsi* appeared to be the immediate precursors of the modern *Micropeplus* species. These fossils are about 5.7 million years old, and Matthews assumed that they had become extinct. However, much younger fossil specimens of *M. hoogendorni* have been found in Middle Pleistocene sites in the English Midlands (Shotton *et al.*, 1993) and the Norfolk coast (Parfitt *et al.*, 2005). Finally, the Russian entomologist, Rjabukhin, has described a Siberian species, *Micropeplus dokuchaevi*, which appears to be synonymous with *M. hoogendorni* (Campbell, personal communication, 1992).

Matthews also described two species of the water-scavenger beetle (Hydrophilidae) genus *Helophorus* from the Lava Camp Mine and another Pliocene age deposit (Beaufort Formation) on Meighen Island in the western part of the Canadian arctic archipelago. The extinct species are the precursors of the extant species *Helophorus tuberculatus*, a species found today in the Arctic. It should be noted that Matthews also found numerous beetle fossils in these Pliocene deposits that match modern species.

The Lost Chicken site in east-central Alaska preserves a Late Pliocene fauna (ca. 3 mya) that includes more than one extinct beetle species. One of these is an ancestral form of the riparian ground beetle, *Asaphidion*. It resembles the modern species *Asaphidion yukonense*, but the elytral microsculpture pattern is different (Fig. 4). It has less pronounced bare (impunctate) patches and more strongly punctate microsculpture than its modern counterpart (Matthews and Telka, 1997). Fossils that are similar to *A. yukonense* have been found in other Late Tertiary assemblages from the Arctic, but some have even more primitive elytral sculpture than the Lost Chicken specimens. These fossils could represent various stages in the evolution of the modern species (Elias, 2007a).

Other evolutionary trends have been spotted in the long fossil records available from Alaska and the Yukon.

Fig. 3. SEM of fossil elytron of the rove beetle, Micropeplus hoogendorni *(from Matthews, 1976b; reproduced with the permission of Natural Resources Canada 2008, courtesy of the Geological Survey of Canada).*

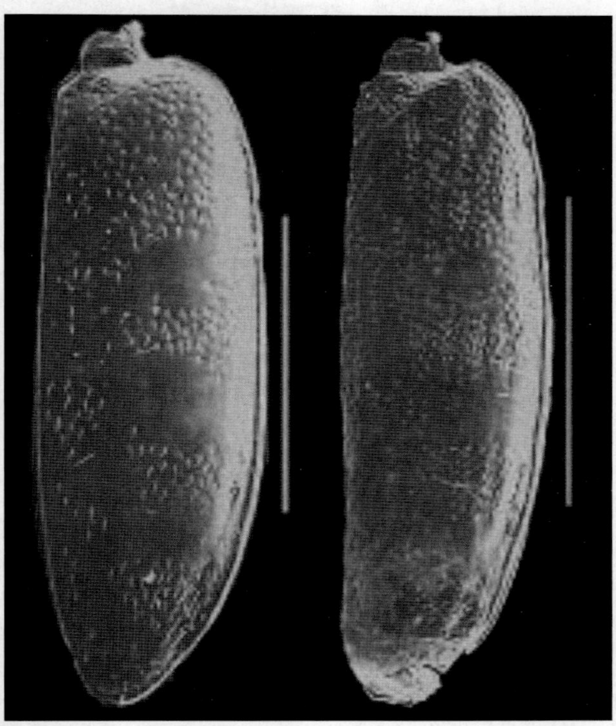

Fig. 4. SEM photographs of elytra of (A) modern species, A. yukonense, *and (B) Late Tertiary specimen of* Asaphidion. *Note the lack of clearly defined impunctate areas in the fossil specimen, and the heavier punctation. Photos courtesy of John Matthews, used with permission.*

One of these is the gradual loss of functioning flight wings. Matthews (1977b) postulated a gradual reduction in wing size in a rove beetle, *Tachinus apterus*, that is now flightless. He assembled a chronological sequence of Alaskan fossils spanning the Quaternary that showed this gradual reduction.

Additional Late Tertiary beetle fossils have been described by Böcher (1989a, 1990) from Kap København, northern Greenland. The fossils probably date to the Pliocene–Pleistocene transition. They are remarkable not only for their excellent state of preservation, but also because they represent a boreal forest environment in northernmost Greenland. The fauna contained 154 insect and arachnid taxa. Nearly all of the insect species (beetles and ants) are extant.

Perhaps the most startling evidence for species longevity comes from the fossil record of the aquatic leaf beetles in the subfamily Donaciinae (Chrysomelidae). Askevold (1990) analyzed donaciine fossils from the Early **Oligocene**-age Florissant shales in Colorado, and discovered that the species described by Wickham as *Donacia primaeva* is indistinguishable from the modern species *Plateumaris nitida* (Fig. 5). This suggests that *P. nitida* has persisted more than 30 million years.

These and numerous other studies provide strong evidence for the constancy of exoskeletal characters in beetles over great lengths of time, but skeptics will ask, ''even if their exoskeletal features show a degree of conservatism, how can you be sure that beetle populations have not become genetically isolated during the Quaternary?'' This is a valid question, because until very recently, paleoentomologists have dealt only with phenotypic evidence in fossils, not with **genotypes**. The developing research on ancient DNA from fossil beetles will be fully discussed in Chapter 15.

Traditionally, there are three lines of evidence that help to resolve this question. First is the physical evidence offered by fossil genitalia. The male **aedeagus** of many beetles, although an internal organ, is heavily sclerotized (Fig. 6). It is therefore resistant to decomposition, and is preserved in some fossil assemblages. If a beetle's abdomen is preserved intact, the genitalia are surrounded and protected from physical abrasion in sediments. The study of fossil genitalia (called by some workers ''paleopornography'') has yielded substantive evidence for constancy of many species through most of the Quaternary and beyond (Coope, 1970). Genitalia are considered the most reliable diagnostic feature in the identification of many beetle genera and families. In fact, there are many beetle genera with species so similar externally that their genitalia are the only reliable means of differentiation.

A second line of evidence on species constancy concerns the stability of ecological requirements of species.

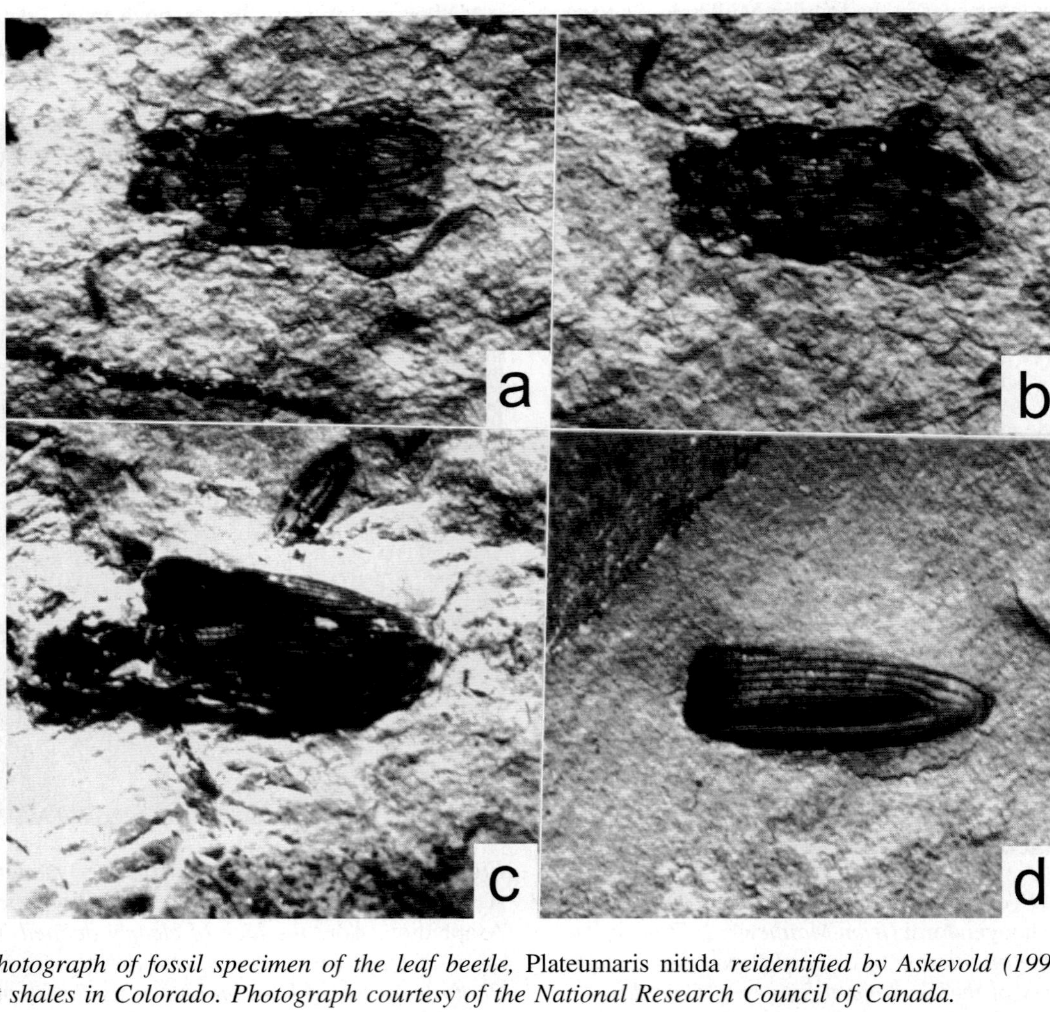

Fig. 5. Photograph of fossil specimen of the leaf beetle, Plateumaris nitida *reidentified by Askevold (1990) from the Florissant shales in Colorado. Photograph courtesy of the National Research Council of Canada.*

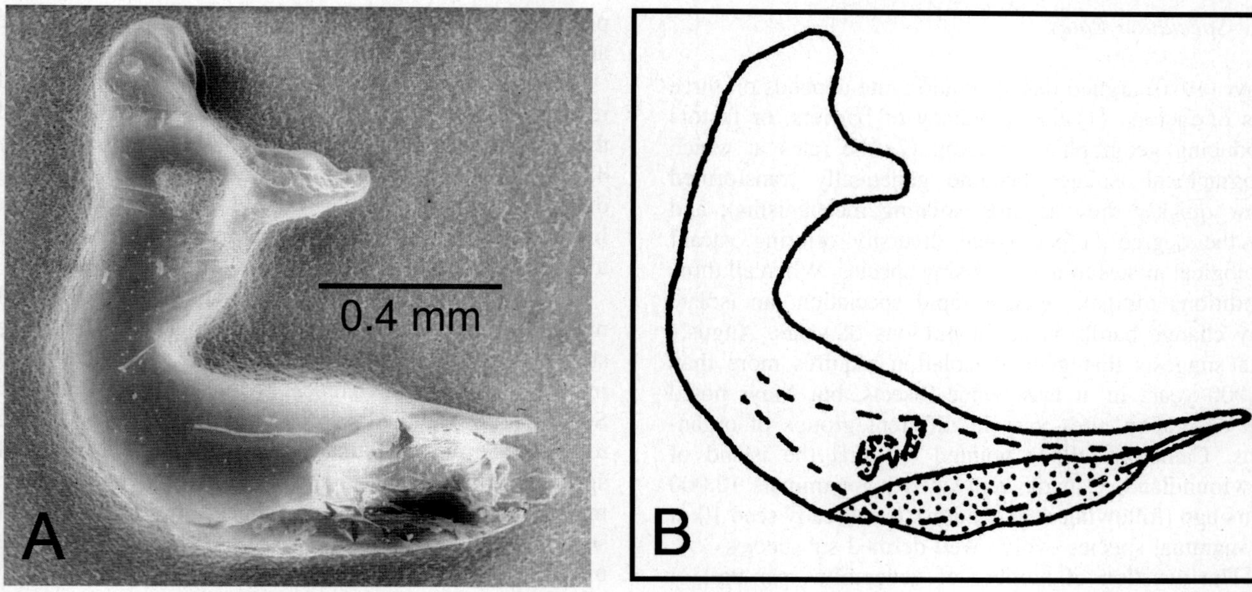

Fig. 6. Fossil (A) and modern (B) aedeagus of the ground beetle, Pterostichus brevicornis *(modern drawing after Lindroth, 1966).*

Again, the skeptic may well ask, ''how can you be sure that physiological changes haven't occurred within those exoskeletons, causing past populations to have different ecological requirements?'' This is another important question, because if a species' ecological requirements change through time, it becomes unreliable as proxy data for reconstructing past conditions, including climatic regime, substrates, host plants, etc. Paleoentomologists are reasonably confident about constancy of ecological requirements in Quaternary beetles because of the consistency of associations of insect species through time.

One of the luxuries afforded the student of abundant, diverse organisms in a fossil record (especially taxa that may be readily identified to the species level) is that many assemblages comprise dozens or even hundreds of species. When the modern habitats and distributions of the species in these assemblages are compiled, a detailed, precise reconstruction of the physical environment and biological community comes into focus. In particular, the spatial overlap in modern distributions of the species in nearly all the fossil assemblages corresponds to a fairly narrow climatic ''envelope'' in which all the species are found living. If hidden physiological evolution were taking place in these animals through the Quaternary, there would be species in a fossil assemblage which are ecologically or climatologically incompatible with the fauna as a whole. This type of discrepancy has rarely been found, though thousands of Quaternary insect assemblages have been examined (Coope, 1978, 1979).

In addition, certain well-studied regions, such as the British Isles, have yielded fossil insect assemblages of nearly identical composition during the series of glacial, interglacial, and **interstadial** climatic episodes. In other words, the warm-adapted faunas of one interglaciation are extremely similar to the warm-adapted faunas of other interglaciations, even though these climatic episodes may be hundreds of thousands of years apart. Likewise, the cold-adapted faunas found during one glacial stage have a great number of species in common with faunas found in previous and subsequent glaciations. If the physiological properties of the species in these faunas were changing through time, then those changes would have to be unilateral in nature, in order to preserve insect communities more or less intact through great lengths of time. The odds of an entire suite of species unilaterally evolving new physiological requirements are too small to merit serious consideration.

Recently, the mutual environmental compatibility of fossil beetle assemblages has been pushed back to the Late Pliocene. Elias and Matthews (2002) used the fossil beetle data from 11 arctic and subarctic sites to reconstruct seasonal temperatures spanning the interval from the Middle Pliocene to the Early Pleistocene. They used the mutual climatic range (MCR) method, discussed in full in Chapter 5. They used 124 species to reconstruct paleotemperatures, by overlapping the climatic tolerance envelopes of all the predatory and scavenging species in a fossil assemblage. In every case, a unique and internally consistent paleotemperature reconstruction emerged from their analyses. The results of their study will be discussed in Chapter 10, but the salient point here is that the ecological compatibility of these many species has remained intact. In one case (Lava Camp Mine, Alaska), the fauna demonstrates this integrity for more than 5 million years; in other cases, it has demonstrably persisted for 2–3 million years.

A third line of evidence for beetle species constancy has been pioneered by Robert Angus. The aquatic scavenger beetle, *Helophorus lapponicus*, was widespread in Western Europe during the cold phases of the last glaciation, but its modern distribution reflects the species' retreat to the remaining cold regions of Europe. These include the northern regions in Scandinavia and mountainous regions as far south as Spain. Angus (1983) captured specimens of *H. lapponicus* from Sweden and Spain; the two populations successfully interbred, despite their genetic isolation during the last 10,000 years.

2.1. Speciation Rates

Mayr (1970) argued that speciation rate depends on three sets of factors: (1) the frequency of barriers, or factors producing geographic isolation; (2) the rates at which geographical isolates become genetically transformed (how quickly they acquire isolating mechanisms); and (3) the degree of ecological diversity offering vacant ecological niches to newly arising species. When all three conditions mitigate against rapid speciation, an isolate may change hardly at all in millions of years. Angus's data suggests that genetic isolation requires more than 10,000 years in at least some insects, but Mayr noted that speciation rates vary in different groups of organisms. Cameron (1958) pointed out that the island of Newfoundland, Canada, reinvaded by mammals 12,000 years ago (following deglaciation), has already seen 10 of 14 mammal species evolve well-defined subspecies.

The question of number of generations per unit of time appears to have little to do with species longevity. Mayr (1970) noted that many insects, with one or more generation per year, have not changed in appearance since the Oligocene, whereas slowly maturing mammals, such as proboscideans, evolved very rapidly in the Quaternary.

Concerning the longevity of geographic isolation necessary for the development of genetically distinct geographic isolates, Coope (1978) has suggested that insect populations responded to large-scale climatic oscillations in the Quaternary by shifts in distribution. These shifts were sufficiently frequent to prevent the genetic isolation of populations that would lead to speciation. Many insects are very mobile. This has been inferred from the fossil record and demonstrated in historic times. However, recent genetic studies have cast doubt on Coope's theory. Reiss *et al.* (1999) and Ashworth (2001) discuss evidence from the cold-adapted ground beetle *Amara alpina*, demonstrating that various North American populations of this species have been genetically isolated from each other for perhaps hundreds of thousands of years, and yet have maintained their morphological constancy throughout the continent (Ashworth, 2004). Additional discussion of this topic is presented in Chapter 15.

3. Colonization of New Regions

Studies of beetle invasion of newly created **polders** in the Netherlands (Haeck, 1971) and newly created land on the Icelandic island of Surtsey (Lindroth, 1971) demonstrate the rapidity of colonization. For instance, carabid beetles appear to be exceptionally well equipped for rapid migration and establishment in new regions. Many species of carabids contain populations with a mix of fully winged (macropterous) individuals and individuals with reduced flight wings (**brachypterous**). Studies of ground beetle invasions have shown that flying carabids make up the vanguard of the invaders. Lindroth (1949) described this group of beetles as a "parachute force," capable of spreading out more rapidly at the edges of the species' range than the flightless "pedestrians." I will discuss this topic more fully in connection with postglacial invasions of deglaciated landscapes in Europe and North America, below.

Insects invest tremendous amounts of energy and resources into dispersal. Heydemann (1967) estimated that on the North Sea coast of Germany, roughly 4,500,000,000 insects per day are lost by aerial drifting during summer. This is equivalent to 270,000 kg of insect **biomass** lost to regional insect communities, a high price to pay for dispersal activity.

Do changes in ecosystems offer vacant ecological niches to newly arising species? Ecological studies have shown how existing beetle populations respond to even minor environmental changes. For instance, Howden and Scholtz (1986) studied the dung beetle fauna of a wildlife refuge in Texas and found a nearly complete turnover of species during a 10-year interval in which precipitation increased roughly 20%. This suggests that niches do not stay vacant for significant lengths of time, because insect mobility is sufficiently great to allow existing species to occupy new habitats as they become available. An obvious exception to this might be on oceanic islands, where invasion by existing taxa is limited to just a few founder species.

It is important to keep in mind that insect habitats are very small, and that they are subject to **microclimates**, rather than **macroclimates**. Part of the secret of insects' success has been that they have been able to secure suitable habitats for themselves in the face of changing environments by readily dispersing across landscapes. Danks (1979) has summarized many of the factors affecting insect distribution and dispersal. His paper examined the Canadian fauna, but many of the same principles apply to other north temperate and arctic regions. Disturbances in more-or-less stable biological communities lead to the development of seral stages of secondary succession. Along with "weedy" plant species, newly disturbed ground is often rapidly colonized by species of beetles and other insects which are preadapted for rapid dispersal. Such species are among the more important pests of agricultural crops. These species are opportunistic, with rapid reproduction and great mobility. Their habit of feeding on the flora of disturbed ground allowed them to take advantage of agricultural habitats created within the last few hundred or thousand years. In the Pleistocene, glacial advances and retreats created newly exposed open ground habitats in many regions. The adults of many of these insect species go through a pre-reproductive dispersal phase. Though many of the dispersing individuals perish, the colonization of newly disturbed open ground habitats is assured for the species as a whole. Vagility, or an intrinsic tendency to disperse, is a reflection of the inherent instability of landscapes.

While there is good evidence for long-distance dispersal in some insects, some groups seem to disperse very little. Lindroth (1957) showed that many of the beetle species accidentally imported into the east coast of North America in ship's ballast have remained in the immediate vicinity of the ports of arrival. This restricted dispersal may be due to the species' lack of ability to compete with the indigenous fauna. Competition between species is an important factor in limiting distributions, but for most groups of organisms, it is little studied, thus far.

Insects exhibit many types of distribution patterns. Some species, though widely distributed, occur only in narrowly defined habitats within their range. For instance, there are several species of beetles which occur today across all of northern North America, but live only in *Sphagnum* bogs. Other species are found in a wide range of habitats (eurytopic species), but are confined to a narrow geographic range. Still other species are cosmopolitan in both their distribution and habitat requirements.

4. Fossil Evidence for Continental-Scale Range Shifts

In the fossil record, we see dramatic evidence of shifts in insect ranges through Quaternary time. The story of the Tibetan dung beetle *Aphodius holdereri* in British Pleistocene assemblages reveals the magnitude of possible distributional change in Eurasia. This type of change has also been shown in the water scavenger beetle, *Helophorus mongoliensis*, a modern inhabitant of Asiatic mountains that was found in 40,000-year-old deposits in England (Angus, 1973), and in the rove beetle *Tachinus caelatus*, known today from Mongolia but found in deposits from the last glaciation in Britain (Taylor and Coope, 1985) and Switzerland (Coope and Elias, 2000).

M. hoogendorni appears to have originated in Alaska in the Pliocene, but its discovery in last interglacial deposits from Britain and in the modern fauna of Siberia shows that interhemispheric distributional shifts have also occurred. In addition, the rove beetle, *Anotylus gibbulus*, has been found in **Sangamon** (**Marine Isotope Stage** 5e) or Early **Wisconsin** age (MIS 4) assemblages from the Scarborough Bluffs in Toronto (Hammond *et al.*, 1979), and from interstadial deposits in the Devensian (last) glaciation in Britain. This beetle has a most peculiar distribution today: it is known only from the Caucasus and from eastern Siberia, north of Vladivostok. These species clearly have been world travelers during the Quaternary. The biogeographic implications of the fossil record are discussed in Chapter 6, but for now, it is sufficient to note the great mobility of insects in response to environmental change.

5. Range Shifts of Alpine Species

Kavanaugh (1979) has suggested that beetle species' longevity as evidenced in the Quaternary fossil record applies only to lowland taxa, since nearly all fossil studies have focused on lowland regions. In his analysis of speciation in the ground beetle genus *Nebria*, Kavanaugh proposed the differentiation of 26 subspecies pairs and three species pairs in post-Wisconsin time (i.e., within the last 10,000 years), and that another 10 *Nebria* species pairs have differentiated earlier in the Quaternary period. In particular, he discussed the *trifaria* species group as showing very recent speciation, arguing that isolation in montane habitats brings about rapid speciation.

However, the fossil record suggests that some beetle species associated with mountain top habitats today exhibited considerable vagility in the past. Late Wisconsin age insect fossil assemblages from Lamb Spring, in the foothills east of the Rocky Mountains of Colorado, contained several species in this distributional category which apparently shifted downslope to the plains during the last glaciation (Elias, 1986; Elias and Nelson, 1989; Elias and Toolin, 1989). One example of this is the water scavenger beetle, *Helophorus splendidus*, which is found today in the Arctic, and on the alpine tundra in Colorado (Smetana, 1985). It occurred in an assemblage from Lamb Spring, dated 17,750 cal yr BP.

6. Tropical, Island and Cave Faunas

Coope (1979) argued that rapid speciation requires environmental stability in a constant geographic location. It is tempting to speculate that the tropics might offer such stability, but even tropical regions seem to exhibit significant environmental fluxes, although the vegetation may have exhibited a dynamic stability in the face of these changes (Colinvaux, 2001). Erwin (1979) hypothesized that changes in Pleistocene precipitation patterns in tropical regions caused the last major impulse of ground beetle evolution. In contrast, the remarkable diversity of insect life in the tropics may be due, at least in part, to a lack of intensity of environmental change over long periods.

Coope (1979) also suggested that isolated geographic situations (i.e., habitat islands) such as equatorial mountain tops, oceanic islands, and caves may represent evolutionary traps from which emigration is denied. Under such conditions, rapid evolution would be a species' only viable response to environmental change. However, beetle species appear to be relatively conservative, even on oceanic islands. On Aldabra, a recent atoll in the Indian Ocean, Basilewsky (1970) found 18 species of ground beetles, none of them **endemic**. All of these species have been collected on Madagascar, and many are also known from the continent of Africa (Thiele, 1977). On the Galapagos Islands, carabid beetles have had roughly 10 million years to evolve, although surveys of carabids on the islands have found surprisingly few species, and Thiele (1977) summarizes carabid speciation on the Galapagos Islands as "not very impressive" in comparison with other groups of animals.

Cave faunas also probably are not as genetically isolated as might be expected. My studies of Late Quaternary insect fossils from packrat middens in the Chihuahuan Desert (summarized in Elias, 1992b) included the discovery of several species of cave beetles in the ground beetle genus *Rhadine* that have shifted their distributions in response to environmental changes in the last 20,000 years. These beetles are flightless and **depigmented**; most have reduced eyes. Their fossil record indicates that they have migrated from caves in the Chihuahuan Desert to caves in central Texas and Oklahoma, even though no subterranean connections are known between these far-distant cave systems. Barr (1960) discusses the possibility of geographic isolation of ancestral stocks of these cave beetles in west Texas during the Late Pliocene or Early Pleistocene, but the fossil evidence indicates that these species are able to migrate substantial distances within at most a few thousand years. In light of this, fossil insect studies should also be carried out on oceanic islands and equatorial mountain tops, in order to test their viability as evolutionary traps.

7. Effects of Long-Term Climate Change

Bennett (1990) discussed the effects of **Milankovitch cycles** on Quaternary biota. Variations in the earth's orbit bring about climatic changes in frequencies varying from 20,000 to 100,000 years. Bennett argues that evolutionary theories, such as the punctuated equilibrium theory (Gould and Eldredge, 1977), overlook the significance of climatic changes on the 20,000–100,000 years scale. Several timescales are at work in the evolutionary history of a species. One of these is the generation time, which for most insects is 1 year or less. A second timescale is the duration of species, which the fossil record now suggests may be up to several million years, at least in Coleoptera. As discussed above, Stanley (1985) surveyed species duration in a wide variety of organisms, and concluded that species endure periods of 1–30 million years, depending on taxonomic group. Species therefore persist much longer than Milankovitch cycles, even though these cycles are thought to have brought massive climatic changes in the Quaternary. These environmental changes have caused the disruption of biotic communities, essentially nonstop, for more than 2 million years. Each species has reacted individually to these changes. Again, the fossil record suggests that insects, as well as many other organisms, have shifted their distributions on regional, continental, and even intercontinental scales in the Quaternary. Bennett argued that macroevolution could scarcely proceed by phyletic gradualism in light of this constant shifting and disruption of populations.

Bennett also noted that most paleontological research deals with timescales of millions or tens of millions of years, too coarse to resolve events at finer intervals. At the other extreme, modern ecological research deals with only a few years at most. The really important evolutionary events fall in between these two scales. However, more detailed study of Quaternary fossil records will illuminate the history of life on timescales between the paleontological and ecological scales, because the type of change that has occurred in the Quaternary has been operating throughout the history of life on earth.

Study of the Quaternary fossil record first began to reveal startling data on the longevity of beetle species about 35 years ago. Unfortunately, these studies have largely gone unnoticed by some beetle systematists and evolutionary biologists. Futuyma (1979) cited a study claiming that poorly flying bog beetles have failed to migrate north of the Wisconsin ice limits in North America since the end of the last glaciation. The fossil record indicates that even flightless species have migrated across major regions of North America, and that these distribution shifts have taken only a few centuries at most (Elias, 1991). This is the type of information that needs to be disseminated outside of the field of paleontology, so that the scientific connections with evolutionary biologists may be strengthened. Interactions between the two fields will be mutually beneficial.

In summary, the fossil record provides evidence for both the longevity of species and the almost constant perturbation of biotic communities and shifting of species' distributions in response to environmental change during the Quaternary. These features of life on a 20,000–100,000-year timescale lend support for the punctuated equilibrium model of evolution, but we are left with some major questions. As Ashworth and Hoganson (1987) pointed out, the problem of trying to assess the precise antiquity of species is that the evidence is inadequate to the task. The fossil data are adequately dated (at least the Late Quaternary assemblages), but lack the precision needed to define close evolutionary relationships. In contrast, **cladistic analysis** may succeed in establishing evolutionary relationships but lacks real temporal control. The only conclusion that emerges is that much remains to be learned. We may begin to resolve some problems in the near future, with the application of ancient DNA techniques to beetle exoskeletons preserved in permafrost environments (see Chapter 15).

8. Sensitivity of Insects to Environmental Change

One of the most important aspects of insects as paleoenvironmental indicators is their sensitivity to environmental change. Species in many families of beetles have demonstrated such sensitivity. This is also true for caddis flies, midges, and other insect groups. In general, predators and scavengers receive the most attention in paleoenvironmental reconstructions, because they are not tied to specific types of vegetation. While some predators and scavengers are eurythermic (adapted to a broad range of thermal conditions), many are stenotherms which are adapted to only a narrow thermal environment. Stenotherms may rapidly colonize a region, as long as the climatic conditions are suitable. When climatic conditions change, stenotherms depart with equal rapidity.

Insect ecologists support the idea that insect abundance and diversity may be controlled by biotic factors (e.g., predation, competition, or parasitism) in the center of a species' range, but that abiotic factors such as climate probably limit populations toward the edges of their range (Price, 1984). It is best, therefore, to study fossil assemblages from ecotones (the edges of ecosystems), rather than assemblages from the centers of past ecosystems. Another advantage to studying the ancient biota of ecotones is that these communities most readily document regional environmental changes. Fossil faunas and floras change as ecotonal boundaries shift across a landscape. These biota are quite literally on the leading edge of change. In contrast, communities from the center of ecosystems tend to be complacent, because ecotonal boundaries seldom move through these regions.

8.1. Experimental Studies on Ground Beetles

A great deal of experimental work on habitat preferences and physiology of beetles has focused on the well-studied family, Carabidae. This is due, in no little part, to the pioneering work of Carl Lindroth (1949–1987) on the ground beetles of Scandinavia. For instance, Thiele (1977) determined the thermal preferences of ground beetles, using controlled experimental environments that provided a gradient of temperatures to the specimens. These tests established a preferred temperature (PT) for a large number of European species. Thiele found no evidence for "temperature races" in carabids (i.e., populations within a species that have differing PTs). In addition, his experiments showed that stenothermic species associated with arctic and alpine environments such as *Nebria nivalis* and *N. gyllenhalli* exhibited PTs of 5 and 8°C, respectively. At the other end of the thermal spectrum, only one species of ground beetle from Western Europe, *Callistus lunatus*, was found to have a PT above 40°C. These results offer substantial evidence linking ground beetle geographic distribution to their thermal requirements.

Additional experiments on metabolic rates of carabids (Thiele, 1977) showed that beetles' optimal metabolisms correspond with their PTs. This phenomenon is strengthened during the process of cold hardening, which takes place in late summer or fall. In preparation for winter cold, ground beetles in northern regions become progressively cold adapted (able to function at cold temperatures). Under these conditions, the beetles' metabolic rates peak at low temperatures, and they may become paralyzed or die if exposed to the temperatures they experience in midsummer.

Surprisingly, terrestrial species are often the most abundant taxa in water-lain sediments. On the whole, they probably reflect regional macroclimates better than aquatic species. However, many aquatic insects (including aquatic beetles, caddis fly larvae, and chironomid larvae) are adapted to waters in a narrow range of temperatures. These water bodies are developed and maintained only within certain climates, so they indirectly reflect macroclimate. Williams (1988) compared fossil caddis fly and beetle paleoclimate reconstructions for sites in the Great Lakes region, and concluded that the two groups respond nearly identically to climate change.

Aquatic insects also provide a basis for inferring detailed limnological data, including types of substrates, water pH, water clarity verses turbidity, and trophic status. Riffle beetles (Elmidae) provide evidence of current speed in streams.

8.2. *Use of Plant-Feeding Beetles in Environmental Reconstruction*

Phytophagous beetles are less useful in paleoclimate reconstructions, but still provide data valuable in paleoecological studies. They are reliable indicators of past vegetation, and host plant-specific beetles occur in many assemblages (although bark beetles are found only in forest assemblages). Bark beetles (Scolytidae) and other tree-associated groups (some weevils, twig borers, etc.) have been used to document ecological changes in forests, and to estimate the position of both latitudinal and altitudinal tree line. Fossil scolytids from Holocene peats at Ennadai Lake, Keewatin, Canada, indicated the continued presence of spruces at the site, even though the mid-Holocene pollen record was lacking in spruce pollen (Elias, 1982a). I have also used changes in the ratio of bark beetle to alpine tundra beetles to infer the history of altitudinal tree line in the Rocky Mountains (Elias, 1983, 1985, 1988b; Elias *et al.*, 1991).

Fossil bark beetles have been used to address other kinds of paleoenvironmental puzzles. British pollen diagrams document a disappearance of elm (*Ulmus*) from the landscape during the mid-Holocene. Was this due to natural or anthropogenic causes, or some combination? The presence of fossils of the elm bark beetle, *Scolytus scolytus*, may explain this elm decline. This problem is difficult to unravel because of the activities of Early **Neolithic** people. However, the beetle evidence suggests that the decline may have been due to Dutch elm disease, which is transmitted by the beetle (Girling, 1988).

There are suites of beetle taxa from sites in northeast Asia and northwestern North America that have been used to indicate the presence of steppe–tundra. This ancient ecosystem has been reduced to tiny relict patches on south-facing mountain slopes in the two regions today, but it once covered whole regions, from the Yukon Territory in the east to northwest Europe in the west. As the name indicates, the dominant vegetation of this ecosystem was a mixture of **steppe** and tundra plants. Steppe–tundra beetle faunas are discussed much more fully in Chapters 9 and 10, but it is worth noting here that all the steppe–tundra beetle faunas shared some important elements. They all contain species of various beetle families that are associated with steppe grasslands, they all contain species of pill beetles (Byrrhidae) that are moss feeders, and they all contain species of predators and scavengers associated with cold, dry climates. Pleistocene faunas from Alaska and the Yukon that are indicative of steppe–tundra always include the weevil, *Lepidophorus lineaticollis*, often contain the sagebrush (*Artemisia*) feeding weevil, *Connatichela artemisiae*, and usually contain the pill beetle, *Morychus*. Steppe–tundra faunas from the other side of the Bering Strait (in northeast Asia) always contain *Morychus* and weevils in the genus *Stephanocleonus*. While there are megafaunal mammals that most people more readily associate with steppe–tundra (saiga antelope, bison, woolly mammoth, woolly rhinoceros, and Pleistocene horses), the steppe–tundra beetle faunas are equally diagnostic, and much more readily recovered from regional sediments than megafaunal mammal bones.

8.3. *Glacial and Periglacial Environmental Indicators*

Some beetle species are extremely useful indicators of particular kinds of habitats, and their presence in a fossil assemblage may be taken as a proxy for the ancient existence of those habitats at study sites. For instance, several species of ground beetles in the genus *Nebria* are highly adapted for preying on insects trapped on the surface of snowfields. Today such snowfield habitats are mostly restricted to arctic and alpine regions, but during

Pleistocene glaciations, snowfields were far more common at low elevations and in the midlatitudes of the Northern Hemisphere. The physiology of snowfield *Nebria* species is geared to operate at temperatures near freezing, and they have unusually long legs, allowing them to run about on the snow without chilling their bodies. These beetles are regularly found in fossil assemblages associated with **periglacial environments**.

Except under the coldest of glacial conditions, snow patches melt away or at least decrease in size during the summer. Pools of meltwater adjacent to snowbeds are the preferred habitat of the water scavenger beetle, *Helophorus glacialis*. As discussed above, this species was widespread in Europe during glacial intervals. Now it is restricted to arctic and alpine regions. Ponel *et al.* (1992) found the fossil remains of this species in a high-altitude fossil assemblage in the French Alps. Mid-Holocene age faunas containing this beetle were interpreted to signify the kind of climatic cooling that would foster the development of snowbeds (and adjacent meltwater pools) at the study site. Another species of *Helophorus*, *H. arcticus*, is associated today with pools of water near the shores of the Arctic Ocean in northeastern North America. During the last glaciation it spread south, probably remaining close to the edge of the ice, where the ice dammed rivers to form large proglacial lakes, and where large meltwater pools would have formed at the ice sheet's edge every summer. At the end of the last glaciation, it is thought to have tracked the proglacial lake habitat north, as the melting ice sheet liberated millions of gallons of water that filled successive basins near the ice margin (Morgan, 1989). If this reconstruction is accurate, then it is safe to say that *H. arcticus* enjoyed a much wider distribution during successive glaciations than it does during interglacial periods, when proglacial lakes vanish. From the perspective of the evolutionary timescale, one could observe that this and other cold-adapted species are currently ''biding their time'' until the next glaciation ushers in much more favorable climatic regimes.

8.4. *Primeval Forest Indicators*

Beside predators and phytophages, other groups of beetles and other insects can play an important role in reconstructing past environments. Fungus beetles provide part of the ecological story of past communities. Many species are associated only with particular types of fungi, such as bracket fungi or mushrooms that grow only in old, well-established stands of trees. There are large numbers of species that feed on the kinds of fungi associated with undisturbed forests, or ''*Urwald*,'' as it is called in German (and now throughout much of the European literature). Many species are part of the so-called saproxylic community. These beetles, or their larvae and/or pupae, feed on dead wood, on fungi that are part of the wood decay process, or on the products of that decay. Whitehouse (2006) documents the history of forest clearance in the British Isles during the Holocene, as marked by the gradual disappearance of the *Urwald* insect fauna from the fossil record. Forty species belonging to this ancient forest fauna have been lost in Britain and 15 species lost in Ireland, during the last few thousand years.

8.5. *Anthropogenic Environmental Indicators*

In archeological studies, dermestid beetles (Dermestidae) and other stored product pests are used to infer human food use, and sanitation (or the lack of it). The beetles that have become successful as stored product pests obviously did not start out in this niche in the Late Tertiary or Early Quaternary; however, Buckland (1981a) noted that human populations have played a major role in their recent dispersal. There is some speculation that these species previously fed on caches of food stored by other animals, such as squirrels and other rodents (Crowson, 1981). This topic is more fully developed in Chapter 7.

Fossils of parasitic insects are used to infer the presence of domesticated animals (livestock) in archeological sites. Even if no cow or sheep remains are preserved in an ancient dwelling, the exoskeletons of the animals' parasites may provide the data needed to infer the practice of animal husbandry (Buckland, 1976a). Fossil remains of human parasites (fleas, lice, and other organisms) document the history of human lifestyles and associated health problems, because insects are associated with human beings from cradle to grave. Various insect genera document the **forensic** pathology of corpses, from desiccated Greenland mummies (Bresciani *et al.*, 1983, 1989) to the unfortunate victims of ritual murders in **Iron Age** Britain (Girling, 1986a).

9. Use of Databases in Reconstructing Environments

The process of building a paleoenvironmental scenario can be time consuming, but it represents the culmination of perhaps months or even years of fossil specimen identification: the building of faunal assemblage lists for a fossil locality. In Britain and adjacent northwest Europe, the development of environmental reconstructions has been greatly aided in recent years by the creation of the BUGS database (Buckland and Buckland, 2006). This database provides modern distribution, habitat, and other ecological data for nearly all species of beetles that have been found in Quaternary fossil assemblages in northwest Europe. By selecting a species name from a drop-down menu, one has instant access to all of these data. For other regions of the world, the investigator must find pertinent taxonomic literature for the species in question. Most taxonomic revisions of insect groups include modern distribution and habitat data. Failing this, such data must be obtained directly from the labels of pinned specimens in major museum collections. This is much more easily done for some regions (e.g., North America and Europe) than for others (much of Asia and nearly all of the Southern Hemisphere). For example, paleoenvironmental reconstructions for fossil beetle assemblages in New Zealand are greatly hindered by the fact that the most extensive collections of the islands' beetle fauna are housed in European museums.

Another difficulty in compiling these kind of modern distributional and ecological data relates to the wide scale human alteration of so many landscapes around the world. We have already examined the problem in Europe, where a significant portion of the *Urwald* fauna has become extirpated from whole regions. In that case, there is at least some historical literature documenting the ecological preferences of such species, or some relict populations in other parts of the continent that can shed light on the species in question. In other parts of the world, the problem is far worse. For instance, the beetle fauna that existed in New Zealand before the coming of European colonists is in some cases nowhere to be found today. Essentially, the entire New Zealand beetle fauna is endemic to those islands, because of their isolation from the rest of the world for many millions of years. A less extreme case concerns the use of modern distribution and habitat data for species living in regions that have become urbanized in recent decades. I compiled such data for desert and semidesert species of beetles found in Late Pleistocene fossil assemblages from the arid southwestern United States. A good deal of beetle collecting and naming of species took place in southern California in the late 19th century. In those days, Los Angeles was a relatively small city; its population was 50,000 in 1890. Collecting localities such as ranches (e.g., Rancho La Brea) and citrus groves that were miles outside the city in 1890 are now fully urbanized, with very different climates and substrates than when darkling beetles were being collected there more than 100 years ago. Therefore, in order to accurately assess the ecological requirements of species living in that region, it is necessary to examine collecting records that predate the human population boom.

As in other macrofossil studies, the ''story'' obtained from a fossil insect assemblage has only local application. A single local reconstruction may not provide a reliable estimate of regional environments. For instance, Ashworth (1977) described a Late Wisconsin beetle fauna from southern Ontario that was indicative of a cold microclimate in close proximity to stagnant ice, but additional regional studies from this period clearly show climatic amelioration (Morgan *et al.*, 1984a). Therefore, to develop an understanding of a region, numerous replicates of local studies are required.

5

Paleoclimatic Studies Using Insects

The transitions from one climatic mode to the other were so sudden at times that the whole terrestrial biota was disrupted....

– Russell Coope (1987a)

Given that the species found in Quaternary fossil assemblages had the same environmental requirements as their modern counterparts, it is then possible to construct a mosaic of environmental conditions for given study sites and time intervals, and then to weave these environmental reconstructions into a regional synthesis depicting macroclimate. The presence of suitable climatic conditions is one of the most important factors determining the geographic range of insects (Coope, 1986).

1. Range Overlap Method

Starting with Coope's (1959) interpretation of Late Quaternary environments at the Chelford site, most climatic interpretations of fossil beetle assemblages have been made on the basis of information gleaned from average temperature and precipitation data from meteorological stations that lie in the modern ranges of the species in the fossil assemblage. The approach taken in most studies has been to plot the region in which the modern distributions of the species in an assemblage overlap, and then to derive a paleoclimate estimate based on the modern climatic parameters within that zone of overlap.

While this method has generally yielded good results, it is not always viable. For instance, during intervals of rapid climate change, insect faunas (as well as the rest of the regional biota) are in such a state of flux that their presence in any one region is very ephemeral. Fossil assemblages representing such faunas often contain mixtures of species for which there is no modern analogue. By this I mean that there is no one region in existence today where all the species represented in the fossil assemblage can be found living.

The range overlap method is also, to a certain extent, subjective. The investigators are called upon to exercise their best judgement about distributional or climatic data that appear out of place with the majority of evidence. Species deemed to be especially climatically sensitive (stenothermic beetles) become ''indicator species'' in an assemblage, and hence receive more weight in paleoclimatic reconstructions than their more cosmopolitan (eurythermic) counterparts. In the 1980s, the MCR method was developed to standardize paleoclimatic interpretations of fossil beetle assemblages, and to tease more climatic information out of the data.

2. Quantitative Approaches to Paleoclimate Reconstruction

After more than 20 years' use of the geographic range overlap method, various investigators sought the development of a more reliable quantitative approach to the task of reconstructing past climates. As discussed below, various methods have been developed in different parts of the world. In Europe and North America, our knowledge of the modern distributions and thermal requirements of most species is reasonably complete, based on more than a century of modern specimen collection in regions with good infrastructure (road and rail networks, the presence of universities and government agencies supporting entomological research, etc.). This knowledge base has fostered the development of methods based on the overlap of climate envelopes for species found in fossil assemblages (see Section 3, below). In less well-studied regions, such as Siberia, New Zealand, Australia, and elsewhere, other methods have had to be developed, because the knowledge base is poorly developed for many species, or absent altogether. Each of these kinds of methods is discussed, in turn.

3. The MCR Method

In 1982, Russell Coope sought the help of paleoclimate modelers Timothy Atkinson and Keith Briffa in the Climatic Research Unit at the University of East Anglia in developing a quantitative method of analyzing the paleoclimatic interpretations of Quaternary insect fossil data. This method avoids the use of indicator species, concentrating on the analysis of entire assemblages, using the presence/absence of species rather than relative abundance (which may vary considerably, depending on depositional environments). The basic principle of the method lies in establishing the range of climates presently occupied by each beetle species found in a fossil assemblage. A species need not necessarily occupy the whole of its potential range, nor need we have a complete picture of its geographic distribution. In order for an adequate climatic envelope to be constructed, all that is needed is the determination of the climatic tolerances of a species, based on the climatic parameters within its known range. This is a great advantage over the

DEVELOPMENTS IN QUATERNARY SCIENCES
VOLUME 12 ISSN 1571-0866

geographical overlay method. In other words, species may live in identical physical environments but in different places. Modern distribution patterns that do not physically overlap may still represent areas of climate space that do overlap. For instance, the modern distribution of the western North American carrion beetle, *Aclypea bituberosa*, does not overlap with that of the eastern North American ground beetle, *Agonum octopunctatum*, but their respective climate envelopes show considerable overlap (Fig. 1)

Once the climatic envelope for each species is established, the climate indicated by the whole assemblage may be taken to be within the area of overlap of the climatic ranges of all the species in the assemblage (Atkinson *et al.*, 1986). Beetles are especially well suited to this technique, because they are a varied group in which many species show clearly defined thermal tolerances (Atkinson *et al.*, 1987). On a hemispheric scale, the distribution of individual species reflects temperature regime, especially summer warmth and degree of **seasonality** (i.e., annual temperature range).

Atkinson *et al.* (1986) have tested the MCR method on modern beetle assemblages from Europe and Asia, and the results of the climate reconstructions match the modern climates associated with the studied modern assemblages. The following method is based on Atkinson *et al.*, 1986.

The first step is the compilation of modern distribution maps for each species. For many species, adequate modern geographic distribution maps have been published (Fig. 2). For others, however, additional literature searches and transcribing of locality labels from museum specimens will have to be performed. Only predators and scavengers are used. Focusing on these groups avoids the problem of having to deal with phytophagous species, the distributions of which may reflect their host plants' ranges, more than climatic parameters. In contrast, predators and scavengers are more or less free to become established in any region of suitable climate, even as part of pioneering communities establishing the first ''foothold'' on bare, mineral soil in recently deglaciated landscapes.

The second step is the determination of climatic range for each species. This has been done in two different ways in Europe and North America. In Europe, the geographic range of each species has been converted into a climatic range with the help of a base map showing locations of meteorological stations. A comparison of the two maps reveals whether given meteorological stations are within a species' geographic range or outside of it. Then the climate data from the two groups of meteorological stations are processed by computer, producing a plot of the two groups of stations on a graph (Fig. 3). The graph's axes are the mean temperature of the warmest month of the year (T_{max}) versus the temperature range between warmest and coldest months (T_{range}). The latter serves as an index of degree of seasonality. **Principal components analysis** of the mean monthly temperature from 495 meteorological stations in the **Palearctic region** shows that over 96% of the variance in temperature regime for the Palearctic is described by two groups of variables, which can be interpreted as (1) summer warmth (T_{max}) and (2) temperature range between the warmest and coldest months (T_{range}) (Atkinson *et al.*, 1987).

There are far fewer meteorological stations in the remote regions of North America, especially in the Arctic. Because of this, species' climate envelopes have been developed in a slightly different way, using a 25-km-grid North American climate database (Bartlein *et al.*, 1994). This database was used to pair climate parameters with the modern beetle collection sites, using the geographically nearest grid location to each collecting site.

For species with modern distributions that include montane sites, care must be taken to match the altitudinal distribution of the species with that of the meteorological stations. The montane species problem illustrates why it is important to find climate data from the nearest meteorological station, and not derive modern climate parameters on the basis of the position of modern beetle collecting localities relative to broad isotherm lines, such as are found in regional climatic atlases. Regional climatic summaries expressed in isotherms often fail to show altitudinal gradients in temperature values, simply because the data are too complex to represent graphically on colored isotherm maps of large regions.

The plotting of species climatic envelopes groups together widely separated geographic locations where a beetle species occurs under similar climates and reveals the climatic homogeneity underlying many species' distributions.

The third step is computer storage and retrieval of species climatic envelopes. In order to facilitate rapid calculation of paleotemperatures by computer, the climatic range envelope for each species is coded and stored in numeric form. In Europe, this has been achieved by superimposing a grid (36 × 60 in 1°C units) on the T_{max}/T_{range} plot and designating each element by a ''one'' or ''zero,'' according to whether it is within or outside the envelope (Fig. 3). In North America, this has been done using a digitizing tablet to draw an envelope around the scatter plot of T_{max} versus T_{range} data for each species (Fig. 4). The envelopes were then stored electronically in a computerized drawing program.

The fourth step is climatic reconstruction from an assemblage of named species. Given a list of species in an assemblage, the computer retrieves and superimposes their numeric envelopes to produce a T_{max}/T_{range} graph of percentage overlap. Visual inspection of this plot allows the area of maximum overlap to be determined, then the values of T_{max}, T_{range}, and T_{min} are taken from that area (Fig. 5)

3.1. MCR Calibration and Accuracy Checking

The MCR method lends itself to rigorous checks by reconstructing modern climates from modern beetle faunas collected within restricted areas, and comparing the results with the mean temperature records from nearby meteorological stations. The results of a test based on species living today at 15 localities in Europe, Iceland, and Siberia are shown in the upper half of Fig. 6. Points representing T_{min} are below and left of the upper boxed point. The vertical bars represent the MCR of all species

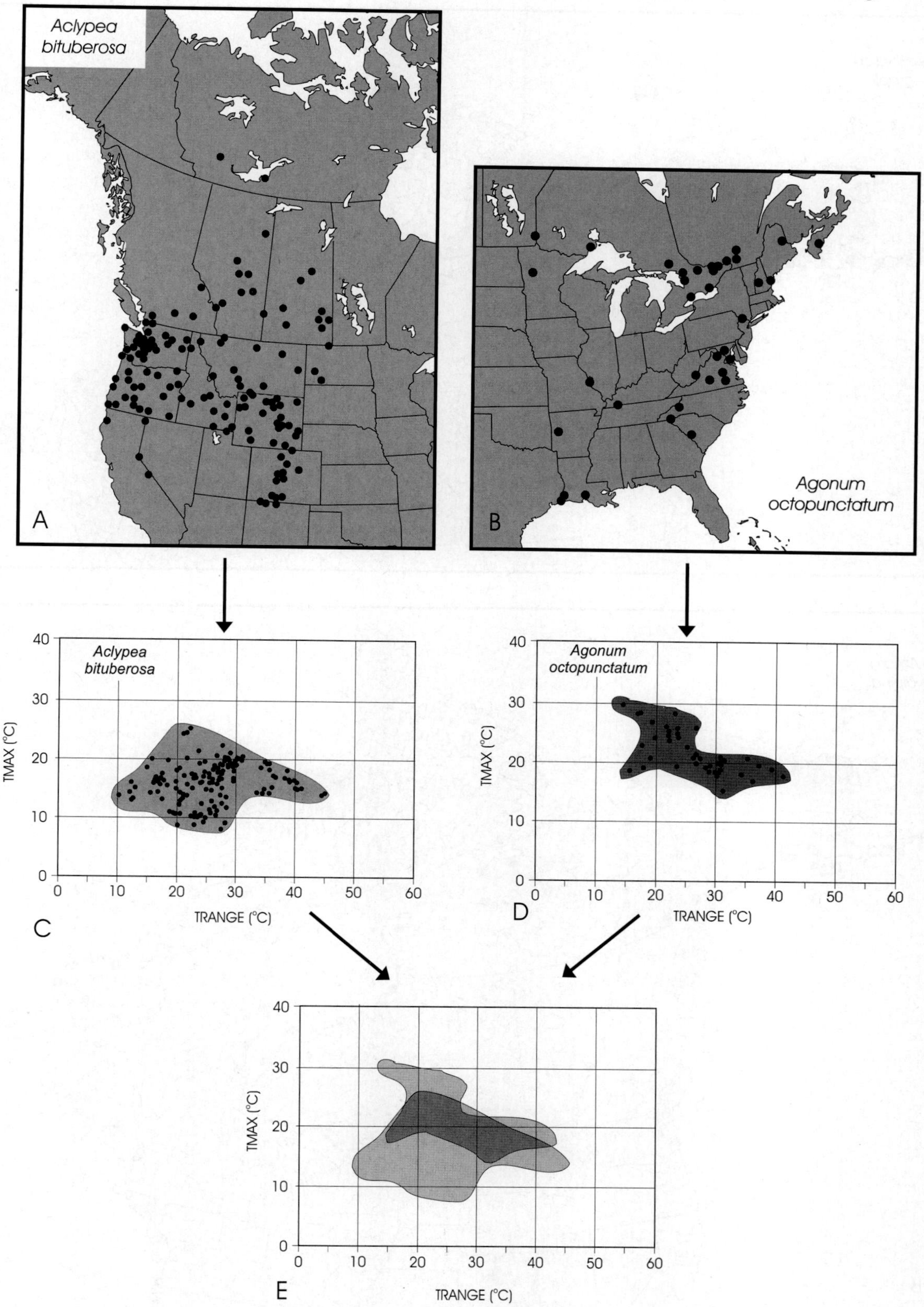

Fig. 1. Modern distribution maps for (A) the carrion beetle (Silphidae) A. bituberosa *and (B) the ground beetle* A. octopunctatum *correspond to climate envelopes for these respective species (C and D) that overlap to produce an MCR (E), even though the modern distributions of these two beetles do not overlap.*

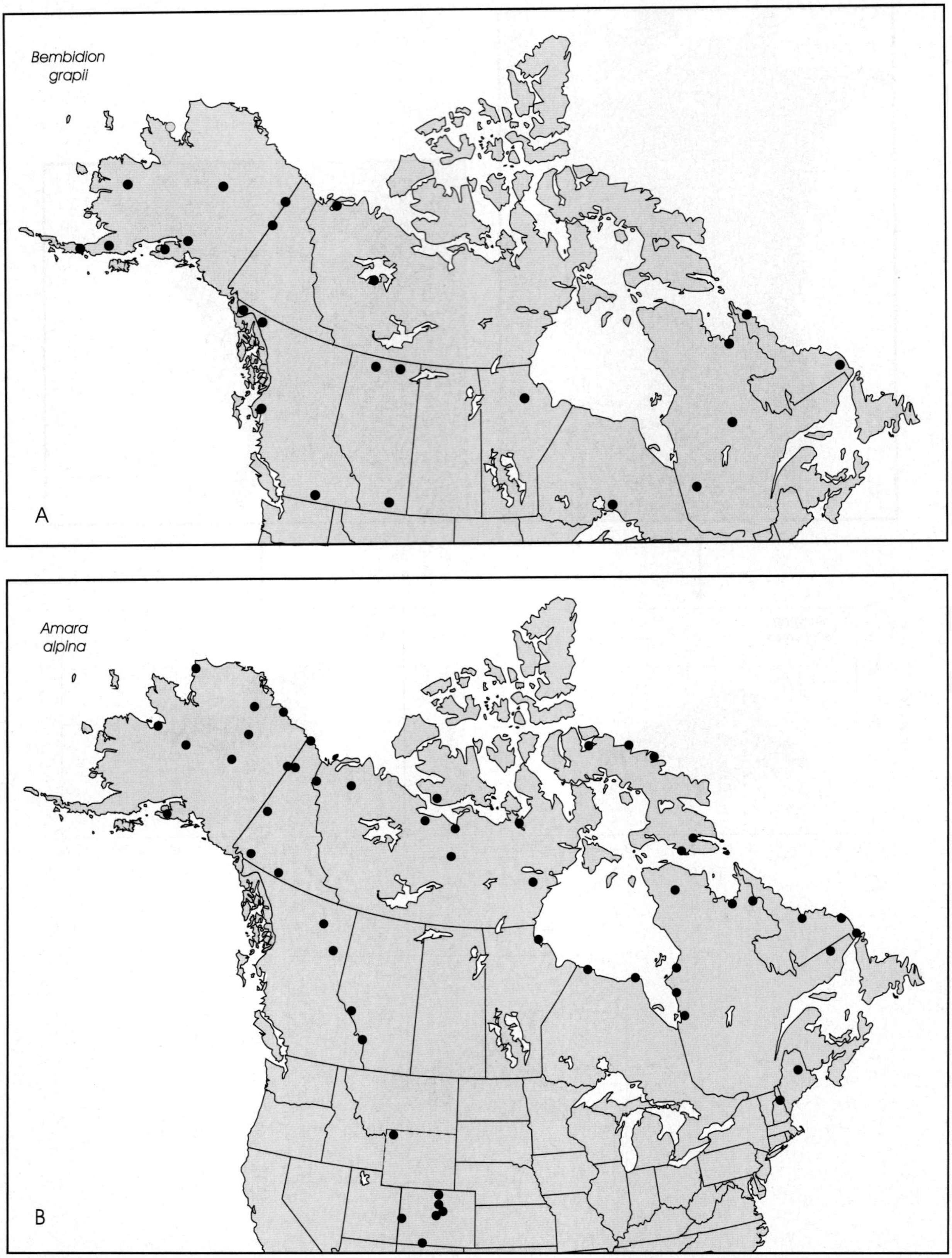

Fig. 2. Known modern North American distributions of the ground beetle species (A) Bembidion grapii *and (B)* Amara alpina*. The distribution pattern of* B. grapii *is typical of boreal species; the distribution of* A. alpina *is typical of arctic and alpine species.*

in each test assemblage. The same test was performed on 35 modern beetle assemblages from North America (lower half of Fig. 6). The results were strikingly similar. These tests suggest that the accuracy of the method is acceptable, using current distributional and meteorological databases. Atkinson *et al.* (1987) published a calibration method, including a correction for overestimation of $T_{\min}$ for colder climates and underestimated

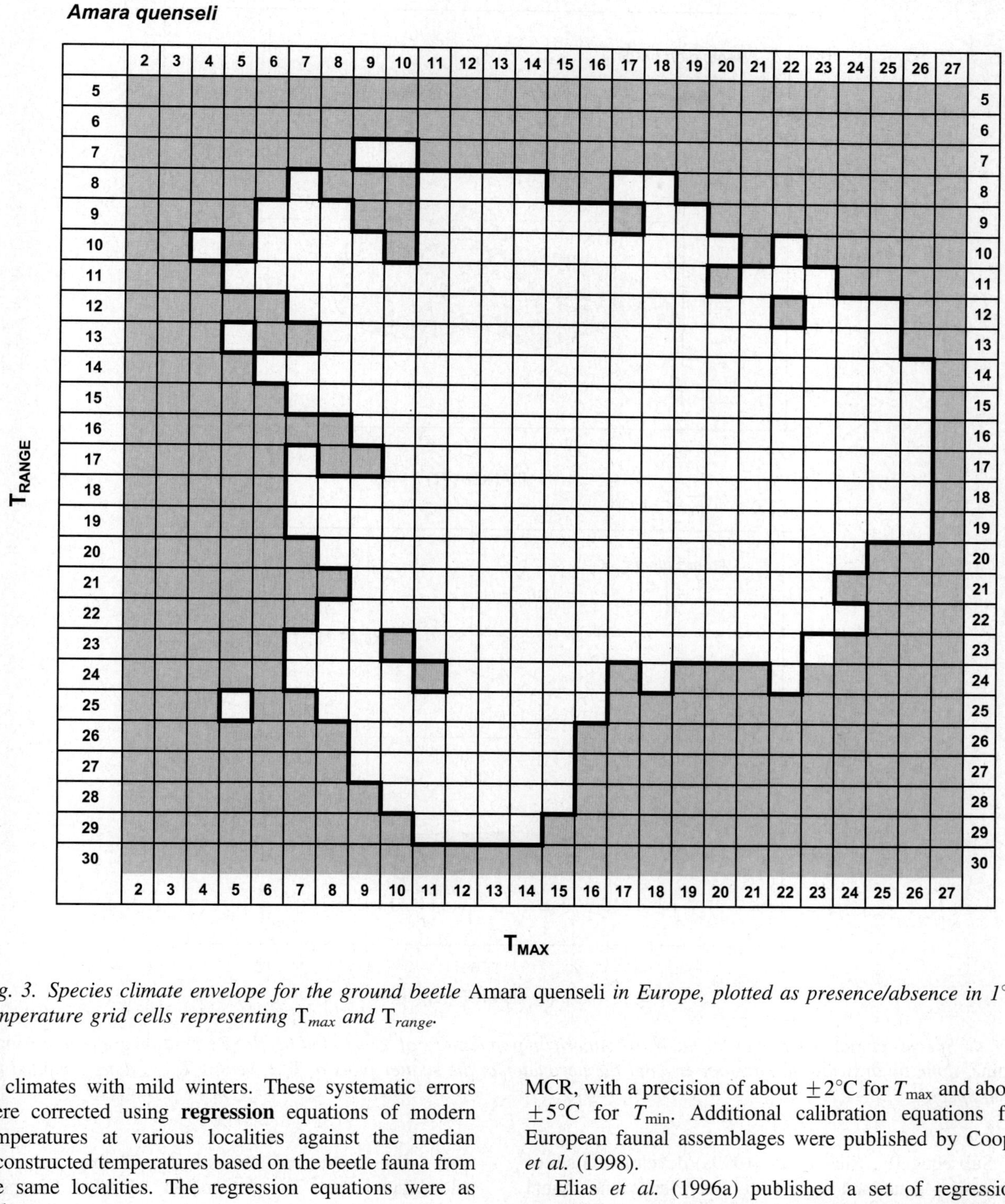

Fig. 3. Species climate envelope for the ground beetle Amara quenseli *in Europe, plotted as presence/absence in 1°C temperature grid cells representing* T_{max} *and* T_{range}.

in climates with mild winters. These systematic errors were corrected using **regression** equations of modern temperatures at various localities against the median reconstructed temperatures based on the beetle fauna from the same localities. The regression equations were as follows:

$$T_{max}(\text{corrected}) = [1.006\, T_{max}(\text{mean})] + [0.0142\, N_{spec}] - 2.96$$
$$r = 0.94; \quad s = 0.83°\text{C}$$

$$T_{min}(\text{corrected}) = [1.416\, T_{min}(\text{mean})] + 1.904$$
$$r = 0.94; \quad s = 2.42°\text{C}$$

where temperatures are in degrees celsius and N_{spec} is the number of species used in the reconstruction. Corrected values obtained using these equations provide unbiased estimates of the most probable paleoclimate within the MCR, with a precision of about $\pm 2°$C for T_{max} and about $\pm 5°$C for T_{min}. Additional calibration equations for European faunal assemblages were published by Coope *et al.* (1998).

Elias *et al.* (1996a) published a set of regression equations to correct for systematic errors in North American MCR estimates, as follows:

$$T_{max}(\text{calibrated}) = (\text{mean predicted } T_{max} \times 0.79) + 3.4$$

$$T_{min}(\text{calibrated}) = (\text{mean predicted } T_{min} \times 0.72) - 4.9$$

These regressions were based on the actual versus predicted values for the 35 modern assemblages. The T_{max} regression yielded an r^2 value of 0.94. The T_{min} regression yielded an r^2 value of 0.82.

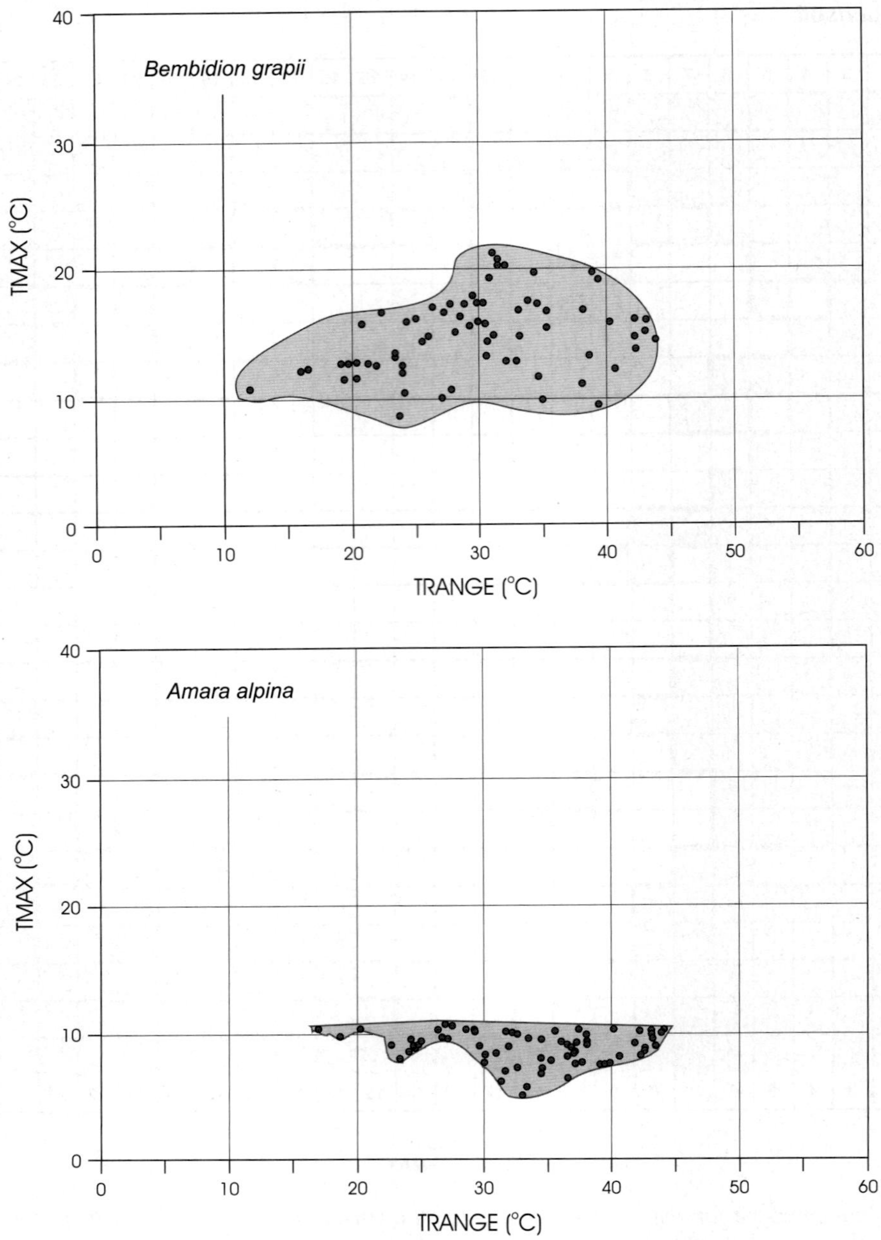

Fig. 4. Species climate envelopes for the North American populations of the ground beetles Bembidion grapii *and* Amara alpina, *done by drawing an envelope around the perimeter of the scatter plots of* T_{max} *versus* T_{range} *data compiled for each species.*

Subsequently, Elias *et al.* (1999a) developed a set of regression equations specifically for Pleistocene faunal assemblages from Eastern **Beringia** (unglaciated regions of the Yukon and Alaska), as follows:

$$T_{\text{max}}(\text{calibrated}) = (\text{mean predicted } T_{\text{max}} \times 0.52) + 6.13$$

$$T_{\text{min}}(\text{calibrated}) = (\text{mean predicted } T_{\text{min}} \times 0.72) - 6.4$$

An example of the original MCR method is shown in Fig. 7. The values of T_{max} and T_{min} are based on time-averaged data. The shaded upper and lower boundaries define the limits of the MCR (paleotemperature estimates lie within these limits). However, the bold line shows what was considered the most probable value of paleotemperatures, as determined by the European regression equations (above). Gaps represent periods without data. The MCR reconstructions for Britain are remarkable, in that they show the rapidity and strength of climatic change at the end of the last glaciation. Extremely rapid warming took place at about 13,000 yr BP, and again at 10,000 yr BP. In between, the marked cooling associated with the **Younger Dryas** is also demonstrated.

When they were developing climate envelopes for species found in Eastern Beringian fossil assemblages, Elias *et al.* (1999a) discovered that the observed modern mean winter temperatures in coastal localities failed to match the temperature estimates made from modern beetle assemblages. The predictions of mean temperatures

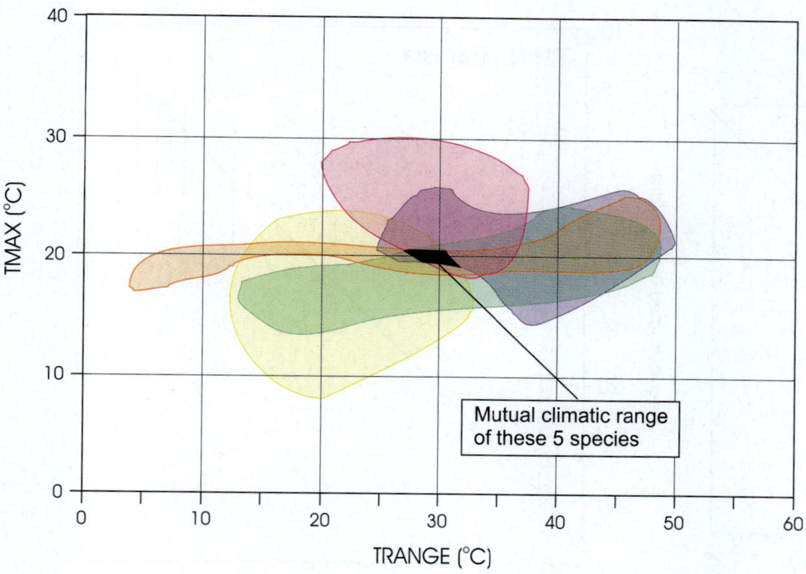

Fig. 5. Hypothetical example of the overlap of five species' climate envelopes, producing an MCR estimate (the black quadrilateral area indicated by the arrow).

for the coldest month (T_{min}) for coastal sites were consistently below the observed values. Thus, the beetle communities living in coastal Alaska today are indicative of colder winter temperatures than are indicated by the observed T_{min}. This phenomenon makes it difficult to apply the MCR method in estimating paleoclimates in Beringia. However, careful study of *daily* winter temperature records for coastal towns in Alaska revealed that T_{min} (the mean temperature of the coldest month) is a poor indicator of the actual climatic variability of these localities. The daily data revealed that incursions of extremely cold arctic air affect even the southern coastal regions of Alaska, almost every year. These ''cold snaps'' may last only a few days each year, but they apparently exert a strong influence on the coastal beetle fauna. There are a number of beetle species that inhabit the Pacific Northwest region of North America that would be able to live in such places as Anchorage or Seward, Alaska, where mean January temperatures are relatively mild. However, they are apparently unable to tolerate the extremes of winter cold that come on a fairly regular basis. A comparison of the lowest mean minimum temperatures recorded for a number of coastal sites in Alaska with modern T_{min} estimates predicted on the basis of the climate envelopes of the modern beetle faunas of those localities shows much closer agreement than a straight comparison of predicted with observed mean January temperatures (Table 1). This study demonstrates that mean monthly winter temperature does not control beetle distributions in these localities, so T_{min} is not a very useful measure of past climate for coastal Alaskan localities. In contrast, most of the Alaskan fossil localities that are situated near the modern coast were well inland during Pleistocene glacial intervals, when the oceanic margins of the Bering Land Bridge were many hundreds of kilometers distant from the modern coastline.

The MCR method devised by Atkinson *et al.* (1986, 1987) and developed for North America by Elias *et al.* (1996a, 1999a) has been shown to be an extremely useful tool for paleoclimate reconstructions. It has allowed the tracking of climatic changes through such oscillations as the lateglacial interval in North America (Table 2). Fossil beetle proxy data were already providing some of the best paleotemperature data, because beetles are sensitive, reliable indicators of thermal regime. It should also be noted that MCR analyses of faunas previously interpreted through the range overlap method have seldom altered those interpretations in any significant way. The paleoclimate curve developed by Atkinson *et al.* (1987) differs little from an earlier curve drawn by Coope (1977a) without the benefit of MCR analysis. However, the addition of the MCR method allows the proxy data from beetle fossils to be quantified, calibrated, and tested against modern communities. MCR also estimates winter temperatures and degree of seasonality (Table 1).

3.2. *Use of MCR for Reconstructing Precipitation in Deserts*

I developed a different set of species climate envelopes for desert-dwelling beetles found in packrat middens from the arid southwestern United States and northern Mexico (Elias, 1998). As discussed above, it is thought that seasonal temperatures are the dominant climatic factors controlling beetle distributions in temperate, boreal and arctic regions. However, the ranges of desert-dwelling insects are more likely to be limited by a combination of temperature and moisture. Desert climates present different types of challenges to insect growth, development, and reproduction, and not all North American deserts are alike.

The North American desert regions share some common climate features, such as hot summers and little annual precipitation. However, there is also considerable heterogeneity in desert climates. For instance, the Sonoran Desert has relatively mild winters: temperatures below 0°C are rare, and just a few consecutive days of freezing temperatures are sufficient to kill some species of Sonoran plants, such as the saguaro cactus. In contrast to this, the Chihuahuan and Great Basin deserts regularly experience below-freezing temperatures in winter. Although all the desert regions of North America receive little precipitation, the seasonal moisture pattern varies

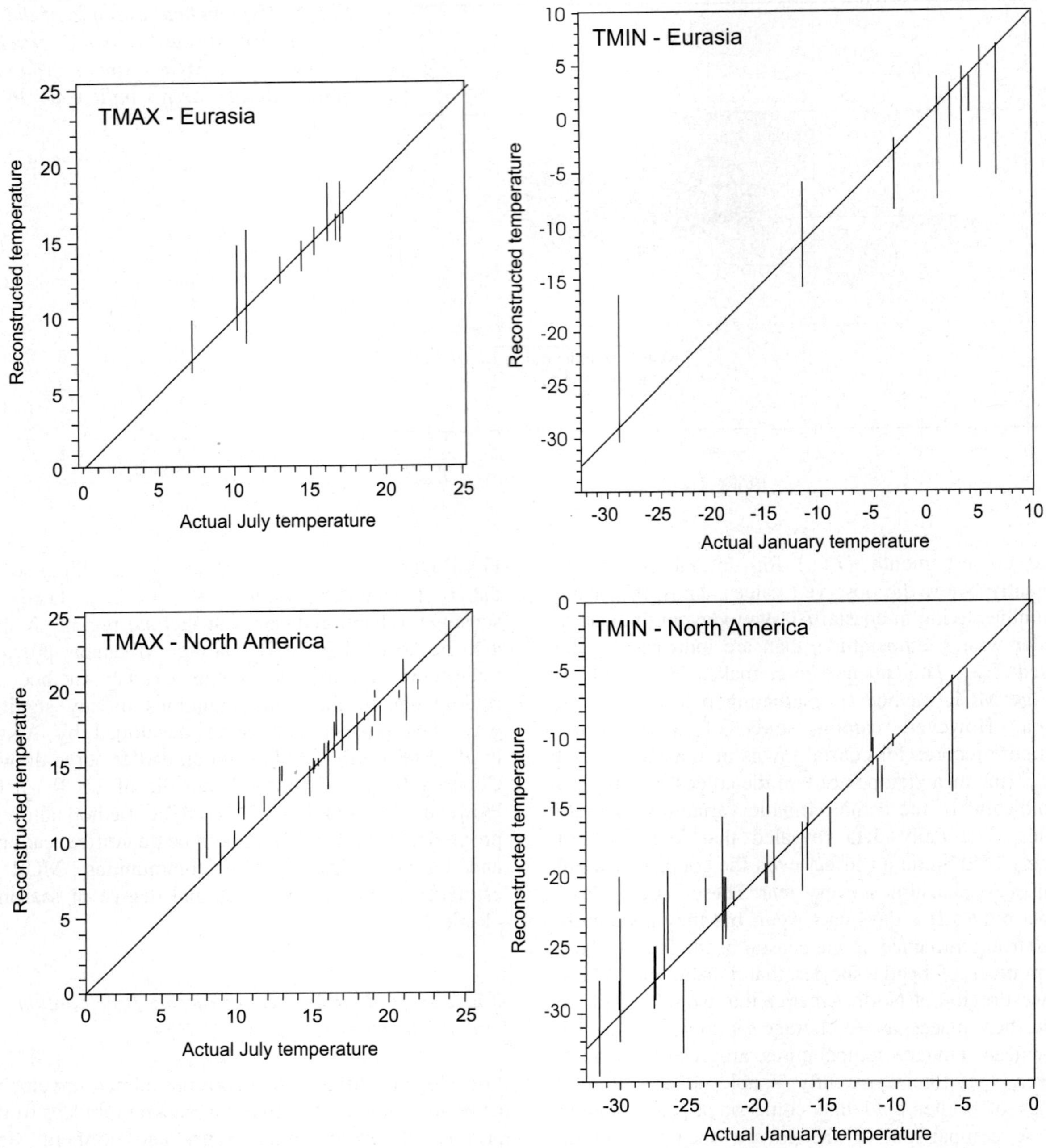

Fig. 6. Graphs depicting the results of tests of predicted versus observed T_{max} and T_{min} for predatory and scavenging species of beetles living in Eurasia (upper two graphs) and in North America (lower two graphs). The vertical bars on each graph indicate the MCR estimate based on species found today at localities for which meteorological data are available. The diagonal lines show where the MCR estimates would fall if the predicted results matched the observed data. Note that for both Eurasian and North American data sets, there is a systematic underestimate in the coldest temperature ranges of both T_{max} and T_{min}, and a systematic overestimation of the warmest temperature ranges. Note also that T_{max} estimates tend to be more tightly constrained than T_{min} estimates, as discussed in the text. The upper two graphs are after Atkinson et al. (1987). The lower two graphs are after Elias et al. (1996a).

considerably from region to region. Differences in atmospheric circulation patterns during the last glaciation caused differences in the seasonality of the precipitation (Swetnam and Betancourt, 1990; Van Devender, 1990a,b).

I worked with Kathy Anderson (INSTAAR, University of Colorado) to create climate envelopes for 50 species of beetles found in packrat midden assemblages from the Chihuahuan Desert, the Colorado Plateau, and the Great Basin. We developed climate envelopes that plotted T_{max} values against mean annual precipitation (MAP) values for the modern collecting localities of these species. Anderson used the Bartlein *et al.*'s (1994) 25-km gridded climate database for North America to derive the climate values employed in the climate envelopes.

To test the validity and calibrate the MCR method for use in North American desert regions, the method was tested using 20 modern locations. Each of these locations was included in the modern distributions of at least four species, so that the climate envelopes of at least four

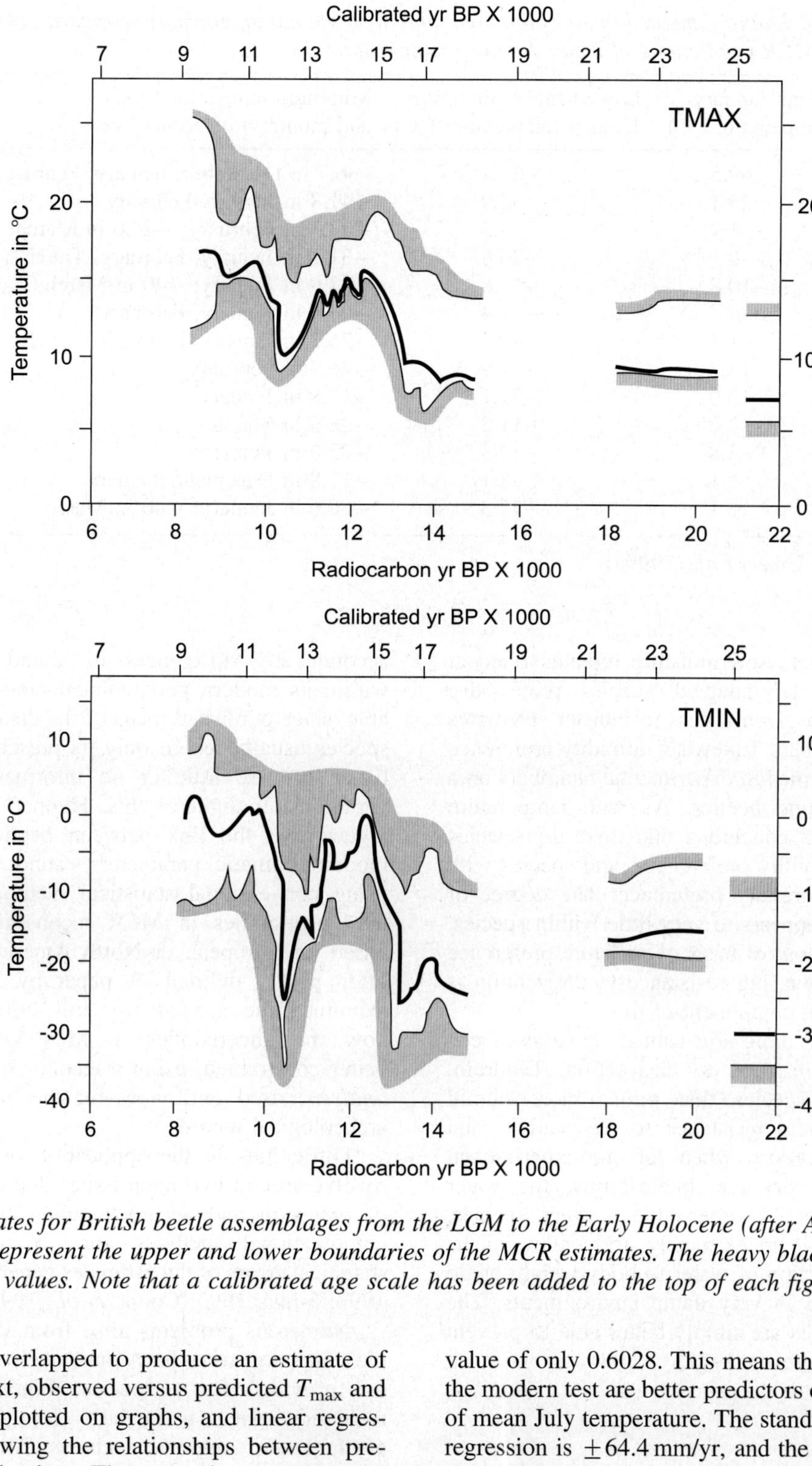

Fig. 7. MCR estimates for British beetle assemblages from the LGM to the Early Holocene (after Atkinson et al., 1987). The shaded areas represent the upper and lower boundaries of the MCR estimates. The heavy black line represents the calibrated estimate values. Note that a calibrated age scale has been added to the top of each figure.

species could be overlapped to produce an estimate of T_{max} and MAP. Next, observed versus predicted T_{max} and MAP values were plotted on graphs, and linear regressions obtained showing the relationships between predicted and observed values. The regression equations are as follows:

$$\text{Mean } T_{max}[\text{calibrated}] = 8.16 + [0.67 \times \text{mean predicted } T_{max}]$$

$$\text{Mean MAP[calibrated]} = 146.5 + [0.712 \times \text{mean predicted MAP}]$$

Surprisingly, the linear regression of predicted versus observed MAP yielded an r^2 value of 0.806, while the regression of predicted versus observed T_{max} yielded an r^2 value of only 0.6028. This means that the beetles used in the modern test are better predictors of MAP than they are of mean July temperature. The standard error of the MAP regression is ± 64.4 mm/yr, and the standard error of the T_{max} regression is ± 1.9°C. The equations suggest that the distribution patterns of beetles in arid regions are more controlled by precipitation than summer temperature. The questions arise: How do beetles experience precipitation? How does it affect them? The answer to these questions lies in soil moisture and relative humidity. A summary of the results of many experiments of soil moisture and relative humidity preferences of ground beetles is provided in Thiele (1977, pp. 191–197). Many species of beetles are found only in habitats with specific soil moisture properties. For instance, some riparian ground beetles live only a given distance from

Table 1. Modern mean January, mean January minimum, and absolute minimum winter temperatures at coastal Alaskan sites, compared with MCR predictions of mean January temperatures.

Site	Mean January temperature (°C)	Lowest mean minimum January temperature (°C)	Minimum temperature (°C) and month(s) of occurrence	MCR predicted T_{min} (°C)
(1) Anchorage	−9.5	−21.4	−36.7 in December, January, February	−25.8
(2) Barrow	−25.1	−34.9	−47.8 in January February	−30.5
(3) Cold Bay	−1.9	−8.3	−23.3 in February; −25.0 in March	−17.7
(7) Homer	−6.1	−26.6	−31.1 in January, February, March	−24.9
(8) Kenai	−10.8	−17.6	−43.9 in January; −40 in March	−15.4
(9) King Salmon	−9.5	−24.4	−44.4 in January, February	−22.8
(10) Kodiak	−1.2	−9.8	−26.7 in January, February, March	−24.7
(11) Kotzebue	−18.3	−31.9	−46.7 in February	−29.0
(12) Nome	−13.9	−31.1	−47.8 in January	−29.9
(14) St. Paul Island	−3.0	−11.2	−28.3 in March	−26.7
(15) Seward	−3.8	−13.3	−22.0 in January	−12.3
(16) Sitka	1.1	−8.1	−17.8 in December, January	−13.6
(17) Unalakleet	−15.1	−27.3	−50.6 in January; −40 in March	−28.6

Source: Data from Elias *et al.* (1999a)

the water's edge, where soil moisture remains fixed at certain percentages. Dry-adapted species prefer dry substrates, even when given access to moister substrates in experimental chambers. Likewise, humidity preference tests have been performed in experimental chambers on a wide variety of ground beetles. As with temperature preference, these tests concluded that there are species with pronounced humidity preferences, and species with little demonstrable humidity preference. The degree of humidity preference appears to vary little within species, however. The physiological basis of moisture preference is difficult to pin down, but resistance to desiccation is probably an important component of this.

Species that prefer more arid habitats have also been found to be more resistant to desiccation (Lindroth, 1949). Conversely, species that prefer more humid environments have less resistance to desiccation, and died within a few hours when left in experimental chambers with very dry air. Interestingly, the water content of ground beetles varies little between species. Even species that live in extremely arid environments have the same proportion of water to body weight as do ground beetles living in very damp environments. The desert-dwelling species are simply better able to prevent desiccation (Boyer-Lefèvre, 1971).

3.3. Ubiquity Analysis

As discussed above, the MCR method assumes that the climatic tolerances of modern beetle species can be adequately defined. A Quaternary MSc dissertation by Dadswell (2003) at Royal Holloway, University of London, sought to examine this assumption, and specifically the statistical steps involved in (a) generating MCR estimates, (b) establishing the link between modern climate and modern beetle distributions, and (c) assessing the errors associated with MCR estimates. One of the criticisms that may be leveled against beetle MCR temperature estimates is that they do not take into account any differences in abundance of a species within its modern geographical range. This is unavoidable, since published maps of the distributions of modern species usually record only its presence or absence, and hence provide little or no information on abundance variations. In light of this, Huppert and Solow (2004) argued that the link between beetle distributions and modern climatic parameters cannot be tested robustly using conventional statistical methods, and hence that the uncertainties in MCR reconstructions, even those based on European or North American beetle data, are often poorly defined. A paper by Bray *et al.* (2006) examined the grounds for this criticism, and showed how the uncertainties in MCR estimates might be better constrained, using a technique known as *ubiquity analysis*, used until now for the analysis of complex archeological records.

Difficulties in the application of the MCR method revolve around two main issues that constrain the degree of precision and accuracy achievable: (a) bias in the computational methods and (b) calculation of the statistical errors of the estimates generated (see Anderson, 1993; Sinka, 1993; Coope *et al.*, 1998; Elias, 2001).

Numerous problems arise from varying adequacy in the collection records. Abundance records are generally not provided for a species. While reasonably high-resolution data, particularly for the Carabidae, are available for regions such as Britain (Luff, 1998), Scandinavia (Lindroth, 1992), Central Europe (Holdhaus and Lindroth, 1939; Horion, 1941), the Iberian Peninsula (Jeanne, 1965 to 1973), and Switzerland (Marggi, 1992), this is not the case for large regions farther east, and the available beetle distribution data cannot match the 0.5° longitude by 0.5° latitude resolution of the climate data currently available for Europe (Hulme *et al.*, 1995). This lack of accurate distribution data is recognized by all Quaternary entomologists, but the problem is currently insurmountable, because it will take many years of research to develop more robust beetle distributional data at the appropriate continental scale. The only way to

Table 2. Summary of lateglacial-age fossil beetle assemblage data from sites in central and eastern North America.

Site	Age (cal yr BP × 1,000)	Late Pleistocene		Modern		Change in temperature		Reference(s)
		T_{max} (°C)	T_{min} (°C)	T_{max}	T_{min}	July ΔT	January ΔT	
(17) Ft. Dodge, IA/I	18.0	12–13	−26 to −20	23.1	−9.4	−11.9 to −10.9	−16.6 to −10.6	Schwert (1992), Elias *et al.* (1996a)
(17) Ft. Dodge, IA/II	17.3	11.5–12.5	−28.5 to −19.5	23.1	−9.4	−11.4 to −12.4	−19.1 to −10.1	Schwert (1992), Elias *et al.* (1996a)
(18) Weaver Drain, MI	17.0	12.5–16.5	−29.5 to −6.5	21.1	−6.2	−8.6 to −4.6	−23.3 to −0.3	Morgan *et al.* (1981); Elias *et al.* (1996a)
(19) Winter Gulf, NY/W08	15.4	17.7–20.5	−10 to −4.5	20.4	−5.2	−2.9 to +0.1	−4.8 to +0.7	Schwert and Morgan (1980); Elias *et al.* (1996a)
(20) Gage St., Ont./G8	15.0–14.5	11.5–13.5	−25.5 to −11.5	19.9	−7.5	−8.4 to −6.4	−18 to −4	Schwert *et al.* (1985), Elias *et al.* (1996a)
(19) Winter Gulf, NY/W06	14.8	16.5–21	−11.5 to −4	20.4	−5.2	−3.9 to +0.6	−8.6 to +1.2	Schwert and Morgan (1980), Elias *et al.* (1996a)
(21) Rostock, Ont.	14.7	12–15	−28 to −20	19.4	−7.8	−7 to −4.4	−20.2 to −12.2	Pilny and Morgan (1987), Elias *et al.* (1996a)
(22) Newton, PA	14.6	10–15	−31 to −9	19.7	−5.6	−9.7 to −4.7	−25.4 to −3.4	Barnowsky *et al.* (1988), Elias *et al.* (1996a)
(23) Norwood, MN/Peat	14.5	16.5–21.5	−17.5 to −3	22.1	−12.3	−5.6 to −0.6	−5.2 to +9.3	Ashworth *et al.* (1981), Elias *et al.* (1996a)
(19) Winter Gulf, NY/W04	14.5	15–21.5	−14 to 0	20.4	−5.2	−5.4 to +1.1	−9.8 to +5.2	Schwert and Morgan (1980), Elias *et al.* (1996a)
(20) Gage St., Ont./G7	14.5–12.8	20–21	−5.5 to −2.5	19.9	−7.5	0 to +1.1	+2 to +5	Schwert *et al.* (1985), Elias *et al.* (1996a)
(23) Norwood, MN/M.S.	13.8	15–20	−25 to −12	22.1	−12.3	−7.1 to −2.1	−12.7 to −0.3	Ashworth *et al.* (1981), Elias *et al.* (1996a)
(21) Rostock, Ont/65–75	13.6	15–20.5	−22.5 to −4.5	19.4	−7.8	−4 to +0.6	−14.7 to +3.3	Pilny and Morgan (1987), Elias *et al.* (1996a)
(24) Two Creeks, WI	13.6	12.5–13.5	−16.5 to −11.5	21.1	−7.3	−8.8 to −7.8	−10.2 to −4.2	Morgan and Morgan (1979), Elias *et al.* (1996a)
(25) Kewaunee, WI	13.5	12.5–17.5	−28 to −5	20.8	−7.8	−8.3 to −3.3	−20.2 to +2.8	Garry *et al.* (1990a), Elias *et al.* (1996a)
(23) Norwood, MN/U.S.	13.4	17–18	−11 to −7	22.1	−12.3	−5.1 to −4.1	+1.3 to +5.3	Ashworth *et al.* (1981), Elias *et al.* (1996a)
(21) Rostock, Ont/55–65	13.3	17–21	−15.5 to −2.5	19.4	−7.8	−2.4 to +1.6	−7.7 to +5.3	Pilny and Morgan (1987), Elias *et al.* (1996a)
(27) Brookside, NS	13.0	12.5–20	−31.5 to 5	18.4	−7.4	−5.9 to +1.6	−24.1 to +12.4	Mott *et al.* (1986), Elias (2007c)
(28) St. Eugene, Que	12.9	10–12.5	−31 to −18.5	17.2	−13.1	−7.2 to −4.7	−17.9 to −5.4	Mott *et al.* (1981), Elias (2007c)
(29) Lockport Gulf, NY/L01	12.8	17.5–21.5	−14 to −3	21.5	−4.8	−4 to 0	−9.2 to +1.8	Miller and Morgan (1982), Elias (2007c)
(30) Johns Lake, ND	12.7	13.5–15.5	−26.5 to −20	20	−17	−6.5 to −4.5	−6.5 to 0	Ashworth and Schwert (1992), Elias (2007c)
(31) Eighteen Mile R., Ont.	12.4	14–15	−16.5 to −10	19.8	−5.8	−5.8 to -4.8	−10.7 to −4.2	Ashworth (1977), Elias (2007c)
(29) Lockport Gulf, NY/L02	12.0	12.5–17.5	−25 to −5	21.5	−4.8	−9 to −4	−20.2 to −0.2	Miller and Morgan (1982), Elias (2007c)
(32) St. Hilaire, Que	11.7	16.5–17.5	−15 to −10	20.7	−10.2	−4.2 to −3.2	−4.8 to +0.2	Mott *et al.* (1981), Elias (2007c)
(26) Nichols Brook, NY-C	11.5–10.1	14.5–18°C	−21 to −12.5	18.1°C	−10°C	−3.5 to −0.1	−11 to −2.5	Fritz *et al.* (1987), Elias (2007c)
(29) Lockport Gulf, NY/LO3	11.4	17–20	−15.5 to −4.5	21.5	−4.8	−4.5 to −1	−10.7 to +0.3	Miller and Morgan (1982), Elias (2007c)
(33) Seibold, ND	11.1	17–21.5	−15.5 to −5	20.5	−14.9	−3.5 to +1	−0.6 to +9.9	Ashworth and Brophy (1972), Elias (2007c)

allow for these inadequacies is to deal with them in a semi-subjective manner. In defining species climate ranges (SCRs) for some difficult taxa, therefore, individual adjustments have to be made, even though they are made on the basis of a thorough understanding of beetle ecology. However, this adds an element of subjectivity to the process, which is one of the features of the prior method of beetle paleoclimate reconstruction that the MCR method was meant to overcome. At the moment, there is no mechanism for ensuring that adjustments such as these are made in a consistent fashion among different researchers (Bray *et al.*, 2006).

The climate data employed to define beetle SCRs has also been quite variable. The most robust publicly available climate data for Europe are arranged on a ½ degree latitudinal by ½ degree longitudinal grid (IPCC Data Distribution Centre). As discussed above, distribution maps for individual beetle species can be overlain onto this climate data grid, either through mathematical coding of the distributions or simply by drawing around printed sheets of the climate data, to define the SCRs. As discussed above in relation to the North American MCR method, data points that appear in the climate database are not necessarily based on locally measured meteorological variables, but are rather interpolated values based on the nearest available meteorological data. This means that the SCRs, and consequently the MCRs derived from them, are based on smoothed climatic gradients, the true regional variance probably being underrepresented in the climatic database. This probably does not cause serious difficulties for sites in Western Europe and much of North America, where there are reasonable densities of climate stations. It becomes a more serious problem in Eastern Europe and Asia, where both beetle collection data and meteorological stations are sparse. Some species that occur in glacial-age faunas in Europe have exclusively Asiatic distributions today, so this problem extends to the interpretation of European fossil assemblages.

The observed deviation between the MCR estimated temperature values and those measured at nearby meteorological stations could, in part, be a reflection of the way in which the SCR envelopes were constructed, but there are additional possible explanations. It could be an inevitable mathematical consequence of the regression method (e.g., Lucy *et al.*, 2002) or an artifact of the climate data themselves. For instance, there could be an underlying skew in the SCRs caused by bias within the raw climate data, the end result of which is to generate skewed MCR data (Bray *et al.*, 2006).

There are statistical problems associated with the use of regression equations to correct systematic deviations in MCR estimates. The point estimates and ranges, which are used in this approach, have no defined probability. In the Atkinson *et al.* (1987) study, the median value of each MCR was adopted as the point estimates plotted in the regressions, although the majority of the calculated MCR ranges plotted on the regression line. The use of the mean point estimate method in MCR would only be appropriate if data exhibit a normal (Gaussian) distribution in climate space. In probabilistic terms, the mean would then be the best estimator of "true" value for Gaussian data. If, however, the MCRs have any other distribution (e.g., log normal, or multimodal) then the method is invalidated. Sinka (1993) explored the possibility that beetle distributions in climate space are normally distributed, and suggested that MCR estimate precision could be improved by weighting against the extreme distributions (the margins of the SCRs) to generate normally distributed data. In all cases, however, the resulting MCRs were less successful at predicting the climate of test sites, which strongly suggests that their distributions within climate space are not Gaussian, and that regression methods are therefore inappropriate for correcting the bias.

3.4. Other Approaches to MCR Analysis

Various attempts have been made to make MCR estimates more reliable. Marra *et al.* (2004) have attempted to improve the procedure for clarifying the climate space distributions of beetle taxa for which the modern ranges are incompletely known and whose populations are geographically restricted (in New Zealand). Huppert and Solow (2004) used a logistic regression approach to fit probability functions to beetle distributions within climate space. This appears to have performed well, but in tests of observed versus predicted temperatures for 602 test sites in North America, there was still significant data scatter, especially in their T_{min} reconstructions. Bray *et al.* (2006) published the results of Dadswell's dissertation project, which used ubiquity analysis in an attempt to refine SCRs for some ground beetle species in Europe. The *ubiquity* of a beetle species is defined by how much of a particular sector of climate space (a particular combination of T_{max} and T_{range}, e.g.) a species actually occupies.

In the test of ubiquity analysis, the distribution of each beetle species was plotted on the geographical climate grid, and the total number of squares of each climate type in which the beetle species is found was determined. For example, if the total number of squares with T_{max} of 15°C and T_{range} of 18°C is 24, and the number of those cells actually occupied by a given species is 18, then the ubiquity of that species can be expressed as 75% ($U = [N_{sq}/SN_{sq}] \times 100$, where U = ubiquity score for each climate square; N_{sq} = number of squares occupied by beetle species in that climate type in Europe; SN_{sq} = total number of squares of that climate type in Europe) (Bray *et al.*, 2006). The ubiquity SCR is thus defined by a plot of the calculated ubiquity of the species in each climate domain. This plot represents the area of climate space occupied by the species, but the ubiquity SCR plot adds a third dimension showing the frequency of occurrence of each species in climate space. Dadswell constructed SCRs for 44 stenothermic species of beetles in European climate space. She discovered that only six of these species have SCRs that are normally distributed in climate space. The remainders have bimodel, tri-model, or other modes of distribution (Fig. 8).

The study published by Bray *et al.* (2006) has forced paleoentomologists to abandon the linear regression models used previously to correct systematic errors in MCR estimates. Additional work in ubiquity analysis is ongoing at Royal Holloway, University of London. More

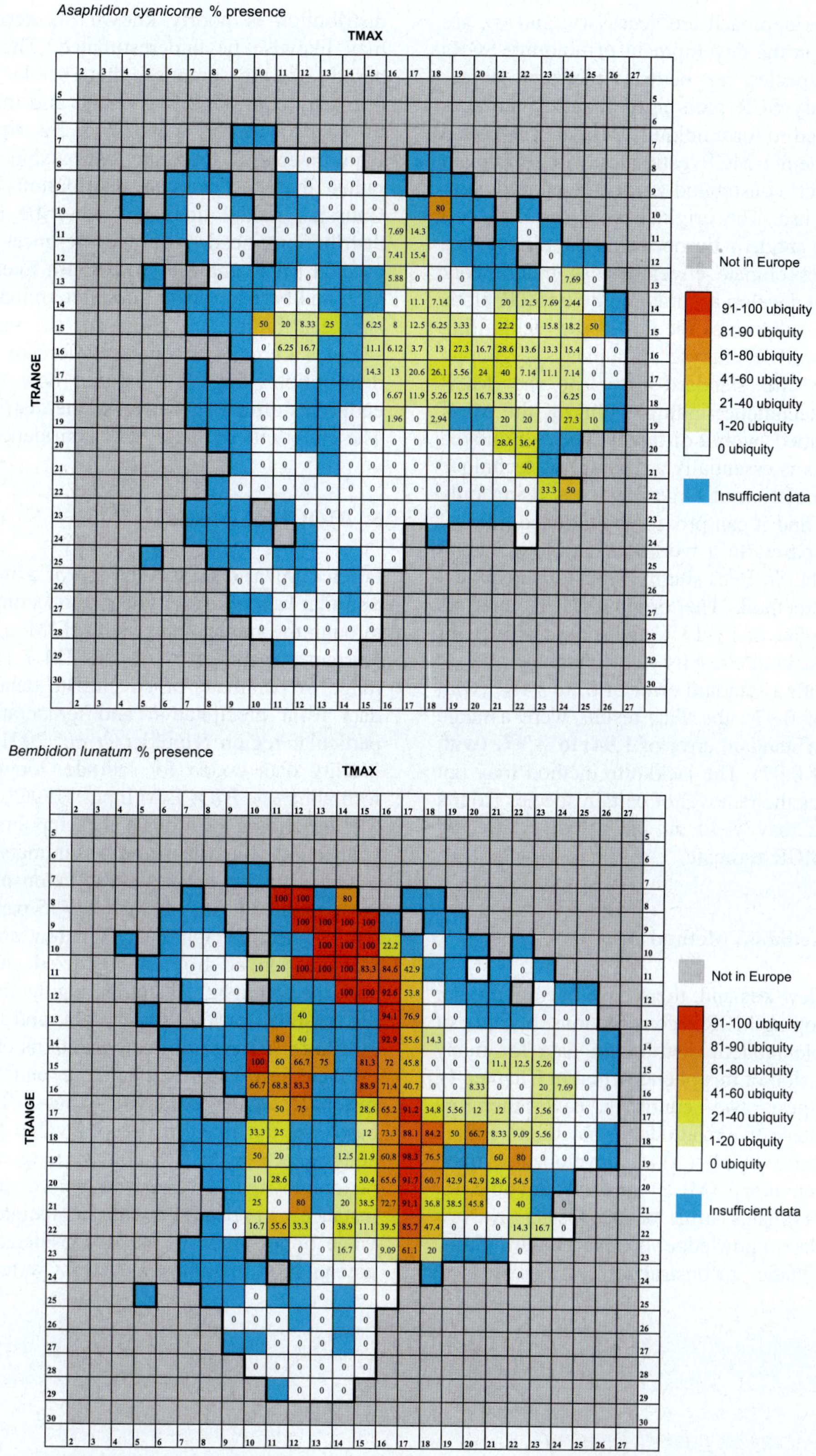

Fig. 8. Ubiquity species climate envelopes for the ground beetles Asaphidion cyanicorne *(above) and* Bembidion lunatum *(below) (European occurrences only). Note the multiple centers of ubiquity (yellow and red grid cells) for each species (after Bray* et al., *2006).*

rigorous tests of the approach are necessary, and we are currently working on the development of ubiquity SCRs for more beetle species, to better understand three-dimensional ubiquity SCR plots as probability surfaces that can be combined to form ubiquity MCRs. It is hoped that these will generate MCR estimates with improved precision and better constrained errors compared with those currently in use. The original work by Dadswell employed only the modern European distributional data to construct species climate envelopes, so much work needs to be done to develop accurate ubiquity SCR plots.

Buckland (2007) discusses the use of the jackknifing statistical method in MCR analyses, as a means of gaining information about the standard error of the MCR estimates. In this technique, multiple runs of the MCR analysis are performed, with a different species removed from each run. This is essentially a resampling technique that allows the simulation of larger data sets using only the available data, and it can provide confidence interval statistics in many cases. In a worked example, a faunal assemblage from the St. Bees site in England was treated by the jackknife method. The original MCR analysis yielded a T_{max} estimate of 12–13°C and a T_{min} estimate of −6 to −3°C. The jackknife results yielded a range of T_{max} values from 12 (with a standard error of 0) to 14°C (with a standard error of 0.97); the T_{min} results were a range from of −8 (with a standard error of 1.94) to −3°C (with a standard error of 0.97). The jackknife method may not work in all cases, as the removal of certain species from a faunal assemblage may yield an inconclusive (i.e., no unique solution) MCR estimate.

4. Maximum-Likelihood Method

In Australia and New Zealand, there have been particular problems in developing MCR reconstructions, because of the lack of reliable modern distribution data for many species (see discussion of this phenomenon in Chapter 4). Researchers in these two countries have therefore developed some alternative methods for paleotemperature reconstruction. Marra *et al.* (2004) developed a maximum-likelihood envelope (MLE) method, suitable for quantifying past climates from beetle fossils in New Zealand. Incomplete knowledge of the modern fauna limits robust climatic reconstruction. If a species' distribution is poorly known, its ecological tolerances may likewise be underestimated. The MLE method is based on the premise that beetle distributions are primarily controlled by climate and that the distribution of each taxon is controlled by finite physiological requirements. It, therefore, uses a sine distribution model rather than a Gaussian distribution model because a Gaussian model does not prescribe absolute limits to distributions needed for the presence–absence approach used in MCR analysis. Maximum likelihood (ML) best-high and best-low values are determined and fitted to the sine distributions to allow for the uncertainty that the samples are realistic representations of the actual climatic distributions. Fig. 9 illustrates how the estimated ML endpoints effectively increase the distribution of a taxon. The endpoints define a 95% confidence distribution.

5. BIOCLIM Method

Porch (2007a) discusses the use of a different approach to a similar lack of modern beetle collecting data in Australia, through the application of BIOCLIM, a computer program (currently included in ANUCLIM 5.1) used to derive a range of parameter-based climate data for point or grid data from precipitation and temperature surfaces of a particular region (Houlder *et al.*, 2001). For each taxon, locality data coded for latitude, longitude, and altitude form the basic BIOCLIM input. BIOCLIM output includes files describing the climate at each point at which the taxon was found (bio files) and a summary of the parameter statistics for the species – the taxon profile (pro). In the Australian context, this includes 35 parameters describing temperature, precipitation, radiation, and moisture indices that are derived from standardized climate surfaces that span the continent. By comparing the relationship between the original profile (raw) data and the data-enhanced profile for a range of taxa, equations can be developed to estimate errors for temperature and precipitation parameters (estimated parameter ranges). These equations are based on the range-per-record value for each parameter and for each taxon. The climatic range of a taxon becomes better known as the range-per-record value diminishes. For instance, in dealing with temperature parameters, a value of less than 0.1°C is considered well known. For precipitation parameters, errors were constructed only

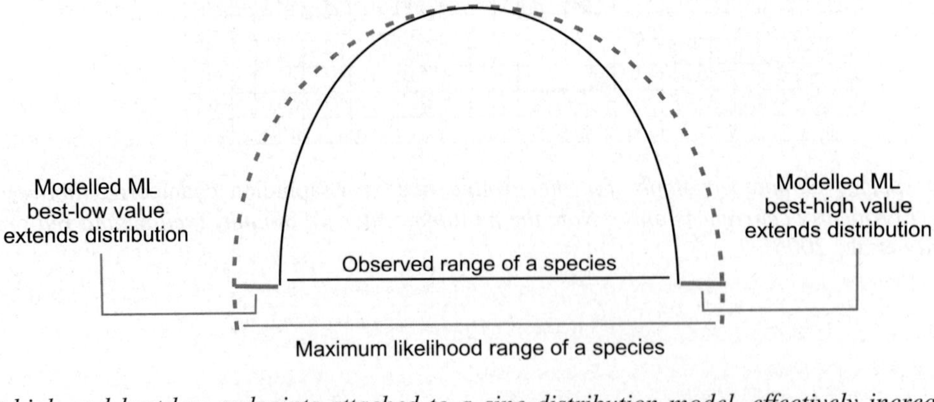

Fig. 9. ML best-high and best-low endpoints attached to a sine distribution model, effectively increasing the climate range of a species (from Marra, 2007; Copyright Elsevier).

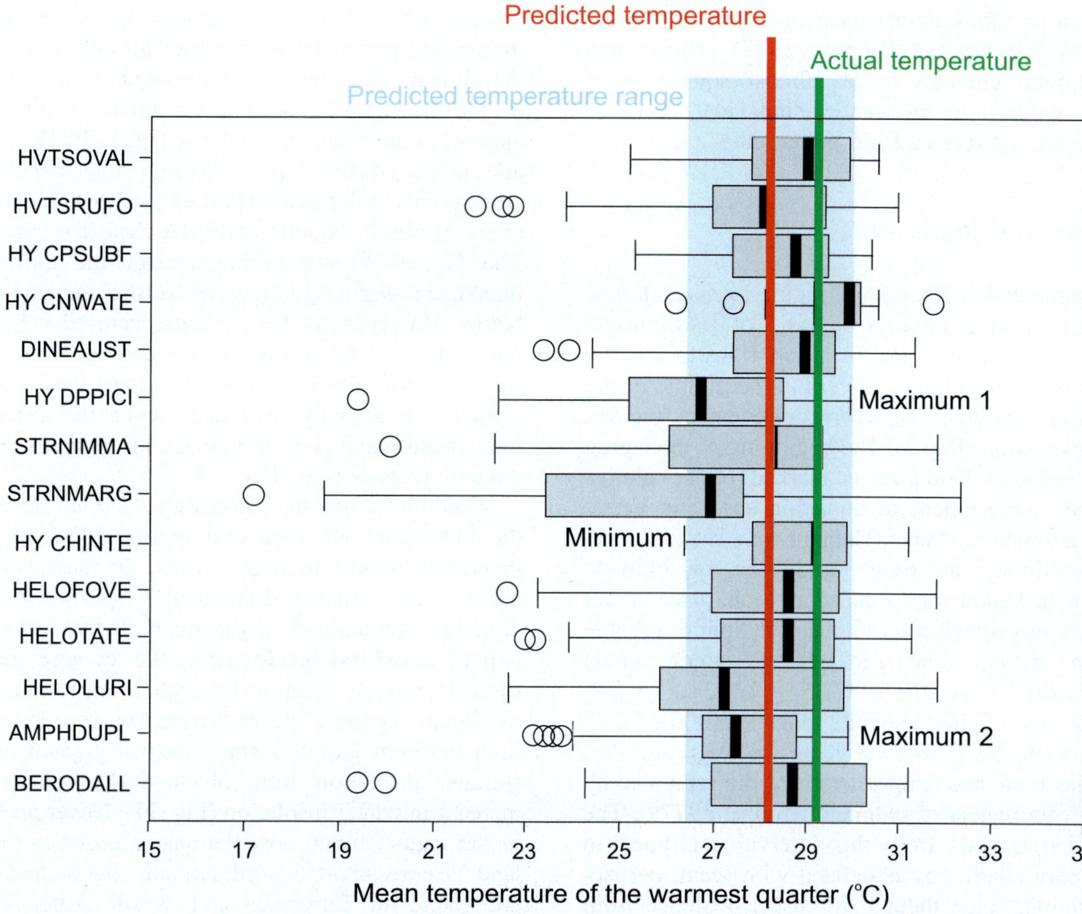

Fig. 10. An assemblage of 14 aquatic beetle species (Dytiscidae and Hydrophilidae – codes listed on left) from a site in northern Australia, showing the derivation of the predicted temperature range (envelope) and the best estimate (predicted temperature). Data for each taxon are displayed as box-plots, which allow examination of the structure of the data underlying the predictions. In this case, a single taxon (labeled ''minimum'') determines the lower (cooler) boundary, and two taxa (labeled maximum 1 and maximum 2) determine the upper (warmer) boundary of the predicted range envelope. This example uses raw data values, rather than estimated parameter range values (figure from Porch, 2007a; Copyright Elsevier).

for minimum values and using the 0–25th percentile spread of values to exclude the influence of the skew to high values with increasing data volumes (Porch, 2007a).

Once the environmental ''envelopes'' for species are developed through BIOCLIM, Porch (2007a) has utilized coexistence methods to reconstruct past Australian climates. For each BIOCLIM parameter, the climatic estimates for modern and fossil assemblages include a range defined by the overlap of the taxa in the assemblage; the median of this range is the best estimate. Using this method, estimation of modern climate at a range of localities across Australia was undertaken to assess the precision and accuracy of beetle-based reconstructions (Fig. 10). Estimation of paleoclimates has used the identical method, based on the taxa in fossil assemblages that have bioclimatic profiles in the modern data set.

6. Insect Evidence for Rapid Climate Change

The previous chapters have shown that insects are in many ways uniquely suited as proxy data for the reconstruction of past terrestrial environments (both on land and in freshwater). They are sensitive to environmental change, and they respond to such changes in ways that can be indicated through fossil assemblages. These phenomena are amply demonstrated in Quaternary insect records.

One of the most important contributions that insect fossil data have made to our understanding of Quaternary environments is the evidence they provide for rapid, intense climate changes. Prior to Coope's work in Britain, nearly all terrestrial paleoclimatic reconstructions were based on studies of past vegetation, interpreted from pollen. Many palynologists assumed that plant communities are reliable indicators of climate change, and that shifts in vegetation patterns are synchronous with climatic fluctuations. Most changes registered in pollen spectra through the Quaternary have been gradual, with transitions between glacial and interglacial episodes lasting hundreds if not thousands of years. However, Coope's work began to cast doubt on paleobotanical reconstructions of climate during intervals of rapid change, especially during the **lateglacial** interval. In fossil insect records, changes between faunas suggestive of major climatic episodes may occur in as little as a few decades, and are often so rapid as to appear instantaneous in the fossil record (Coope and Brophy, 1972). These data suggest climatic changes in an almost square wave pattern, as opposed to the gentle, sinusoidal curve of climatic change interpreted from pollen data.

6.1. Evidence from British Studies

I will examine two episodes of rapid, intense climate change, first inferred through analysis of fossil insect assemblages in Britain. One is a rapid warming (the Upton Warren Interstadial Complex) in the middle of the last glaciation. The other is the series of events at the end of the last glaciation (Fig. 7). The beginning of the Upton Warren Interstadial Complex is marked by an almost instantaneous replacement of an arctic–subarctic beetle fauna with a temperate fauna (Coope *et al.*, 1961; Coope, 1977a). The timing and exact rapidity of this climatic change remain unknown, because it took place about 47,000 years ago, which places it near the limit of reliable radiocarbon dating. The age of the arctic fauna (44,300 $\pm$ 1,600 ^{14}C yr BP, or 47,852 $\pm$ 2,103 cal yr BP) overlaps the age of the temperate fauna (43,000 $\pm$ 1,200 ^{14}C yr BP, or 46,733 $\pm$ 1,841 cal yr BP). Stratigraphically, however, the arctic and temperate faunas are separated by only a few centimeters of sediment (Coope, 1977a). The British pollen records from this interval continued to indicate tundra conditions associated with scant, herbaceous vegetation. Even though the insect evidence from this episode indicates conditions sufficiently warm to allow the establishment of trees in Britain, palynological evidence indicates a total lack of trees at this time. This is probably a reflection of a rapid and intense climatic amelioration which was responded to more rapidly by the beetle fauna compared with the trees. The beetles show evidence for a subsequent climatic cooling that meant that in the British Isles, the trees had no time to register this climatic event.

Evidence for the timing of interstadial warming during marine isotope stage (MIS) 3 has been found from different time intervals in different regions (Fig. 11). The two best-studied regions, Britain and France, have MCR reconstructions showing interstadial warm peaks at 47,000–46,000 and 39,000 cal yr BP, respectively. The British records fail to show a warming at 39,000 cal yr BP, and likewise the French record at Grand Pile (Ponel, 1995) fails to show a warming between 47,000 and 46,000 cal yr BP. However, there is a gap in the Grand Pile MCR reconstructions between about 44,000 and 48,000 cal yr BP, so this may explain the missing earlier interstadial signal in the French reconstruction.

The **oxygen isotope** records from Greenland show numerous oscillations attributed to interstadial events (Fig. 12). No single terrestrial fossil record shows all of these interstadials. In particular, the fossil beetle records discussed here fail to show any interstadial warming between 40 and 36 ka and between 27 and 21 ka, although the ice core oxygen isotope records indicate two distinct warming events, and the North Atlantic foraminifera record also shows ameliorations during this interval. The North Atlantic marine fossil and Greenland ice core records both indicate what appears to be a long, intense interstadial period between about 50 and 44 ka, whereas the climatic warming in the subsequent interstadial peak in the North Atlantic and Greenland record (43–41) appears to have been less intense (Fig. 12). The calibrated age range of the Upton Warren Interstadial interval corresponds to the middle part of the 50–44 ka interstadial event in North Atlantic and Greenland ice core records. The Upton Warren event matches the timing of an interstadial warming in eastern North America, discussed below. However, as these events occurred very near the outer limits of reliability of radiocarbon dating, it could well be that either or both beetle-derived interstadial signals are actually correlated with the peak of the interstadial event seen in Greenland ice cores and North Atlantic records (Fig. 12).

Could it be that the paleoclimate signals derived from the Greenland ice core and marine fossil records are inherently noisy? In other words, do these two sets of proxy data respond dramatically to relatively minor climatic fluctuations? If the main factor driving global climate is **orbital forcing**, then the ice core and marine fossil records do seem to be responding to other aspects of climate, because the insolation curves for the mid-to-high northern latitudes show only a gradual decline in summer insolation from 50 to 20 ka, and a gradual increase in winter insolation (Fig. 11). Nevertheless, some of the major climatic oscillations recorded in the Greenland ice cores and the North Atlantic marine fossil records are found in European and North American beetle records. However, as we have seen, there is no single interstadial event that is recorded in all terrestrial regions. The ''spotty'' nature of interstadial occurrences across the Northern Hemisphere may argue for more regional expression of interstadial warming, perhaps driven by shifting ocean currents and Heinrich events (Elias, 1998). As always, we will need more fossil data before we can come to grips with these issues.

The lateglacial sequence from Britain is better understood than the Upton Warren Interstadial Complex, mainly because lateglacial events (LGEs, 17,000–11,500 cal yr BP) have been reliably dated by radiocarbon assay. Insect data indicate that the British lateglacial warming was sudden and intense, beginning 15,900 years ago. It is signaled by a replacement of arctic and subarctic species by temperate species. Beetle evidence points to a rise in summer temperatures of 7°C and a rise of winter temperatures of perhaps as much as 20°C. These changes constitute a replacement of cold, continental climate by mild, oceanic climate in the space of a few decades.

Following the Younger Dryas oscillation, warming of similar magnitude and rapidity took place (Ashworth, 1973; Osborne, 1974a; Bishop and Coope, 1977). This reconstruction of climatic changes is considerably different from the traditional view (based on palynological evidence) of European climatic changes at the close of the last glaciation (Coope, 1987a). The pollen record from the early lateglacial in northwestern Europe shows continuity with vegetation from the previous glacial interval. As Coope and Brophy (1972) stated, the palynological interpretation of tundra conditions during

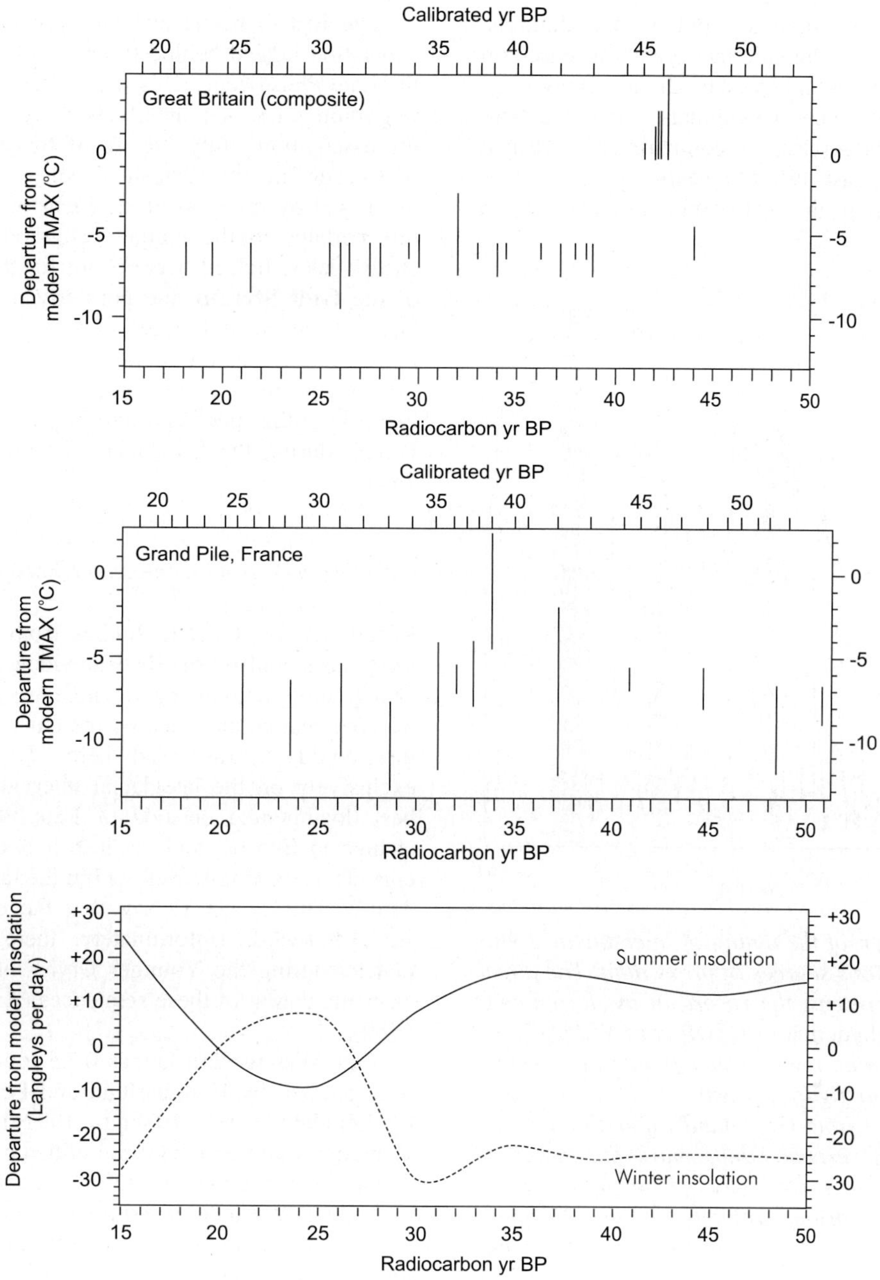

Fig. 11. MCR reconstructions from a suite of sites in Great Britain (data from Coope, 1977a; Coope et al., 1997, Gao et al., 1998) and the Grand Pile site in France (data from Ponel, 1995) compared with changes in insolation at 65°N (data from Berger, 1978 and figure after Elias, 1999).

the early phase of the lateglacial is based largely on negative evidence, that is, on an impoverished flora and a lack of trees. But the vegetation from this time is composed largely of pioneering plant species. Their presence should be viewed as evidence for a lack of competition in an impoverished flora, rather than as an index of arctic conditions.

Independent paleoclimatic reconstructions based on snow accumulation rates reconstructed from Greenland ice cores offer corroboration of the British paleoclimate scenario based on insect data (Lowe *et al.*, 1995). Likewise, the climatic changes inferred from oxygen isotope data from Greenland ice cores corroborate the timing and intensity of changes seen in British lateglacial beetle assemblages. Dansgaard *et al.* (1989) inferred climatic warming of 7°C within 50 years after the termination of Younger Dryas cooling. This is precisely the same estimate of amelioration that Coope made for terrestrial environments in Britain at this time. More recently, Rasmussen *et al.* (2006) published a revised chronology for last glacial termination section of the NGRIP ice core. When this is compared with calibrated radiocarbon ages from British beetle fossil assemblages, the timing and intensity of the two records match almost exactly (Fig. 13).

Moore (1986) compared British paleoclimatic reconstructions based on mammals with those based on insects, and concluded that "beetles are better climatic indicators than bears." Two problems are inherent in the mammalian fossil record. One of these is the paucity of specimens in comparison with invertebrates, pollen, and diatoms. A second problem is that climatic interpretations based on mammalian fossils are complicated by the relatively large range of climatic tolerance in mammals, compared with **poikilothermic** invertebrates. In comparing the British fossil record of the last 120,000 years, Moore (1986) concluded that ''perhaps the beetles will still have the last word.''

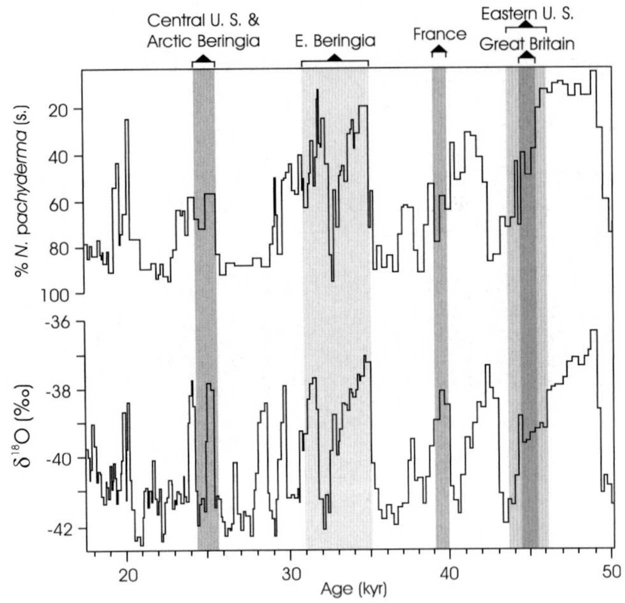

Fig. 12. Comparison of the timing of interstadial events as indicated by various sources of proxy data. The upper half of the figure presents the record of the foraminifer Neoglobigerina pachyderma *in DSDP core VEMA-23-81 from the North Atlantic region (after Bond et al., 1993). The lower half of the figure presents the $\delta^{18}O$ record of the Summit ice core from Greenland (after Bond et al., 1993). The shaded vertical bands mark the timing of interstadial warming events marked in fossil beetle assemblages from various regions of the Northern Hemisphere, as discussed in the text (figure after Elias, 1999).*

The British beetle evidence argues strongly that the vegetation lagged behind the rapid changes in climate in the Late Pleistocene. This lag is due most likely to slow migration rates of plants, especially trees. This will be discussed more fully, below. If this phenomenon were only seen in the British fossil record, it might be dismissed by many as an accident of biological history, due perhaps to the unique geographic position of the British Isles. Indeed, were it not for the warming effects of the **Gulf Stream**, the northerly position of much of Great Britain would dictate climatic conditions similar to those in central Labrador. The shifting of the Gulf Stream's position in the North Atlantic may have played a key role in the speed and intensity of climatic changes in Britain during the lateglacial (Ruddiman and McIntyre, 1981).

6.2. Evidence from Continental Europe

Results of fossil beetle studies from a wide variety of landscapes confirm the British evidence for rapid climatic change quite convincingly. There are numerous episodes of rapid change indicated by the Quaternary insect record that could be discussed here. This section focuses exclusively on the lateglacial interval, because it is the best documented episode of Late Pleistocene climatic change in Europe, and because it is also the best dated episode, since it falls well within the range of radiocarbon dating. This is not to say that the dating of LGEs is straightforward. Unfortunately, there is a radiocarbon plateau during the Younger Dryas interval that complicates the dating of these sequences (Ammann and Lotter, 1989).

Late **Weichselian** faunas from southern Sweden have been studied by Hammarlund and Lemdahl (1994), and by Lemdahl (1985, 1988a,b, 1991a), who have documented a series of lateglacial climatic changes (Fig. 14). These faunas show an initial lateglacial amelioration somewhat later than in Britain (ca. 14,900 vs. 15,800 cal yr BP), and of slightly less intensity, probably because of the regional cooling influence of the ice sheet. The

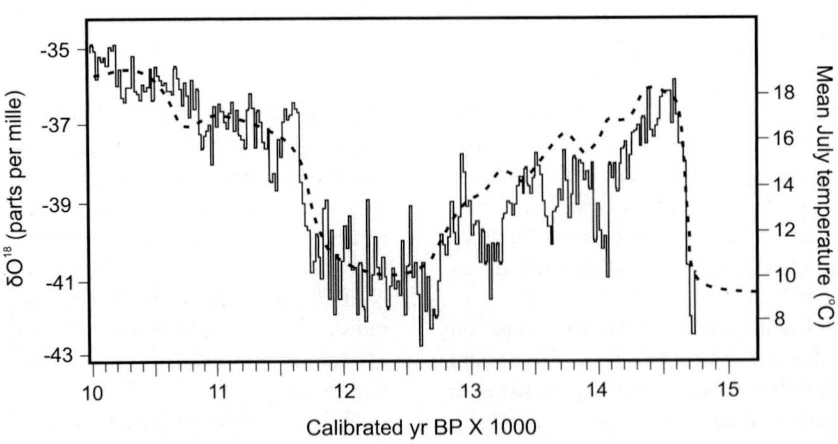

Fig. 13. Comparison of the chronology of lateglacial and Early Holocene climatic changes seen in the oxygen isotope record of the NGRIP ice core from Greenland (solid line) (data from Rasmussen et al., *2006) and in the composite MCR reconstruction for the British Isles (dashed line) (data from Atkinson* et al., *1987), plotted on the calibrated time scale.*

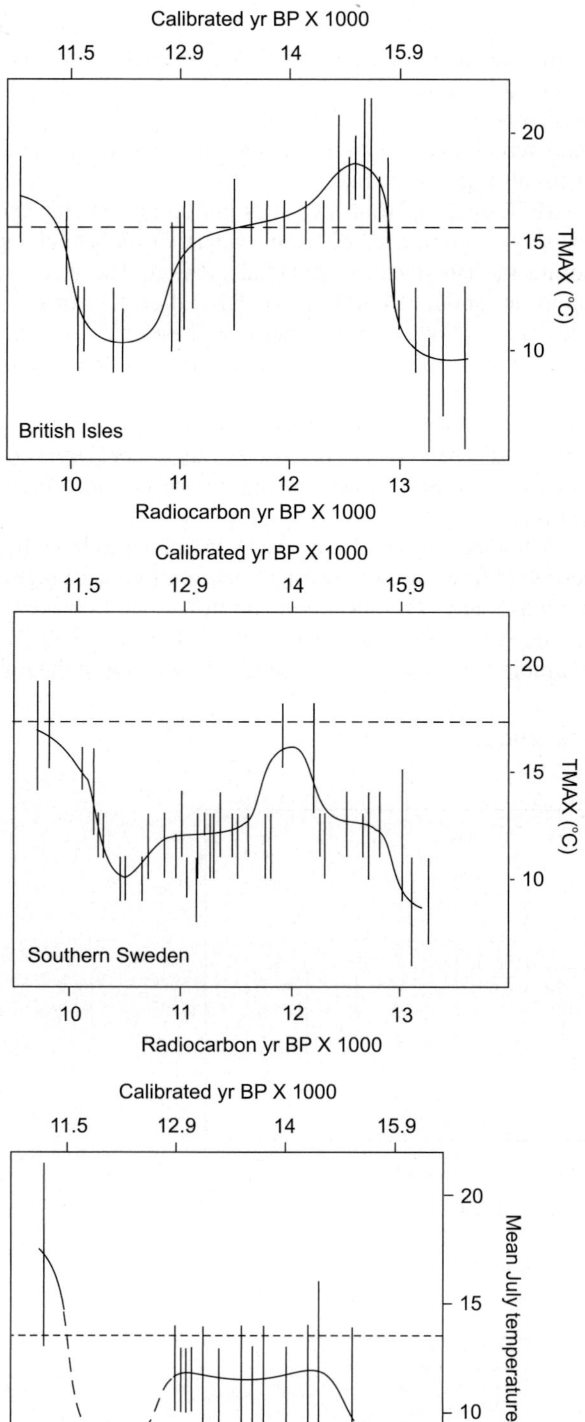

Fig. 14. MCR estimates of T_{max} for the lateglacial interval in (top) the British Isles, (center) southern Sweden, and (bottom) western Norway. The dashed horizontal line on each graph indicates the modern mean July temperature in the study region (data from Coope and Lemdahl, 1995).

Swedish vegetation shows a gradual change in the early lateglacial, or **Bølling pollen zone**. Even though the insect fauna is comprised of species from the boreal zone, the regional vegetation was open ground tundra with shrub birch (Iversen, 1954; Berglund *et al.*, 1984). Birch forest arrived in southern Sweden during the **Allerød pollen zone**, nearly 1,000 years after the initial climatic warming.

Following a Younger Dryas cooling, another rapid warming began at about 11,900 cal yr BP. The temperate vegetation signature of the **Preboreal pollen zone** lagged behind this amelioration by about 500 years (Lemdahl, 1985, 1991a). An Early Holocene beetle assemblage from a site at Hano Bay on the Baltic Sea (Gaillard and Lemdahl, 1994a) yielded an MCR T_{max} estimate of 14–18°C, essentially bracketing the modern T_{max} in this region. This fauna is dated about 9,500 cal yr BP (8,480 $\pm$ 80 ^{14}C yr BP), and demonstrates that full Holocene interglacial climates had become established by that time.

Coope and Lemdahl (1995) have compared the lateglacial climatic signals generated by MCR reconstructions from Britain, western Norway, and southern Sweden (Fig. 14). They concurred with Lemdahl's (1991a) conclusion that the climate of southern Sweden was cooled by its proximity to the waning Fennoscandian ice sheet, delaying the onset of interstadial warming. The climate of western Norway remained only just above full-glacial temperatures until after 11,500 cal yr BP, because of the cooling effects of the ice sheet.

A limited number of MCR estimates from Poland show amelioration by 15,900 cal yr BP, as in Britain. This is to be expected, as the Polish fossil sites were well south of the Fennoscandian ice sheet during the lateglacial interval (Coope and Lemdahl, 1995). Lemdahl (1991b) described a lateglacial insect faunal sequence from Zabinko in western Poland. Even though the insects from this site indicate summer temperatures of 14–15°C, subarctic vegetation was associated with the initial warming signaled by the beetle fauna (Tobolski, 1988).

A reconstruction of temperature gradients across northern Europe was done by Coope *et al.* (1998) for the transition period between the last glaciation and the Holocene, spanning the interval from 17,600 to 10,300 cal yr BP (14,500–9,000 ^{14}C yr BP). They found that during some intervals, such as 17,600–15,900 and 12,900–11,500 cal yr BP, the MCR evidence suggests climatic uniformity across much of northern Europe. However, during other intervals, such as 15,900–14,800 cal yr BP, there appear to have been steep temperature gradients – far steeper than modern gradients in this region. Coope *et al.* (1998) drew isotherm lines across the map of northern Europe, based on the MCR data. These lines trended northwest–southeast during colder episodes, but changed to west–east or northeast–southwest during warmer intervals, perhaps reflecting shifts from oceanic influence during the cold intervals to more continental climates during warm intervals. The remnant Scandinavian ice sheet continued to exert local cooling effects during the warmer episodes of the lateglacial interval, a phenomenon also observed in lateglacial beetle faunal assemblages from eastern North America (see below).

Paleotemperature reconstructions from sites in France have revealed variable patterns of oscillations during the lateglacial interval. Ponel and Coope (1990) studied a lateglacial insect faunal sequence from La Taphanel in the Massif Central region of southern France (Fig. 15). The timing and intensity of climatic changes was very similar to the British reconstruction, with rapid amelioration by 15,800 cal yr BP, and a marked dip in temperatures at about 14,000 cal yr BP, followed by a marked Younger Dryas cooling just after 13,000 cal yr BP, and another rapid warming before 11,500 cal yr BP. Interestingly, the reconstruction of climatic events at La Taphanel based on insect data is much more synchronous with the vegetation record than in northern Europe. Presumably, this is because the temperate plant species that survived the last glaciation in refugia in southern Europe were able to migrate rapidly into the Massif Central region when climate ameliorated. However, the climatic cooling at about 14,000 cal yr BP (corresponding to a synchronous cooling in Britain) is not seen in regional pollen records (de Beaulieu *et al.*, 1982, 1984). Even when plant migration routes are short, it appears that some insect groups (e.g., ground beetles) that are commonly found in fossil assemblages do a better job of signaling rapid or short-term climate change than do most plants represented in fossil pollen spectra.

MCR analyses from two sites in northern France (Fig. 15) show a somewhat different pattern of climate change during the lateglacial interstadial interval. These records begin at about 14,500 cal yr BP, after the onset of lateglacial climatic amelioration. These records show little or no evidence for cooling during the Allerød interval. In this aspect, the climatic reconstruction for northern France is similar to that for Switzerland (see below). There is a slight hint of an Older Dryas (ca. 14,000 cal yr BP) cooling in one of the northern France records.

A number of lateglacial insect assemblages have been described from the southwestern and northwestern regions of Switzerland. The sites in the northwest are Lobsigensee on the Swiss Plateau (Elias and Wilkinson, 1983) and Champreveyres, adjacent to the Jura Mountains at the Lake

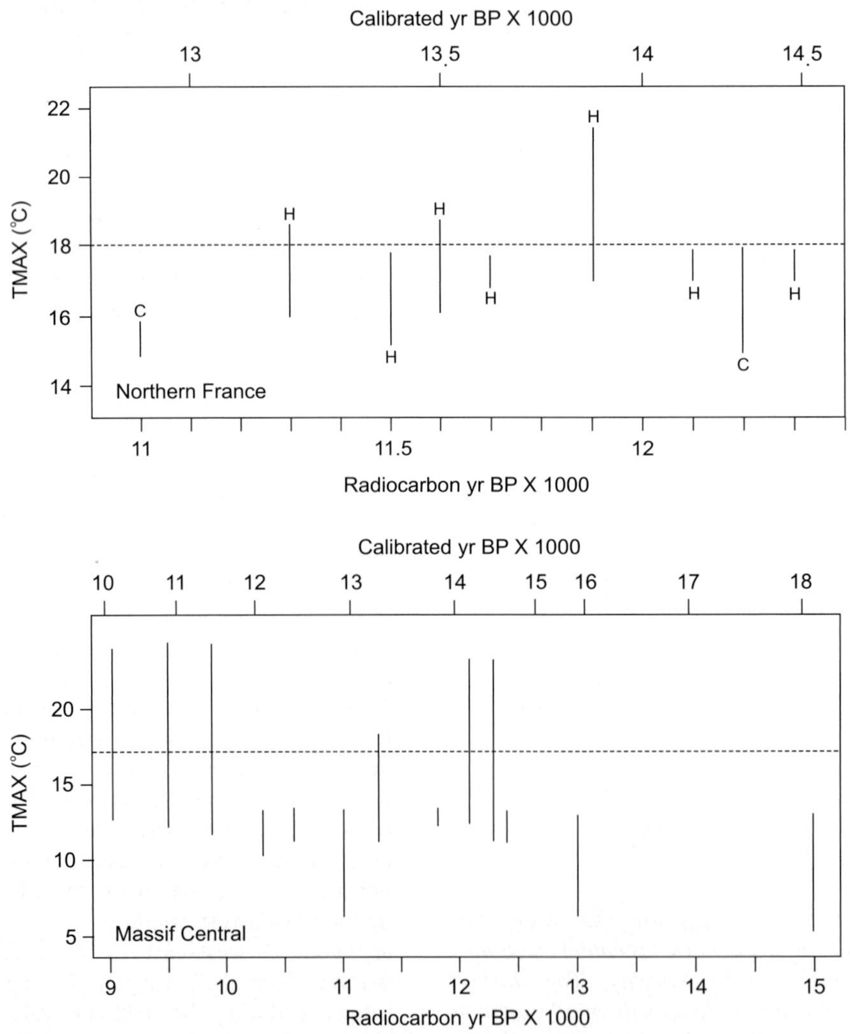

Fig. 15. MCR estimates of T_{max} for the lateglacial interval in (top) northern France, and (bottom) the Massif Central region of France. The letter "C" indicates MCR data from the Conty site; the letter "H" indicates MCR estimates from the Houdancourt site (data for northern France is from Ponel et al. (2005) and data for the Massif Central region is from Ponel and Coope (1990)).

of Neuchâtel (Coope and Elias, 2000). The Lobsigensee record contains an abrupt change from an arctic and alpine fauna to a temperate, boreal fauna at about 15,800 cal yr BP. As in La Taphanel, changes in the vegetation and insect records from Lobsigensee appear to be synchronous at the beginning of lateglacial warming. The shift from arctic and alpine to temperate insect assemblages coincides with pollen spectra including tree birch, willow, and juniper in the Bølling pollen zone (Ammann *et al.*, 1983). However, while the insect fauna in the Bølling is indicative of climatic conditions found today in central Europe, available evidence suggests that the modern vegetation of that region was not developed at Lobsigensee until 2,500 years later. This pattern is often repeated in lateglacial and Early Holocene paleoecology: insect faunas suggest an abrupt, early warming to levels very near modern parameters, and vegetation appears to develop slowly from open ground through shrubs, and pioneering trees (different taxa, in different regions), finally establishing the more or less modern plant communities.

The earliest insect faunal assemblage from the Champreveyres site (Coope and Elias, 2000) has been dated at about 15,250 cal yr BP. This assemblage reflects temperate conditions, so the timing of the transition from last-glacial climates predates this assemblage. The presence of **thermophilous** beetles in early lateglacial assemblages suggests that the climate was warm enough to have sustained mixed deciduous forest by at least 14,800 cal yr BP, although regional forests of modern composition seem not to have been established until the Holocene, more than 2,000 years later. Climatic warming alone is not sufficient to induce forest establishment. All of the proper conditions must first be developed, including soil chemistry and organic content, seasonality, and quality of moisture, as well as other factors. Trees would have had to migrate a distance of at least 600 km from their southern refuge in Italy.

In southwestern Switzerland, Gaillard and Lemdahl (1994b) reconstructed lateglacial environments at Grand Marais de Boussens, near Lausanne. Here, the timing of the shift from full glacial to interstadial climates was reflected in changes in beetle faunas dated about 15,000 cal yr BP, although there is a gap in faunal assemblages between that time and about 16,000 cal yr BP, so the faunal shift may well have come just after 15,900 cal yr BP, as it did further north in Europe. In a separate paper, Gabus *et al.* (1987) documented a warm-adapted beetle fauna from a terrace of Lake Geneva that was dated approximately 12,600 ^{14}C yr BP (14,900 cal yr BP). This fauna yields a T_{max} estimate of 15–16°C, about 2–3°C cooler than T_{max} in this region today.

Taken together, the MCR reconstructions for the Swiss lateglacial (Fig. 16) reveal a pattern of rapid warming by 15,900 cal yr BP, followed by relatively stable climate until 13,700 cal yr BP – the start of the Younger Dryas episode. Evidence from the Gerzensee and Zeneggen sites shows that during the Younger Dryas, mean July temperatures were depressed by 5–8°C, and mean January temperatures were depressed by 10–12°C (Lemdahl, 2000a). These are relatively high altitude sites, and the effects of the Younger Dryas appear to have been more pronounced at high altitude than in the lowlands.

6.3. Evidence from North America

Morgan *et al.* (1984a) did the first summary of lateglacial insect fossil research in North America. The most intensively studied region includes the central and eastern United States and southeastern Canada. A series of insect faunas indicate changing conditions at the end of the last (Wisconsin) glaciation. The Longswamp site in southeastern Pennsylvania (Morgan *et al.*, 1982b) contained a basal fauna indicative of conditions immediately postdating the maximum extent of the **Laurentide ice sheet**, at about 18,250 cal yr BP. This early fauna is comprised of species associated with open ground habitats from within the boreal zone. In other words, the beetles live in the boreal climatic zone, but their specific habitats include meadows, stream banks, and other openings in the forest. The presence of coniferous bark beetles in the assemblage offers evidence for the proximity of coniferous trees. In contrast to this, Watts (1979) reports ''clear evidence for tundra vegetation'' in pollen spectra from this basal horizon at Longswamp. However, more recent paleobotanical analyses indicate that spruce (*Picea*) was present in sites just south of the Laurentide ice sheet as early as 18,250 yr BP (Jackson *et al.*, 1997).

A fauna dated at 14,800 cal yr BP was sampled from a kettle pond deposit at Brampton, near Toronto, Ontario (Morgan *et al.*, 1984a). Again, this fauna is characterized by species living in open ground situations within the boreal zone of Canada. Paleobotanical evidence from Brampton is indicative of open ground tundra, including *Dryas integrifolia*, *Vaccinium uliginosum*, and dwarf birch. Regional plant macrofossil and bark beetle evidence indicates that conifers did not colonize the site until about 500 years later.

This 500 years lag in forest response to climatic warming in the Great Lakes region was also demonstrated from the Gage Street site in Kitchener, Ontario (Schwert *et al.*, 1985). Insect assemblages from 15,900 cal yr BP onward reflected climate characterized by mean summer temperatures greater than 10°C. Today, regions in Canada which experience summer temperatures greater than 10°C support boreal forest, whereas regions with summer temperatures colder than 10°C support arctic tundra (Bryson, 1966). So the insect data suggest that southern Ontario was warm enough to support the growth of conifers from 15,900 cal yr BP onwards, but park-like tundra vegetation continued at the site until about 14,800 cal yr BP. However, according to Jackson *et al.* (1997), spruce, fir (*Abies*), and larch (*Larix*) all migrated rapidly into the recently deglaciated terrain south of the Laurentide ice sheet after about 18,000 cal yr BP, based on mapped time slices of plant macrofossil and pollen records.

MCR reconstructions for lateglacial beetle assemblages in eastern and central North America were published by Elias *et al.* (1996a). The results of this analysis are shown in Fig. 17. Faunal assemblages representing ice-proximal sites (within 100 km of the Laurentide ice sheet margin) are shown in blue. These assemblages reflect T_{max} values about 5°C cooler than contemporaneous assemblages from sites well away from the ice margin. The earliest sign of warming following the LGM comes from Weaver Drain,

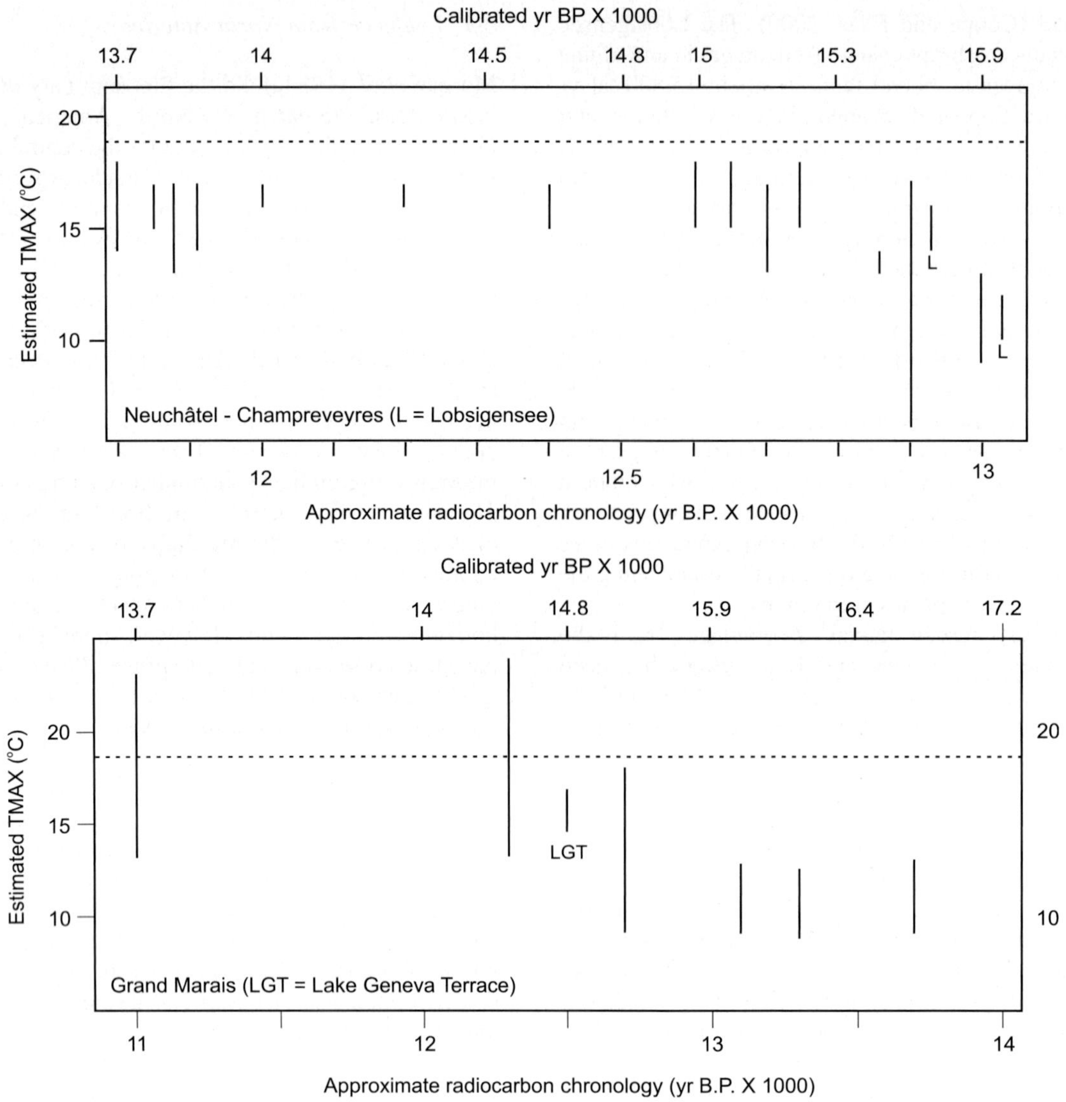

Fig. 16. MCR estimates of T_{max} for the lateglacial interval in (top) the Neuchatel region of northwestern Switzerland, and (bottom) the Grand Marais site in southwest Switzerland. In the upper figure, the letter ''L'' is used to indicate MCR estimates from the Lobsigensee site (data from Elias and Wilkinson, 1983). MCR data from Champreveyres comes from Coope and Elias (2000). In the lower diagram, the letters ''LGT'' indicate an MCR estimate from the Lake Geneva Terrace site discussed in Gabus et al. (1987). MCR data from the Grand Marais site is from Gaillard and Lemdahl (1994b). In both diagrams, the dashed line represents the modern mean July temperature of the study region.

Michigan. This site was ice-proximal at about 17,000 cal yr BP, but MCR analysis shows that T_{max} (4.6–8.6°C cooler than present) had started to rise above the levels reconstructed from a 17,200 cal yr BP assemblage at Fort Dodge, Iowa. Ice-proximal sites showed the effects of local cooling until the ice margin retreated and pockets of stranded dead ice melted, after 12,900 cal yr BP. Summer temperatures were apparently influenced by stagnant ice in close proximity to the Norwood, Minnesota, site at 13,600 cal yr BP.

By about 15,300 cal yr BP, substantial warming in both T_{max} and T_{min} had occurred at Winter Gulf, New York. T_{max} values were less than 3°C cooler than modern, and T_{min} values were less than 5°C cooler than modern. This degree of warming was not seen in the oldest assemblages from Gage St., Ontario. There, T_{max} from about 15,100 to 14,600 cal yr BP was 6–8°C cooler than modern. During the interval 14,600–12,800 cal yr BP, T_{max} values at Gage St. were as much as 1.1°C warmer than present and T_{min} values were 2–5°C warmer than present. At Nichols Brook, New York, T_{max} values were as much as 2.9°C warmer than present from 13,400 to 11,500 cal yr BP, and T_{min} values were as much as 6°C warmer than present. This level of warming was also suggested by assemblages from Rostock, Ontario.

T_{min} values between 12.4 and 10.8 ka indicate steady, substantial warming of winter temperatures in both ice-proximal and ice-distal sites. Proximity to ice apparently did not affect winter temperatures as much as summer. However, T_{min} values from most sites did not reach modern parameters until after 11,500 cal yr BP.

Between about 13,000 and 11,500 cal yr BP, a third set of sites (St. Eugene and St. Hilaire, Quebec; Eighteen Mile River, Ontario; and Lockport Gulf, New York) was situated near large bodies of glacial meltwater (proglacial

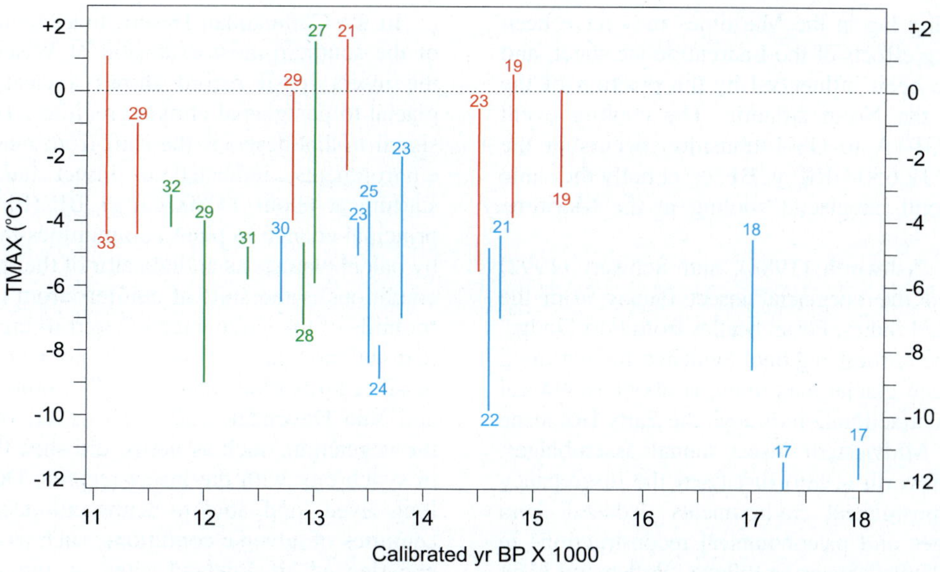

Fig. 17. MCR reconstructions for lateglacial beetle assemblages in eastern and central North America. Red bars indicate MCR estimates from ice-distal sites; blue bars indicate MCR estimates from ice-proximal sites; green bars indicate MCR estimates from sites near proglacial lakes (data from Elias et al., 1996a).

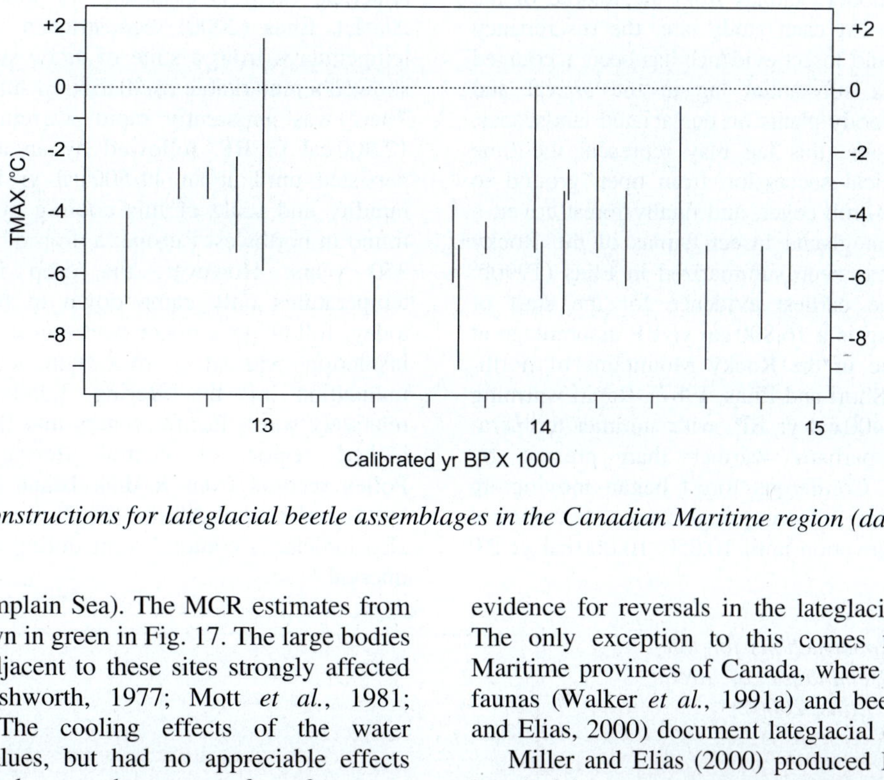

Fig. 18. MCR reconstructions for lateglacial beetle assemblages in the Canadian Maritime region (data from Miller and Elias, 2000).

lakes and the Champlain Sea). The MCR estimates from these sites are shown in green in Fig. 17. The large bodies of chilled water adjacent to these sites strongly affected local climates (Ashworth, 1977; Mott *et al.*, 1981; Morgan, 1987). The cooling effects of the water depressed T_{max} values, but had no appreciable effects on T_{min}, presumably because the water surfaces froze over in winter. From 13,000 to 11,500 cal yr BP, Elias *et al.* (1996a) reconstructed a warming trend at these sites. It is difficult to discern whether this trend represents the warming of the meltwater bodies or whether regional warming overwhelmed the cooling effect of proglacial meltwater.

After 11,500 cal yr BP, the ice sheet margin had retreated well north of the study sites in eastern and central North America, and all the fossil assemblages reflect T_{max} values near modern parameters.

Unlike the European records, the North American insect fossil faunas do not provide any significant evidence for reversals in the lateglacial warming trend. The only exception to this comes from sites in the Maritime provinces of Canada, where chironomid larval faunas (Walker *et al.*, 1991a) and beetle faunas (Miller and Elias, 2000) document lateglacial cooling episodes.

Miller and Elias (2000) produced MCR estimates of lateglacial temperatures from lateglacial faunal assemblages from 11 sites in New Brunswick and Nova Scotia (Fig. 18). The analysis showed marked declines in both T_{max} and T_{min} from 14,000 to 13,500 cal yr BP, and after 12,600 cal yr BP. The former interval corresponds to the Older Dryas cooling in northwest Europe; the latter corresponds to the Younger Dryas cooling. A comparison of lateglacial climate change in Maritime Canada with the GRIP ice core record from Greenland shows several similarities. In particular, the warming event marking the **stadial**–interstadial transition from GS-2 to GI-1e begins about 14,700 GRIP yr BP in the ice core record. This warming took place in the maritimes about 150–200 years later. The warming lag in the Maritimes may have been due to the cooling effects of the Laurentide ice sheet, and it may also have been influenced by the position of the **Polar Front** in the North Atlantic. The cooling event designated the GI-1A to GS-1 transition occurs in the GRIP ice core at 12,650 GRIP yr BP, essentially the same time as the second lateglacial cooling in the Maritime beetle record.

Schwert and Ashworth (1988), and Schwert (1992) have summarized the lateglacial insect faunas from the Midwestern United States. Fossil beetles from Fort Dodge, Iowa, provide the earliest regional evidence for warming after the Wisconsin glacial maximum, at about 18,400 cal yr BP. Additional amelioration through the Early Holocene is indicated by Midwestern insect faunal assemblages. Schwert and Ashworth (1988) discussed the discrepancy between early postglacial environments deduced from insect assemblages and paleobotanical reconstructions in the Midwestern United States as follows. Pollen and plant macrofossils that accumulated in ice marginal deposits just after deglaciation reflect tundra-like communities with sedges and *Dryas*, and are lacking in trees and shrubs. However, the insect faunas are consistently different; they are analogous to modern faunas from the middle of the boreal forest zone. At each study site, the discrepancy between the plant and insect evidence has been accounted for by invoking a substantial lag in the arrival and establishment of woody plants on deglaciated landscapes. As in central Europe, this lag may represent the time needed for ecological succession from open ground to herbaceous cover, shrub cover, and finally forest cover.

My work on lateglacial insect faunas of the Rocky Mountain region has been summarized in Elias (1990b, 1991, 2007c). The earliest evidence for the start of postglacial warming is a 16,800 cal yr BP assemblage at the Mary Jane site in the Rocky Mountains of north-central Colorado (Short and Elias, 1987). Rapid warming occurred after 13,400 cal yr BP, with summer temperatures becoming perhaps warmer than present by 11,900 cal yr BP. Coniferous forest began moving up slope in the Rockies following deglaciation, but did not reach its present elevation until 10,850–10,000 cal yr BP (Short, 1985).

In the Chihuahuan Desert, more than 1,000 km south of the southern-most extension of Wisconsin ice sheets, the insect fossil record shows a clear transition from glacial to postglacial climatic regimes. The chief climatic signal in this desert is the shift from mesic to xeric (dry) environments, indicated in insect faunal assemblages starting at about 14,800 cal yr BP (Elias, 1992b). The principal change in plant communities that has been used by paleobotanists as an indicator of the shift to postglacial conditions is the shift of conifers from the lowlands and foothills of the Chihuahuan Desert to higher elevations in regional mountains. This shift is not generally recorded in fossil records older than 13,400–13,000 cal yr BP (Elias and Van Devender, 1992). However, other elements of the vegetation, such as herbs, did shift their distributions in synchrony with the insect records. Desert conifers are long lived, and able to withstand decades or perhaps centuries of adverse conditions, such as they would have experienced in lowland sites at the end of the last glaciation. This makes them relatively poor indicators of climatic shifts from cool and moist to hot and dry.

Although there is no indication of a Younger Dryas cooling in fossil beetle records throughout most of North America, there is evidence for this cooling in arctic Alaska. Elias (2000) reconstructed lateglacial seasonal temperatures from a suite of arctic sites, and the results showed a remarkable oscillation in summer temperatures. There was apparently rapid warming from 13,500 to 12,900 cal yr BP, followed by an abrupt cooling that persisted until about 11,500 cal yr BP (Fig. 19). The rapidity and scale of this cooling event matched those found in northwest Europe: a drop of 7°C within at most 350 years. However, the drop in Alaskan arctic temperatures only came down to the same levels as today, following a major warming event in which arctic insolation was at a maximum, combined with the inundation of the Bering Land Bridge, bringing relatively warm Pacific waters into the previously land-locked region of central Beringia (Elias, 2000). Pollen records from Kodiak Island (Peteet and Mann, 1994) and central Alaska (Bigelow and Edwards, 2001) also indicate a cooling event during the Younger Dryas interval.

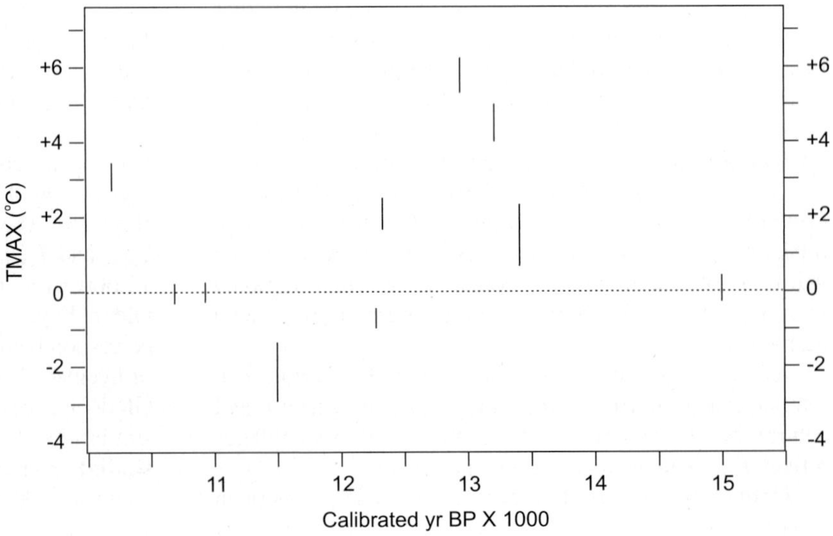

Fig. 19. MCR reconstructions for lateglacial beetle assemblages in arctic Alaska (data from Elias, 2000).

One of the most puzzling examples of differences between insect and plant response to postglacial warming comes from southwestern Alaska. Postglacial warming began in this region by 14,800 cal yr BP (Lea *et al.*, 1991). The insect faunas from this time reflect open ground habitats within the boreal forest. The insect fossils offer evidence of early postglacial warming to near-modern levels, indicating summer temperatures warm enough to support the growth of spruce forest (Elias, 1992c). However, spruce forest did not arrive in southwestern Alaska until the mid-Holocene (Short *et al.*, 1992), or about 8,000 years after the initiation of postglacial warming. Edwards *et al.* (2005) argued that this apparent lag in vegetation was at least partially due to shifting plant structural types in postglacial times. Early postglacial vegetation in Alaska was a unique mix of taxa, dominated in some regions by deciduous trees such as alder (*Alnus*), birch (*Betula*), and poplar (*Populus*). They argue that the shift from shrub tundra to deciduous forests could have happened rapidly in Alaska as the result of individual or population scale responses to warming. Species within these genera of deciduous plants are capable of growing as prostrate shrubs under cold climatic conditions, and they can grow as tall shrubs or trees under warm climates. For instance, shrub birch (*Betula glandulosa*) can be a prostrate shrub (less than 50 cm tall) or a robust, multistemmed shrub nearly 2 m tall. It can also hybridize with tree birch (*Betula neoalaskana*) to form a tree reaching several meters in height (Edwards *et al.*, 1991). The pollen from these tall shrub and tree forms of birch would be quite similar to the pollen of the dwarf shrub, so the change from shrub tundra to deciduous woodland would be virtually invisible in the pollen record.

7. Plant Migration Lag

Ever since the fossil insect data first began to offer evidence of rapid response to climate change, the issue of plant migration lag in response to climate change has been the subject of vigorous debate in the literature and at scientific meetings (e.g., Cole, 1985; Markgraf, 1986). Differential response times between pollen and other proxy indicators of climate change, such as beetles, have been explained in several ways. Huntley and Webb (1989) circumvented the problem of "supposed lags" by proposing that vegetation changes are consistent with climate change "if viewed at the appropriate spatial and temporal scales." On the largest scales (thousands of years and subcontinental regions), vegetation may well be in equilibrium with climate, although the word "equilibrium" seems inappropriate in this context. This scale may be useful for establishing broad outlines of change, but the implication seems to be that pollen is inadequate as proxy data for rapid change. However, the most recent computations of vegetation response to climate change since the LGM in eastern North America (Jackson *et al.*, 1997) indicate little or no vegetation lag in response to climate change, but they created snapshots of past vegetation at 3,000-radiocarbon year intervals.

The question of vegetation–climate equilibrium has some important implications beyond isolated studies of insect versus pollen data. One of the most popular trends in paleobotany in recent years has been the application of **multiple regression statistical analyses** of pollen data as a means of paleoclimatic reconstructions. These reconstructions formally assume that vegetation patterns (albeit at the continental scale) in the Late Quaternary (at time scales of 500–1,000 years) have been in equilibrium with climate change. This assumption appears incompatible with the results of some detailed studies on tree migration. One of these is the investigation of postglacial tree migrations in the Great Lakes region (Davis *et al.*, 1986). They reported evidence that the geographic limit of beech (*Fagus*) was in disequilibrium with climate in southern Wisconsin during the interval 7,800–6,800 cal yr BP, and that hemlock (*Tsuga*) was also in disequilibrium with climate prior to 5,800 cal yr BP. In another study, Johnstone and Chapin (2003) examined migration rates for lodgepole pine (*Pinus contorta*) in the northern boreal forest regions of western Canada. Pollen records indicate that this species only arrived in the central Yukon region within the past millennium (MacDonald and Cwynar, 1986, 1991). Johnstone and Chapin found that lodgepole pine is continuing to expand northwards in Canada – a process that began in the Early Holocene. Their study has implications for how we interpret models of vegetation response to climate change, as it provides an example of a species whose current range limits are apparently not in equilibrium with modern climate. The authors hypothesized that dispersal dynamics and regeneration niche are factors likely to constrain lodgepole pine migration dynamics.

Concerning the responses of trees to climate change, Brubaker (1986) noted that population responses of trees to climate are difficult to document, because of their long life span. Brubaker's ecological studies also concluded that tree population changes may lag behind climatic shifts on the order of decades or centuries. Clark *et al.* (1998) acknowledged that tree species move at different rates and in different directions as a result of climate change.

An example of the importance of soil development in vegetation establishment was described by Pennington (1986). The postulated intense warming in lateglacial Britain, discussed above, preceded the expansion of tree birch by 500–1,500 years. Small populations of tree birch were represented by macrofossils early in the period, so the explanation of the delayed regional expansion appears to lie in differing rates of soil development on a heterogeneous landscape. Precipitation levels may also have been limiting in the early lateglacial (Prentice, 1986). These various studies highlight the need for better understanding of the ecological requirements of the various species of trees that have been used by palynologists as hallmarks of climate change.

Genetic studies, in combination with fossil data, are providing useful insights into the history of various tree species at the population level (see Petit *et al.*, 2008 for a review). This is the most biologically interesting level to study, as it is populations that actually respond to changing regional climates. Recently, genetic studies of modern tree populations have shown that small populations of key tree species may have survived *in situ* in regions previously thought to be devoid of trees during glacial intervals. For instance, Anderson *et al.* (2006) have produced genetic evidence that strongly indicates white spruce (*Picea glauca*) survived in unglaciated regions of Alaska during the last glaciation. Its postglacial spread in Alaska was therefore not the result of long-distance migration from populations south of the Laurentide and **Cordilleran ice sheets**. These results suggest that rates of tree migration estimated solely from fossil records may be far too high. Davis *et al.* (2005) used more recent fossil and genetic evidence to show that the postglacial migration of some deciduous trees may be less than 100 m per year.

6

Insect Zoogeography in the Quaternary

The patterns of distributions of animal and plant species is the result of a complex history that cannot be understood completely without the evidence of this fossil record. To ignore the fossil data would be like trying to reconstruct the plot of a film from a study of its last frame.

– Russell Coope (1990)

1. Early Work

The aim of this chapter is to demonstrate some of the remarkable changes in insect distributions over the last few thousand years, as shown in the fossil record. The fossil record is making some very important contributions to the field of **biogeography**, by establishing rates and patterns of distributional shifts through time. The science of biogeography (including zoogeography, the study of animal distribution patterns, and phytogeography, the study of plant distribution patterns) began in earnest in the 19th century, including the work of Charles Darwin, whose observations on the distributions of species, past and present, played a key role in the development of his evolutionary theory. Systematic entomologists began considering insect zoogeography at about the same time. Notable among these is the American entomologist, John LeConte, who made significant strides in this developing field, as early as 1859. However, fossil evidence from the Quaternary record had little impact on insect zoogeography until much more recently. Van Dyke (1939) offered his considerations on the origin and distribution patterns of the North American beetle fauna. In his paper, he complained of the lack of Pleistocene fossil data, outside of the California asphalt deposits and Scudder's specimens from Toronto. To his credit, Van Dyke noted that the available Pleistocene fossils were ''usually identical'' to the modern species.

Given the lack of fossil data, Van Dyke proceeded to postulate the origins of modern beetle distributions based solely on evidence gleaned from their modern ranges. As Coope (1990) pointed out in the quotation above, it seems intuitively obvious that it may be impossible to discern the history of species movements over time based only on where they have settled down most recently. Nevertheless, numerous zoogeographers have attempted to reconstruct insect distributional histories and to erect biogeographic and evolutionary theories without consideration for (or perhaps in spite of) the fossil evidence. For instance, biogeographers have suggested that the evolution of populations in refugia may be an important mechanism for speciation and the production of geographic distribution patterns (Noonan, 1985). Noonan (1988) suggested that Pleistocene glaciations may have had the general effect of causing extinction of many of the more sedentary insects of North America, and that Pleistocene environmental stresses on North American insects probably selected for relatively vagile taxa. As we have seen, there is little or no fossil evidence of insect extinctions due to Pleistocene glaciations, even in the more sedentary insect groups (Coope, 1978). Crowson (1981) suggested that many flightless beetles have very limited powers of dispersal, to the extent that they have practically no power to extend their range across even narrow belts of unfavorable terrain. This may be true for some taxa, but the fossil record suggests that many flightless species have shifted distributions dramatically over Quaternary time.

As Morgan and Morgan (1980) pointed out, with the accumulation of fossil records it is now becoming possible to raise knowledge of faunal histories and speciation above the realm of speculation. Let me issue one caveat before proceeding. While the fossil record broadens our knowledge considerably, the records of most individual species cover only a small proportion of a species' life span. The fossil record of a given beetle species may cover 100,000 years, but this would represent only 5% of the total life span of a species, which may have been 2,000,000 years. Van Dyke (1939) anticipated this problem when he stated:

> Most of our orders of insects were well established in Permian times, at the end of the Paleozoic. Thus, the insects being of infinitely older stock than the higher vertebrates, we are justified in concluding that the history of their earlier migrations and distribution goes back very much further in geological times than does that of the mammals and birds.

An important constraint in postulation of zoogeographic events is the dispersal power of insects. This was touched upon in Chapter 5, as it related to insect response to climate change. In a general summary of dispersal mechanisms in beetles, Crowson (1981) recognized several categories of beetles in respect to their powers of dispersal, ranging from the very vagile to the extremely sedentary. As indicated below, even some supposedly sedentary groups have moved about considerably during the Late Quaternary.

DEVELOPMENTS IN QUATERNARY SCIENCES
VOLUME 12 ISSN 1571-0866

2. Pleistocene Records from Europe

The British fossil record provides the best documented history of distributional shifts in response to climate change. Fossil assemblages from the last (Ipswichian) interglaciation and interstadials within the last glaciation provide the basis for inferring invasion of Britain by species which today have a Mediterranean range (Coope, 1990). These beetles have a variety of ecological requirements, and some have such restricted modern distributions that they have been labeled endemic Mediterranean species.

Ground beetles such as *Bembidion grisvardi* (modern range shown in Fig. 1) live in open habitats with sparse vegetation. Their presence in Britain at the end of the last glaciation (ca. 15,900 cal yr BP) is indicative of sudden climatic warming and an open, poorly vegetated landscape with poor soil development. This environment scarcely exists in modern Britain, but must have been much more widespread at the end of the last glaciation (Coope, 1990). *B. grisvardi* was also identified from one of the lateglacial assemblages at Champreveyres, Switzerland (Coope and Elias, 2000).

Another thermophilous ground beetle found in last interglacial deposits in Europe is *Oodes gracilis*. This species lives in aquatic environments, particularly reed swamps, where adults crawl on submerged plant stems. Interglacial-age fossils of *O. gracilis* have been found in Britain (Coope, 1990) and Byelorussia (Nazarov, 1991). In both instances, the presence of *O. gracilis*, in combination with other thermophiles, indicates climatic conditions warmer than present.

During past episodes of warm climate in Britain, the dung beetle fauna was enriched by a number of thermophilous species. Among these are the Mediterranean beetles *Aphodius bonvouloiri* and *Onthophagus massai* (Fig. 1). *A. bonvouloiri* was one of the most abundant dung beetles in warm interstadial assemblages in Britain (Girling, 1974a). At the termination of the Upton Warren Interstadial, *A. bonvouloiri* was replaced in the British fauna by *Aphodius holdereri* (Fig. 2), the dung beetle discussed in Chapter 1 that today is found only on the Tibetan Plateau (Coope, 1973). Perhaps the most exotic thermophilous dung beetle in the British Pleistocene faunas is *O. massai*, a species found today only on the island of Sicily. Here modern distribution might suggest endemism, but the fossil record refutes that hypothesis. As Coope (1990) pointed out, such modern geographic isolation could just as well be the species' last stand as its place of birth.

Another scarab beetle found in last interglacial deposits in Britain is the genus *Drepanocerus*. This group is known today only from Africa south of the Sahara Desert, the Indian subcontinent, and Southeast Asia (Coope, 1979). Its modern range is a fragmented remnant of a once wide distribution that was apparently disrupted during the last glaciation.

Even host-specific phytophages can provide useful zoogeographic clues to past environments. The bark

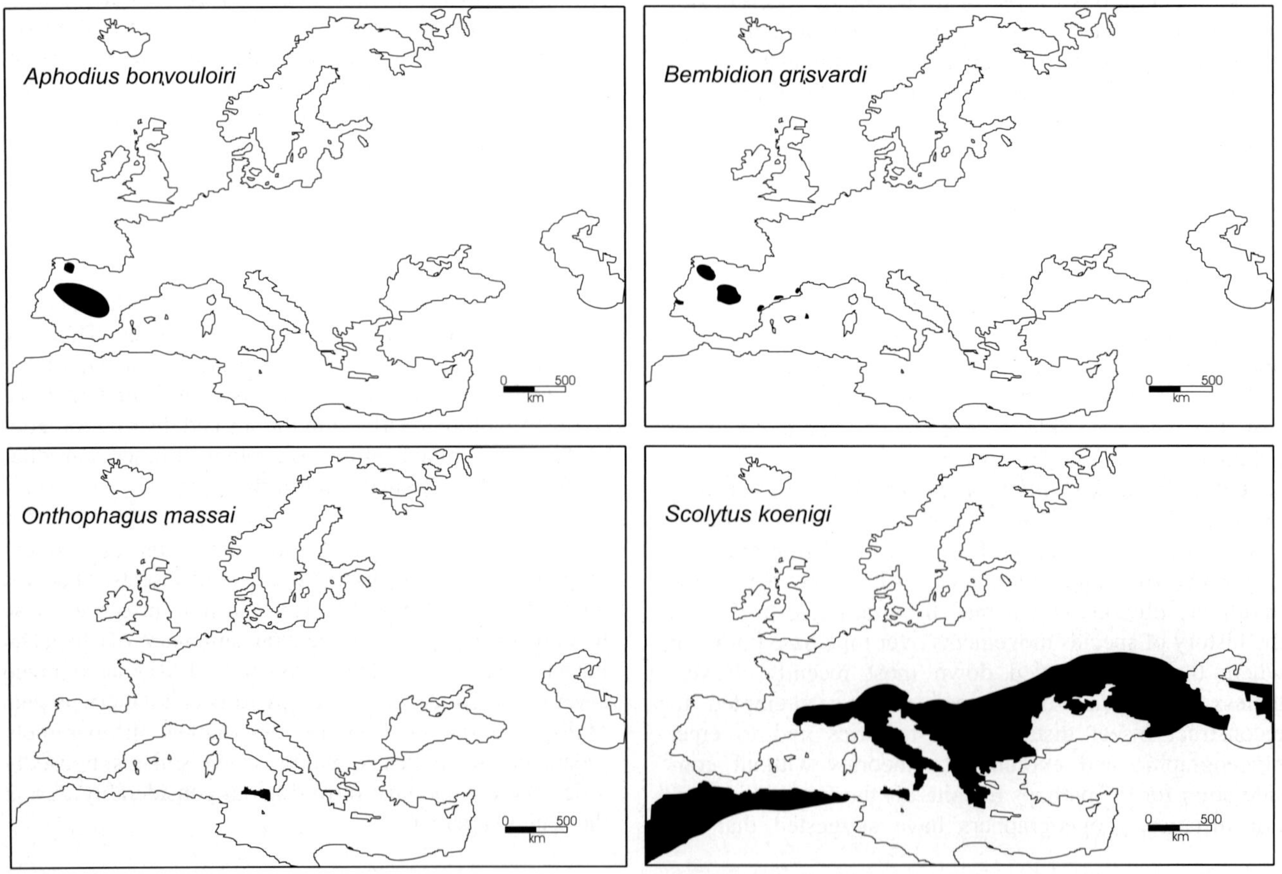

Fig. 1. Modern distributions of thermophilous species of beetles found in British interstadial and interglacial deposits (after Coope, 1986).

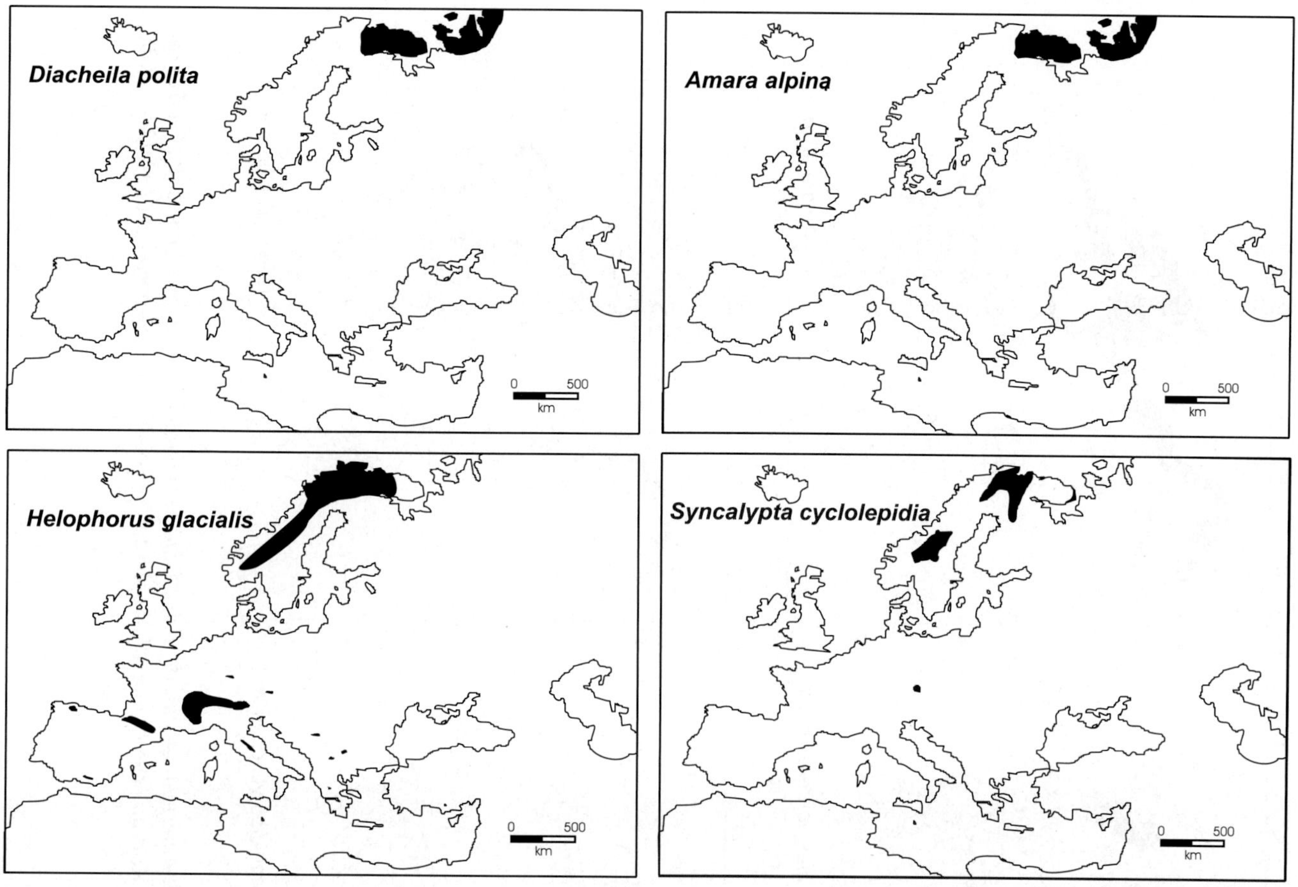

Fig. 2. Modern European distributions of cold-adapted species of beetles found in British glacial deposits (data from Lindroth, 1960).

beetle, *Scolytus koenigi*, lives under the bark of maple (*Acer* spp.) and allied trees, and is found today across southern Europe and northwestern Africa (Fig. 1). It was found in British fossil assemblages from the last interglacial, as were pollen and macrofossils of the host plants (Coope, 1990).

Cold-adapted faunas replaced these thermophilous species during glacial cycles in Britain. Beetles with arctic, or arctic and alpine affinities were the most abundant elements in the glacial assemblages. *Diacheila polita* is an arctic ground beetle found in British fossil records as well as in glacial assemblages from Eastern Europe and North America south of LGM ice sheets. In Eurasia, *D. polita*, a flightless species, is found today only in arctic regions from the Kola Peninsula eastward. In North America, it is found today only in arctic and subarctic regions of Alaska and the Yukon Territory, with isolated populations on alpine tundra in the Alaska Range.

Another ground beetle, *Amara alpina*, is an arctic and alpine species today, with populations extending south along mountains in Scandinavia (Fig. 2). It is the northernmost ground beetle in the modern North American fauna (Fig. 3), but isolated populations are found on mountain tops as far south as northern New Mexico in the Rocky Mountain chain, and on top of the few high mountains in the northern Appalachians. It has been found in numerous glacial-age deposits in Britain (Coope, 1977a), as well as in Late Weichselian assemblages from southern Sweden (Lemdahl, 1988b, 1991a).

The pill beetle, *Syncalypta cyclolepidia*, was a widespread inhabitant of Europe during the last glaciation. It has been found in British fossil faunas as well as from glacial-age assemblages near the end of the last glaciation at Lobsigensee, Switzerland (Elias and Wilkinson, 1983). Today this moss-feeding beetle occupies arctic and alpine regions of Scandinavia, and has isolated populations in the Alps of Austria (Fig. 2).

Another cold-adapted beetle that was apparently widespread in Europe during the last glaciation is the water scavenger beetle, *Helophorus glacialis*. The modern distribution of this species in Europe reflects its retreat to cold climate regions, both in the north and on mountain tops throughout the rest of Europe (Fig. 2). In the last glaciation, the fossil record indicates that it lived in Sweden (Lemdahl, 1991a), Britain (Coope, 1977a), France (Ponel and Coope, 1990), and Switzerland (Elias and Wilkinson, 1983; Coope and Elias, 2000).

The relative severity of various cold stages has been clarified by the presence of certain beetle species in the British fossil record. For instance, the water scavenger beetle, *Ochthebius kaninensis*, has been found in deposits from a cold period just prior to the Upton Warren Interstadial. Today, it is known only from the Kanin Peninsula in the Russian Arctic (Coope, 1990).

The carabid species in the subgenus *Cryobius* (genus *Pterostichus*) are cold adapted, and occupied Europe during the last glaciation. This group is no longer found anywhere in Western Europe. Most species live in eastern

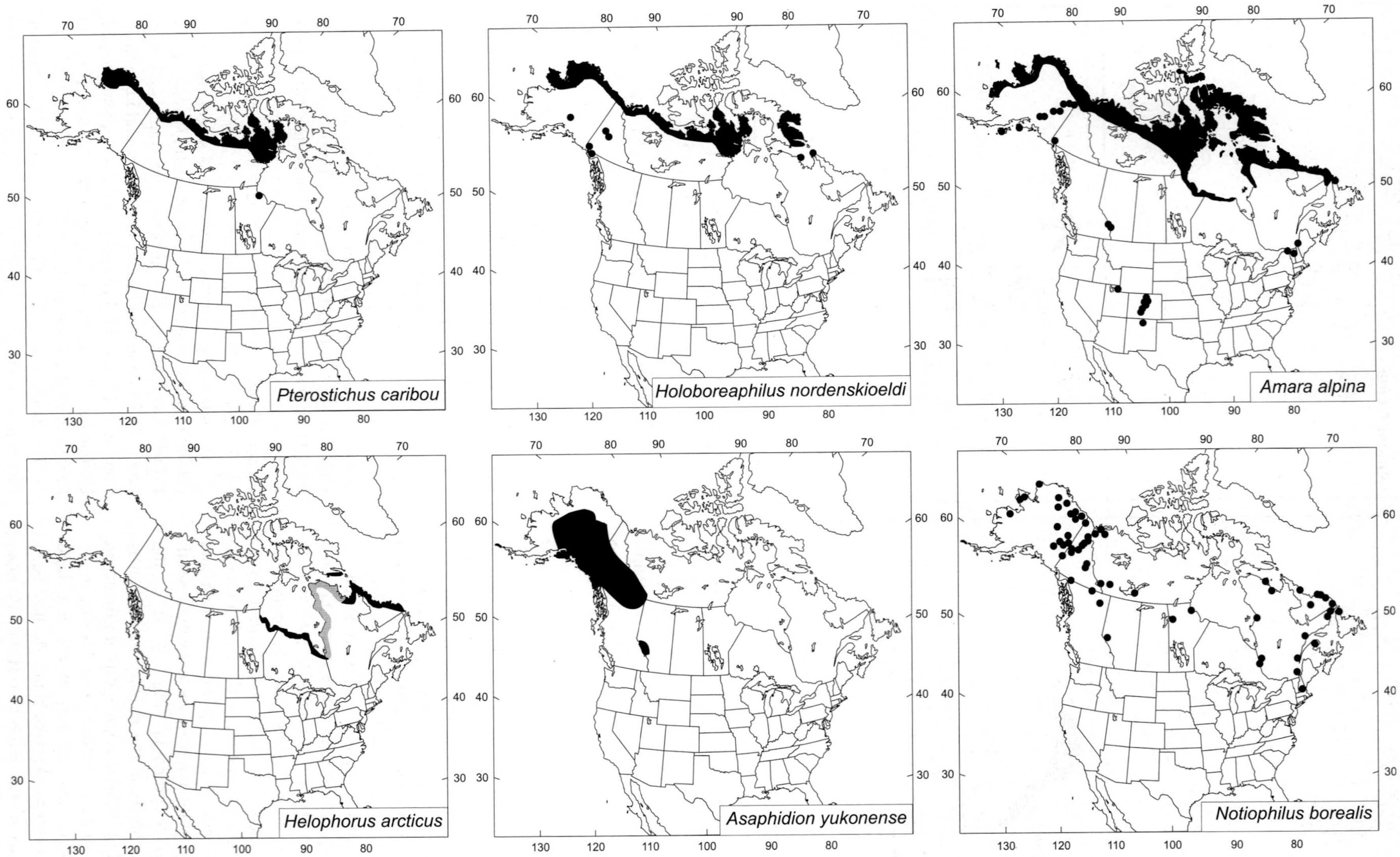

Fig. 3. Modern North American distributions of cold-adapted species of beetles found in North American glacial deposits. Data for Pterostichus caribou *from Ball (1966), map of* Holoboreaphilus nordenskioeldi *after Morgan et al. (1984b), map of* Helophorus arcticus *after Morgan (1989), map of* Asaphidion yukonense *after Morgan et al. (1984a), data for* Amara alpina *from Lindroth (1968), and data for* Notiophilus borealis *from Lindroth (1961).*

Siberia and arctic North America, with many species known today only from Alaska (Ball, 1966). Coope (1990) points out that the concentration of *Cryobius* species in a given area today (such as in Siberia and Alaska) may be more indicative of their common ecological requirements than of an original center of dispersal. *Cryobius* fossils have been found in glacial deposits from Britain (Coope, 1977a), and mid-Weichselian assemblages at the Niederwenigen site in Switzerland (Coope, 2007a). Nazarov (1991) reported *Cryobius* fossils from middle Pleistocene (ca. 440,000 yr BP) deposits in Byelorussia, in conjunction with *D. polita* and other indicators of arctic conditions. These European examples show that it may be unwise to describe any beetle species as endemic to a given region until its fossil history has been determined.

Dramatic distributional shifts have also been documented elsewhere in the Pleistocene beetle fauna of Eurasia. For instance, Kiselyov (1973) described distributional changes in Late Pleistocene beetle faunas in the Ural Mountain region. His work showed that the Transuralian region was once home to a unique mixture of **xeric-adapted** species found today in steppe regions of Kazakhstan and arctic/subarctic species found today in eastern Siberia and Mongolia.

Likewise, Pleistocene insect assemblages from eastern Siberia contain mixtures of steppe and arctic tundra elements (Kiselyov and Nazarov, 1984; Sher *et al.*, 2005). These species have undergone continental-scale shifts in distribution within the last glacial/interglacial cycle. For instance, the weevils in deposits from the Kolyma lowland include the genera *Coniocleonus* and *Stephanocleonus*, confined today to steppe- and montane-steppe regions well to the south in Siberia and Mongolia. Some steppe-associated weevils still live in isolated habitats on south-facing slopes in Yakutia, the Taimyr Peninsula, western Chukotka and Wrangell Island (Korotyayev, 1977).

The phenomenon of Beringian steppe–tundra and its insect fauna is treated in detail in Chapters 9 and 11. The arctic species found in many of these assemblages include *A. alpina* and several species of *Cryobius*, although species of the latter group occur only in intervals of more mesic climate.

3. North American Studies

Fossil evidence indicates that North American insects underwent large-scale distributional shifts in the Quaternary. For instance, the last glaciation forced the dispersal of cold-adapted insects into two major refugia (Morgan *et al.*, 1984a). Unlike Europe, not all high latitude regions of North America were glaciated during the Pleistocene. The lowlands of Alaska, the Yukon Territory, and westernmost edge of the Canadian Northwest Territories were free of ice for much of the Quaternary. Moreover, as sea levels lowered during glaciations, the shallow continental shelf regions between Alaska and Siberia became dry land, connecting the ice-free regions on either side of the Bering Strait. This region served as a refugium for arctic biota. Eric Hultén put forward the idea of a Beringian refuge in 1937, based solely on the modern distribution patterns of arctic plants.

A second refugium for cold-adapted biota was south of the ice sheets, and extended from the state of Washington through Montana and the Dakotas into southeastern Wisconsin, Illinois, northeastern Pennsylvania, and New York. These two refugia have been well-established in the fossil record. A third refugium on the exposed continental shelf regions off the east coast of North America has been postulated. As yet no insect fossils have been recovered (Morgan *et al.*, 1984a) to support the existence of this refugium. However, vertebrate fossils have been dredged up from the continental shelf region.

3.1. Dispersal of Beetles from the Southern Refugium

The insect faunas of Late Wisconsin age from the southern refugium are composed of species with a variety of modern distributions. No arctic style climate occurs today in the refugial region, except on mountain tops in the Rocky Mountains and isolated peaks in northern Appalachians. Accordingly, the cold-adapted fauna that inhabited the refugium during the Wisconsin glaciation was extirpated from lowlands south of the ice sheets when climate warmed at the end of the last glaciation. A few cold-adapted beetle species became established on alpine tundra regions in the aforementioned mountain regions, following local deglaciation of mountain ice caps. One of these is the ground beetle species, *A. alpina* (Fig. 3). This species did not become established in the European Alps at the end of the last glaciation. It is adapted to cold, xeric conditions, such as are found in the high arctic. Similar conditions persist on some high mountains in North America, but perhaps the Alps are too moist on the whole to support *A. alpina* today.

Most other cold-adapted beetle species were extirpated from the southern refugial regions (Schwert and Ashworth, 1988). The retreat of the Wisconsin ice sheets was slow, and the ice-proximal regions became too warm to support species such as *D. polita* (Fig. 4). In some regions, boreal forest encroached on the retreating ice margins. Following deglaciation, Beringian populations of cold-adapted beetles have recolonized the central and eastern arctic regions of North America.

Not all beetle groups have been equally successful in becoming established in previously glaciated landscapes. Predaceous beetles (especially Carabidae and Staphylinidae) were relatively effective in recolonizing Canada. One reason for this is that they exhibit good dispersal abilities; hence they were able to rapidly reinvade newly exposed regions following deglaciation (Campbell, 1980).

One of these invaders is the ground beetle species, *Asaphidion yukonense*, known today from central Alaska, south through the Yukon Territory and northern British Columbia, with isolated populations in the Rocky Mountains of Alberta (Fig. 3). *A. yukonense* lives in open areas with scarce vegetation consisting of small moss patches. It is easy to envision this type of habitat in recently deglaciated landscapes, and in fact fossils of this species have been found in this depositional context from

Fig. 4. "Diacheila polita *in Iowa, 12,000 yr BP" (after Faukse et al., unpublished data, with permission).*

sites in Midwestern and eastern North America (Morgan *et al.*, 1984a).

Species richness of the modern arctic beetle fauna declines markedly from west to east. This appears to be due to migrational lag following deglaciation and probably not because of differences in productivity in the respective landscapes (Schwert, 1992). The beetle fauna of the eastern arctic has yet to achieve its potential diversity. One of the factors curtailing the easterly dispersal of tundra insects from Beringia has been northward expansion of trees in the Mackenzie River Valley during the Holocene. However, strong westerly winds prevalent across the northern tundra regions should have facilitated eastward spread of insects through aerial dispersal, despite the presence of the Mackenzie forest barrier (Danks, 1981).

Hudson Bay has also apparently been an effective barrier in eastward dispersal of beetles in the Holocene. The ground beetle species *Pterostichus (Cryobius) caribou* is one of many arctic beetle species that exhibit a modern distribution including northern Alaska and northern Canada, with an abrupt termination on the western shores of Hudson Bay (Fig. 3). Another factor limiting the postglacial reinvasion of the eastern arctic is that the central part of the Labrador-Ungava peninsula was not fully deglaciated until after 6,800 cal yr BP (Andrews and Dyke, 2007), whereas much of the western arctic was deglaciated by 11,500 yr BP.

In contrast to arctic species, the boreal insect fauna that survived the last glaciation south of the ice sheets was probably able to disperse northwards along the margin of receding ice. An example of this type of dispersal is provided by the boreal ground beetle species, *Notiophilus borealis* (Fig. 3). The northward shift of the boreal fauna has been documented in several Midwestern fossil assemblages (Morgan, 1987; Schwert and Ashworth, 1988).

The modern distributions of some of the beetles that have recolonized arctic North America suggest modes of dispersal that are quite intriguing. The rove beetle species, *Holoboreaphilus nordenskioeldi*, is known today from arctic Canada and Alaska, including the northern tip of the Labrador-Ungava peninsula and southeastern Baffin Island (Fig. 3). It has been found in only two Wisconsin-age deposits from south of the glacial margins. One of these is an ice-proximal deposit from Titusville, Pennsylvania (Totten, 1971), dated roughly between 44,500 and 40,000 yr BP. The other occurrence is from a 13,300 yr BP assemblage at Marias Pass, Montana (Elias, 1988a). This species is also common in British fossil assemblages during several of the cold periods. Its modern presence on Baffin Island and northern Ungava is puzzling. Morgan *et al.* (1984b) proposed three hypotheses to explain its dispersal to the eastern Canadian arctic. The first is easterly postglacial dispersal from population centers in Alaska. The second is dispersal on drifting pack ice from the last vestiges of disintegrating Laurentide ice in eastern Keewatin, across Hudson Bay to Baffin Island. The third is dispersal by drifting on pack ice from the northern tip of Labrador, following local deglaciation (ca. 9,500 yr BP). Perhaps additional fossil studies on the Labrador coast will help elucidate the postglacial movements of *H. nordenskioeldi*.

Fossils of the water scavenger beetle species, *Helophorus arcticus*, are known from a number of Wisconsin age fossil sites in central and eastern North America, south of the Laurentide ice sheet. It has also been found in British cold-stage fossil assemblages (Coope, written communication, April 2009). It is found in fossil assemblages generally indicative of climatic conditions associated with northern treeline (i.e., mean July temperatures at or near 10°C). Morgan (1989) postulated the movements of *H. arcticus* from the Sangamon interglaciation onwards. With the onset of the Wisconsin glaciation, northern Canadian populations were forced south in eastern North America. Lowered sea level probably provided substantial habitats along the exposed continental shelf from Labrador south to New England. Inland, *H. arcticus* inhabited the shores of proglacial lakes, a habitat not currently in existence, but one that most certainly has been available throughout much of the Quaternary history of North America. In this context, it is important to keep in mind that glaciations account for the majority of Quaternary time, and interglaciations are relatively short, of the order of 10,000 years. With the onset of deglaciation, *H. arcticus* migrated north, following the shores of huge proglacial lakes, formed by meltwater from the receding Laurentide ice sheet. This dispersal route allowed *H. arcticus* to migrate eastward in the Great Lakes region, onto the southern shores of the lateglacial Champlain Sea in Quebec, and north from there as the ice continued to recede into Labrador. Other than *H. arcticus*, the only cold-adapted beetle to have succeeded in postglacial dispersal into northern Quebec along eastern coastal tundra is *A. alpina* (Schwert, 1992).

3.2. Dispersal of Rocky Mountain Insects

The Rocky Mountains have been a refuge for many cold-adapted species of both plants and animals. Unlike Midwestern and eastern North America, where most cold stenotherms were extirpated after the Wisconsin glaciation, some of the Wisconsin fauna that lived in close proximity to the Rocky Mountains was able to track suitable habitats up slope to the alpine tundra zone. Hence, several cold-adapted species that lived on the plains south of Denver toward the end of the last glaciation have retreated to high elevations in the Colorado Front Range (Elias, 1986). Nevertheless, numerous **extirpations** of cold-adapted species have occurred in the Rocky Mountain region.

Thirteen species of beetles that are no longer living in the Rocky Mountain region have been identified from regional fossil assemblages dating to the interval from 21,000 to 2,700 yr BP. While additional modern collecting in the region may reveal extant populations of some of these beetles, it appears likely that most of them are gone. Six of the eight species eliminated before the Holocene have shifted their distributions to the north. These are all cold-adapted animals, found today either in the arctic or in the boreo-arctic regions of Canada and Alaska.

Of the species extirpated from the Rocky Mountain region during the Holocene, only one has shifted its distribution to the north. The riparian ground beetle species *Bembidion rusticum* lives now only in British Columbia and northward, more than 2,600 km from the fossil site at Lake Emma, Colorado (Elias *et al.*, 1991). Two other species in this category have modern ranges in eastern North America, one in the Appalachians, and one in the grasslands of the Midwest. These distributional shifts appear not to be due to postglacial warming, although some extirpation occurred during the interval of **Altithermal** warming in the Holocene. They may be related to changes in moisture.

The trends observed from the Rocky Mountain fossil record also hold true for fossil insect faunas from the Chihuahuan Desert. During the lateglacial interval, most regional extirpations were of species which live to the north of that desert today, whereas in the Holocene, most of the regional extirpations were of species which live either to the east or to the west of the Chihuahuan Desert (Elias, 1992b). Many temperate species of beetles were able to take advantage of cool, moist habitats established in the Chihuahuan Desert during the last glaciation, the spine of the Rocky Mountains serving as a corridor to the desert regions.

In his biogeographic analysis of the carabid genus *Harpalus*, Noonan (1990) noted that the species in western North America occur in lowland habitats in the north, and shift progressively upward into mountains to the south, as desert and other xeric environments replace mesic lowland habitats. This ability to exist in montane habitats facilitates broad distributions by enabling species to expand their ranges southward. *Harpalus* species have persisted in the southern Rocky Mountains at least in part because of their ability to track suitable habitats up and down slope as climates changed.

3.3. *Postglacial Shifts in the Beetle Fauna of the American Southwest*

The study of insect fossils from packrat middens preserved in the arid climates of southwestern deserts is beginning to bring to light the insect faunal history of this region. While some fossil insect research has been undertaken for sites in the Great Basin (Elias *et al.*, 1992b), the most intensively studied region (191 midden insect fossil samples from 27 sites) is the Chihuahuan Desert (Elias and Van Devender, 1990, 1992; Elias, 1992b; Elias *et al.*, 1995). Late Pleistocene faunas comprised mixtures of temperate and desert species not seen in any one region today. What has become of the Late Pleistocene beetle fauna of this region? Some species live elsewhere in the Chihuahuan Desert; others live now in other regions. In fact, an examination of the modern distributions of several species permits the postulation of movement in every possible direction: several species now live to the west, into the Sonoran, Mohave, and Great Basin deserts; some now live in the east, on the plains of Texas and Oklahoma; others now live to the north and northeast, on the prairie regions on the east flank of the Rockies; a few species now live to the south, in Mexico.

Sufficiently detailed regional analyses of Late Quaternary Chihuahuan Desert insect faunas are available to

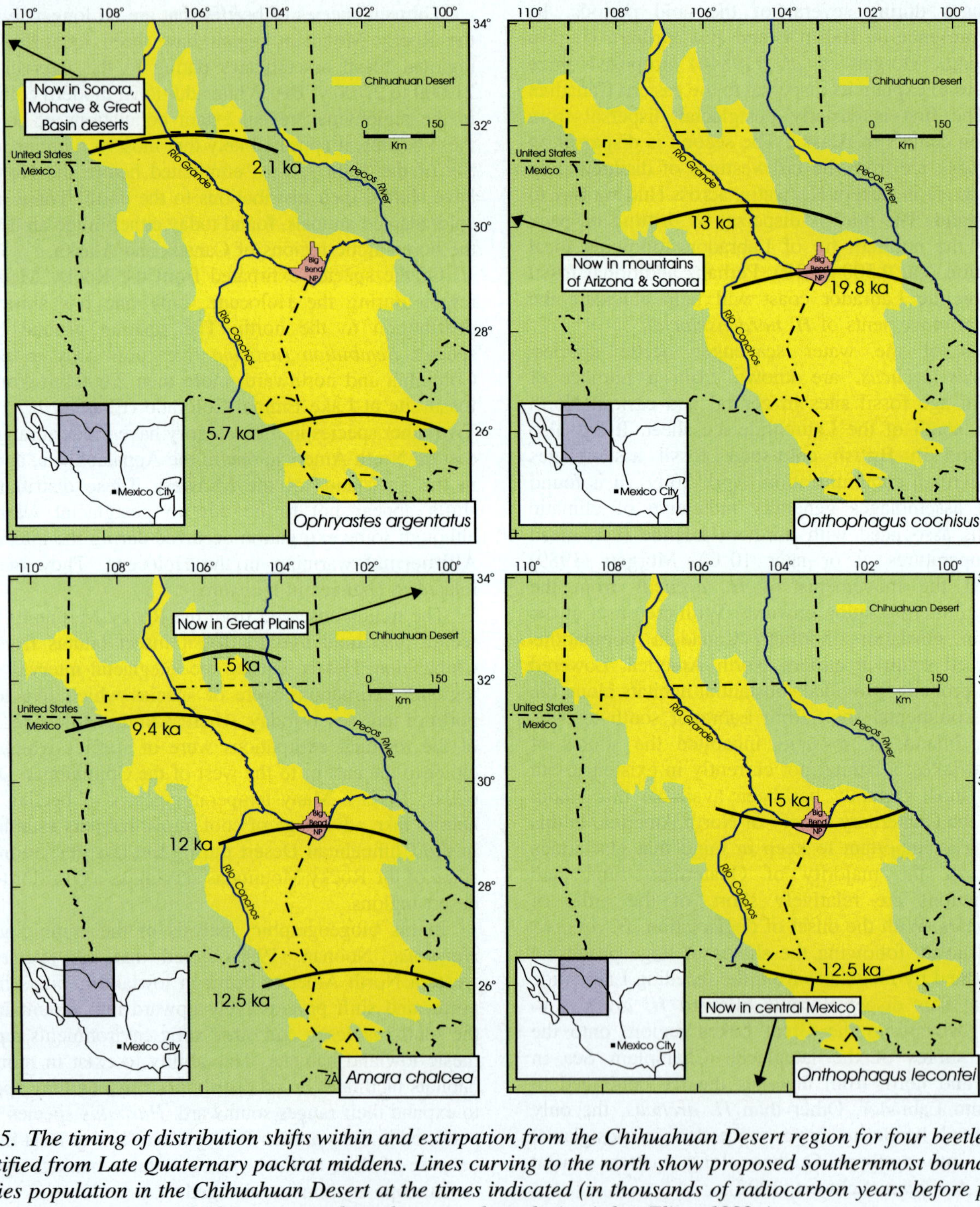

Fig. 5. The timing of distribution shifts within and extirpation from the Chihuahuan Desert region for four beetle species identified from Late Quaternary packrat middens. Lines curving to the north show proposed southernmost boundaries of species population in the Chihuahuan Desert at the times indicated (in thousands of radiocarbon years before present). Lines curving to the south show proposed northernmost boundaries (after Elias, 1992a).

allow inferences on the timing of their distributional changes (Elias, 1992b). These shifts follow no single pattern (Fig. 5). Some species, such as the dung beetle *Onthophagus lecontei* may have begun migrating as early as 18,000 cal yr BP, and was last recorded from sites in the southern Chihuahuan Desert at 14,800 cal yr BP. Its range is now restricted to central Mexico. Another dung beetle, *Onthophagus cochisus*, appears to have begun moving northward from the Big Bend region at 23,700 cal yr BP. It was last recorded from the Trans-Pecos region at 13,000 yr BP, and is now found only in the higher elevations (1,500–2,400 m) of the Chiricahua Mountains of southeastern Arizona and in the mountains of northern Sonora, Mexico (Howden and Cartwright, 1963). The ground beetle species, *Amara chalcea*, apparently departed the southern Chihuahuan Desert by 14,800 yr BP, but remained in the central part of the desert until 10,600 cal yr BP, with at least one population surviving in the hills of Last Chance Canyon, New Mexico, until 1,400 cal yr BP. Today this species is found only in the grasslands of the Great Plains region, on dry, open, sandy ground (Lindroth, 1968).

Some of the most puzzling distributional changes of Pleistocene Chihuahuan Desert beetle species are those of the cave-dwelling ground beetles in the genus *Rhadine*. These beetles are dorsoventrally flattened, which enables them to maneuver through narrow cracks in caves. They are flightless and depigmented; most have reduced eyes. Pleistocene fossils have been identified from assemblages in the Chihuahuan Desert (Elias and Van Devender, 1990, 1992), but modern records of some of these species are limited to caves in central Texas and Oklahoma. Needless to say, there are no known subterranean connections between these far-distant cave systems. Another puzzling shift involves *Rhadine longicolle*, known today only from Carlsbad Caverns, New Mexico, and nearby caves within Carlsbad Caverns National Park (Barr, 1960). Based on the modern evidence alone, one might suppose that *R. longicolle* has always been restricted to the caves of the Carlsbad region, but fossil specimens were identified from the Big Bend region of Texas, about 400 km to the southeast (Elias and Van Devender, 1990). Barr (1960) discussed the possibility of geographic isolation of ancestral stocks in west Texas during the Late Pliocene or Early Pleistocene, but the fossil evidence indicates that these beetles are able to migrate substantial distances within at most a few thousand years.

The fossil insect record of the Chihuahuan Desert indicates that sedentary, flightless beetles (such as the heavy-bodied weevils in the genus *Ophryastes*) have undergone marked distributional shifts in the American southwest within the space of a few centuries. Moreover, even highly specialized cave dwellers such as the ground beetles of the *Rhadine* group have somehow managed to move from one cave system to another in response to changes in Late Quaternary environments. It appears certain that packrats are not agents of dispersal of beetle remains from one cave to another. Insects are not an important part of packrat diets (Armstrong, 1982). Moreover, modern studies with packrats have shown that beetle fragments found in middens, including those representing cave-dwelling species, have not been consumed by rats (Elias, 1990a). Cave-dwelling insects probably become trapped in the sticky mass of rat feces and urine as they forage near the mouth of a cave, where the rat makes its nest.

While the fossil record sheds new light on some important questions, it has also generated a new batch of questions that are at least as exciting as the previous ones. Hopefully, the results reviewed here will stimulate zoogeographers and paleoentomologists to cooperate more fully in future.

7

The Use of Insect Fossils in Archeology

Entomological evidence has seldom been collected in archaeological investigations. Although its usefulness seems promising.

– Samuel Graham (1965)

Well-preserved insect fossils have been found in many archeological settings. Along with other types of biological proxy data (plant macrofossils, vertebrate remains, pollen, etc.), fossil insects are now making an important contribution to the reconstruction of both natural and anthropogenic environments associated with archeological sites. Evidence is accumulating about human lifestyles and living conditions. The biological evidence sometimes reveals far more about these matters than can be obtained from artifacts.

Ignorance of the ubiquity of insects in both water-lain deposits and in arid environments remains a problem among nonpaleontological scientists. A recent article by geneticists working on the recovery of ancient DNA from fossil fleas from Peruvian mummies (Dittmar *et al.*, 2003) made the following statement:

> Generally, arthropod remains are rarely found among archaeological material. Arthropods, like any other organic material are subjected to post mortem decay. The size and the fragility of most arthropods greatly limit the possibility of retrieval at archaeological sites (Kenward, 1978a). Most records of arthropods from archaeological sites are limited to records of subfossil mites from soil samples of archaeological deposits (Girling, 1978; Kenward, 1978b). Traditionally those samples have been used for paleoenvironmental and paleoclimatic analyses. Occasionally, arthropods are found in coprolites, contributing to the knowledge of dietary components in historic societies. (Bryant, 1974; Panagiotakopulu, 1999).

I find this level of ignorance particularly troubling because the authors have quoted from at least three authors (Kenward, Girling, and Panagiotakopulu) who, between them, have written more than 250 articles on these ''rarely found'' fossils.

As with paleontological studies, the ''home'' of research on insects from archeological sites is the British Isles. I will discuss the British record in more detail, below. For now, I discuss the types of deposits that have yielded good insect results.

1. Types of Anthropogenic Deposits Yielding Insects

Chief among these are waterlogged sediments at sites of human occupation, such as are common in Britain and elsewhere in northern Europe. Insect exoskeletons in waterlogged sediments show excellent preservation, and are sufficiently abundant at some sites to nearly overwhelm the investigator. For example, a medieval moat surrounding a manor house at Cowick, England (Fig. 1, No. 39) yielded 224 taxa of beetles and caddisflies (Girling and Robinson, 1989). The insect fauna from the trackway peats at Thorne Moor (Fig. 1, No. 142) comprised 340 taxa in 73 families (Buckland, 1979). As discussed in Chapter 2, insects as well as other organic detritus, are readily preserved in situations where normal decomposition is retarded, due to the lack of oxygen. Numerous insect studies have been based on these types of sediments, either from archeological sites themselves, or from localities adjacent to (and the same age as) such sites.

Examples of saturated organic deposits yielding insect fossils include anthropogenic features, such as ancient trackways through bogs and other wet lowlands, organic detritus from wells, and ancient occupation horizons below the current water table. Some nonanthropogenic organic repositories, such as ponds, lakes, and peat bogs, are in close proximity to archeological sites, and may contain insect faunas that are contemporaneous with nearby human occupation.

Domestic debris from under floors, in trash middens, latrines, sewers, borrow pits, etc., have also yielded abundant insect remains. The garbage and sewage of past generations offers many clues to their lifestyles, sanitation, animal husbandry, and land-use (Moore, 1981; Osborne, 1983).

Preservation of insects is also excellent both in cold (arctic/subarctic) and hot, dry environments (such as desert caves). Again, this is a function of retarded decomposition in these environments.

Insect fossils have been recovered from a variety of archeological contexts. Insect remains have been sampled from mummies from Greenland (Bresciani *et al.*, 1983, 1989), Egypt (Alfieri, 1931; Audouin, 1835; Curry, 1979; David, 1984), the Aleutian Islands of Alaska (Horne, 1979), and Chile (Baker, 1990). Insect fossils have also been removed from the remains of ''bog people'' (some of these were victims of ritual murders whose bodies were thrown into bogs). The exoskeletons of stored product pest

DEVELOPMENTS IN QUATERNARY SCIENCES
VOLUME 12 ISSN 1571-0866

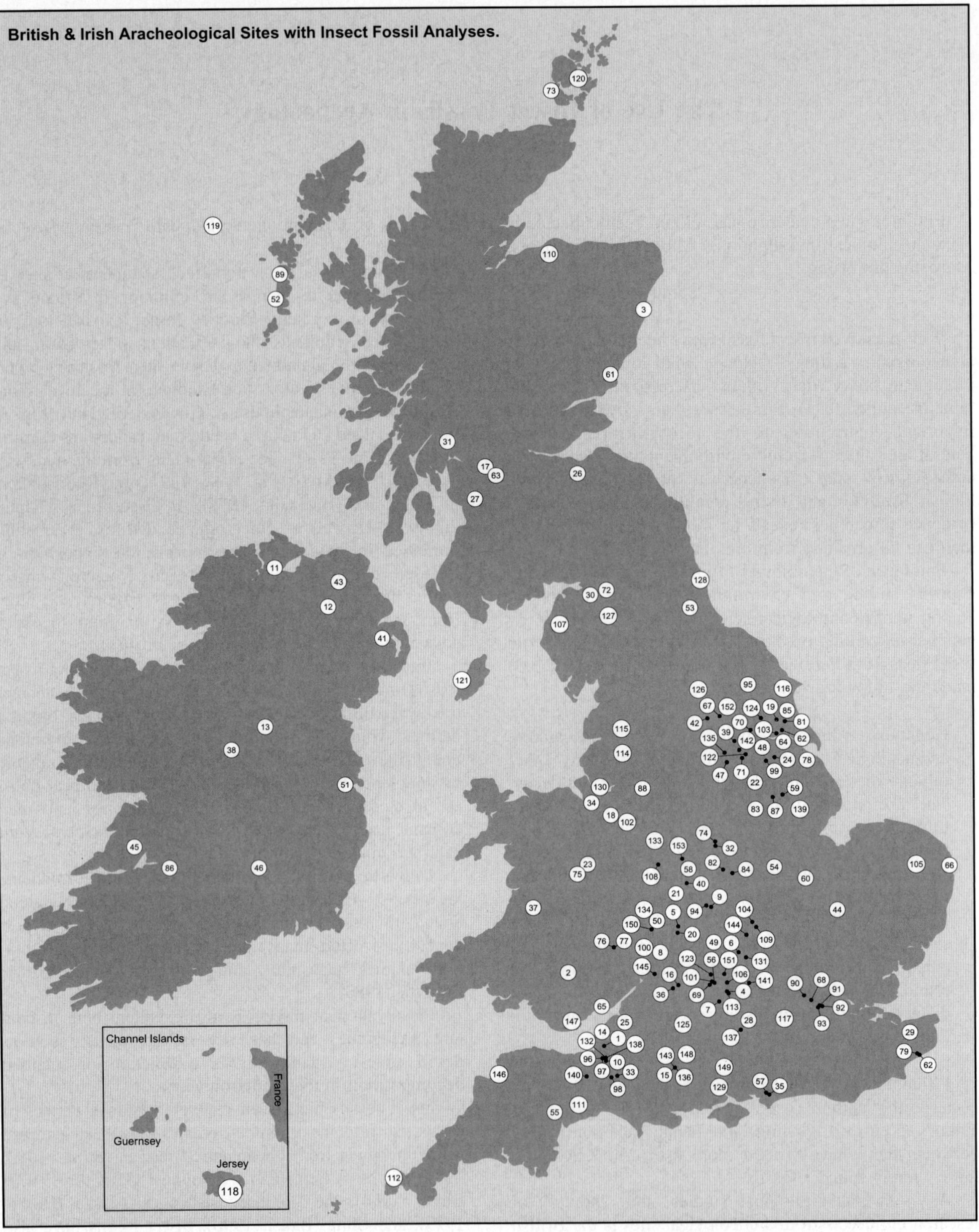

Fig. 1. Map of the British Isles, showing locations of archeological sites from which insect fossils have been analyzed. Site numbers are keyed to list of sites in Table 1.

insects have been recovered from Pueblo Indian cliff dwellings (Graham, 1965), the tombs of Egyptian pharaohs (and the tombs of less regal Egyptians) (Solomon, 1965), and from numerous medieval dwellings in Europe.

2. Environmental Indications from Insects

The ecological sensitivity of insects makes them useful indicators of past environments, both natural and anthropogenic. Evidence from Pleistocene assemblages paved the way for archeoentomological studies. In Europe, the majority of Holocene paleoclimate interpretations based on insects as well as other biological proxy data is confounded by anthropogenic effects on landscapes, especially forest clearing (Osborne, 1976). Buckland (1979) has asserted that the transformation of most of Europe from a wholly forested landscape to artificially maintained meadows, farmer's fields and urban environments represents the most dramatic nonclimatic environmental change of the Quaternary. Hence, most British archeological sites contain a mixture of **synanthropic** and nonsynanthropic biota, and it is often difficult to make clear separations between these two components (Kenward, 1975a, 1976, 1978a, 1982). Kenward (1985b) studied indoor and outdoor insect taxa found in modern buildings, which helped to define potential sources of insects deposited in ancient dwellings (Fig. 2). Smith (1996; 2000c) studied the beetle fauna associated with hay, straw, haystacks, and deep litter beds in a modern barn, to try to determine the accuracy of reconstruction of such environments based on fossil assemblages. He concluded that it is not possible to differentiate these materials on the basis of their resident beetle faunas, because of depositional problems. Straw and hay move around a farm in the course of the agricultural year, and even though specific beetles may be attracted to either hay or straw at some stage, if they die *in situ*, their exoskeletons may remain in the vegetable matrix, even after the hay moves from the field to the barn loft to the deep litter bed on the barn floor.

Still, insect remains can provide a wealth of archeological detail, and may yield information on human activities and environmental conditions not brought out by other types of data. As noted above, stored product pests, as well as insect scavengers and phytophages, provide unique documentation of human food, food storage and use patterns, food types, building methods, and other details of domestic life, farming and animal husbandry. Human parasites and synanthropes offer evidence of sanitation conditions, health, crowding, and human dispersal patterns. Carrion beetles, flies, and other insects provide forensic data on human and animal corpses (Erzinclioglu, 1983; Smith, 1986). I will develop each of these themes on a regional basis, below.

3. Special Methods for Archeological Sites

Buckland (1976a) provides a useful first account of methods for the sampling of insects in archeological contexts. More recent treatments are found in Coope

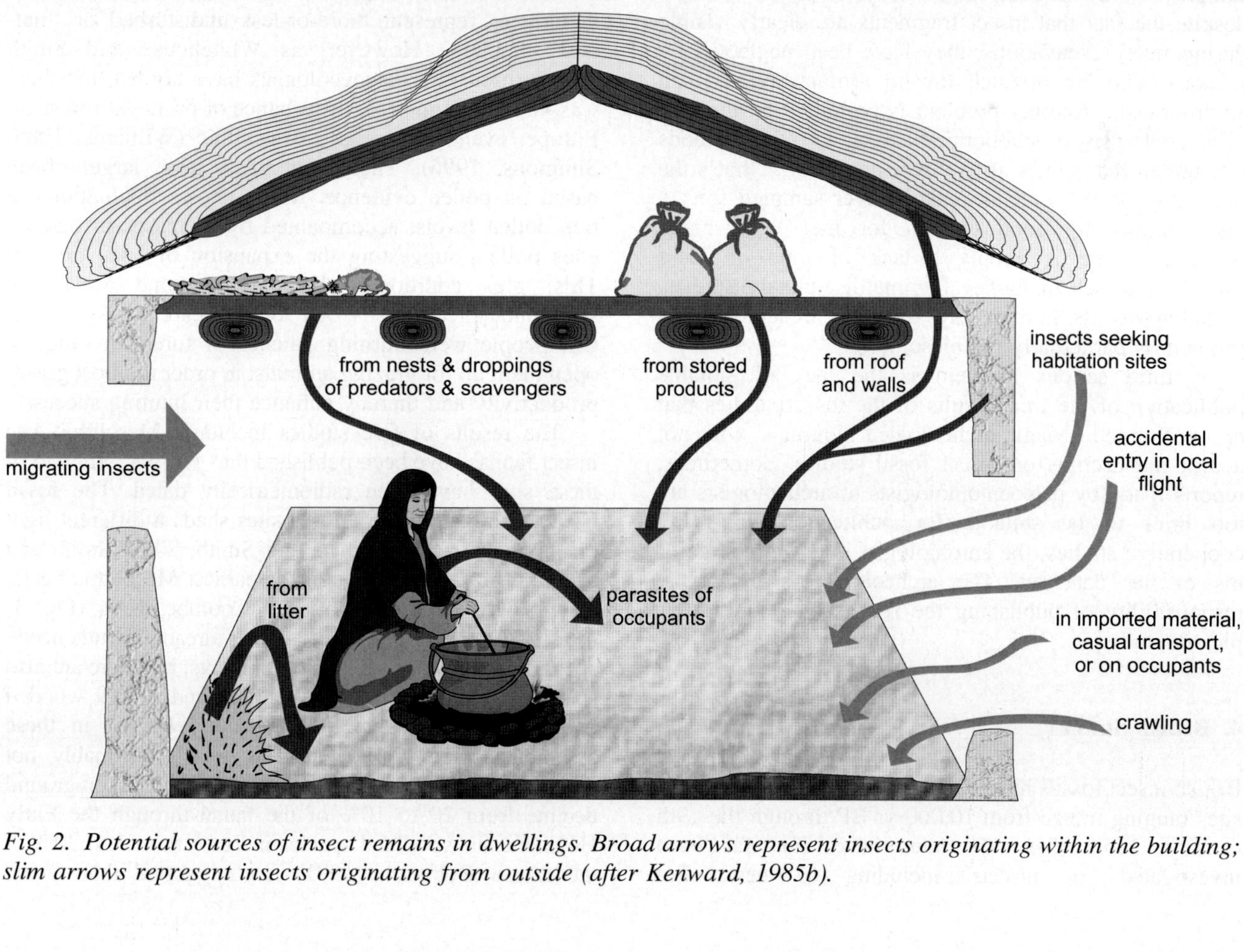

Fig. 2. Potential sources of insect remains in dwellings. Broad arrows represent insects originating within the building; slim arrows represent insects originating from outside (after Kenward, 1985b).

(1986) and in Buckland and Coope (1991). In general, sediments containing insect fossil samples should be greater than 5 cm thick, to provide adequate numbers of specimens per sampling horizon. However, this is not practical in all situations. For instance, some cultural horizons are only a few millimeters thick. Several kilograms of sediment at one site may yield little, while relatively small samples at another site may yield enormous numbers. Osborne (1969) extracted a huge insect fauna from a small sample from the Wilsford Shaft in southern England (Fig. 1, No. 148), but I have obtained only twenty identifiable insects after screening several thousand kilograms of sediment from the Lubbock Lake site in Texas. So, there are no hard and fast rules for sampling intervals, but the paleoentomologist and archeologist should work together to develop sampling strategies.

Pay careful attention to disconformities, whether natural or anthropogenic. Even a single event (such as posthole digging) may confound a sedimentary sequence beyond recognition. Whenever possible, sample from newly exposed horizons to avoid modern contamination. Needless to say, it is best if the fossil insect worker does the sampling. If this is impossible, then the archeologist should take as much material as possible and seal it in plastic bags for delivery to the paleontologist (Buckland, 1976a).

Some problems are inherent in **archeoentomology**. Perhaps the chief among these is the lack of sampling for insects by archeologists. This can be due to ignorance (''you mean there might be beetles in my sediment samples?''), or neglect. Buckland (1976a) noted that despite the fact that insect fragments are clearly visible during many excavations, they have been neglected by workers who are oriented toward artifacts rather than environments. Another problem is prejudice on the part of archeologists, or stubborn adherence to older methods (''a pollen diagram is all we ever need'' or, ''that's the only thing Professor Glockenspiel ever sampled for ... his methods are good enough for me!''). In many instances, there is simply a lack of time or more especially a lack of money for interdisciplinary studies, including insects. In these days of rescue archeology, this problem is particularly troublesome.

A third serious problem is the lack of suitable publication of the final results of the insect studies that are performed. Many archeological journals will not accept manuscripts on insect fossil studies. Sometimes, reports made by paleoentomologists to archeologists are too brief to be suitable for publication. In many cooperative studies, the entomologist is not in charge of his or her data set. The archeologist is given the responsibility of publishing the results of a multidisciplinary study.

4. British Studies

British insect fossils have been studied from archeological sites ranging in age from 10,000 yr BP through the 19th century. More than 150 sites have thus far been investigated for insects, including 67 new sites investigated since *Quaternary Insects and Their Environments* was published in 1994. Table 1 summarizes the publications of these studies that are more-or-less readily available to the interested reader, and that include faunal lists. It is interesting to note the dominance of Roman-Age and Medieval archeological sites in this list (Fig. 3). Roman archeology is generally well funded, because of great public interest. Most of the Medieval archeology comes from urban settings, such as York and Oxford, where Medieval deposits can be found beneath modern streets and buildings, and where there are resident paleoentomologists available to study the insect remains.

The **Upper Paleolithic** period ended in Britain about 9700 ^{14}C yr BP (11,000 cal yr BP) (Mithen, 2003). Only one Upper Paleolithic site has been discovered in which insects are preserved in an archeological context. This is the Messingham site (Fig. 1, No. 99) in north Lincolnshire (Buckland, 1982b, 1984). Coversand deposits yielded stone artifacts, in conjunction with a nonsynanthropic fauna indicative of a harsh, treeless landscape at ca. 12,100 cal yr BP.

The British **Mesolithic** is generally considered to span the interval from 9,700 to 5,300 ^{14}C yr BP, or 11,000 to 6,100 cal yr BP) (Parker-Pearson, 1995). This was the last era of hunter-gatherer societies in Europe. This is important from an entomological standpoint, because the human inhabitants of Britain did not alter the landscape significantly during this period. There was no large-scale clearing of trees for agriculture, for instance, that has characterized all subsequent periods in Britain, as elsewhere. So Mesolithic-period beetle assemblages might be thought to represent more-or-less undisturbed or ''natural'' habitats. However, as Whitehouse and Smith (2004) discussed, paleoecologists have argued that there was, in fact, large-scale disturbance of primeval forests in Europe, even in the Early Mesolithic (Williams, 1985; Simmons, 1996). These arguments have largely been based on pollen evidence, which shows fluctuations in tree pollen levels, accompanied by increases in herbaceous pollen, suggesting the expansion of open ground. This, plus additional charcoal in some Mesolithic sedimentary records, has led some researchers to suggest that people were burning patches of forest in order to open them up for grazing animals, in order to boost grazer productivity and thereby enhance their hunting success.

The results of five studies including Mesolithic-Age insect faunas have been published thus far, but only two of these sites have been radiometrically dated. The fossil beetle record from these three sites sheds a different light on this debate (Whitehouse and Smith, 2004; Smith and Whitehouse, 2005). Some of the earliest Mesolithic beetle assemblages are from Holywell Coombe, Kent (Fig. 1, No. 79) (Coope, 1998). These faunas already include beetle species indicative of forest environments, but there are also species found today in open ground and lightly wooded habitats (Fig. 4). There are no dung beetles in these assemblages, so grazing animals were presumably not common, at least locally. Species indicative of open ground decline from 20 to 10% of the fauna through the Early Holocene, as woodland species increase from 3 to 30% of the fauna. Robinson (1991) has suggested that faunal

Table 1. British and Irish archeological sites with insect fossil analyses.

Site	Reference(s)
(1) Abbotts Way	Girling (1976a)
(2) Abercynafon	Panagiotakopulu (2004a)
(3) Aberdeen	Kenward and Hall (2001)
(4) Abingdon	Robinson (1979a)
(5) Alcester	Osborne (1971a), Girling (1986b)
(6) Alchester	Giorgi and Robinson (1985), Robinson (1975, 2001c)
(7) Appleford	Robinson (1981a)
(8) Aston Mill	Whitehead (1989)
(9) Baginton	Osborne (1975)
(10) Baker	Girling (1980b)
(11) Ballyarnet	O'Neill *et al.* (2007), Whitehouse (2007)
(12) Ballymacombes Moor	Whitehouse (2006)
(13) Ballywillin Crannog	O'Brien *et al.* (2005), Selby *et al.* (2005)
(14) Banwell Moor	Smith (2000a)
(15) Barford St. Martin	Smith (1999)
(16) Barnsley Park	Coope and Osborne (1968)
(17) Bearsden	Dickson *et al.* (1979)
(18) Beeston Castle	Girling (1993)
(19) Beverley	Hall and Kenward (1980), Kenward (2004), Allison *et al.* (1996)
(20) Bidford on Avon	Osborne (1988)
(21) Birmingham	Osborne (1981a)
(22) Bole Ings	Dinnin (1997), Brayshay and Dinnin (1999), Dinnin and Sadler (1999)
(23) Breiddin Hill Fort	Musson *et al.* (1977), Girling (1991), Buckland *et al.* (2001)
(24) Brigg	Buckland (1981c)
(25) Bristol	Jones (1998)
(26) Brixwold	Kenward *et al.* (1997)
(27) Buiston Crannog	Kenward *et al.* (2000)
(28) Burghfield	Robinson (1992a)
(29) Canterbury	Girling (1981b)
(30) Carlisle	Kenward (1984, 1990), Kenward and Carrott (2006)
(31) Carrick Castle	Warsop and Skidmore (1998)
(32) Castle Donnington	Smith and Howard (2004)
(33) Catsgore	Girling (1984b)
(34) Chester	Jaques *et al.* (2004)
(35) Chichester	Girling (1989a), Grove (1995)
(36) Cirencester	Osborne (1982)
(37) Copa Hill	Mighall *et al.* (2002)
(38) Corlea	Reilly (1996)
(39) Cowick	Girling and Robinson (1989), Hayfield and Greig (1989)
(40) Croft	Smith and Howard (2004), Smith *et al.* (2005a)
(41) Crossnacreevy	Girling (1974b)
(42) Dalton Parlours	Sudell (1990)
(43) Deer Park Farms	Kenward and Allison (1994a), Allison *et al.* (1999)
(44) Denny Abbey	Robinson (1980)
(45) Derragh	Whitehouse (2007)
(46) Derryville Bog	Caseldine *et al.* (2001), Whitehouse (2006)
(47) Doncaster	Smith (1978; 1989), Kenward (2004)
(48) Dragonby	Buckland (1996)
(49) Drayton	Robinson (2003a)
(50) Droitwich	Osborne (1977, 2006)
(51) Dublin	Coope (1981), O'Connor (1979, 1987), Reilly (2003)
(52) Dun Vulen	Roper (1999)
(53) Durham	Kenward (1979a)
(54) Empingham	Buckland (1981b)
(55) Exeter	Kenward (1975b), Straker *et al.* (1984)
(56) Farmoor	Robinson (1979a)

(*Continued*)

Table 1. (*Continued*)

Site	Reference(s)
(57) Fishbourne	Osborne (1971b)
(58) Fisherwick	Osborne (1979)
(59) Fiskerton	Osborne (2003)
(60) Flag Fen	Robinson (1992b, 2001d)
(61) Flodden Hill, Millfield	Kenward (2004)
(62) Folkestone: Amsterdam shipwreck	Hakbijl (1987)
(63) Glasgow	Buckland (2002)
(64) Glebe Farm	Carrott *et al.* (1993), Kenward (2004)
(65) Goldcliff	Smith *et al.* (2000)
(66) Great Yarmouth	Jones (1976)
(67) Grims Ditch	Kenward and Large (2001)
(68) Hampstead Heath	Girling and Greig (1977, 1985), Girling (1989b)
(69) Hardwich-with-Yelford	Robinson (1993)
(70) Hasholme Hall	Holdridge (1987)
(71) Hatfield Moors	Buckland and Sadler (1985), Buckland and Dinnin (1997), Dinnin (1997), Boswijk and Whitehouse (2002), Whitehouse (2004)
(72) Hayton	Jaques *et al.* (2000), Kenward (2004)
(73) Heldale Water	Sadler and Buckland (1998)
(74) Hemington	Smith and Howard (2004)
(75) Hen Domen	Greig *et al.* (1982)
(76) Hereford	Girling (1985a), Kenward (1985a)
(77) Hereford City Arms	Girling (1985a)
(78) Hibbalbstow	Hodder *et al.* (2005)
(79) Holywell Coombe	Coope (1998)
(80) Hull	Kenward (1977, 1979b), Miller *et al.* (1993a)
(81) Kingston upon Hull	Kenward and Carrott (2006)
(82) Kirby Muxloe	Smith (2000)
(83) Langford	Smith and Howard (2004)
(84) Leicester	Girling (1981a)
(85) Leven-Brandesburton	Hall *et al.* (1994), Kenward (2004)
(86) Limerick	Reilly (2003)
(87) Lincoln	Dobney *et al.* (1998), Kenward (2004)
(88) Lindow	Girling (1986a), Dinnin and Skidmore (1995)
(89) Loch Druidibeg	Dinnin (1996)
(90) London: Mowbray	Stafford (1971)
(91) London: Copthall Ave.	Allison and Kenward (1987)
(92) London: Southwark	Girling (1979c), Tyers (1988)
(93) London: Trig Lane	Kenward (1975b)
(94) Lunt	Osborne (1975)
(95) Malton	Buckland (1982a)
(96) Meare East	Caseldine (1987)
(97) Meare Heath	Girling (1976b, 1982)
(98) Meare Lake Village	Girling (1979a)
(99) Messingham	Buckland (1984)
(100) Midsummer Hill	Osborne (1981b)
(101) Mingies Ditch	Robinson (1993)
(102) Nantwich	Colledge (1981)
(103) North Ferriby	Buckland *et al.* (1990)
(104) Northampton	Keepax *et al.* (1978, 1979), Robinson (1983)
(105) Norwich: Fishergate	Kenward and Allison (1994b)
(106) Oxford	Brown and Robinson (1985), Robinson (1979b, 1981b, 1984, 1986, 2001b, 2001c, 2003b)
(106) Oxford Castle	Varley (1980)
(106) Oxford: St. Johns	Robinson (1992c)
(107) Papcastle	Kenward *et al.* (1988)

(*Continued*)

Table 1. (*Continued*)

Site	Reference(s)
(108) Penk	Smith (2000)
(109) Piddington	Simpson (2001)
(110) Pluscarden	Buckland (1994), Skidmore (1994)
(111) Pomeroy Wood	Robinson (1999)
(112) Porth Meare Cove	Osborne (1976)
(113) Radley	Robinson (1996)
(114) Red Moss	Ashworth (1972)
(115) Ribchester	Carrott *et al.* (2000), Kenward and Carrott (2006)
(116) Rudston	Buckland (1980)
(117) Runnymede	Robinson (1991, 2000a)
(118) Saint Helier, Jersey	Holdsworth (1976)
(119) Saint Kilda	Allison and Kenward (1996)
(120) Saint Ola, Orkney	Smith (1981)
(121) Saint Patrick's Isle	Tomlinson *et al.* (2002)
(122) Sandtoft	Samuels and Buckland (1978)
(123) Shakenoak Farm	Robinson (1978)
(124) Shiptonsthorpe	Palmer and Whitehouse (2006)
(125) Silbury Hill	Robinson (1997)
(126) Skellgarths, Ripon	Rowland *et al.* (2000), Kenward (2004)
(127) South Dyke	Kenward and Large (2001)
(128) South Shields	Osborne (1994)
(129) Southampton	Buckland *et al.* (1976), Kenward and Allison (1987), Kenward and Girling (1986)
(130) Speke Hall	Kenward and Tomlinson (1991)
(131) Steeple Claydon	Smith (1999)
(132) Stileway	Girling (1985b)
(133) Stone	Moffett and Smith (1997)
(134) Stourport	Osborne (1996)
(135) Sutton Common	Roper and Whitehouse (1997)
(136) Sutton Mandeville	Smith (1999)
(137) Swalecliffe	Tyers (2003)
(138) Sweet Track	Girling (1979)
(139) Tattershall Thorpe	Chowne *et al.* (1986)
(140) Taunton	Osborne (1984)
(141) Thame	Smith (1999)
(142) Thorne Moor	Buckland (1979), Buckland and Kenward (1973)
(143) Tisbury	Smith (1999)
(144) Towcester	Girling (1983)
(145) Upton Saint Leonards	Smith (1999)
(146) Westward Ho	Girling and Robinson (1987)
(147) Whitton	Osborne (1981c)
(148) Wilsford	Osborne (1969, 1986)
(149) Winchester	Kenward (1975)
(150) Worcester	Colledge and Osborne (1980), Osborne (1980d)
(151) Yarnton	Greig *et al.* (2004)
(152) York	Allison *et al.* (1996), Buckland *et al.* (1974), Hall *et al.* (1983, 1992), Kenward (1975, 1987, 1988a, 1998), Kenward and Hall (1995)
(152) York: Bedern	Kenward *et al.* (1986), Hall *et al.* (1993)
(152) York: Church Street	Buckland (1976c)
(152) York: Coney Street	Hall and Kenward (1976), Kenward and Williams (1979)
(152) York: Colonia	Hall and Kenward (1990)
(153) Yoxall Bridge	Smith *et al.* (2001), Smith and Howard (2004)

assemblages containing at least 20% forest-associated species should be considered indicative of closed woodland in Britain. When open ground and dung beetle species constitute 10% or more of a fauna, then Robinson considers this good evidence for a pastoral landscape. So the Holywell Coombe fauna gives indications for the presence of both closed forest and open ground habitats nearby. Whitehouse and Smith (2004) argue that early successional woodlands, such as existed in Britain in the Early Holocene, would necessarily have been open woodlands. The pioneering trees in early postglacial Britain included birch, pine, and hazel – all trees adapted to open woodland environments. This, plus the poorly developed soils of the Early Holocene, would have meant that open woodlands were the dominant ecosystem of that era.

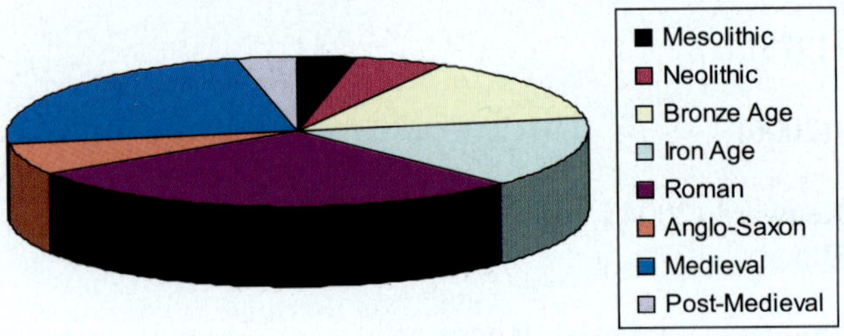

Fig. 3. Percentage composition of the age of archeological sites in Britain and Ireland from which insect fossil assemblages have been studied.

The second dated set of Mesolithic beetle assemblages comes from a site called Bole Ings (Fig. 1, No. 22), in the Lower Trent region of Nottinghamshire (Dinnin, 1997; Brayshay and Dinnin, 1999; Dinnin and Sadler, 1999). The dating chronology of this site is not very well established, but the faunal trends still provide insights into changes in the regional vegetation cover. Indicators of primary forest cover dominate all the samples from Bole Ings (Fig. 4), comprising at least 20% of each faunal assemblage. However, the percentage of open ground taxa fluctuates at this site, starting with 7% in the oldest fauna (>9,300 cal yr BP), then decreasing to about 2% around 5,800 cal yr BP, then increasing again after 3,800 cal yr BP. Whitehouse and Smith (2004) argue that the open ground at this site was maintained by the floodplain of the Trent, but that the later opening of the forest was tied to anthropogenic activities. Greenwood and Smith (2005) summarized the paleohydrologic reconstructions of the Trent River system, based on insect fossil evidence. Floodplain habitats are inherently unstable, creating new patches of open ground as the river channels shift, and frequent flood disturbances add another dimension of habitat instability. Open ground faunas of probable Mesolithic Age are also associated with floodplain habitats on the Thames at Runnymede (Robinson, 1991, 2000b).

The insect faunal evidence suggests that any single cause for the opening up of Mesolithic forests, whether anthropogenic or otherwise, is probably too simplistic.

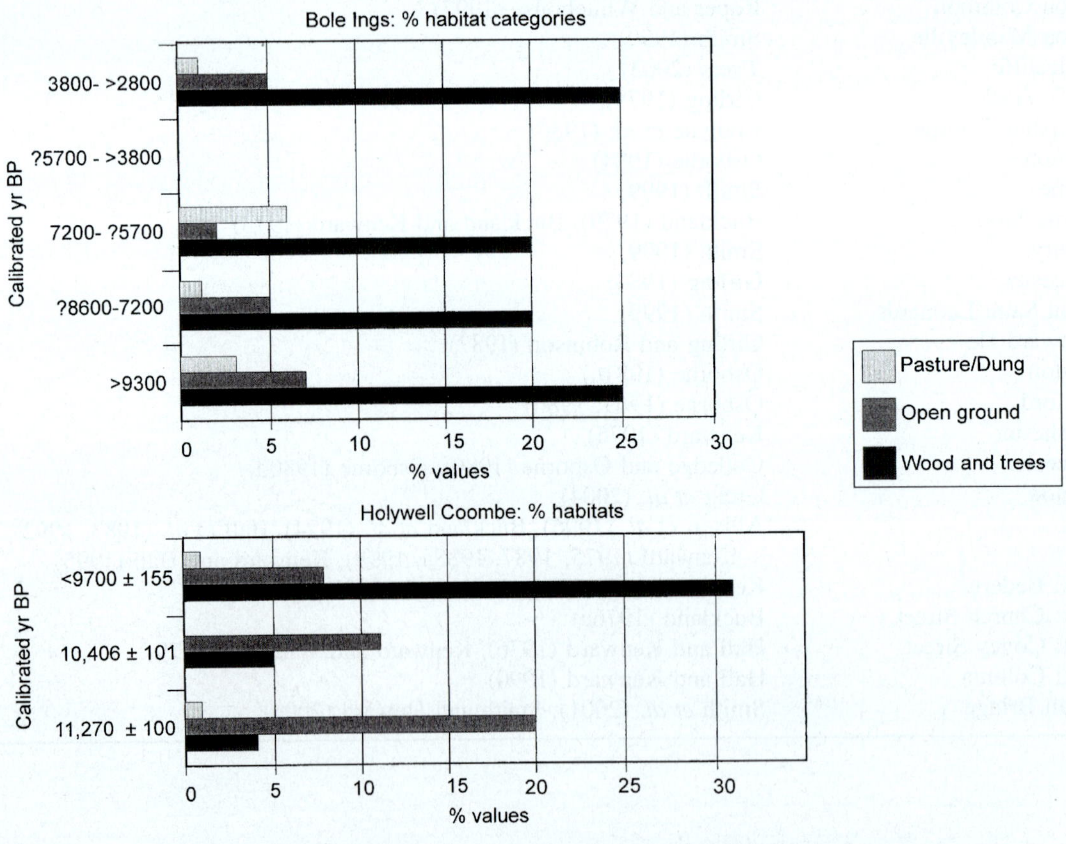

Fig. 4. Percentage composition of fossil beetle assemblages at the Bole Ings and Holywell Coombe sites, showing changes in the proportions of pasture/dung, open ground, and wood/tree-associated species through the Holocene (after Whitehouse and Smith, 2004).

Natural disturbances, such as floods, high winds, and forest fires certainly each played a role in creating openings in British and other European forests during the Early to mid-Holocene, and indeed, many of the beetle species associated with forest habitats from this era actually rely on the presence of a mosaic of habitat types, ranging from open ground to closed canopy forest. The fossil beetle record clearly has an important role to play in unraveling such paleoecological conundra. The development of a better understanding of the types of faunas associated with "open" and "closed" forest habitats will be critical to these investigations. Toward this end, Kenward (2006) performed a study of modern beetle assemblages in forested and open ground habitats in Britain, and concluded that while the presence of tree-associated species in an assemblage can be taken as a clear indication for the proximity of woodland, the rarity of such species in an assemblage cannot be taken as evidence of open ground, because some woodland-region faunas that he studied contained very few tree-associated beetles.

Only a handful of Neolithic period sites have been studied for insect remains. At Hampstead Heath, London (Fig. 1, No. 68), the insect fauna registered a change from natural forest to cleared sites used for cultivation and grazing (Girling and Greig, 1977, 1985). The insects from this assemblage include lime feeders, ivy feeders, holly feeders, and oak leaf-miners. The Neolithic elm decline was noted at the site, as was the presence of the elm bark beetle, *Scolytus scolytus* (Girling and Greig, 1985). A bog site at Abercynaton, Wales (Fig. 1, No. 2) recorded a succession of wetland beetle faunas, but had only one indicator of trees, the oak-feeding weevil, *Rhynchaenus signifer* (Panagiotakopulu, 2004a). Sweet Track (Fig. 1, No. 138), a Neolithic trackway across a wetland in the Somerset Levels region, yielded several insect faunal assemblages. The plant feeding species in these faunas document that the trackway was made of the branches of ash, elm, and lime trees, set over a reed swamp with pools of standing water. The beetle faunas of the Neolithic Period need considerably more study than they have received up till now.

Twenty-six Bronze-Age insect faunal assemblages have been published from the UK and Ireland. One of the best documented of these insect faunal successions has been described from Thorne Moor in South Yorkshire (Fig. 1, No. 142) (Buckland and Kenward, 1973; Buckland, 1979; Buckland and Johnson, 1983; Roper, 1996). A trackway, dated at 3090 yr BP, was discovered in a peat bog. This date corresponds to the time of regional forest clearing and the rise of cereal pollen in the regional vegetation. The trackway peat contained beetles from open water and bogs, which suggest that local flooding may have killed the forest. As with other trackway sites, waterlogging of the landscape eventually overwhelmed local farming. This may have been because of hydrologic changes associated with the rivers of the Humber Basin. Beetle species that are now extinct or extremely rare in England provide evidence of old, mature forest. This type of primeval forest and its associated insect fauna do not exist in Britain today, and are rare elsewhere in Europe.

In a somewhat different environmental setting, Whitehouse (2007) discussed Bronze-Age beetle assemblages from Ballyarnet Lake in County Derry, Northern Ireland (Fig. 1, No. 11). The earliest assemblages predate human occupation of the site, and include many ancient woodland species no longer found in Ireland or elsewhere in the British Isles, such as *Ryncolus sculpturatus*. Younger faunas indicate the gradual increase in meadows and open ground, at the expense of woodland. This has been interpreted as forest clearance and increasing animal husbandry, as dung beetle numbers also increased. This trend away from woodland communities persisted through the sequence, to the point at which the only remaining wood indicators were species such as the furniture beetle *Anobium punctatum* (Fig. 5), which probably fed on timbers in buildings. Additional regional studies have been done in Northern Ireland at Corlea (Fig. 1, No. 38) (Reilly, 1996) and Derryville Bog (Fig. 1, No. 46) (Caseldine *et al.*, 2001). At Derryville, **Bronze-Age** and Iron-Age archeological sites situated along a lakeshore were investigated. One side of the lake had more-or-less undisturbed forest that persisted through the Bronze Age, although some local sites had apparently been cleared for pasture or crop cultivation.

The history of the systematic destruction of primeval forest in Britain and Ireland is being reconstructed using

Fig. 5. The furniture beetle Anobium punctatum *(photo courtesy of K. V. Makarov, Zoological Institute, Russian Academy of Sciences).*

several methods. Apparently, nearly all primeval forest was cleared from the English Midlands by the end of the Roman period. Secondary forests, needed for firewood, were preserved for hunting by the nobility (including that much maligned conservationist, William the Conqueror), but even these forests were constantly under pressure (Buckland, 1979). Many species of beetles that require undisturbed woodland for habitat are saproxylic species: beetles that are dependent on the dead or dying wood of moribund or dead trees (standing or fallen), or on wood-inhabiting fungi, during part or all of their life cycle. Many of these beetles have poor mobility, and as primeval forests became fragmented into increasingly smaller patches, these poorly dispersing species eventually died out (Buckland and Dinnin, 1993; Smith and Whitehouse, 2005; Whitehouse, 2006). Peat bog expansion and the loss of particular kinds of woodlands, such as pine forests, also played a role in the regional extirpation of the primeval forest beetle fauna (Whitehouse, 1997, 2000, 2004, 2006; Boswijk and Whitehouse, 2002). Holocene climate change may also have played a role (Dinnin and Sadler, 1999; Whitehouse, 2006), although this factor is always difficult to tease apart from habitat modification in Europe, at least during the last 5,000–6,000 years.

Deforestation also had a deleterious effect on the rivers and streams of Britain, beginning at least in the Bronze Age. Increasing sediment loads, associated with soil erosion in newly deforested regions, choked once-clear running streams, eliminating parts of the insect fauna (Osborne, 1988; Smith, 2000b; Smith and Howard, 2004).

A series of Bronze- and Iron-Age trackways have been excavated in the Somerset Levels region (Fig. 1, Nos. 1, 10, 96–98, 132, 138). Extensive work on the insect faunas from these sites was done by Maureen Girling (Girling, 1976a,b, 1977a,b, 1978, 1979a,b, 1980b, 1982, 1984a,c, 1985b; Caseldine, 1987; Coles, 1987). Prehistoric farmers tried to develop agriculture in boggy lowlands with the aid of an extensive series of wooden trackways, but these measures eventually failed and the farms were abandoned. The beetle evidence shows that local woodlands were not completely cut down to make the trackways. A mixed deciduous forest persisted at sites such as Stileway, right through the Neolithic and Bronze Age. These woodlands included oak, hazel, and willow, based on the host-specific weevil and bark beetle species found in the fossil assemblages.

A Bronze-Age well at Wilsford, Wiltshire (Fig. 1, No. 148), was investigated by Osborne (1969, 1989). The samples of organic detritus came from the bottom of the well (a depth of 30 m); 23 L of sample produced a beetle fauna of 2,600 individuals in 138 taxa and 27 families! The insect horizon dates to 3,330 yr BP, and suggests a climate similar to modern, or slightly warmer (Osborne, 1969). Contrary to archeological expectations that this was a ritual well, the insects indicate that the well was an everyday farmer's well, used for watering cattle (based on a super-abundance of dung beetles in the fossil fauna). Quantities of cut grasses were brought near the well, probably as fodder or bedding. This was suggested by beetle species that live in decaying grasses. Evidence from wood boring beetles suggests timbers were erected over the mouth of the well.

A Bronze-Age field at Tower's Fen, Thorney, Peterborough was investigated by Green *et al.* (2008). During the Middle and Late Bronze Age, the land around the fen supported a mosaic of woodland, scrub, arable fields, meadow, and short grazed grassland. Dung beetles such as *Aphodius distinctus* suggest the presence of horses or cattle. The nearby presence of agricultural fields is suggested by the presence of the ground beetle, *Trechus quadristriatus*, and by the wheat shoot beetle, *Helophorus nubilus*. The larva of this beetle attacks wheat stems (Jones and Jones, 1974). The archeological and paleoenvironmental evidence from this site suggests that the Bronze Age agricultural landscape developed piecemeal and was based upon a mixed arable and pastoral economy.

Iron-Age records abound in Britain. One curiosity from the Iron Age is Lindow Man, a corpse preserved in a sphagnum bog in Cheshire (Fig. 1, No. 88). The insects associated with this "bog man" are a part of the forensic evidence of how the man died (Girling, 1986; Skidmore, 1986). Taxonomic and ecological data gathered by forensic entomologists for use in criminology has also been put to use in archeological contexts. Forensic entomological research has documented the insect faunal succession on human and other corpses, allowing more precise reconstructions on the timing of deaths, modes of burial, and other circumstances surrounding death and burial (Erzinclioglu, 1983; Smith, 1996). A procession, or "waves," of insect groups are attracted to corpses during the various stages of decomposition. These carrion insects include a variety of maggot (fly larvae) species, and both larvae and adults of beetles in more than 16 families, including Carabidae, Silphidae, Leiodidae, Staphylinidae, Histeridae, and Dermestidae. The same insects have been attracted to carrion for many thousands of years, so the fossil insects found with ancient human and domesticated animal remains have likewise provided good evidence on the timing and mode of ancient deaths and burials. More of these types of studies are discussed below.

Another Iron-Age bog body has been found in Ireland, at Oldcroghan in County Offaly. Insect remains from peats surrounding the body show that the victim was thrown in a pool in a raised bog. There are indicators of anthropogenic activity in and around the bog, so the ritual murder was apparently not done in an isolated place, but rather where people were frequently around (Plunkett *et al.*, 2009).

Iron-Age sites from Britain contain abundant evidence of human modifications of the landscape. Land continued to be cleared for agriculture, and trackways were built across wetlands. Several of these trackways were studied from the Goldcliff site on the Severn Estuary in Wales (Smith *et al.*, 1997). These investigations also encompassed insect faunas from the floors of nearby Iron-Age houses. Smith *et al.* (1997) found that a substantial synanthropic insect fauna was established in these houses (Table 2). As these dwellings were built either in fen woodland or on a raised bog surface, later inundated by seawater, it seems unlikely that the synanthropic fauna

Table 2. Synanthropic beetle species from the Iron-Age Goldcliff site, Wales.

Family	Species	Habitat
Hydrophilidae	*Megasternum bolitophagum*	Decaying plant matter
Hydrophilidae	*Cryptopleurum minutum*	Decaying plant matter
Staphylinidae	*Trogophleus bilineatus*	Urban archeological sites
Staphylinidae	*Xylodromus concinnus*	Mouldering hay
Mycetophagidae	*Typhaea stercorea*	Mouldering hay
Lathridiidae	*Enicmus minutus*	Mouldering hay
Colydiidae	*Aglenus brunneus*	Deep layers of decomposing organic matter
Histeridae	*Acritis nigricornis*	Preys on insects in decaying plant matter
Histeridae	*Paralister purpurascens*	Preys on insects in decaying plant matter
Histeridae	*Paralister carbonarius*	Preys on insects in decaying plant matter

Source: Data from Smith *et al.* (1997)

built-up slowly here, as has been hypothesized by Kenward (1997) for the development of synanthropic faunas at human occupation sites in rural Ireland.

Roman occupation and colonization also left its marks on the British landscape, and the Roman period is the best studied of all archeological intervals, in terms of fossil insect research. The Romans built many forts and other strongholds, as well as houses, baths, and other buildings. In one of the northern outposts at Bearsden, Scotland (Fig. 1, No. 17), insect fossils from a ditch included stored product pests associated with grain. These pests were abundant in samples taken from a Roman warehouse at York (Fig. 1, No. 152) (Hall and Kenward, 1976; Kenward and Williams, 1979). At Bearsden, they were found in cereal debris, which may have entered the ditch as sewage, presumably after both the grain and the pests had been consumed by Roman soldiers. The Romans imported quite a few stored product pests into Britain, so much so that the presence of some of these beetles, such as *Sitophilus granarius* (Fig. 6A), *Palorus ratzeburgi* (Fig. 6B), *Oryzaephilus surinamensis* (Fig. 6C), and *Cryptolestes ferrugineus*, have been considered diagnostic of Roman-Age faunas at certain sites (Kenward and Carrott, 2006). The maintenance of a large standing army probably contributed to the accidental importation of these pests, hiding in large quantities of stored grain.

Agricultural development proceeded apace during the Roman period. At Dragonby, Lincolnshire (Fig. 1, No. 48), the beetle evidence indicates a landscape wholly cleared of trees, given over locally to animal husbandry (Buckland, 1996). A large number of Roman wells have been excavated, and their preserved organic detritus has yielded abundant beetle remains. Animal husbandry was the primary reason for many of these wells, so the associated faunas are often rich in dung beetles, beetles associated with hay or straw, and beetles that lived in the timbers used to construct the wells and their coverings.

Urban Roman archeology has been particularly well documented from York, where more than 5 m of waterlogged sediments accumulated from Roman through medieval times, leaving layer upon layer of refuse. All of the insect assemblages from York reflect human habitations. The preservation is excellent because of waterlogging. Based on the quantity and quality of debris, initial research suggested that the Romans that occupied York were somewhat more hygienic than the subsequent Anglo-Danish inhabitants (Buckland, 1976c). Personal hygiene in Roman Britain was not all that it could have been, however, as attested by the presence of pubic lice remains (*Pthirus pubis*) in Roman sites (Kenward, 1999).

Reconstructions of medieval life at Anglo-Danish sites in York have included detailed research on insect fossils from sediments taken beneath Lloyd's Bank and subsequently at Coppergate (Buckland, 1973, 1974, 1976b; Buckland *et al.*, 1974; Addyman *et al.*, 1976; Hall and Kenward, 1976; Kenward *et al.*, 1978; Hall *et al.*, 1983; O'Connor *et al.*, 1984; Kenward, 1988a; Addyman, 1989; Hall and Kenward, 1990). Denford (1978) has also examined fossil mites from the Coppergate site. The overwhelming impression from the insect evidence at these sites is one of squalor in Anglo-Danish time. Rush and reed flooring materials were rarely changed; a well-developed compost fauna lived in every building studied. This lifestyle was not unique to York, however. Coope (1981) documented the same type of floor debris in a medieval house from Dublin (Fig. 1, No. 51). He hypothesized that the heat from fermenting plants on floors might have helped warm the house in winter. In summarizing what the fossil insect record can tell us about medieval living conditions in Britain, Osborne (1997) compared the insect faunas found in human living accommodations to those found living today in compost heaps.

In contrast to this, a group of Saxon pits at Southampton (Fig. 1, No. 129) indicate that mammal dung was removed from buildings and buried outside (Buckland *et al.*, 1976; Hall *et al.*, 1988). At a tannery site in York, dating to 1000 AD, the remains of thousands of flies (houseflies and biting stable flies) were found. The investigators also found chicken feathers and egg shells. Chicken manure was probably used in tanning hides. Among the chicken manure, a beetle was found that is common in dung-covered floors of chicken coops. There is no evidence of stored grains at the tannery, just mealworms that were probably brought in with some flour.

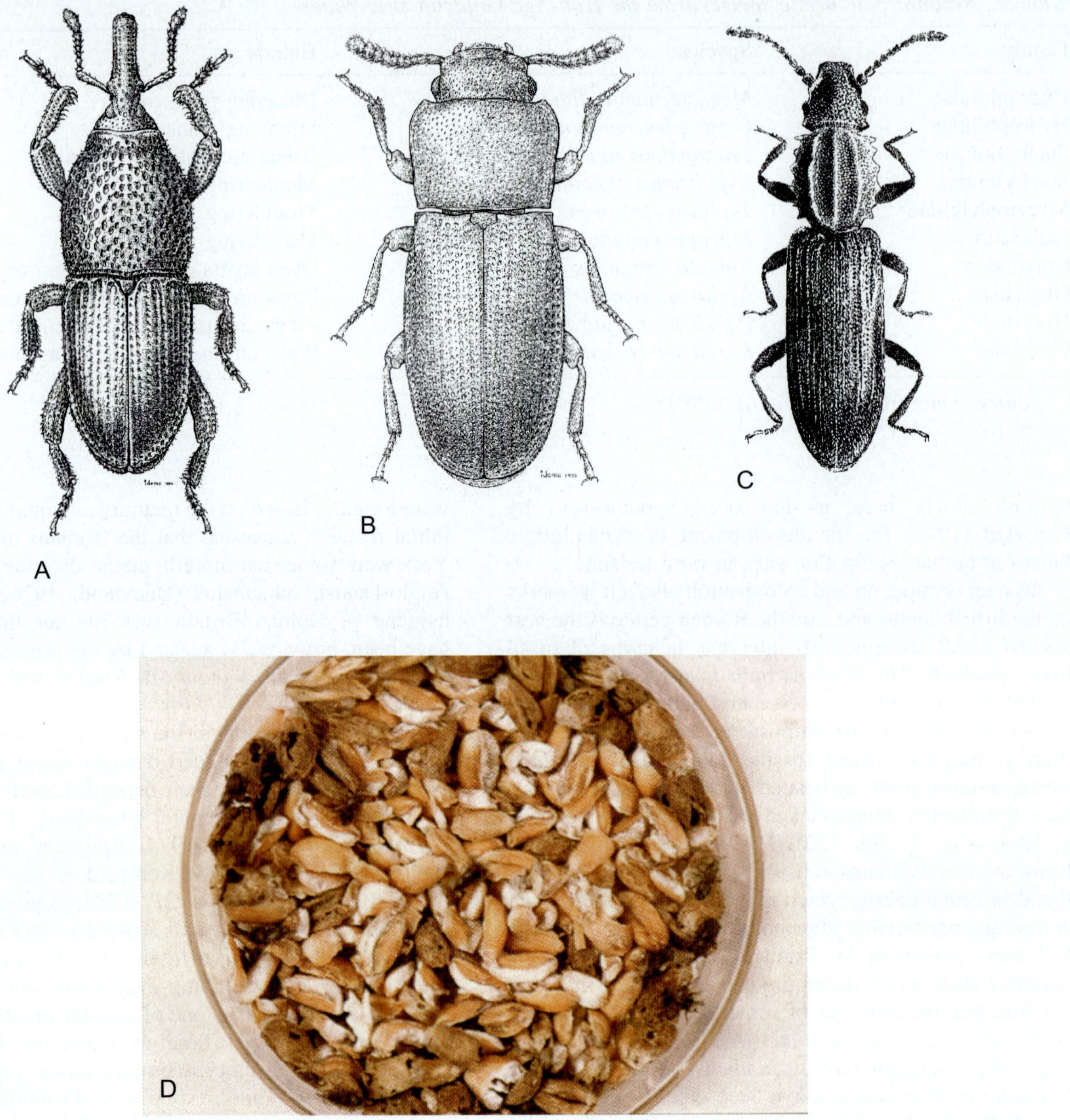

Fig. 6. Stored product pest species typical of Roman-site faunas in Britain; (A) the granary weevil, Sitophilus granarius*; (B) the small-eyed flour beetle,* Palorus ratzeburgi*; (C) the sawtoothed grain beetle,* Oryzaephilus surinamensis*; (D) a sample of grain infested by* S. granarius *(photos (A) and (B) from Bousquet, 1990; (C) courtesy of Hein Bijlmakers; (D) courtesy of Paul Buckland).*

In a York house, a fossil insect assemblage provides evidence of heath plants used as bedding, a practice still used in parts of Ireland. Unfortunately for the would-be slumberers, the heath insect fauna includes numerous biting ants.

Timbers in York buildings were infested with deathwatch and powderpost beetles. Although little is left of the original walls, some clues to their construction are found in beetle remains. The presence of dung beetles (dung was used to make daub), in combination with willow stem-eating beetles (from the wattle) indicate the use of wattle and daub architecture. Nettle feeding beetles were also found in the house. Buckland suggested that ''perhaps they wandered in from the back yard'' of this bijou residence.

The evidence from beetle species brought in from natural habitats (riverbanks, peat moss, heather, etc.) suggests that the climate of medieval York was similar to modern (O'Connor *et al.*, 1984). The apparent affluence of the people based on their artifacts (jewelry, imported pottery) contrasts sharply with the squalor inferred from the biological evidence. This modifies our image of medieval York considerably from what it was, based on previous archeological studies.

Medieval sites in and around Dublin, Ireland likewise shed light on human activities there. The processing of animal hides was the chief occupation of people living at a site on Newmarket Street in Dublin, as attested by the fossil insect fauna (Whitehouse, 2007). Timbers used in the construction of medieval buildings and furniture in Dublin contained the remains of locally extinct beetles associated with wood. Reilly (2003) suggested that the wood and associated beetles were transported to Dublin from forest stands elsewhere in Ireland or beyond.

Post-medieval deposits from Dublin document the introduction of certain insect pests, including the bedbug (*Cimex lectularius*) and the oriental cockroach (*Blatta orientalis*), in the early 18th century (Reilly, 2003). The cockroach appears to have been introduced repeatedly from southern Europe. Its first fossil record comes from a Roman-Age site in Lincoln (Dobney *et al.*, 1998).

Research on an early Christian archeological site at Deer Park Farms, Antrim, Northern Ireland (Fig. 1, No. 43) has forced a reevaluation of the ways in which the characteristic urban fauna developed, because there are species from this site (an isolated, rural site) which were formerly regarded as slow colonizers, needing long-lived settlements in order to become established (Kenward and Allison, 1994a; Allison *et al.*, 1999). Clearly, much remains to be learned about how insects have become adapted to anthropogenic environments (Kenward, 1997). The work of these British paleoentomologists constitutes a first major step.

The variety of habitats available for British insects is steadily declining, as has been documented by the fossil record spanning the Holocene. Osborne (1997) summed up the state of affairs quite eloquently:

> The final nails in the coffin of the British insect fauna are still being driven home, such as the deluge of agricultural insecticides, vast monocultures of such diverse plants as wheat and conifers, incessant trimming of hedges by machine, and the digging up and reseeding of grassland. These factors would appear to be likely to have a more deleterious effect on the insects than all the events of the last ten thousand years.

5. Studies in Western Europe (Fig. 7; Table 3)

The use of insect remains in paleoenvironmental reconstructions from archeological sites has grown considerably in Europe since the publication of *Quaternary Insects and Their Environments* in 1994. This review of European work proceeds from north to south from Scandinavia to the Mediterranean and North Africa.

5.1. Svalbard

Van Wijngaarden-Bakker and Pals (1981) reported on the presence of a stored product pest on board the Dutch merchant ship *Smeerenburg*. The wreck of this ship was discovered in Svalbard. It was a 17th century vessel, used to supply Dutch whalers working in arctic waters. Only one species of beetle was found, a bean weevil named *Acanthoscelides obtectus*. This is a stored product pest, often imported in peas, beans, and lentils from warmer parts of world (Harde, 1984).

5.2. Norway

Insects have been studied from medieval sites at Oslo, Norway (Fig. 7, No. 48) (Kenward, 1980, 1988b). The rich faunal assemblages contained mixtures of species that lived out-of-doors in the regional environment, such as a variety of ground beetles, water beetles, and rove beetles. There were indications that Medieval Oslo remained in close proximity to woodlands. For instance, the bark-gnawing beetle, *Nemozoma elongatum* (Trogossitidae) lives in woodlands and woodland margins.

There were also ''mixed'' (indoor–outdoor) species. These included *Latridius minutus*, a species found today in both synanthropic habitats (such as stored products) and natural habitats (in wasp nests, dung piles, etc.), as well as the sap beetle *Omosita colon* found in decaying vegetable matter that might have been either indoors or out. Evidence for poor sanitary conditions were ubiquitous. Rotting plant matter must have been abundant around the site, as attested to by three species of the flat bark beetle genus *Monotoma*, the cylindrical bark beetle *Aglenus brunneus*, and the silken fungus beetle species *Ephistemus globulus*, all of which inhabit piles of rotting straw and hay, or other decaying grasses or animal dung. Kenward found *Typhaea stercorea* and *Mycetaea subterranea* that feed on moldy grains and other moldy stored products, as well as the stored product pests, *Ptinus fur*, *Dermestes lardarius*, and the darkling beetle, *Tenebrio obscurus*. The *Ptinus* species feeds on animal skins, bird feathers, and a variety of stored food products. The dermestid species feeds on stored animal products, such as ham, bacon, or cheese. The darkling beetle species is a pest in stored grain. Its larva is called ''the dark mealworm.'' The carrion beetle, *Necrobia violacea* and the scarab, *Trox scaber* were also present, indicating the proximity of carrion or the hides or bones of dead animals. Kenward also found the remains of *Anobium denticollis*, a beetle that breeds in dead deciduous wood, and likely was attacking the wooden beams and furniture of dwellings. In addition to the beetle fauna, he also identified the human flea, *Pulex irritans* (Fig. 11B), and the remains of a honeybee (*Apis mellifera*).

The evidence, when taken together, indicates that the inhabitants of medieval towns in Scandinavia, such as Oslo, lived in houses with decaying straw underfoot, built with timbers infested with woodworm. They ate moldy, insect-infested grains and meats. The bones, hides, feathers, and dung of domestic animals were part of the household refuse that was rarely cleared away. It is perhaps not surprising then, that these people were also infested with parasites, such as fleas. As discussed throughout this chapter, these living conditions were essentially the norm for the vast majority of Medieval Europeans.

5.3. Sweden

In Sweden, Mesolithic and Early Neolithic sites at Skateholm in the south (Fig. 7, No. 60) have yielded insect faunas indicative of natural lacustrine, riparian, and coastal marine environments (Lemdahl, 1982, 1988a;

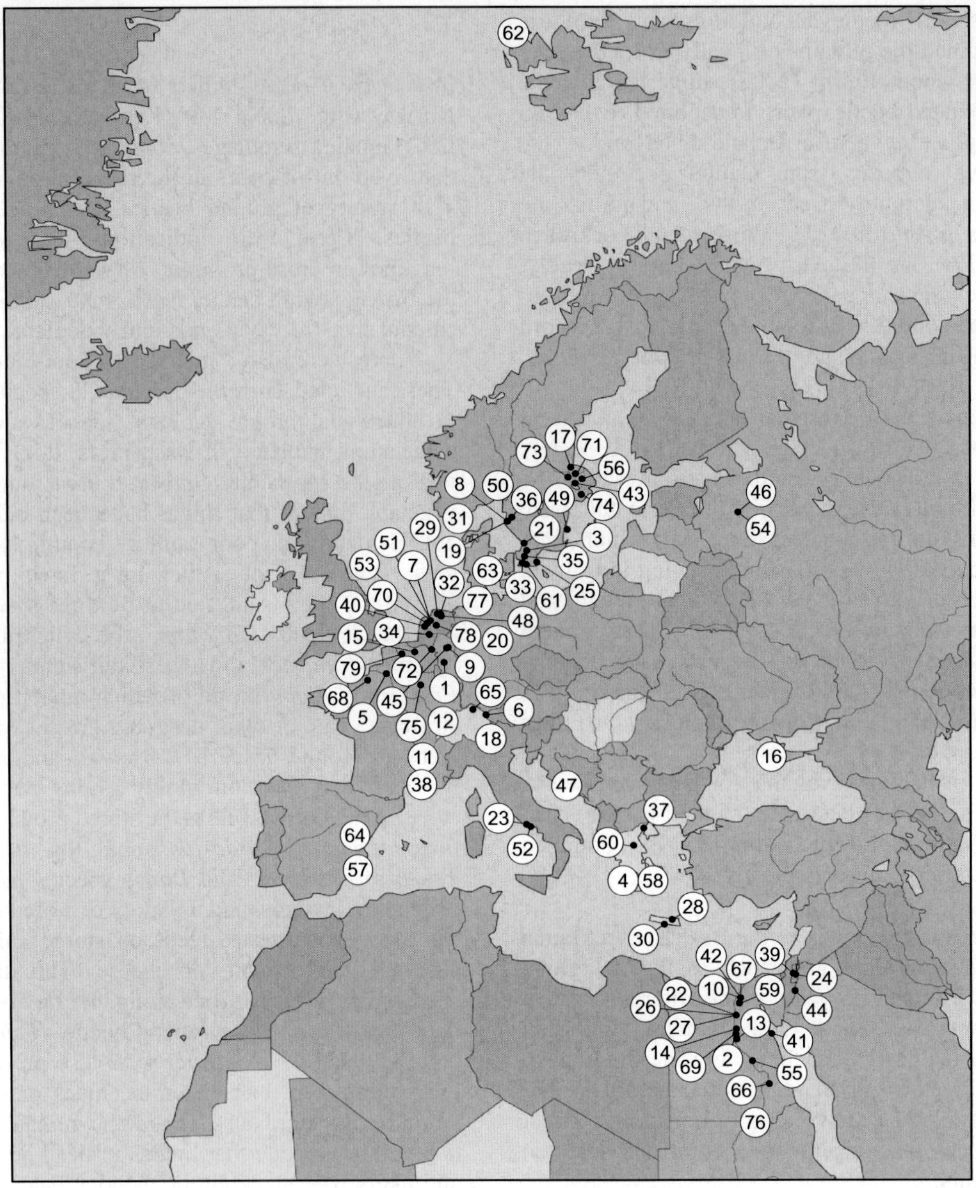

Fig. 7. Map of Europe, the Middle East, and North Africa, showing locations of archeological sites from which insect fossils have been analyzed. Site numbers are keyed to list of sites in Table 2.

Lemdahl and Thelaus, 1989). The insect fauna is consistent with the development of a lagoon at the site, caused by a **marine transgression**. Roughly contemporaneous insect assemblages from the Ageröd V bog site (Fig. 7, No. 3) are likewise indicative of natural environments adjacent to a village. The fauna includes thermophilous species, which have subsequently been extirpated from Sweden. A Late Iron-Age well and pit from Järrestad (Fig. 7, No. 25) yielded an extensive fauna (71 insect taxa), mainly indicative of an open, pastoral landscape where cows and sheep were grazed. Surprisingly, the well fauna did not contain abundant aquatic insect remains, but instead was dominated by the open ground fauna (Lemdahl, 2003).

Insect assemblages from the 15th century Halmstad site in southwestern Sweden (Fig. 7, No. 20) also include species that have since been regionally extirpated. These shifts in distribution have been attributed to either changes in land-use patterns or changing climate.

Medieval faunas have been studied by Lemdahl (1991c) from a site in the city of Lund (Fig. 7, No. 33); Hellqvist and Lemdahl (1996) studied a medieval site in Uppsala (Fig. 7, No. 70); Andersson (1992a) investigated a medieval site at Gothenburg (Fig. 7, No. 18). The faunal samples were recovered from a number of different archeological contexts, including soil samples, refuse layers, wells, and barrels. The Uppsala faunal assemblages included 81 taxa (mostly beetles), indicative of open ground habitats in an agricultural landscape where wheat, barley, and cabbage were grown. By the beginning of the 15th century, the local economy appears to have shifted mainly to animal husbandry. Interestingly, even though all but one of the fossil assemblages came from out-of-doors sites, they all included beetle species

Table 3. Continental European, Middle Eastern, and North African archeological sites with insect fossil analyses.

Site	Reference(s)
(1) Aachen, Germany	Koch (1970, 1971)
(2) Abydos, Egypt	Fletcher (1994, 1998), Schlüter and Dreyer (1984)
(3) Ageröd, Sweden	Lemdahl and Nilsson (1982), Lemdahl (1983)
(4) Akrotiri, Greece	Panagiotakopulu (2000), Panagiotakopulu and Buckland (1991), Panagiotakopulu *et al.* (1995, 1997)
(5) Amiens, France	Yvinec (1997)
(6) Arbon Bleiche, Switzerland	Lemdahl (2004)
(7) Assendelver Polders, The Netherlands	Hakbijl (1989)
(8) Bjorsjoas, Sweden	Andersson (1992a)
(9) Büderich, Germany	Koch (1970, 1971)
(10) Cairo Museum mummies, Egypt	Ruffer (1914)
(11) Chalain, France	Ponel (1997a)
(12) Champreveyres, Switzerland	Coope and Elias (2000)
(13) Deir el Medina, Egypt	Levinson and Levinson (1994)
(14) El Amarna, Egypt	Ryder (1983), Panagiotakopulu (1999)
(15) Erps-Kwerps, Belgium	Lentacker *et al.* (1992)
(16) Eupatoria, Ukraine	Antipina *et al.* (1991)
(17) Falun, Sweden	Hellqvist and Lemdahl (1990)
(18) Fiave, Italy	Osborne (1985)
(19) Gothenburg, Sweden	Andersson (1986, 1992b)
(20) Göttingen, Germany	Büchner and Wolf (1997)
(21) Halmstad, Sweden	Andersson (1999), Lemdahl and Thelaus (1989)
(22) Hawara, Egypt	Curry (1979), David (1978)
(23) Herculaneum, Italy	Capasso and Di Tota (1998)
(24) Horbat Rosh Zayit, Israel	Kislev and Melamed (2000)
(25) Järrestad, Sweden	Lemdahl (2003)
(26) Kahun, Egypt	Panagiotakopulu (1998)
(27) Karanis, Egypt	Blaisdell (1927)
(28) Knossos, Crete	Jones (1983)
(29) Kolhorn, The Netherlands	Hakbijl *et al.* (1989), Schelvis (1989)
(30) Kommos, Crete	Shaw and Shaw (1995)
(31) Langanes 1, Norway	Buckland *et al.* (2006a)
(32) Leeuwarden, The Netherlands	Schelvis (1987)
(33) Lockarp, Sweden	Buckland (2007), Eliasson and Kishonti (2003)
(34) Lower Rhine River, The Netherlands	Klink (1989)
(35) Lund-Apotekaren, Sweden	Lemdahl (1991c)
(36) Lundbacken, Sweden	Hellqvist (1999)
(37) Mandalo, Greece	Valamoti and Buckland (1995)
(38) Marseille, France	Ponel and Yvinec (1997)
(39) Masada, Israel	Kislev and Simchoni (2001)
(40) Midden-Delfland, The Netherlands	Schelvis and Koot (1995)
(41) Mons Claudianus, Egypt	Panagiotakopulu and Van der Veen (1997)
(42) Mummy of Han-Em-Kem-Esi, Egypt	Strong (1981)
(43) Myrby Bog, Sweden	Hellqvist (1996)
(44) Nahal Hemar Cave, Israel	Zias and Mumcuoglu (1991)
(45) Neuss, Germany	Cymorek and Koch (1969), Koch (1971)
(46) Novgorod, Russia	Hellqvist (1999)
(47) Okrugo, Croatia	Smith *et al.* (2006)
(48) Oldeboorn, The Netherlands	Schelvis (1990a)
(49) *Oskarshamn* shipwreck, Sweden	Lemdahl *et al.* (1995)
(50) Oslo, Norway	Kenward (1980, 1988b)
(51) Oudenburg, The Netherlands	Schelvis and Ervynck (1992)
(52) Pompeii, Italy	Meyer (1980)
(53) Raised Bog at Oostzaan, The Netherlands	Van Geel *et al.* (2003)
(54) Rjurikovo Gorodisce, Russia	Hellqvist (1999)

(*Continued*)

Table 3. (*Continued*)

Site	Reference(s)
(55) Rifeh, Egypt	David (1978), Levinson and Levinson (1994)
(56) Samnan, Sweden	Hellqvist (1999)
(57) Santa Pola, Spain	Moret and Martín-Cantarino (1996)
(58) Santorini, Greece	Panagiotakopulu and Buckland (1991)
(59) Saqqarah, Egypt	Chaddick and Leek (1972), Levinson and Levinson (1994), Solomon (1965)
(60) Servia, Greece	Hubbard (1979)
(61) Skateholm, Sweden	Lemdahl (1990)
(62) Smeerenburg, Svalbard	van Wijngaarden-Bakker and Pals (1981)
(63) Svendborg, Denmark	Jørgensen (1986), Noe-Nygaard (1982)
(64) Teruel, Spain	Compte and Perales (1984)
(65) Thayngen-Weier, Switzerland	Nielsen (1989), Nielsen *et al.* (2000), Guyan (1981)
(66) Thebes, Egypt	Andres (1931)
(67) Tomb Offerings, XI Dynasty, Egypt	Chaddick and Leek (1972)
(68) Touffréville, France	Ponel *et al.* (2000)
(69) Tuna el-Gebel, Egypt	Boessneck (1988), Levinson and Levinson (1994)
(70) Uitgeest, The Netherlands	Van Geel *et al.* (2003)
(71) Uppsala, Bryggaren, Sweden	Lemdahl (1991d), Hellqvist and Lemdahl (1996)
(72) Valkenburg, The Netherlands	Hakbijl (1988)
(73) Vallby-Tillberga, Sweden	Hellqvist (1999)
(74) Valsgarde, Sweden	Hellqvist (1999)
(75) Ville-sur-Returne, France	Buckland *et al.* (2006b)
(76) Wadi Halfa, Sudan	Armelagos (1969)
(77) Wilhelmshaven, Germany	Lemdahl (1990)
(78) Woerden, The Netherlands	Pals and Hakbijl (1992)
(79) Ypres, Belgium	Schelvis (1999)

considered to be synanthropic in Scandinavia, such as the ground beetle *Trechus rubens*, the water scavenger beetles *Cercyon unipunctatus* and *C. quisquilius*, and species in the family Lathridiidae. As in British studies, the Swedish studies document the presence of species that are either rare or extirpated from the study regions today. In Sweden, these species include *Enicmus brevicornis*, *Stereocorynes truncorum*, *Xyleborus monographus*, and *Aphodius subterraneus*. The first three of these require large, old trees; the latter species requires a certain type of grazed meadow, also not found in southern Sweden today.

The most recent fossils studied from Swedish archeological sites come from Falun (Fig. 7, No. 4), where 300-year-old buildings have been excavated. Soil from within the houses yielded 43 insect taxa (Hellqvist and Lemdahl, 1990).

The fauna from Gothenburg (Andersen, 1992) included many of the same synanthropic species found in Medieval Oslo, including *Tenebrio molitor*, *Trox scaber*, four species of the dung beetle *Aphodius*, *Anobium punctatum*, *Omosita colon*, *Mycetaea subterranea*, and *Sitophilus granarius*. Clearly domestic conditions were no more hygenic in Gothenburg than they were in Oslo.

Shipwrecks have yielded some fascinating glimpses into the ancient seafaring trade of Scandinavia. The shipwreck of the frame-timbered vessel *Oskarshamn* (Fig. 7, No. 47) yielded insect fossils studied by Lemdahl (1991e). This was a 13th century shipwreck off the Swedish coast. The fossil insect fauna was extracted from between the ribs and planking of the ship, in association with barley and weed seeds. The fauna included numerous species associated with rotting or fermented straw or hay, such as *Cilea silphoides* and *Cartodere nodifer*. Several other species were found that feed on mold or mildew, such as *Latridius*, *Thes bergrothi*, *Dienerella filiformis*, and *Tytthaspis sedecimpunctata*. The stored product pests on board the ship included *Ptinus fur*, known to eat a wide variety of animal and plant products, as well as the grain weevil, *Sitophilus granarius*.

Another Swedish ship, the East Indiaman, *Gotheborg* (Fig. 8), ran aground in Gothenburg harbor in 1745. It was fully laden with trade goods from China. In 1986, Andersen published a short list of beetle remains found in the excavation of the ship's contents that had been dredged up from the harbor. The ship's hold was found to contain *Tenebroides mauritanius*, *Cryptolestes ferrugineus*, *Rhizopertha dominica*, and *Alphitobius diaperinus* – all pests known to infest stored grain throughout the world today. Also in residence onboard the ship was the darkling beetle, *Uloma culinaris*, and inhabitant of rotting wood.

5.4. Denmark

Little paleoentomological work in connection with archeology has yet been done in Denmark. A series of samples from a Medieval urban setting were studied from a site in Svendborg, Denmark (Fig. 7, No. 62) (Noe-Nygaard,

Fig. 8. The reconstructed sailing ship, Göteborg, modeled exactly on the 18th century East Indiaman that sank in Gothenburg harbor (photo by Åke Fredriksson, used with permission).

1982; Jørgensen, 1986). The earlier sample has been dated by dendrochronology to AD1188–1218. The younger fauna, from a sample of charred barley, dates to the early 13th century. These samples contained a few stored product pests, including the grain weevil, *Sitophilus granarius*.

5.5. Germany

Similarly, little work of this kind has yet been done in Germany. Lemdahl (1990) investigated insects from an Iron-Age settlement at Butjadingen, near Wilhelmshaven (Fig. 7, No. 76), but did not publish a faunal list. Koch (1970, 1971) studied insects from Roman and medieval sites from Aachen and Büderich in the Rhineland (Fig. 7, No. 52). The faunal assemblages are small, and most of the species identified are from the surrounding environments (woods, meadows, standing freshwater). The Aachen fauna includes the rove beetle *Coprophilus striatulus*, a species associated with decaying vegetation, such as hay or straw. It also had the stored product pest, *Ptinus fur*, and six species of *Aphodius* dung beetles. The Büderich fauna included the rove beetle *Bisnius puella*, an inhabitant of cow dung or decomposing vegetation.

5.6. The Netherlands

Studies in the Netherlands have been more extensive, ranging from Neolithic to post-Medieval. The Neolithic site at Kolhorn yielded insect remains from a well, studied by Hakbijl *et al.* (1989). The insect fauna presented a **taphonomic** conundrum to the authors. Unlike most British studies of ancient wells, the well from Kolhorn did not contain rich concentrations of beetles associated with the watering of livestock. The lack of water beetle remains suggests that the deposit was not washed into the well. The inclusion of relatively heavy plant remains indicates that the deposit was not wind-borne. Based on the combined evidence of the insect and plant remains, it appears that this well was back-filled with a mixture of burnt and unburnt *Atriplex* and *Phragmites*, along with associated insects. As with some other Dutch sites, the presence of salt-tolerant beetles indicates the proximity of salt marsh to the site in Neolithic times. However, the authors concluded that the immediate region had been desalinated, and that the sea was some distance away. Perhaps the well ceased to be used shortly after being dug, because of contamination of brackish groundwater.

Hakbijl (1989) has investigated insect remains from an Early Iron-Age site at the Assendelver Polders, west of Amsterdam (Fig. 7, No. 6). More than 175 beetle and other insect taxa were identified, including substantial numbers of salt marsh and other saline environmental indicator species. Excavations of sandy floor level samples yielded the largest percentage of **halobionts**. Most of these insects show a preference for some type of water-land transitional environment, such as the strand of a salt marsh creek or the edge of a tidal pond. They suggest that house floors in the settlement included layers of sand from salt marsh levees. The insect remains also led Hakbijl to conclusions about the gathering and indoor use of various food items, and while some evidence indicates that the houses were kept relatively free of rubbish, poor sanitary conditions are suggested by the presence of human fleas, abundant houseflies, and stable flies.

Sometimes the presence of a host-specific parasite in a fossil assemblage can be used to establish the presence of the host at the site. Such was the case when Schelvis and Koot (1995) identified the sheep louse, *Damalinia*, in a fossil assemblage from an Iron-Age deposit at Midden-Delfland in the Netherlands. There was some confusion among the archeologists as to whether goats or sheep were being kept by the Iron-Age farmers, but the presence of this louse confirmed that sheep were being kept, because *Damalinia* only parasitizes sheep.

Van Geel *et al.* (2003) studied biological remains from a Roman-Age settlement at Uitgeest, the Netherlands. The fossils were extracted from ancient sods used to build the walls of a well at the site. Beetle species recovered from this site include *Aphodius* and *Anotylus*, both associated with dung, as well as *Enicmus transversus*, that feeds on fungi, and the aquatic beetles *Cercyon convexiusculus* and *Microcara testacea*. Fossil mites from this site occur today in wet meadows, in both brackish- and fresh-water environments. The overall reconstruction of the site suggests animal husbandry around the site of the well, in a treeless landscape that included fresh- and brackish-water meadows.

Schelvis (1987, 1990a,b, 1997) has used fossil oribatid mites to interpret the environment of archeological sites. The presence of brackish-water adapted mites in a Roman-Age site at Oudenburg, West Flanders, helped

Schelvis and Ervynck (1992) to reconstruct regional sea level changes. Fossil mite research from the Dutch medieval site of Oldeboorn (Fig. 7, No. 46) documented types of local landscapes, ranging from dry litter and moss covered soil through perennially wet moorland (Schelvis, 1990b). Schelvis (1987) also investigated medieval mite fossils from Leeuwarden (Fig. 7, No. 30). Schelvis (1999) used the remains of sheep ectoparisites to infer wool processing at medieval sites in Flanders.

Hakbijl (1986, 1987) has also delved into insect fossils in marine archeology. A wrecked hull of the Dutch East-Indies ship *Amsterdam*, found in the English Channel near Folkestone (Fig. 1, No. 3), yielded a container of "Spanish Fly," the ground-up exoskeletons of the blister beetle (Meloidae), *Lytta vesicatoria*. However, closer examination showed that the shipment had been adulterated with similarly colored bodies of the green chafer (Scarabaeidae), *Cetonia aurata*. Apparently the adulteration of controlled substances with cheap fillers has gone on for quite some time. Hakbijl and De Groot (1997) also studied insect remains taken from the improvised dwelling of Willem Barents' 1596 Arctic expedition, forced to overwinter on the Russian island of Novaya Zemlya in the late 16th century (Fig. 9). Along with the remains of human parasites such as lice and fleas, they once again found the remains of *Lytta versicatoria*. Apparently Dutch sailors were treated by naval surgeons for various disorders, using ground-up exoskeletons of this beetle, which contain cantharidin that blisters the skin.

Klink (1989) compiled a large body of insect fossil data concerning anthropogenic changes in the lower Rhine River in the Netherlands (Fig. 7, No. 32). A total of about 15,000 insect remains in 167 taxa were recovered from 52 sediment core samples from the Rhine. Most were aquatic larvae, including flies, caddisflies, and stoneflies. Klink's study indicates that the insect fauna of the Rhine has undergone a complete change during historic times, with 80–100 insect species disappearing during the last few centuries. Numerous deformed head capsules of aquatic fly larvae were found. Such deformities are thought to have been caused by the dumping of heavy metal compounds into the river during the last 200 years.

Fig. 9. Drawing of the hut built by Willem Barents' men to overwinter in Novaya Zemlya (after De Veer, 1598). Perhaps the man on the bed is being treated with cantharidin?

5.7. Switzerland

Only three Swiss studies have focused on insect remains from archeological sites. The Swiss lake village of Champreveyres at Neuchâtel (Fig. 7, No. 11) has been the subject of an interdisciplinary archeological/environmental study, beginning with **Magdalenian** occupation during the lateglacial interval. The paleoenvironmental work was based on lake sediments adjacent to the occupation site (Egloff, 1987; Coope and Elias, 2000). Insect fossils were described from sediments spanning the interval 12,670–11,820 yr BP. The site yielded 171 beetle taxa in 36 families. The only evidence of human disturbance is from a mixture of temperate and arctic beetles in one horizon. Perhaps people mixed the sediments through digging, although such disturbances were not evident in the stratigraphy of the site. No synanthropic insects were found, and the Paleolithic inhabitants seem to have had little impact on local landscapes.

Nielsen *et al.* (2000) studied an insect fossil assemblage from a Neolithic barn, built on an island in a shallow lake at Weier in northeast Switzerland. The assemblage has been radiocarbon dated to about 3600 BC. It yielded 54 arthropod taxa, including 37 beetles, which are dominated by Staphylinidae. The faunal assemblage is mostly indicative of dung piles and foul, decaying straw. The presence of puparia of the housefly, *Musca domestica* indicates that decaying, fermenting organic matter or dung was present, as this is the main breeding medium of this species. The cattle that lived in this byre apparently stood in thick accumulations of their own dung and decaying fodder and straw for much of the year. The combination of dung, decaying organic matter, and associated fungi provided a rich set of habitats for beetles, mites, and spiders, forming a food web (Fig. 10). The arthropods that fed on organic detritus, dung, and fungi were preyed upon by predatory spiders, beetles, and mites.

Lemdahl (2004) also studied insects from a Neolithic site, Arbon Bleiche (Fig. 7, No. 6). The ancient dwellings here were built on piles over the water on the southern shore of Lake Constance. The five samples came from organic deposits associated with floors of houses, and an alley between houses. An abundance of aquatic insects demonstrated the proximity of these dwellings to the water, but dung beetles in the fossil fauna indicate that the adjacent uplands were being used for grazing cattle.

5.8. France

Most archeoentomological work in France has focused on Roman-Age sites. An exception to this has been the study of insect remains from a Neolithic occupation site dated ca. 5300 BP at Chalain, in the Jura region (Fig. 7, No. 10). Here, Ponel (1997a) found a mixed "natural" and synanthropic fauna. The study region was apparently situated close to more-or-less primeval forest that contained firs, pines, oak, and elm, based on the bark

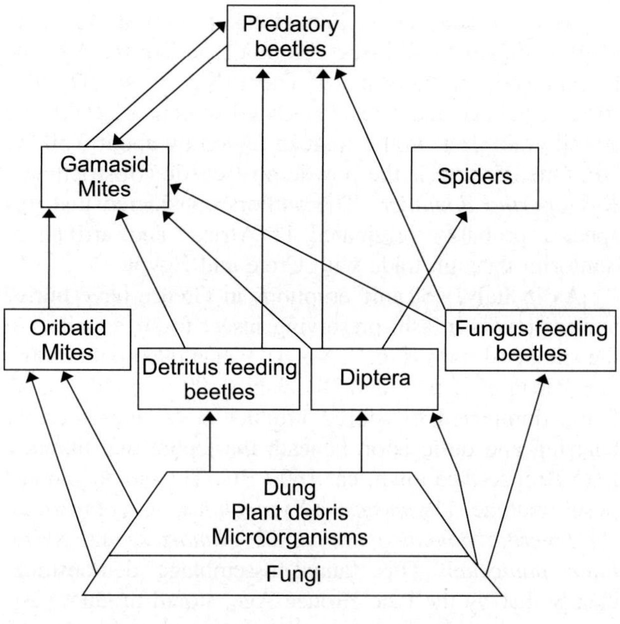

Fig. 10. Simplified food web of the detritus, dung, and fungal feeding arthropods at the Neolithic Weier barn site in Switzerland (after Nielsen et al., 2000).

beetle and weevil species found in the fossil assemblage. For instance, the flat bark beetle, *Uleiota planata*, is associated with ancient broad leaved forests (Hyman, 1992). The domestic side of the story is surprisingly similar to much younger (e.g., Medieval) archeological sites. Already in the Neolithic, French farms were apparently infested with beetles feeding on animal carcasses, such as *Dermestes frischii* and *Trox scaber*. The presence of moldy or mildewed hay on their floors is attested to by such beetles as the flat bark beetle, *Monotoma longicollis,* and the minute brown scavenger beetle, *Latridius minutus*. The presence of animal dung, specifically sheep, horse, and cow dung, is documented by three species of *Aphodius* dung beetles, and two species of *Onthophagus*.

Yvinec (1997) studied insect remains from charred grain taken from a second century AD Roman granary at Amiens (Fig. 7, No. 5). Not surprisingly, this fauna included grain pests, such as *Oryzaephilus surinamensis* and *Sitophilus granarius*. Many of the beetle remains were charred, suggesting that the granary was deliberately burned down to get rid of the insect-infested grain, as was inferred from a similar Roman granary site in Britain.

Buckland *et al.* (2006b) studied a series of three Roman-Age faunal assemblages taken from a deposit in a river valley at Ville-sur-Returne (Fig. 7, No. 73) in the Champagne region. The deposit comes from sediments over a Roman road, dated after AD177–178. The deposit was sealed by a later Roman road. These assemblages are essentially of synanthropic species, and represent the natural fauna of the region. There are numerous beetles associated with standing freshwater ponds or lakes with emergent vegetation along the margins. A diverse dung beetle fauna may be indicative of large mammal (horses, oxen, and sheep) traffic along the Roman road. The limited tree-associated species indicate the proximity of sandy pine woodlands along a river floodplain.

Ponel *et al.* (2000) studied a series of beetle faunal assemblages from a ditch, a shaft, a well, and a sedimentary basin at Touffréville, in the Calvados region (Fig. 7, No. 66). This is the richest set of archeological site faunas thus far studied from France, with 159 identified insect taxa. The faunal assemblages range in age from Iron-Age to Roman. Unlike many other European assemblages from this interval, the Touffréville faunas have very few synanthropic taxa, only *Stegobium paniceum, Nicobium castaneum, Anobium punctatum, Sitophilus granarius, Aglenus brunneus, Mycetaea hirta,* and *Enicmus minutus*. Furthermore, the numbers of individuals of these species are relatively low, in comparison with some British Roman and Medieval faunas.

The ditch assemblage contains beetle species that feed on herbaceous vegetation, several water beetles associated with standing water, and the seed weevil, *Bruchus pisorum* that feeds mainly on cultivated pea (*Pisum sativum*). The shaft fauna is dominated by species associated with decaying plant matter, and surprisingly few aquatic species. Some stored product pests occur in this assemblage. The well samples contain aquatic species, but these decrease upward, giving way to species associated with decomposing vegetation. The overall impression left by the insect and plant remains at the site is of an agricultural site that was abandoned toward the beginning o the 3rd century AD. There is clear evidence of waste ground, occasionally visited by humans, but choked with weeds and pools of stagnant water. The well at the site was apparently reused as a refuse pit. The synanthropic beetle fauna indicates that at least some domestic buildings (e.g., houses and barns) were located at the site or very close by.

5.9. Italy

Osborne (in Greig, 1985) studied a small insect fauna from a Bronze-Age lake dwelling at Fiave, Italy (Fig. 7, No. 17). He found the remains of aquatic insects living in still waters at the site, as well as stored product pests and dung beetles that may have fed on the dung of domesticated animals.

The eruption of Vesuvius in AD 79 preserved the remains of insects in nearby communities. In ancient Herculaneum, near Naples (Fig. 7, No. 22), Dal Monte (1956) found the stored product pests, *Oryzaephilus* and *Sitophilus granarius* in charred cereals beneath AD 79 tephra. Also at Herculaneum, Capasso, and Di Tota (1998) identified the remains of nits of the human head louse (*Pediculus humanus capitis*), preserved in the hair of a victim of the eruption. Not surprisingly, stored product pests were also found in charred grain samples from Pompeii (Meyer, 1980).

5.10. Spain

Only one archeological site has been studied for insect fossil remains thus far in Spain.

This is a Roman well, dated from the 1st–4th centuries AD, at Santa Pola, on the southern coast (Fig. 7, No. 55).

Here, Moret and Martín-Cantarino (1996) found two species of stored product pests: the darkling beetle *Blaps gigas*, and the grain weevil, *Sitophilus granarius*. The other species found include dung beetles and darkling beetles found today in dry steppe regions of southern Spain and the Mediterranean margins. The dung beetles are indicative of cow, sheep, and horse dung. One unusual find was the metallic wood borer, *Melanophila cuspidata*. This genus of beetles is attracted to recently burnt woodlands.

5.11. Greece

Greek studies range from Neolithic to Bronze Age, and they have been particularly useful in the development of our knowledge of the origins of stored product pests. There are over 600 documented species of beetles in the modern world that are known to be pests of stored products (Bousquet, 1990). Only about 18 of these are of primary economic importance, but the problem of stored product pests is probably at least as great now as it was in antiquity, if not greater. Today these pests damage about 10% of the world's annual food production. A more detailed analysis of various crops in various regions of the world is given by Parkin (1956). In the United States, the annual loss due to stored product pests was estimated to be about $1 billion in 1981.

The Greek fossil record is helping elucidate the origins of some of these pests. For instance, the charred head of a specimen of *Oryzaephilus surinamensis* from a Late Neolithic burnt grain deposit at Mandalo, Macedonia, has been described by Valamoti and Buckland (1995). This is one of the earliest finds of this species, with a calibrated age of 4340–4350 BC. It was found in charred emmer wheat. Until recently, the earliest find of *Oryzaephilus* was from the mid-second millennium BC, from the Aegean island of Santorini (Panagiotakopulu and Buckland, 1991).

While the *Oryzaephilus* specimen is clearly associated with the charred grain, the wheat does not bear any evidence of insect attack, and only one specimen was found, so there is little evidence of a full-blown infestation of the grain.

Fossil insect evidence from around the Mediterranean basin suggests that large-scale shipment of grain may have begun by the middle Bronze Age (Panagiotakopulu and Buckland, 1991). Older records of stored product pests such as *Oryzaephilus* probably represent their presence in low numbers, where they fed on broken grain, chaff, or crop residues (Valamoti and Buckland, 1995). What were the origins of these stored product pests? What did they feed upon prior to the human invention of stored products? Bell (1991) considered *Oryzaephilus surinamensis* to be of tropical origin, but there are also records of this species from birds' nests (Woodroffe, 1953), rotting wood and hay (Horion, 1960), fungi (Joy, 1932), and leaf litter (Crowson, 1958) in various parts of Europe. The transition from detritus feeder to synanthropic stored product pest was apparently already underway by the fourth millennium BC, as documented from the Mandalo site. Here it had already become closely associated with human habitations.

As discussed above, Panagiotakopulu and Buckland (1991) studied fossil insects from a Late Bronze-Age site at Santorini, on the island of Thera (Fig. 7, No. 57). The fossil evidence indicates the establishment of field and stored grain pests in the Aegean region by about 3,500 yr BP. One of these is the powderpost beetle (Bostrichidae) *Rhyzopertha dominica*. The authors considered that this species probably originated in Africa, and arrived at Santorini through trade with Crete and Egypt.

As in Italy, volcanic eruptions in Greece have buried ancient towns in ash, preserving insect fossil remains. At the site of Akrotiri (Fig. 7, No. 4), Panagiotakopulu (2000) and Panagiotakopulu and Buckland (1991) found a small fauna dominated by stored product pests, preserved by charring and desiccation beneath the tephra that buried a Late Bronze-Age town, ca. 1600 BC. The stored product pests include *Dermestes*, *Oryzaephilus*, *Cryptophagus*, *Rhyzopertha dominica*, *Sitophilus granarius*, and *Stegobium paniceum*. This faunal assemblage demonstrates clearly that by the Late Bronze Age, stored product pests were gaining in both numbers and diversity in the Mediterranean region. There were also deciduous woodland indicators at this site, such as the weevil, *Trogloryhnchus anophthalmus* and the seed weevil, *Bruchus rufipes*.

The inhabitants of Santorini apparently did not sit idly by and let stored product pests devour their foodstuffs. Another study from the Akrotiri site concerns the development of natural pesticides, to combat stored product pests. Panagiotakopulu *et al.* (1995) discussed a wide variety of methods that were apparently used to kill insects. These include the use of airtight storage containers, and the use of various plant and animal substances. Dried residues indicate that some foodstuffs were stored in either olive oil or vinegar, either of which would repel insect pests. The leaves of various plants were placed in stored food containers, apparently as insecticides. These include laurel leaves and *Thymelaea hirsuta*, and *Echium*.

Panagiotakopulu *et al.* (1997) also discussed intriguing entomological evidence that the inhabitants of Akrotiri and elsewhere on the island of Santorini were raising native silk moths, in support of an indigenous silk trade. A fossil cocoon of a moth, thought to be from either *Saturnia pyri* or *Pachypasa otus*, was found inside a ceramic cylinder, dated to the middle of the 2nd millennium BC. This is some debate about the source of silk used in the Mediterranean region in antiquity. Some argue for Chinese silk, brought to Egypt or Greece by very early traders. The fossil evidence from Santorini argues for the utilization of silk from European silk moths.

6. North African and Middle Eastern Studies

A number of insect fossil studies have been done in recent years, concerning archeological sites in arid regions of the Middle East and North Africa. Many of these studies involve desiccated insect remains from tombs or ancient dwellings. As in packrat midden studies, the dried exoskeletons of insects from desert and semidesert regions of the Old World can be remarkably well preserved, and these fossils are shedding new light on ancient human lifestyles from these vitally important archeological regions.

6.1. Egypt

Interest in Egyptian antiquities launched a great deal of archeological research, as far back as the early 19th century, and this included investigations of insect fossils from the tombs of the pharaohs. In 1814, a French entomologist named Millin described beetle remains from an unwrapped mummy, taken by scientists who accompanied Napoleon on his Egyptian military campaign (Alluaud, 1908). The full range of fossil insect finds from Egyptian archeological sites is discussed by Panagiotakopulu (2001a), and summarized in Table 4. When archeologists unwrap the bandages from ancient mummies, they frequently find numerous carcass beetle remains – notably the adults and larvae of dermestid beetles, such as *Dermestes maculatus* and *Necrobia violacea* (Fig. 11D,I). These beetles are attracted to dried animal remains, and in spite of the best efforts of the ancient Egyptians who prepared bodies for mummification, sealed them in seemingly impregnable sarcophagi, and buried them deep inside seemingly airtight stone tombs, the beetles managed to find their prey. Mummies described by Pettigrew (1834) had their flesh eaten away by *Dermestes frischii*. Even the mightiest of pharaohs was not immune. The mummy of Rameses II was found to contain the larvae (Fig. 11F) and adults of the carrion-feeding beetle, *Thylodrias contractus* (Alluaud, 1908).

Foodstuffs left as offerings in Egyptian tombs, as well as stored products found in other archeological contexts, have yielded the usual crop of stored product pests. The grain weevil *Sitophilus granarius* has been found in barley left in a tomb at Saqqarah (Fig. 7, No. 57). Its earliest appearance in Egyptian sites comes from the tomb of Queen Ichetis, dated ca. 2323–2150 BC. *Oryzaephilus surinamensis* has only been found in one location: the Roman site of Mons Claudianus (Fig. 7, No. 39) (Panagiotakopulu and Van der Veen, 1997). The remains of the lesser grain borer *Rhyzopertha dominica* have been more widely found in ancient Egyptian sites, including Kahun (Fig. 7, No. 24), Mons Claudianus, Tutankhamun's tomb, and the workman's village at El Amarna (Fig. 7, No. 13) (Panagiotakopulu, 2001a). Another widespread pest in ancient Egypt was the biscuit beetle *Stegobium paniceum*. The remains of this species have been found in Kahun, in Tutankhamun's tomb, and in the tomb of Kha at Deir el Medina (Fig. 7, No. 12).

One of the more unusual stored product pest fossil records from Egypt is that of *Aglenus brunneus*. This is a flightless, blind beetle that has now become cosmopolitan but was apparently much more restricted in ancient times. In Egyptian antiquity, it has only been found from the Roman site of Mons Cladianus. Panagiotakopulu (2001a) suggested that it may have been introduced there by the Romans – a stowaway in stored grain that came from the European side of the Mediterranean. One of the more exotic pests preserved in Egyptian antiquity is a bark beetle *Coccotrypes dactyliperda* that is known to attack the stones in palm dates. A fossil of this species was recovered from inside a doum palm (*Hyphnaea thebica*) from the Greco-Roman site of Berenice, on the Red Sea coast (Panagiotakopulu, 2001b). Apparently, the grain weevil, *Sitophilus granarius* infested grain supplies throughout the Roman Empire. Charred remains of this pest were found in an Iron-Age deposit from Jordan (Fig. 12).

The ancient Egyptians suffered from the same host of human parasites found elsewhere in the world. Remains of bedbugs (*Cimex lectularius*) (Fig. 11A) were found in El Amarna, as was the human flea (Fig. 11B) (Panagiotakopulu and Buckland, 1999; Panagiotakopulu, 2001b). Panagiotakopulu (2004b) has proposed that rat fleas from Egypt may have been the original vectors of bubonic plague. The primary host for this flea (*Xenopsylla cheopis*) is the Nile rat (*Arvicanthis niloticus*). Panagiotakopulu believes that people living along the Nile in ancient Egypt came in contact with these rats, especially when the Nile was in flood. The first plague epidemic may have happened at the workman's village at Amarna, where large numbers of people lived in filthy, squalid conditions and rats abounded. The black rat, *Rattus rattus* was accidentally introduced into Egypt via trade vessels from India (perhaps via Mesopotamia). This rat could have become the new vector for plague, as *Xenopsylla cheopis* was able to switch hosts. As is so often the case, the new host species would not have been as resistant to the plague bacteria as was the original African rat host. According to Panagiotakopulu's theory, outbreaks of plague ensued as black rat populations expanded in urban settings across the Old World, leading to occasional pandemics (Fig. 13).

Buckland and Sadler (1989) postulated that the human flea is of New World origin, in South America. They note the lack of clear historical or archeological evidence for *Pulex irritans* in the Mediterranean world (i.e., classical Egypt, Greece, or Rome) or elsewhere in Europe until Medieval times. Their theory is that the human flea may have originally evolved as a parasite of guinea pig or peccary, shifting to humans as they entered South America at the end of the last glaciation. From there, fleas could have spread rapidly through human populations, northwards to Alaska, and across the Bering Strait to Asia. The findings of *Pulex* sp. on mummified guinea pigs from high elevation sites in the Peruvian Andes demonstrates the presence of this flea genus in South America prior to European contact, thus strengthening the hypothesis of its origin in the New World (Dittmar *et al.*, 2003).

Head lice (*Pediculus humanus capitis*) and their eggs (nits) are not as dangerous to humans as plague-ridden fleas, but lice have also been found preserved in a wide range of mummies. Lice fossil records extend south to ancient Nubia, where 40% of the scalp and hair samples from human remains dating from AD 350 to 550 at Wadi Halfa (Fig. 7, No. 74) were found to contain lice (Armelagos, 1969). The oldest louse record comes from across the Red Sea in what is now Israel, from Nahul Hemar Cave. These lice fossils date back to the Early Holocene, ca. 6900–6300 BC (Zias and Mumcuoglu, 1991). Fossil lice from the First Century AD have also been described from Masada, Israel (Mumcuoglu *et al.*, 2003).

Table 4. Fossil insects found in Egyptian archeological sites.

Species	Egyptian Records
Carrion and carcass beetles	
Dermestes ater	On human mummies, including Han-Em-Kem-Esi
Dermestes carnivorus	On human mummies
Dermestes frischii	On human mummies, including Ramesis II, Minya mummy, Han-Em-Kem-Esi; on ox and ibis mummies
Dermestes maculatus	On human mummies, including Bristol mummy, Graeco-Roman at Thebes
Necrobia rufipes	On human mummies, including Pum IV, Ramesis II, Bristol mummy
Necrobia violacea	On human mummies, including Ptolomaic mummy from Thebes
Thylodrias contractus	On human mummies, including Pum IV and Ramesis II
Stored product pests	
Aglenus brunneus	At Mons Claudianus (introduced in stored products from Roman empire?)
Anthrenus coloratus	In workman's village, Amarna
Attagenus astacurus	In workman's village, Amarna
Attagenus unicolor	In workman's village, Amarna
Bruchidius spp.	In *Acacia* seeds from Kahun; in pulses from Tukankhamun's tomb; at Amarna
Coccotrypes dactyliperda	In date (*Hyphaena thebica*) stones from Graeco-Roman site at Berenice
Cryptolestes turcicus	At Mons Claudianus and Amarna
Gibbium psylloides	At Mons Claudianus, on human mummies
Latheticus oryzae	At Amarna
Oryzaephilus surinamensis	In storage vessel from Mons Claudianus
Palorus ratzeburgi	In dried pig dung from workman's village at Amarna (pigs fed infested grain)
Palorus subdepressus	In cereal stores from Amarna
Rhyzopertha dominica	In plant remains from Kahun; in offerings in Tukankhamun's tomb; at Mons Claudianus; at workers village at Amarna
Sitophilus granarius	Barley in tombs at Saqqarah
Stegobium paniceum	At Kahun; in offerings in Tukankhamun's tomb; on mummy of Rameses II
Thorictodes heydeni	Roman town site at Karanis
Tribolium castaneum or *T. confusum*	In 3rd Millennium BC tomb; in pot dated 1000 BC; at Mons Claudianus
Trogoderma granarium	In material from Byzantine site of Kom-el-Nana
Human parasites	
Cimex lectularius (bedbug)	At Amarna
Pulex irritans (human flea)	At Amarna
Pediculus humaus capitis (human head louse)	On human mummies in Cairo Museum; from Abydos and Hierakonpolis; from Manchester Museum mummy; from Nahal Hemar Cave (6900–6300 BC)

Source: Data from Panagiotakopulu (2001a)

7. Studies in the North Atlantic Region (Figs. 14 and 15; Table 5)

In recent decades, insect fossil studies from archeological sites in the North Atlantic region have flourished. With the exception of the Asummiut, Qeqertasussak, and Qilakitsoq Inuit sites in Greenland (Fig. 14, Nos. 1, 12, and 13), all of the sites discussed here from Iceland, Greenland, and the Faeroe Islands are necessarily young, falling within the time of Norse settlement in about the last thousand years. These studies are contributing a great deal to our understanding of how the Norse peoples lived in these relatively extreme environments.

In southern Iceland, Buckland *et al.* (1986a) provided records of pre- and post-settlement insect faunas from trenches dug through peat associated with an archeological site at Ketilsstadir farm (Fig. 15, No. 8). At this site, the *Landnám* tephra was deposited at the time of the first Viking landings, about 900 AD, and hence provides a convenient marker horizon. *Landnám* is the Old Norse term for the taking or settling of land. Altogether, 46 beetle samples indicate a fauna of a few species that became more diverse as humans modified the landscape (especially the draining of bogs and overgrazing of hillsides by sheep). Among the synanthropic species found were a dung beetle, *Aphodius lapponum* that feeds on the dung of large herbivores, and several species associated with stored hay.

Another study in southern Iceland focused on insect faunas from a farm at Holt (Fig. 15, No. 7) (Buckland *et al.*, 1991). This paper discussed the ecological effects of the Norse colonization of Iceland, which began in the 9th century AD. These impacts were investigated through comparisons between the pre- and post-Landnám insect faunas and floras of this and other sites. The farm at Holt was settled early, and appears to have been continuously occupied for more than 1,000 years. By sampling organic deposits from a series of ditches, the authors were able to obtain fossil assemblages in stratigraphic sequence.

Fig. 11. Insect species found preserved in ancient Egyptian tombs. (A) bedbug, Cimex lectularius*; (B) human head lice (*Pediculus humanus capitis*): fossil specimens from Amarna, Egypt, and modern specimen (inset); (C) human flea,* Pulex irritans*; (D) the hide beetle,* Dermestes maculatus*; (E) the smooth spider beetle,* Gibbium psylloides*; (F) larva of the odd beetle,* Thylodrias contractus*; (G) the longheaded flour beetle,* Latheticus oryzae*; (H) the small-eyed flour beetle,* Palorus ratzeburgi*; (I) the blacklegged ham beetle,* Necrobia violacea. Photo Credits*: (A) Centers for Disease Control, USA; (B) (modern specimen) Dr. K. Baldwin, Monmouth College, Illinois; (fossil specimens) Dr. Eva Panagiotakopulu, University of Edinburgh, used with joint permission of Antiquity Publications Ltd.; (D) courtesy of Joyce Gross, University of California, Berkeley; (E) photo courtesy of M. E. Smirnov, Zoological Institute, Russian Academy of Sciences; (F) courtesy of Dr. Jarmo Holopainen, University of Kuopio, Finland; (G) courtesy of K. V. Makarov, Zoological Institute, Russian Academy of Sciences; (H) and (I): courtesy of the Canadian National Collection of Insects, Ottawa.*

Fig. 12. Remains of the grain weevil, Sitophilus granarius *from an Iron-Age archeological site in Jordan (photo by Eva Panagiatakopulu, University of Edinburgh, used with permission).*

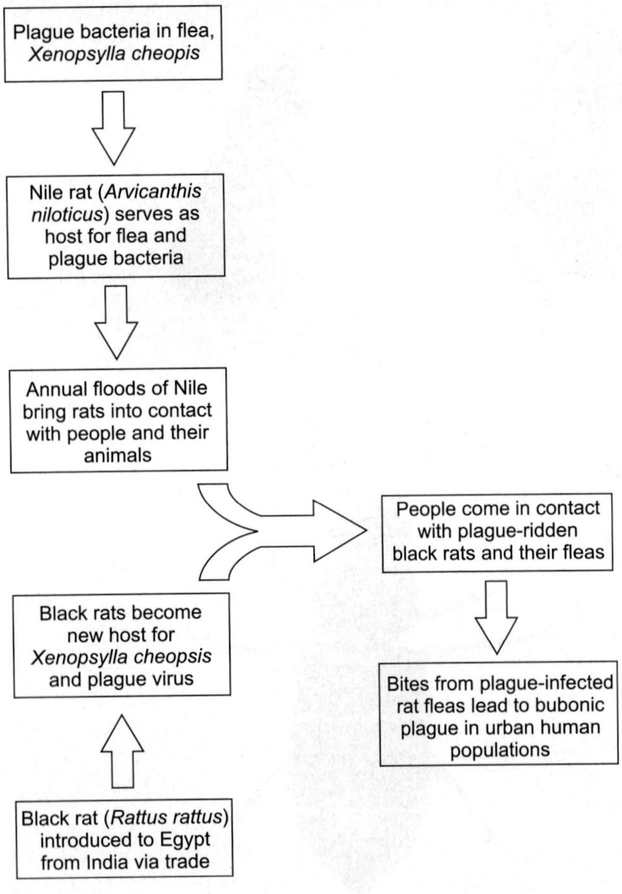

Fig. 13. Diagram illustrating the hypothesized development of bubonic plague in Egypt (after Panagiotakopulu, 2004b).

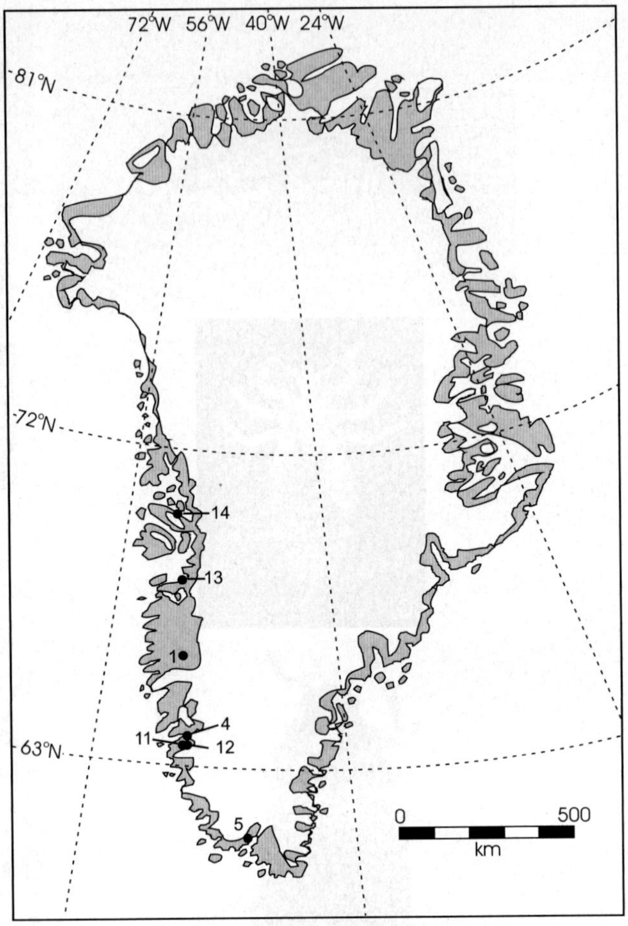

Fig. 14. Map of Greenland, showing locations of archeological sites from which insect fossils have been analyzed. Site numbers are keyed to list of sites in Table 5.

The pre-*Landnám* insect fauna from Holt was comprised of subarctic species. The post-*Landnám* fauna included several synanthropic insects, including *Omalium excavatum*, *Xylodromus concinnus*, *Xylodromus depressus*, *Atomaria* cf. *apicalis*, *Mycetaea hirta*, *Corticaria elongata*, and *Typhaea stercorea*. The data obtained from the Holt study provided the basis for a regional synthesis of paleoenvironments. This showed that the pre-*Landnám*, subarctic biota of Iceland represented a fragile ecosystem, which faired very poorly after the Norse settlement of the island. Human occupation has been ecologically devastating to much of Iceland. This process began with the earliest settlers, as indicated by the fossil study.

Perry *et al.* (1985) discovered a mixture of ecological groups in the fossil insect fauna from the late medieval farm at Storaborg, Iceland (Fig. 15, No. 15). The authors used **cluster analysis** to help refine their interpretation of the faunal elements. Two principal groups were distinguished: a midden component and a house floor component. Taken together, the evidence of insects and plant macrofossils pointed to domestic areas that had a warm, foul, well-trod covering of cut vegetation containing large quantities of decomposing garbage. In contrast, the outdoor midden-type layer, exposed to the elements, had a cool, less foul, less stable environment.

Following the colonization of Iceland, the Vikings proceeded west to establish settlements on Greenland. Cereal agriculture was precluded by the severe climate, and by about 1350 AD, the settlements in western Greenland were abandoned (Fredskild, 1988). The work of archeologists and paleoecologists has given us a glimpse of what life was like for these hardy folk.

Table 5. North Atlantic Archeological sites with insect fossil analyses.

Site	Reference(s)
(1) Asummiut, Greenland	Böcher (1998)
(2) Bessastadir, Iceland	Amorosi *et al.* (1992)
(3) Engihlid, Iceland	Buckland and Sadler (1991)
(4) Eqalugialik, Greenland	Sadler (1987)
(5) Garðar, Greenland	Buckland *et al.* (2009a)
(6) Garden under Sandet, Greenland	Buckland *et al.* (1998), Buckland (2000, 2007)
(7) Godataettur, Iceland	Buckland *et al.* (1995)
(8) Holt, Iceland	Buckland *et al.* (1991)
(9) Ketilsstadir farm, Iceland	Buckland *et al.* (1983a, 1986a)
(10) Nesstofa, Iceland	Amorosi *et al.* (1994)
(11) Niaqussat, Greenland	McGovern *et al.* (1983), Sadler (1987)
(12) Nipaitsoq, Greenland	Buckland *et al.* (1983b), McGovern *et al.* (1983)
(13) Qeqertasussak, Greenland	Böcher and Fredskild (1993), Skidmore (1996)
(14) Qilakitsoq, Greenland	Bresciani *et al.* (1989)
(15) Reykholt, Iceland	Buckland *et al.* (1992)
(16) Storaborg, Iceland	Sveinbjarnardóttir *et al.* (1981), Perry *et al.* (1985), Buckland and Perry (1989);
(17) Tjornuvik, Faroe Islands	Buckland and Dinnin (1998)
(18) Toftanes, Faroe Islands	Edwards *et al.* (1998), Vickers *et al.* (2005)

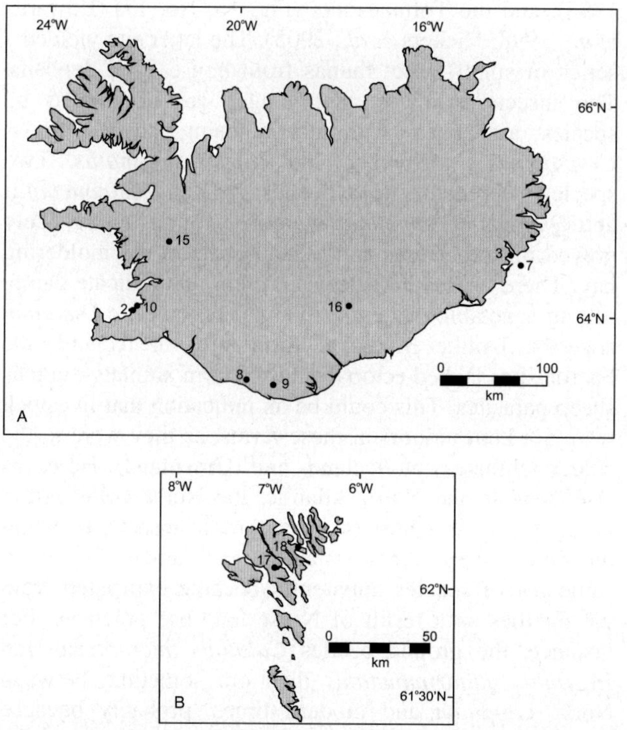

Fig. 15. Map of Iceland (A) and the Faroe Islands (B), showing locations of archeological sites from which insect fossils have been analyzed. Site numbers are keyed to list of sites in Table 5.

Archeological investigations of the remains of farms from Niaqussat and Nipaatsoq (Fig. 14, Nos. 10 and 11), on the southwestern coast, were part of the Inuit-Norse Project (McGovern *et al.*, 1983). The farms date from ca. AD 1000–1350. Insect remains in detritus layers on floors of farm buildings indicate refuse dumping from the living rooms and barns. Several species associated with damp, moldering hay died out in Greenland when the Vikings colonies were abandoned. The midden also contained some native Greenland beetles, including subarctic ground beetles and weevils. In living rooms, abundant byrrhids provide evidence that mosses were brought in, perhaps to fill bedding. Abundant remains of fly puparia (species often associated with human feces) were also found in living rooms. Such flies are found in many unhygienic environments, in latrines, sewers, etc. McGovern *et al.* (1983) considered that ''In a place where it was essential to conserve as much heat as possible, it is probable that visits to the outside world were kept at an absolute minimum during winter, and the use of one room for defecation, even if it was not strictly a latrine, is not surprising.'' The living rooms also showed evidence of rotting meat and feces, covered over by fetid hay.

Historians have pieced together accounts of life in Norse Greenland, although these were mostly not first-hand accounts, because few of these exist, and even fewer discuss the squalid living conditions. Life in 18th century Iceland may not have been too different, as the following account relates: ''The floor was bare, made of packed earth which turned to mud when the roof leaked during rainstorms. The walls were covered with a grayish form of mold, and a greenish, slimy liquid dripped constantly down them, particularly in winter'' (Jónsson, 1892–1894). It is tempting to equate the squalor with a low standard of living, but the decaying vegetation on the floor probably insulated against the permafrost. The relationship between poor hygiene and poor health was not noted in medieval times; Greenland farms were no more squalid than city dwellings in Norse cities, such as Dublin, York, and Oslo.

In the so-called Eastern Settlement, the climate was somewhat milder than in the Western Settlement, which was situated further north. At the bishop's seat of Garðar in the Eastern Settlement, there is evidence for both manuring and irrigation of fields to increase hay production. The fossil insect fauna from this site was described by Buckland *et al.* (2009a). It contained species indicative of wet **eutrophic** peatland with small pools of water, combined with a rich "indoor" (anthropogenic environment) fauna indicative of stored hay and dung.

The demise of the Norse colony in southwest Greenland remains an archeological mystery, as there is no record of whether the colonists died out *in situ*, or went back to Iceland. There is a clear signal in the insect faunal record, however, as the synanthropic fauna was completely replaced by native Greenland species in the abandoned houses and barns (Panagiotakopulu *et al.*, 2007). Buckland *et al.* (1996a) discovered an intriguing succession of fossil flies preserved in bedroom sediments at Nipaatsoq. In lower layers, the dominant fly was *Teleomarina flavipes*, a species of housefly that required warm quarters and was accidentally introduced by the Norse to Greenland. In the penultimate layer, however, *T. flavipes* disappeared, perhaps as room fires died, and two cold-tolerant indoor carrion species came in. It is unknown whether these carrion insects were feeding on human remains, or on animal carcasses. In the top layer, which likely accumulated during a 2-year period encompassing the abandonment of the farm, an outdoor fauna of flies predominated, as if the farmhouse roof had caved in. McGovern (quoted in Pringle, 1997) reconstructed the scenario as follows: a late-winter disaster, in which they eat the cows and they eat the dogs and then the flies get them."

Böcher's (1988b) analysis of fossils from the Inuit archeological site of Qeqertasussuk (Fig. 14, No. 12), occupied from about 4,400–3,200 yr BP, yielded insect assemblages reflecting faunas more diverse than those occupying the site today. Several species of beetles in the fossil assemblages have modern ranges significantly farther south in Greenland. These include the ground beetle, *Trichocellus cognatus* and the rove beetles *Micralymma brevilingue* and *Atheta islandica*. Based on the insect assemblage data, the climate at 3,500 yr BP at Qeqertasussuk was warmer than present. The only arthropod indicators of human disturbance at the site were abundant remains of litter-inhabiting sheet-web spiders (Linyphiidae), and abundant fly pupae, suggesting that maggots fed on decaying meat or other garbage.

The mummified corpses of six Inuit women and two children, dating to the 15th century AD, have been described from the Qilakitsoq site in western Greenland (Fig. 14, No. 13) (Hansen, 1989). Many lice and their eggs were found on these mummies, especially on one woman. Some lice were even in her intestines, indicating their ingestion (apparently not uncommon among Greenlanders) (Bresciani *et al.*, 1983, 1989). Like the human corpses, the lice essentially were freeze-dried, and in an excellent state of preservation. The discovery of lice remains associated with humans or their habitations is not isolated to Greenland, however. Lice have also been found in **coprolites** of Paleoindian Age in Utah (Fry, 1976), and from 16th century Aleutian Island mummies (Horne, 1979). Evidence from the abandoned farms in Greenland indicates that the Vikings also had lice, but they probably brought them from Scandinavia, since the Vikings had little contact with the Inuit.

Norse farms on Greenland, Iceland, and Orkney have yielded flea remains. The Norse Greenland farms had *Pediculus humanus humanus* (Sveinbjarnardóttir and Buckland, 1983; McGovern *et al.*, 1983; Sadler, 1990), the louse that is found on the trunk of humans. It requires temperatures of 29–30°C, whereas the head louse, *P. humanus capitis*, can withstand lower temperatures. These two parasites, in addition to the crab louse, *Phthirus pubis*, are known in antiquity. Lice infestations brought on by squalor and lack of personal hygiene have been ubiquitous for a very long time. They may contribute greatly to the spread of disease, but this association was not well known in medieval times. Poor living conditions were seldom commented on in medieval literature (why state the obvious?) except that authors from one country would comment on unsanitary conditions in another. The economic failure of Norse farms on Greenland was probably the cause of their demise, but louse-borne disease cannot be ruled out.

Another North Atlantic Norse colonization took place in the 9th century AD, on the remote Faeroe Islands, southeast of Iceland. Two Norse archeological sites on the Faeroes have yielded insect faunal assemblages: the Tjornuvik site (Fig. 15, No. 16) (Buckland and Dinnin, 1998), and the Toftanes site (Fig. 15, No. 17) (Edwards *et al.*, 1998; Vickers *et al.*, 2005). The latter site yielded a series of small insect faunas from house floor deposits. The insects in these assemblages are dominated by species associated with decomposing hay, such as *Cryptophagus*, *Atomaria* and *Latridius minutus*. Two species of predatory rove beetles, *Xylodromus concinnus* and *Quedius mesomelinus* were also found. These likely preyed on the beetles and other insects in the moldering hay. There were other species present that indicate damp, rotting vegetable matter, such as the rove beetle *Omalium rivulare*. Unlike Medieval farm sites in Iceland, the Faeroe sites lacked ectoparasites of farm animals, such as sheep parasites. This could be an indication that livestock were not kept indoors at these farms, as they were in the colder climates of Iceland and Greenland. Here, as elsewhere in the North Atlantic, the Norse colonization brought with it a host of synanthropic insects, boosting the diversity of remote island insect faunas. However, some native species apparently became extirpated from the Faeroes as a result of Norse land-use practices. For instance, the ground beetles *Calathus micropterus* and *Harpalus quadripunctatus* died out sometime between Norse *Landnám* and modern times, probably because their habitats have greatly diminished. These beetles prefer shaded landscapes, and the native vegetation that would offer such shade – stands of heather and juniper – has largely been destroyed by sheep grazing.

8. Insect Fossils in New World Archeological Sites (Table 6; Fig. 16)

Most insect fossil studies from New World archeological sites deal with contemporaneous faunas from more-or-less natural environments in the vicinity of the archeological site. In this respect, New World studies differ from most Old World studies, which concern insect

Table 6. North American archeological sites with insect fossil analyses.

Site	Reference(s)
(1) Aubrey Spring, Texas, USA	Elias (1997)
(2) Bamert Cave, California, USA	Nissen (1973)
(3) Big Bone Cave, Tennessee, USA	Faulkner (1991)
(4) Bighorn Basin, Wyoming, USA	Chomko and Gilbert (1991)
(5) Bonnet Plume Basin, Yukon Territory, Canada	Hughes *et al.* (1981)
(6) Boston, Massachusetts, USA	Baine (1998)
(7) Cahokia, Illinois, USA	Pauketat *et al.* (2002)
(8) Danger Cave, Utah, USA	Fry (1976)
(9) Dirty Shame Rockshelter, Oregon, USA	Hall (1977)
(10) False Cougar Cave, Montana, USA	Elias (1990b)
(11) Glen Canyon, Utah, USA	Fry (1976)
(12) Hogup Cave, Utah, USA	Fry (1976)
(13) Huntington Canyon, Utah, USA	Elias (1990b)
(14) Lamb Spring, Colorado, USA	Elias (1986), Elias and Nelson (1989), Elias and Toolin (1989)
(15) Leavenworth, South Dakota, USA	Gilbert and Bass (1967)
(16) Lovelock Cave, Nevada, USA	Heizer (1970)
(17) Lubbock Lake, Texas, USA	Elias and Johnson (1988)
(18) Mesa Verde, Colorado, USA	Evans and Baldwin (1977), Graham (1965)
(19) Ocampo Caves, Mexico, USA	Callen (1970)
(20) Old Crow, Yukon Territory, Canada	Matthews (1975b), Matthews *et al.* (1990a), Morlan and Matthews (1978)
(21) Plainview, Texas, USA	Elias (1995b)
(22) Quebec City, Quebec, Canada	Bain (2001)
(23) Roberts Creek, Iowa, USA	Baker *et al.* (1993)
(24) San Fernando de Vellicata, Baja California, Mexico	Essig (1927)
(25) San Vincente Ferrer, Baja California, Mexico	Essig (1927)
(26) Snowflake, Arizona, USA	Hevley and Johnson (1974)
(27) Tehuacan Valley, Mexico	Callen (1967)

assemblages from ancient human habitations. A few American studies have archeological connections, however. Graham (1965) examined insect fossils from a number of localities in the famous Anasazi Indian sites at Mesa Verde, Colorado (Fig. 16, No. 18). A sequential study of insect assemblages, spanning the interval from the Basketmaker culture through the Pueblo culture, indicates that the synanthropic insect fauna remained virtually unchanged. Stored product pests of corn were found both in dried corn samples and in human coprolites.

While human parasites such as lice were not found in occupation sites at Mesa Verde, louse eggs were observed on mummified corpses. Some buried human remains were heavily infested with various kinds of blowflies (Diptera: Calliphoridae), suggesting that the fly eggs were laid while the body was being prepared for burial, presumably during the summer months (the flies' active period). Other corpses and skeletons were devoid of fly larvae; these were probably buried in winter.

Bark beetle evidence from timbers cut for use in cliff dwellings reveals a number of aspects of this activity. The Anasazi cut down trees from the top of the mesa, and stripped them of bark before placing them in walls. This bark removal probably took place in July or early August, because at this time, phloem-wood boring beetles enter the wood and thereafter feed only on wood. After a year or more, these beetles emerge as adults, cutting a characteristic exit hole. These exit holes were for the most part missing from Mesa Verde timbers, indicating that the bark had been removed in midsummer while the larvae of phloem-wood borers were still mining between the bark and the wood. Furthermore, it appears that the Indians mostly felled the wood in the early summer, and then left it on the ground, allowing the phloem-wood feeding bark beetles to perform the arduous task of loosening the bark.

Dermestid **larval exuviae** were identified by Evans and Baldwin (1977) from a salt cake recovered from an Anasazi jar at Oak Tree House at Mesa Verde. Apparently, the insect remains had been mixed accidentally with loose, granular salt as it was moistened and molded into a cake.

8.1. Coprolite Studies

The study of ancient human feces, preserved in dry caves as coprolites, has yielded some interesting information on insects and their interactions with Amerindians. Some arthropods were consumed as food. Others, such as fleas, lice, and mites, were external parasites that were either accidentally or deliberately consumed.

Stiger (1977) found the remains of cicadas and grasshoppers in Anasazi coprolites from Mesa Verde.

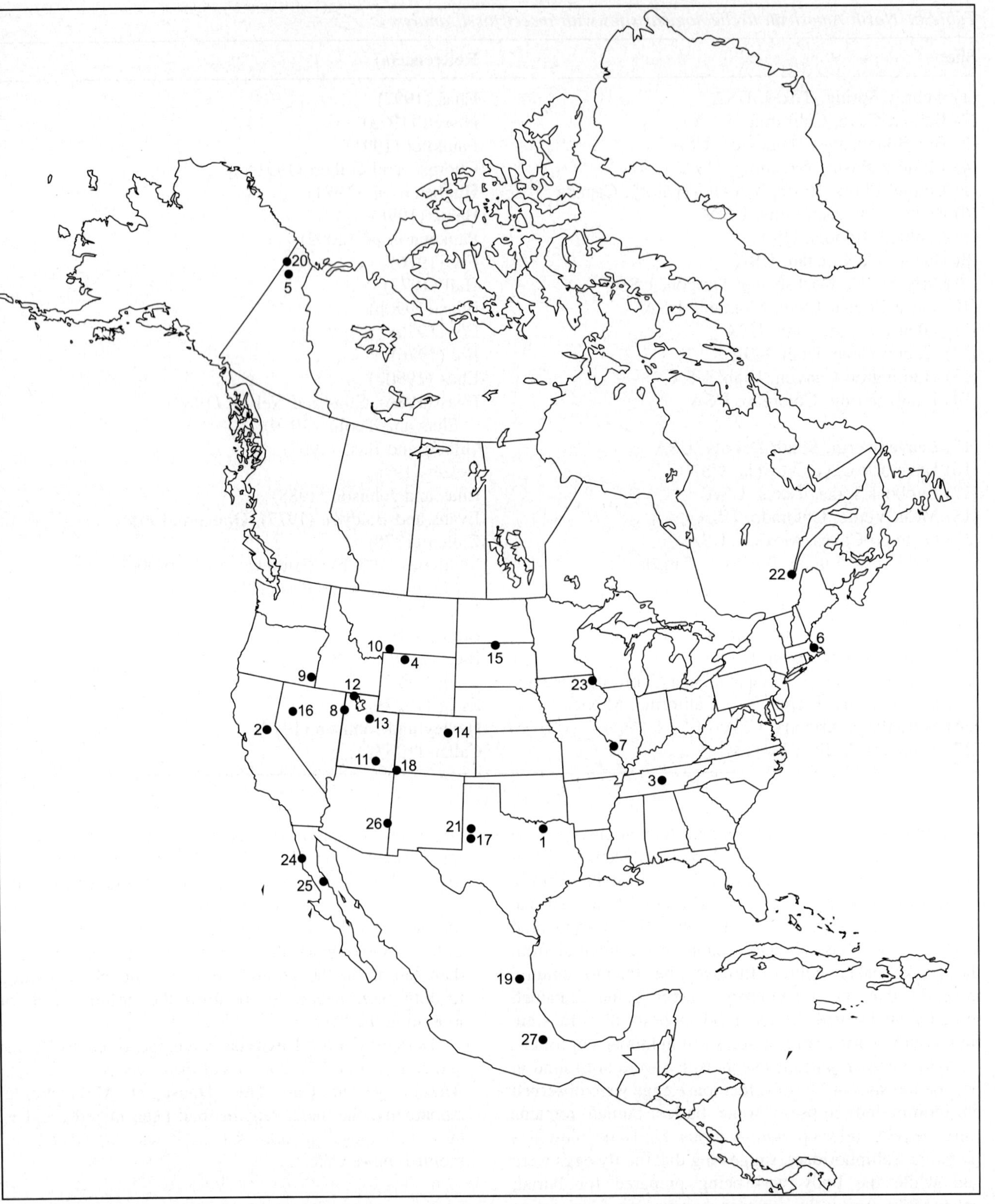

Fig. 16. Map of North America, showing locations of archeological sites from which insect fossils have been analyzed. Site numbers are keyed to list of sites in Table 6.

He suggested that insects became increasingly important in the Anasazi diet through time, and that regional environmental change may have played a role in this sequence. As shrub-grassland became dominant, the number of grasshoppers would have increased. Evidence of increased numbers of turkey bones in coprolites also led Stiger to suggest that domesticated turkeys were used to control grasshoppers on cultivated lands. Grasshoppers were apparently an important part of the diet of ancient peoples in the Great Basin. Madsen and Kirkman (1988) found the remains of enormous numbers of grasshoppers in caves near Great Salt Lake. They concluded that native peoples consumed dried, salted grasshoppers that had drowned in the lake, and had been washed up on the shore in their millions.

A number of mites were found in human coprolites from Mesa Verde (Morris, 1986). These have not been specifically identified, but some belong to the genus *Trombicula*, known as chiggers in the American southwest. The larvae of these mites parasitize humans by burrowing beneath the skin to feed on blood and interstitial fluid.

Paleoindian coprolites from the Dirty Shame Rockshelter, Oregon (Fig. 16, No. 9), have been dated at 9,500 yr BP. Some contained abundant remains of red ants (Hymenoptera: Formicidae: *Formica* sp.); others were comprised almost completely of indigestible parts of the termite species *Reticulitermes* cf. *tibialis* (Isoptera) (Hall, 1977). Termites comprised 78% of one coprolite, which indicates that Paleoindians sometimes made a complete meal of termites. Consumption of insects by late prehistoric Amerindians from western North America was documented by European historians who first made contact with Amerindians, and corroborated by fossil evidence (Stewart, 1938; Madsen and Kirkman, 1988). For example, Nissen (1973) found remains of craneflies (Diptera: Tipulidae) in Amerindian coprolites from Bamert Cave, California (Fig. 16, No. 2).

In the Great Basin region of western North America, many dry caves contain human coprolites. In Utah, Fry (1976) analyzed coprolites found in Danger and Hogup Caves, as well as from caves in the Glen Canyon region (Fig. 16, Nos. 8, 12, and 11). A 2,000-year-old coprolite from Danger Cave contained an egg case (nit) of *Pediculus humanus*, the human louse, still attached to a human hair. A 5,000-yr-old coprolite from Danger Cave contained the complete exoskeleton of a mosquito. The ingestion of the mosquito must surely have been accidental; it would take a great many mosquitoes to make a meal! Fry (1976) also found that human mummies from caves in Glen Canyon had louse-infested scalps.

Heizer (1970) summarized the results of insect fossil studies from coprolites in Lovelock Cave, Nevada (Fig. 16, No. 16). Some samples, dating between 1200 and 150 yr BP, contained the exoskeletons of the large predaceous diving beetle, *Cybister*. The heads were missing, which suggests that, being less palatable, they were removed before the beetles were swallowed.

Human coprolites dated at about 2,000 yr BP have been studied from Big Bone Cave, Tennessee (Fig. 16, No. 16). Faulkner (1991) examined eight samples in order to reconstruct diet, disease and lifestyle. In addition to food items in the feces, fleas, weevil larvae, and fly fragments were found. The weevil larvae were probably derived from weevil-infested grain in the diet. The weevil species is in a group known to attack seeds of legumes and other plants that could be stored for food. The fleas found were woodrat (*Neotoma*) fleas, and may represent post-depositional invasion of the feces in the cave. In contrast, they may have been eaten by the Indians. Non-insect parasites, such as pinworm eggs, roundworm eggs, hookworm eggs, and *Giardia* cysts were also found in the fecal samples.

In Mexico, coprolites from cave deposits spanning much of the Holocene have been investigated for insect remains. Callen (1970) studied coprolites from caves near Ocampo, Tamaulipas (Fig. 16, No. 19). Some of the coprolites were riddled with exit holes and pupal cases from blowflies and flesh flies. Larval exuviae of the latrine fly, *Fannia scalaris*, were also abundant, in addition to the remains of the housefly, *Musca domestica*, and fungus gnats. Several species of dermestid beetles were identified. Among these, the larvae of the odd beetle, *Thylodrias contractus*, (Dermestidae) is the most remarkable. This beetle, a scavenger on dried animal material and fur, is thought to have been carried on dried animal skins by humans migrating over the Bering Land Bridge into the New World (Callen, 1970).

Another set of sites with preserved coprolites was discovered in the Tehuacan Valley of central Mexico (Fig. 16, No. 27). Insect fossils have been identified from coprolites ranging in age from 8,700 to 450 yr BP. Early coprolites (>5,400 yr BP) contained ant heads, dung beetle larvae, caterpillars, and ticks. Coprolites dating from 2,900 to 2,200 yr BP contained dung beetles, friutfly (*Drosophila*) larvae, latrine fly larvae, and chewing lice (Mallophaga). The lice were probably bird parasites, accidentally ingested by people. Coprolites dating from 2,200 to 1,300 yr BP contained sucking lice (Anoplura), fleas, ticks, flies, beetles, ants, and pseudoscorpions. The beetles were sufficiently broken up to demonstrate that they had been chewed. Beetle exoskeletons formed the bulk of several individual coprolites. Some of the fossil ants resembled honey ants (*Myrmecocystus* sp.), whose abdomens store honeydew, a sugary excretion harvested by the ants from the abdomens of aphids. These ants may have been deliberately consumed for their sweet taste.

8.2. *Insect Remains from Prehistoric Buildings*

In Arizona, insect remains have been extracted from debris middens in rooms in a Prehistoric period pueblo near the modern town of Snowflake (Fig. 16, No. 26) (Hevley and Johnson, 1974). A small assemblage of darkling beetles, ants, and ichneumon wasps was found. The authors speculated that these scavenging insects were attracted to food resources in the Pueblo, and that the parasitic wasps then preyed on the scavengers. They noted that insects are found in many such midden deposits, but that this line of research had yet to be pursued.

Indian pueblo buildings have traditionally been made of adobe, a mixture of clayey mud and straw, sun-baked into bricks, which are made into walls that are then plastered over with more mud to form a smooth finish. When the Spanish came to the American southwest, they adopted the use of adobe in making their forts, houses, and mission churches. Insect fossils retrieved from old adobe walls from Spanish mission churches of San Fernando de Vellicata and San Vincente Ferrer (Fig. 16, Nos. 25 and 24) in Baja California (Essig, 1927) comprise species that lived in the streambank mud that was used to make the adobe, mud wasps that made nests in the fresh adobe mud, and stored product pests including the granary weevil (*Sitophilus granarius*) and the rice weevil

(*Sitophilus oryzae*). Essig speculated that these pests were introduced into the mission communities by priests who brought them in bags of contaminated grain. Since the fossil record has shown that these species have Old World origins, his speculation seems to have been correct.

8.3. *Insect Remains from Human Burials, Food Caches, and Refuse Pits*

The remains of fly maggots and their pupal exuviae (the skin cast from the larva as it pupated) have been used in some New World sites to help determine mode and timing of burial. Gilbert and Bass (1967) studied fly remains from a Late Prehistoric Arikara Indian burial site at Leavenworth, South Dakota (Fig. 16, No. 15). Exuviae of many flies were found, placing the time of burial between late March and mid-October (the flies' active season in South Dakota).

In the Bighorn Basin of Wyoming (Fig. 16, No. 3), a Late Prehistoric bone pit was analyzed by Chomko and Gilbert (1991). The contents of a bone-filled pit were thought to be the remains of a meal cooked in a skin container. Then the residue was dumped in the pit and covered with a slab of stone. Abundant remains of blowfly puparia found at the bottom of the bone mass provide evidence that the bone mass was formed outside the pit, because the flies could not have burrowed that deeply through bones to lay their eggs at the bottom. Flies are attracted to rotting meat, so it must have been exposed a few days before burial. Development of maggots and maturation through several **instars** until pupation requires at least 21 days in warm weather. So the contents of the container were exposed at least 3 weeks before being dumped in the pit. This series of events probably took place between June and September, the active season for blowflies in Wyoming. If it had occurred in winter or spring, secondary carrion beetles such as dermestids would have been attracted to the bone mass after it thawed in spring. These beetles feed on dried carrion. Because of their niche in natural environments, dermestids were predisposed to become pests in dried meats and animal skins.

The Mound Builders of the Mississippi Valley region developed a fascinating, highly advanced agricultural society in the centuries immediately before European contact. A massive ceremonial plaza was constructed at the Cahokia site in Illinois (Fig. 16, No. 7). Monks Mound was the central focus of this ceremonial center. It is the largest man-made earthen mound in North America, standing 30 m high, with a base of 316×241 m. The surrounding plaza was 19 hectares in size. The creation of these massive features took tremendous human effort. Insect fossil analyses of a nearby refuse trench helped archeologists determine the time of year in which this work was carried out (Pauketat *et al.*, 2002). I studied seven bulk samples from six sampling zones at the Cahokia ditch site, dated ca. AD 1050–1100. The samples yielded beetles, fly puparia, and ants. Nearly all of the submound pit zones contained sizable quantities of insects that are active during the warm months between April and October. The insect and other evidence indicate that the pit was dug, sat open for at least a year, and was then in-filled rapidly, probably over a 2–10 year span. The heterogeneous textures and subzones within the final zone probably represent several filling episodes, the additional fill being packed into the pit as it settled. Otherwise, as evidenced by the zone contents, the former borrow pit was filled as part of large-scale, detritus-rich, single-event dumping activities. The pit's proximity to the plaza probably accounts for the pit's rapid infilling.

8.4. *Insect Remains from European Colonial Sites*

The youngest fossil insect assemblages studied from archeological sites in North America come from 17th century Boston (Fig. 16, No. 6) and 19th century Quebec City (Fig. 16, No. 22). The Boston fauna is dated to about AD 1650 – less than 30 years after the Pilgrims landed at Plymouth Rock. This is a most remarkable fauna, in that it contains 24 species, 22 of which are Old World introductions (Table 7). Many familiar species names appear on this list – synanthropic beetles that have previously been mentioned in this chapter, in connection with European and Egyptian archeological sites. These include the rove beetle, *Quedius mesomelinus*, a predator on fly maggots and other insect larvae in rotting vegetation and animal carcasses, *Dermestes lardarius* found in animal carcasses and cured meats (bacon and ham), and the stored product pests, *Tenebroides mauritanius*, *Ptinus fur*, *Trox scaber*, and *Sitophilus granarius*. The two New World species on the faunal list are also associated with synanthropic habitats, and have become cosmopolitan in the last few centuries. These are the rove beetle, *Carpelimus obesus*, and the flat bark beetle, *Nausibius clavicornis*. The former species lives in rotting vegetation and dung, where it preys on small insects. The latter species is found in dried foods of all kinds. It probably originated in the tropical regions – perhaps the Caribbean Islands – and has since become established in indoor environments throughout the world, and in outdoor habitats throughout the tropics (Halstead, 1993).

Bain's (2001) paper on the insect remains she recovered from late 19th century insect assemblages from Quebec City mainly concerns sanitation issues. She was investigating sanitation practices by studying insect remains from organic deposits taken from house sites, warehouses, privies and wells, and outbuildings. Along with dung- and carrion-feeding beetles and stored product pests, she found the remains of *Lytta vesicatoria*, Spanish Fly.

Baker *et al.* (1993) and Schwert (1996) discussed the results of a study of the impacts of agricultural development in Iowa on the native beetle fauna. While not an archeological study *per se*, their research documents the deterioration of woodland and freshwater habitats in the American Midwest, following clearance of trees and ploughing of fields for agriculture. This land use change lowered the diversity of ground beetles and essentially eliminated riffle beetles (Elmidae) from the study site at Roberts Creek. The introduction of Old World species took place here, even in a rural setting. Ten species of beetles, including rove beetles, dung beetles, a leaf beetle, and four weevils, were introduced into the region in the late 19th or early 20th century.

Table 7. Beetle species found in 17th century faunal assemblage from colonial Boston, Massachusetts.

Species	Family	Origins and habitat
Bembidion tetracolum	Carabidae	An Old World species, introduced before 1834; Almost always on cultivated fields, waste ground, or near human habitation
Cercyon haemorrhoidalis	Hydrophilidae	Old World species found in cow dung
Cercyon terminatus	Hydrophilidae	Old World species found in cow dung and decaying grass
Cercyon analis	Hydrophilidae	Old World species found in rotting vegetation
Carpelimus obesus	Staphylinidae	Native North American species since become cosmopolitan in rotting vegetation and dung
Oxytelus sculptus	Staphylinidae	Old World species found in cow dung and decaying grass
Anotylus rugosus	Staphylinidae	Old World species found in dung and decaying grass (preys on maggots)
Philonthus politus	Staphylinidae	Old World species found in dung and decaying grass, fungi, carrion (preys on maggots)
Creophilus maxillosus	Staphylinidae	Old World species found in dung and carrion (preys on maggots)
Quedius mesomelinus	Staphylinidae	Old World species found in variety of synanthropic habitats (preys on insect larvae)
Dermestes lardarius	Dermestidae	Old World species common in bacon, ham, cheese, animal carcases
Tenebroides mauritanicus	Ostomidae	Old World species that lives in stored grain
Carpophilus hemipterus	Nitidulidae	Old World species that lives in dried fruits
Omosita colon	Nitidulidae	Old World species found in carrion, dried seaweed, decaying vegetable matter
Monotoma picipes	Cucujidae	Old World species found in grass heaps, decaying vegetation
*Nausibius clavicornis**	Cucujidae	New World species found in stored food products
Oryzaephilus surinamensis	Cucujidae	Old World species that lives in stored grain
Latridius minutus	Lathridiidae	Old World species that lives in stored grain
Ptinus fur	Ptinindae	Old World species that lives in wide variety of stored products
Gnatocerus cornutus	Tenebrionidae	Old World species that lives in stored grain, milled flour
Trox scaber	Scarabaeidae	Old World species found in carrion and rubbish
Aphodius granarius	Scarabaeidae	Old World species found in cow dung
Sitophilus granarius	Curculionidae	Old World species that lives in stored grain
Sitophilus oryzae	Curculionidae	Old World species that lives in stored grain

Source: Data from Bain (1998) and Buckland and Buckland (2006)
**N. clavicornis* is a New World species-carried to all parts of the World by commerce, becoming established in warmer regions. Afrotropical region, Oriental, Australasian, Nearctic and Neotropical (Halstead, 1993)

8.5. Insects from Nonanthropogenic Deposits at Archeological Sites

Insect fossils from nonanthropogenic environments (i.e., undisturbed, natural habitats) provide data useful in building an environmental framework in which to place archeological interpretations of a site. Insects, along with other biological proxy data, provide inferences on climate, soil, and regional vegetation.

I have studied a number of insect fossil assemblages from this type of deposit, including samples from the Aubrey Spring Clovis site near Denton, Texas (Fig. 16, No. 1). Abundant insect assemblages were recovered from pond sediments. The insects range in age from 17,250 to 16,200 cal yr BP. This predates a Clovis Paleoindian occupation on a hill adjacent to the ancient pond. Unfortunately, the pond sediments dating to the time of Clovis occupation contained few insects.

The older sediments yielded an abundant, diverse fauna indicative of open ground environments surrounding a sedge-lined pond. The modern distributions of the insects suggest a climate with summers about 10°C cooler than present at 17,250 cal yr BP, then warming 2°C by 16,400 cal yr BP. This scenario agrees with a contemporaneous grassland pollen spectrum, but adds some additional climatic definition (Elias, 1997).

I have also examined insect fossils from a Clovis archeological site at False Cougar Cave, Montana (Fig. 16, No. 10) (Elias, 1990b). The insects were recovered from cave floor sediments, which also contained Paleoindian artifacts. The age of the fossil insect sample is 10,500 yr BP. The fauna is small but provides an environmental reconstruction that is internally consistent. The insects in the assemblage came from outside the cave (there are no cave dwellers). Based on the insects, the environment was very similar to modern conditions by 12,400 cal yr BP.

At the Lubbock Lake site in west Texas (Fig. 16, No. 17), I have sampled insects from an organic horizon including bison bones butchered by Clovis Indians. The age of this sample is ca. 12,100 cal yr BP. Only a few beetle species were identified (Elias and Johnson, 1988).

One of these is the ground beetle species, *Calosoma porosifrons*, a caterpillar hunter. Its modern range is limited to mountain slopes in the state of Durango, Mexico. The climate there is cooler and wetter than modern day Lubbock. Younger (nonwaterlogged) strata at Lubbock Lake apparently are oxidized too heavily for good preservation of insects.

At the Plainview site in west Texas (Fig. 16, No. 21), I recovered insect fossils from organic horizons in a sand and gravel pit. The samples range in age from 11,500 to 10,000 cal yr BP, contemporaneous with Paleoindian archeology at Plainview but not in direct association with artifacts. A limited beetle fauna shows climatic conditions similar to modern.

Lamb Spring, Colorado (Fig. 16, No. 14), is a multicomponent archeological site, with a basal bone bed including many extinct megafaunal mammals. Stanford *et al.* (1982) described possible Pre-Clovis artifacts at the site. A radiocarbon date on mammoth bone was 13,100 ± 1,000 yr BP (15,650 ± 1,382 cal yr BP); plant macrofossils yielded an age of 12,750 ± 150 (15,164 ± 387 cal yr BP). I sampled insect fossils from the bone bed clay horizon (Elias, 1986). This fauna was ecologically mixed, including both prairie and alpine tundra species. AMS dating of fossil head capsules, thoraces, and elytra of selected prairie species yielded an age of 17,850 ± 550 (21,402 ± 760 cal yr BP); AMS dating of specimens of alpine tundra species yielded an age of 14,500 ± 500 (17,741 ± 592 cal yr BP) (Elias and Nelson, 1989; Elias and Toolin, 1989). These first published AMS dates made directly from insect fossil exoskeletons indicate subtle reworking of the Lamb Spring sediments by spring activity, although this reworking was not observed in the stratigraphy.

Huntington Canyon, Utah (Fig. 16, No. 13), is a mammoth site with possible archeological connections, and is the highest elevation mammoth site yet found in North America (2,950 m above sea level, asl). A projectile point (Pryor Stemmed Point) was found at the site, in close proximity to a mammoth skeleton. As with the Lamb Spring site, there are dating problems. Spruce wood from below the mammoth skeleton was dated at 9,700 yr BP (11,000 cal yr BP), but mammoth bone collagen has yielded an AMS date of 11,200 yr BP (13,100 cal yr BP) (Gillette and Madsen, 1992). I extracted insect fossils from the mammoth's skull cavity, and selected insect exoskeletons have yielded an AMS date of 10,500 yr BP (12,400 cal yr BP). The insects are indicative of climate similar to modern conditions, and reflect a shallow, sedge-lined pool in which the mammoth died.

Another set of archeological sites originally believed to be of Pre-Clovis Age are the Old Crow and Bonnet Plume Basin sites in the Yukon Territory (Fig. 16, Nos. 20 and 5). Initial excitement over very old bone "artifacts" led to extensive archeological and paleontological investigations of these sites, including insect fossil analyses (Morlan and Matthews, 1978), although the insect fossils did not come from horizons containing artifacts. At present, most if not all of these bones are no longer considered bona fide artifacts, and other bones clearly modified by humans have been shown to be of Holocene Age. These artifacts were not found *in situ*, but had weathered out of the bluffs at Old Crow. The age range of insects studied from Old Crow is from > 40,000 to 30,000 ^{14}C yr BP (Matthews, 1975b; Morlan and Matthews, 1983; Matthews *et al.*, 1990a). Regardless of archeological problems, the Yukon sites in the Old Crow and Bluefish Basins remain some of the most valuable paleontological sites in Eastern Beringia. These sites are discussed further in the chapter dealing with Beringian paleoecology.

9. South American Archeology

At Monte Verde, Chile, an archeological site dating to ca. 13,000 yr BP (15,900 cal yr BP), described by Dillehay (1986, 1989), contains abundant wooden artifacts in peat. Peat samples were taken by Ashworth *et al.* (1989) for fossil insect analysis (Fig. 17). Samples from units MV-5 and MV-6 consisted of organic-rich sands and peats, coeval with the 15,900 cal yr BP artifacts. The identified beetle taxa represent two types of aquatic habitats: shallow, vegetation-rich water and riffles in a shallow stream. In addition, riparian beetles were found which

Fig. 17. The Monte Verde archeological excavation. An excavated hearth-like structure is visible in the foreground. The peat MV-5 that contains abundant fossil beetles and is dated between 11 and 810 to 10,860 yr BP is visible at the base of the cut on the far wall of the excavation (photograph by A. C. Ashworth, from The Encyclopedia of Quaternary Science, ©Elsevier, 2007).

represent open habitats. Rain forest-inhabiting species in the fauna are dominated by *Nothofagus* (southern beech)-associated taxa, including beetles that attack dead branches, decaying trunks, rotting wood, and leaf litter.

The environmental reconstruction based on the insect fauna is a shallow creek with sparsely vegetated sand bars and some boggy margins, flowing through rain forest. This reconstruction is consistent with that based on plant macrofossils, sediments and diatoms (Dillehay, 1989). The lateglacial paleoclimate was interpreted as very similar to modern; an interpretation in agreement with all of the other lines of evidence except for the pollen interpretation. Heusser (1989) concluded from pollen studies that the climate was in transition from glacial to postglacial at 13,000 yr BP, with substantially colder climates during the oldest human occupation (14,000–13,500 yr BP). In rebuttal, Ashworth *et al.* (1989) stated:

> The stratigraphic arrangement of samples in the base of MV-5, shown by Heusser, was interpreted to fit a preexisting regional pollen stratigraphy; only then do the cold-indicator species appear to be restricted to the basal portions of MV-5. Because of the broad temporal span (about 1500 years) represented by the indicator species, we do not believe that Heusser's argument for a climatic change is particularly convincing.

No synanthropic insect species were found in the Monte Verde assemblages. This is explained by Ashworth *et al.* (1989) through the following arguments:

(1) European sites with synanthropes are all sites of accumulation of dung and rotting meat (wells, middens, etc.), whereas no such accumulations were found at Monte Verde.
(2) Waterlogged meat scraps and dung in the bog would not attract these insects.
(3) There may have been no species preadapted to synanthropy available in Chile in lateglacial times.

Any or all of these factors may have played a role in precluding synanthropic insects from the site.

10. Comparison with European and other Old World Studies

Most New World studies are from the Late Pleistocene or earliest Holocene, while nearly all of the Old World Studies are Mid- to Late Holocene in age. Differences in types of human occupation and lifestyles between these times are important. In Europe, the more-or-less sedentary lifestyles from the Bronze Age through Medieval times led to the accumulation of organic debris (in wells, cesspits, latrines, trash heaps, etc.) and associated insect faunas. In contrast to this, Paleolithic hunter-gathers leading a transhumance lifestyle left very little organic debris on the landscape. Regrettably, the only substantial artifact records of these people are their stone tools. Their temporary shelters did not foster synanthropic insects, so that coeval insect faunas generally reflect undisturbed habitats.

Old World sites contain abundant synanthropic insect species; New World sites contain few, and most of these are human parasites in feces (most of which are not insects, anyway). There appears to have been a lack of adaptation of New World insects to synanthropic lifestyles until much later in the Holocene. This was probably because of a lack of suitable human-induced habitats, at least in western North America, until late in prehistory. As we have seen, only a handful of late prehistoric and early historic archeological sites have been analyzed in the New World.

The lack of human influence on natural environments until the Late Holocene has both advantages and disadvantages. It allows researchers studying most New World sites to examine natural biotic responses to macroclimate without the confounding effects of humanity. In contrast, insects studied thus far from New World archeological sites rarely shed much light on human living conditions and activities.

8

European Studies

The last word has not been said and much further work has still to be done.

– van Geel *et al.* (1989)

The study of Quaternary insect fossils is most fully developed for sites in Western Europe, especially Great Britain. These studies have made an impact on regional Quaternary studies, especially in the study of climate change. As elsewhere, most European research has focused fossil assemblages dating to the last glacial–interglacial cycle, but numerous others of Middle Pleistocene age have also been investigated. This chapter presents an outline of these studies.

1. British Research

The knowledge of Quaternary insects has developed furthest in the British Isles. In the 1994 book, *Quaternary Insects and Their Environments*, I reported that more than 50 studies have been carried out in the UK and Ireland. The number has more than doubled to at least 125 studies for which full faunal lists have been published (Table 1), mostly covering faunas from the last interglaciation onwards. The only important gap in the British record of the Late Quaternary is the latter half of the Holocene. While there is no lack of fossil material, it is nearly impossible to discern whether these insect assemblages reflect natural or anthropogenic changes to the landscape during the last several thousand years.

I am sure that my British colleagues would argue, quite rightly, that there is still much to be done there. However, because of its depth and clarity, the British record provides some unique insights, not only into the timing and intensity of regional environmental changes, but also into the responses of insect species and communities to those changes. When combined with other paleontological studies, the British Quaternary insect record also reveals the different ways that these ecosystem components have responded to change.

As Coope (1988) pointed out, during the Late Quaternary, the British Isles have been on the receiving end of some of the greatest fluctuations in sea surface temperatures anywhere in the world. This is because at various times, Great Britain either has basked in the warmth of the Gulf Stream (as it does today) or has been chilled by the icy waters north of the Atlantic polar front (see Ruddiman and McIntyre, 1981). The effects of ocean currents, combined with Milankovitch cycles in solar radiation through the Late Quaternary, have created climatic episodes in Britain varying from Mediterranean warmth to Arctic cold (Coope, 1987b).

In Britain, as elsewhere, it is very difficult to assign a reliable age to most fossil assemblages older than the last interglaciation. Many events that occurred within the last 40,000 years can be dated by radiocarbon (given suitable quantity and quality of organic preservation), but beyond this are hundreds of thousands of years that fall in a gap between radiocarbon and other isotopic dating methods. Some sediments and fossils in this age range have been successfully dated through **thermoluminescence** (Forman, 1989), **amino acid racemization** of proteins in shells and bone (Miller and Brigham-Grette, 1989; Geyh and Schleicher, 1990; Preece and Penkman, 2005), and **tephrochronology** (Walter, 1989). However, many pre-last interglacial deposits have not been dated accurately. For instance, Briggs *et al.* (1975b) described an interglacial deposit containing insect fossils, plant macrofossils, pollen, molluscs, and vertebrates from Sugworth, England (Fig. 1, No. 104). Based on stratigraphic position and correlation with other floras and faunas, this site was placed in the Cromerian interglacial (several interglaciations prior to the last interglaciation). But, because there is no radiometric dating of the site, this remains a tentative assignment. See Fig. 2 for chronostratigraphic correlations between British and Continental European sequences, and Marine Isotope Stages.

Coope (2006, 2007b) has reviewed his work on Middle Pleistocene insect faunas associated with sites of Ancient Human Occupation of Britain (the AHOB project; Stringer, 2005). These faunas could have been discussed in Chapter 7, but there appears to have been no appreciable anthropogenic influence on these faunas, so they are essentially paleontological assemblages, not archeological. He studied five Pre-Anglian (i.e., older than MIS 12) faunas in the English Midlands and East Anglia, including Waverley Wood, Brooksby, and High Lodge in the Midlands (Fig. 1, Nos. 115, 21, and 58), and Pakefield and multiple assemblages from Happisburgh site 1 on the coast of East Anglia (Fig. 1, Nos. 80 and 56). The three inland sites were situated in the valley of the ancient Bytham River, before glacial ice forced an alteration in its course. The precise age of these faunas remains unknown, but the Pakefield fauna stands out, because the beetle fauna is indicative of warmer-than-modern temperatures, and it includes several thermophilous species not found in Britain today. Based on correlations with similar faunas, particularly the Sugworth fauna, Coope (2006) has assigned the Pakefield fauna to the Cromerian Interglacial Complex (MIS 13). The faunas

DEVELOPMENTS IN QUATERNARY SCIENCES
VOLUME 12 ISSN 1571-0866

Table 1. Locations and references of British Quaternary insect faunas.

Site	Reference(s)
(1) Abingdon	Aalto *et al.* (1984)
(2) Aghnadarragh	McCabe *et al.* (1987)
(3) Alcester	Shotton *et al.* (1977)
(4) Ardleigh	Coope (2007b)
(5) Armthorpe	Gaunt *et al.* (1971), Buckland (2001)
(6) Aston Mill	Whitehead (1989)
(7) Austerfield	Gaunt *et al.* (1972)
(8) Aveley	Coope (2001)
(9) Barling	Bridgland *et al.* (2001)
(10) Barnwell Station	Coope (1968a)
(11) Baston Fen	Briant *et al.* (2004)
(12) Beckford	Briggs *et al.* (1975a)
(13) Beeston	Coope (2000a, 2007b)
(14) Bigholm Burn	Bishop and Coope (1977)
(15) Birmingham	Osborne (1980c)
(16) Bobbitshole	Coope (1974a)
(17) Brandon	Coope (1968b, 1989), Osborne and Shotton (1968), Maddy *et al.* (1994)
(18) Brighouse Bay	Bishop and Coope (1977)
(19) Brimfield	Shotton *et al.* (1974)
(20) Brimpton	Bryant *et al.* (1983)
(21) Brooksby	Coope (2006)
(22) Broomfield	Coope (2006)
(23) Burnhead	Coope (1962a)
(24) Cassington	Maddy *et al.* (1998)
(25) Chelford	Coope (1959), Worsley *et al.* (1983)
(26) Church Stretton	Osborne (1972)
(27) Clettnadal, Shetland	Whittington *et al.* (2003)
(28) Coleshill	Coope and Sands (1966)
(29) Colnbrook	Coope (1982)
(30) Colney Heath	Pearson (1967)
(31) Corstorphine	Coope (1968c)
(32) Croyden	Peake and Osborne (1971)
(33) Davenham	Hughes *et al.* (2000)
(34) Deeping St. James	Keen *et al.* (1999)
(35) Derryvee, Ireland	Colhoun *et al.* (1972)
(36) Dimlington	Penny *et al.* (1969)
(37) Dowdeswell	Hutchinson and Coope (2002)
(38) Drumurcher, Ireland	Coope *et al.* (1979)
(39) Earith	Coope (2000b)
(40) Elsing	Taylor and Coope (1985), Coope (2001)
(41) Falcon's Hall	Taylor and Coope (1985)
(42) Farmoor	Coope (1976)
(43) Farnham	Moseley (1983)
(44) Fladbury	Coope (1962b, 1994), Ullrich and Coope (1974)
(45) Fliquet Bay, Jersey	Coope *et al.* (1980)
(46) Flixborough	Buckland (1982b, 1984)
(47) Folkestone	Coope (1980)
(48) Four Ashes	Morgan (1973)
(49) Frog Hall Pit	Keen *et al.* (1997)
(50) Glanllynnau	Coope and Brophy (1972), Coope (1977b)
(51) Glen Ballyre	Coope (1971a)
(52) Gransmoor	Walker *et al.* (1993)
(53) Great Billing	Morgan (1969)
(54) Great Totham	Bridgland *et al.* (1994)
(55) Hackney, London	Green *et al.* (2006)
(56) Happisburgh	Coope (2006)
(57) Hawks Tor	Coope (1977a)
(58) High Lodge	Coope (2007b)
(59) Holywell Coombe	Coope (1998)

(*Continued*)

Table 1. (Continued)

Site	Reference(s)
(60) Kempton Park	Gibbard *et al.* (1981)
(61) Kirkby on Bain	Girling (1980a)
(62) Isleworth	Coope and Angus (1975)
(63) Itteringham	Coope (2001)
(64) Kilmaurs	Bishop and Coope (1977)
(65) Latton	Lewis *et al.* (2006)
(66) Lea Marston	Osborne (1973, 1974)
(67) Lea Valley	Coope and Tallon (1983)
(68) Leeds: Oxbow	Gaunt *et al.* (1970)
(69) Low Wray Bay	Coope (1977a)
(70) Marlow	Coope *et al.* (1997)
(71) Marsworth	Green *et al.* (1984)
(72) Mathon	Coope *et al.* (2002), Coope (2007b)
(73) Maxey	Davey *et al.* (1991)
(74) Messingham	Bateman *et al.* (2001), Buckland (1982b, 1984)
(75) Minworth	Coope and Sands (1963, 1966)
(76) Nechells	Shotton and Osborne (1965)
(77) Needham	Taylor and Coope (1985)
(78) North Lechlade	Briggs *et al.* (1985)
(79) Northmoor	Coope (1976)
(80) Pakefield	Coope (2007b)
(81) Pools Farm	Coope (2007b)
(82) Porthmeare	Osborne (1976)
(83) Queensford	Briggs *et al.* (1985)
(84) Radwell	Rogerson *et al.* (1992)
(85) Red Moss	Ashworth (1972)
(86) Redkirk Point	Bishop and Coope (1977)
(87) Roberthill	Bishop and Coope (1977)
(88) Rodbaston	Ashworth (1973)
(89) Saint Aubins Bay, Jersey	Coope *et al.* (1985)
(90) Saint Bees	Pearson (1962), Coope and Joachim (1980)
(91) Sandy	Gao *et al.* (1998)
(92) Sanquhar	Bishop and Coope (1977)
(93) Scrooby	Gaunt and Girling (1996, 1997)
(94) Shortalstown	Coope (1971b)
(95) Shropham	Coope (1995, 2001)
(96) Shustoke	Kelly and Osborne (1965)
(97) South Kensington	Coope *et al.* (1997)
(98) Sproughton	Rose *et al.* (1980)
(99) Stafford	Morgan (1970)
(100) Standlake Common	Briggs *et al.* (1985)
(101) Stanton Harcourt	Briggs *et al.* (1985)
(102) Star Carr	Dinnin and Welsh (2001)
(103) Stoke Goldington	Green *et al.* (1996)
(104) Sugworth	Briggs *et al.* (1975b), Osborne (1980a), Coope (2007b)
(105) Syston	Bell *et al.* (1972)
(106) Tame Valley	Coope and Sands (1966)
(107) Tattershall Thorpe	Coope (2001)
(108) Teith Valley	Merritt *et al.* (1990)
(109) Tiln	Howard *et al.* (1999)
(110) Torrie Borehole	Merritt *et al.* (1990, 2003)
(111) Trafalgar Square	Franks *et al.* (1958), Coope (2001)
(112) Twyning	Whitehead (1992)
(113) Upper Strensham	de Rouffignac *et al.* (1995)
(114) Upton Warren	Coope *et al.* (1961)
(115) Waverley Wood	Shotton *et al.* (1993), Keen *et al.* (2006)
(116) West Bromwich	Osborne (1980b)
(117) West Drayton	Coope (1982)
(118) West Runton	Coope (2000a)

(Continued)

Table 1. (*Continued*)

Site	Reference(s)
(119) Westwoodside	Bateman *et al.* (2001)
(120) Whitacre Heath	Coope and Sands (1966)
(121) Wilden	Shotton and Coope (1983)
(122) Wilsford	Osborne (1969)
(123) Woodston	Horton *et al.* (1992)
(124) Woolpack Farm	Gao *et al.* (2000)
(125) Wretton	Coope (1974b)

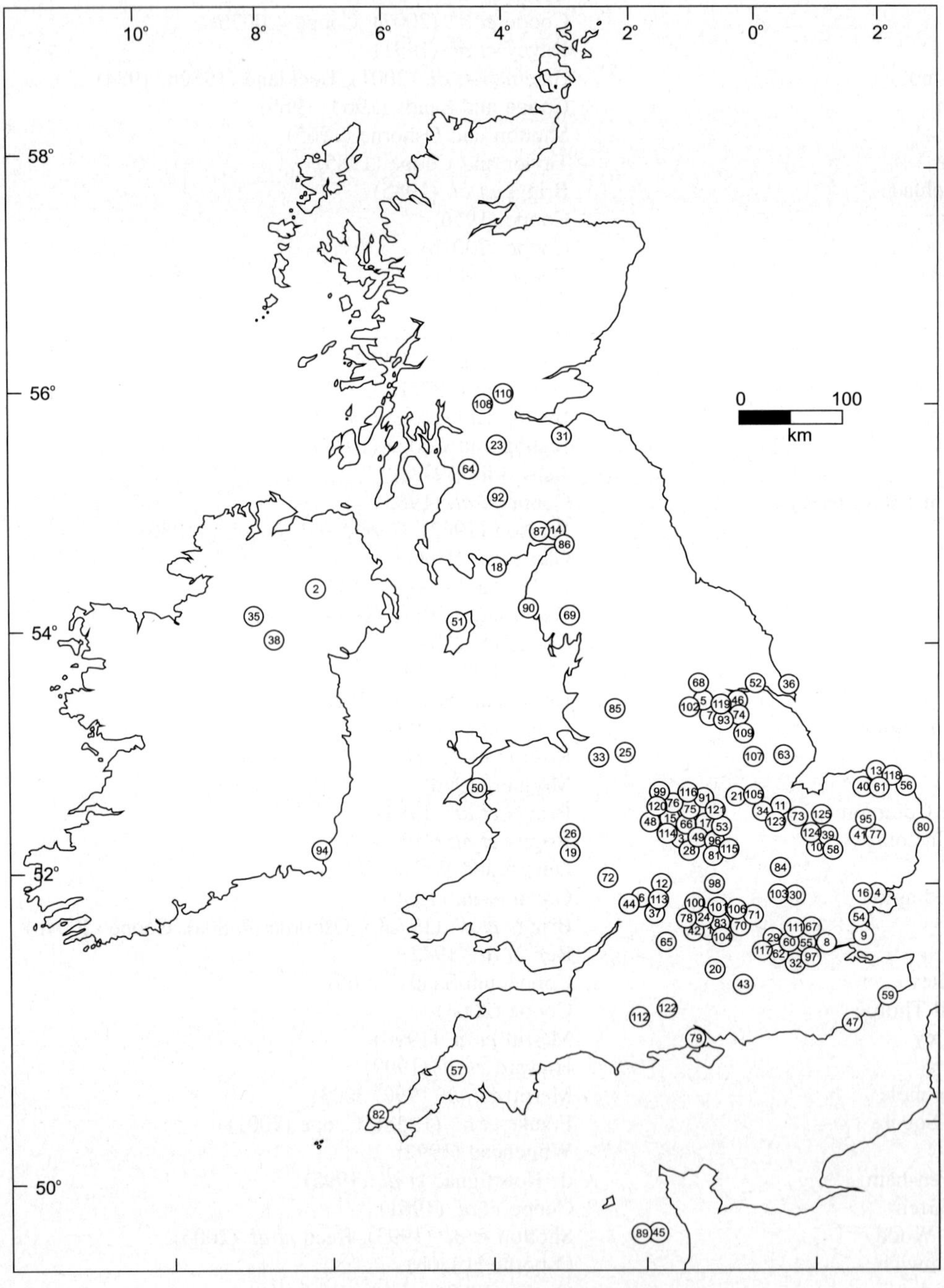

Fig. 1. Map of British Isles, showing Quaternary insect fossil localities. See Table 1 for site names.

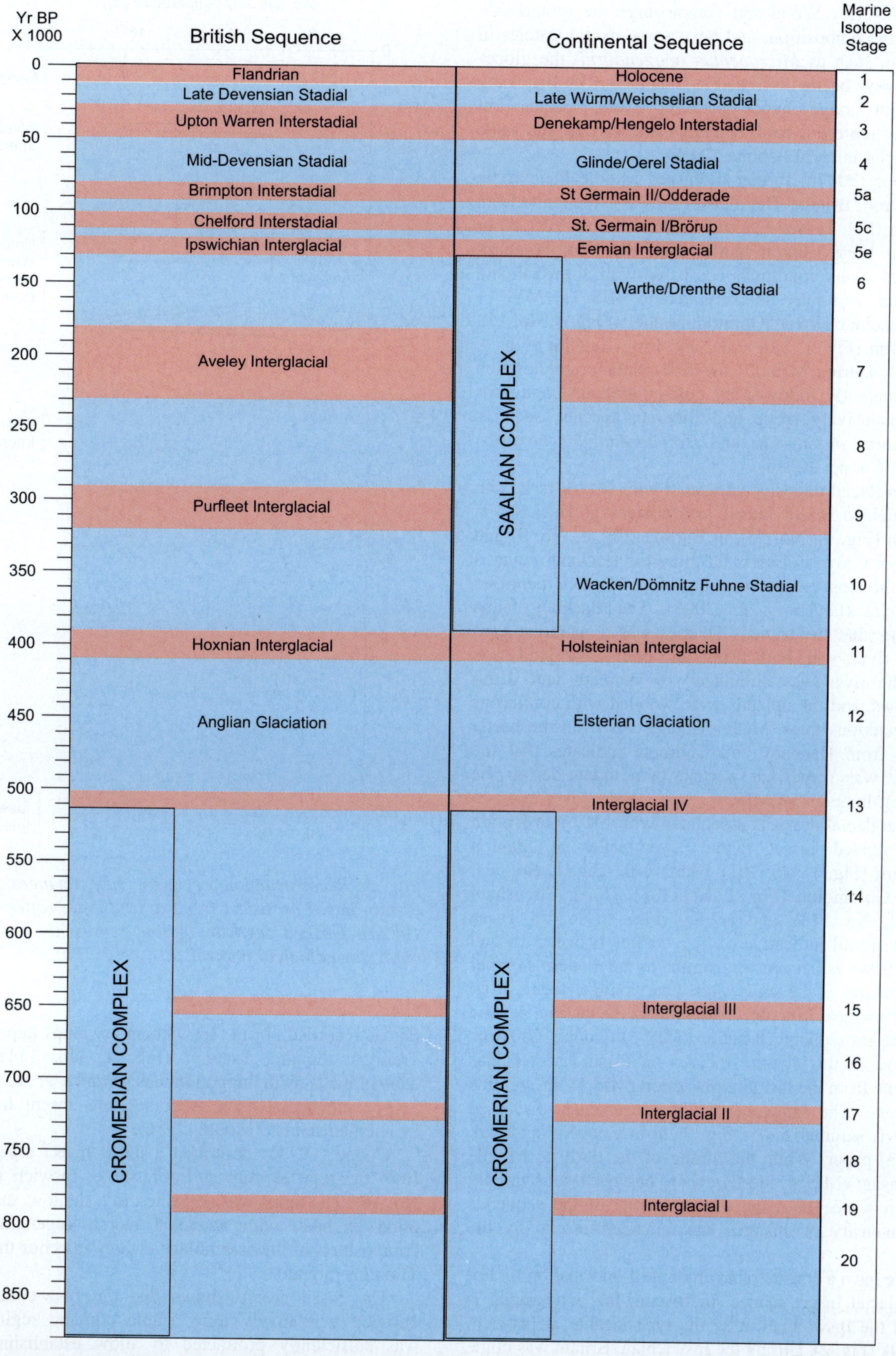

Fig. 2. Chronostratigraphic correlations between British and Continental European sequences, and Marine Isotope Stages. Continental sequence after Gibbard et al. (2005) (modified 2007). British sequence after Stringer (2005, p. 300).

from Waverley Wood and Happisburgh are remarkably similar in composition, and have some exotic species in common, such as *Micropeplus hoogendorni*, the globe-trotting rove beetle discussed in Chapter 4. The Brooksby and High Lodge faunas lack this exotic species, but probably represent cool-temperate intervals in this same interglacial interval (Coope, 2006).

Coope (2007b) further discussed Middle Pleistocene faunas from Britain. The oldest of these is the fauna from Beeston (Fig. 1, No. 13), which may date to MIS 18 or 20. The West Runton fauna, from the Norfolk coast, indicates cooler conditions compared with Sugworth and Pakefield, and may be correlated to MIS 13. MIS 11 faunas include Hoxne, Quinton, Nechells (Fig. 1, No. 16), Woodston (Fig. 1, No. 123), and Frog Hall Pit (Fig. 1, No. 49). During MIS 11, the first climatic amelioration was apparently followed by cold-continental conditions with exclusively arctic and Siberian species, such as *Helophorus obscurellus* and *Holoboreaphilus nordenskioeldi* (Coope, 2007b).

Interglacial deposits associated with MIS 9 in Britain have yielded insect faunal assemblages at Hackney in London (Fig. 1., No. 55), at Barling (Fig. 1, No. 9) and Cudmore Grove in Essex. Of these the Hackney fauna is by far the most extensive, comprising 181 taxa identified to species (Green *et al.*, 2006). The Hackney fauna indicates that the ancient Thames was a mature meandering river, with both riffles and pools. The landscape near the river was dominated by marshes and damp meadows, and the uplands were wooded with coniferous and deciduous trees. MCR evidence based on the beetle faunas from Hackney and Cudmore indicates that the climate was somewhat warmer than today during the warmest intervals of MIS 9.

Interglacial deposits associated with MIS 7 in Britain have yielded insect faunal assemblages at Stanton Harcourt (Fig. 1, No. 101), Marsworth (Fig. 1, No. 71), Stoke Goldington (Fig. 1, No. 103), Upper Strensham (Fig. 1, No. 113), and Aveley (Fig. 1, No. 8). These faunas are all indicative of T_{max} values between 16 and 18°C; these values are very similar to the modern summer temperatures at the study sites. One younger fauna, from Balderton, near Lincoln, clearly marks the thermal decline into the subsequent glaciation (MIS 6) (Coope, 2007b).

Most of this chapter focuses on fossil insect faunas that date from the last interglaciation (MIS 5) through the Holocene. The sequence of climatic changes discussed below is summarized in Fig. 3, from Coope's landmark (1977a) paper. While the timing of the Early to Middle Devensian (MIS 4 and 3) events has been substantially revised since 1977, the interpretations of the sequence and intensity of changes has remained essentially the same.

We have a considerable amount of information on last interglacial insect faunas. In Britain, this interglacial is called the Ipswichian, after the type locality at Ipswich, Suffolk (Fig. 2). During the Ipswichian, Britain was home to stenothermic ground beetle species, which today live in southern Europe. This indicates that the climate at the height of the last interglacial (MIS 5e) was warmer than modern (Coope, 1990). Other lines of evidence point in the same direction. For instance, the thermophilous insect

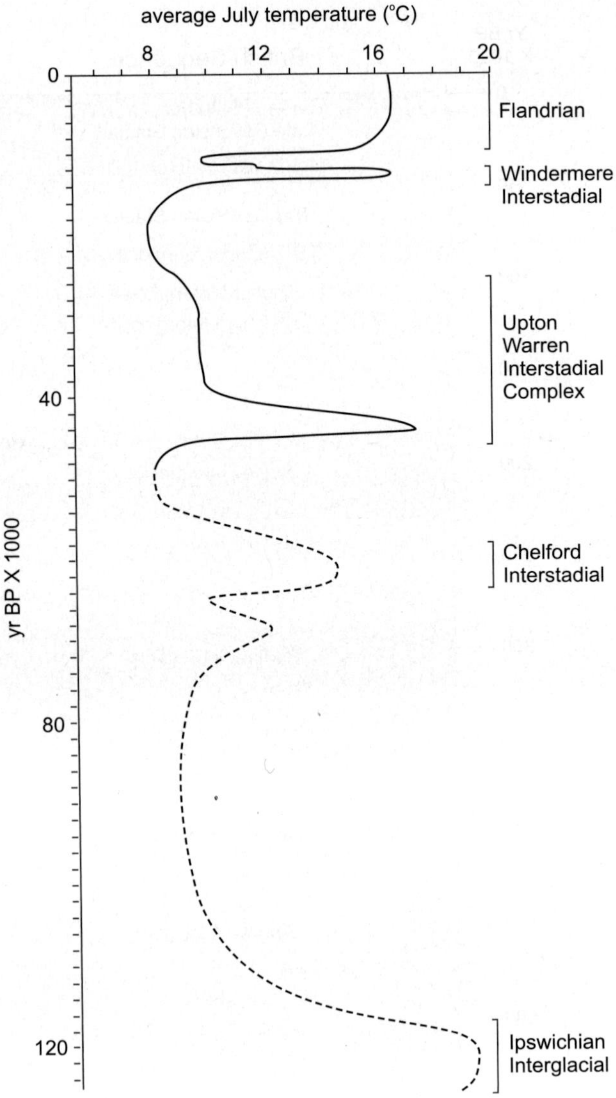

Fig. 3. Reconstruction of mean July temperatures in Britain based on insect fossil assemblages (after Coope, 1977a). Dashed segment of line represents interval of poor chronological control.

fauna recorded from Ipswichian swamp deposits at Trafalgar Square, London (Fig. 1, No. 111), were coinhabitants with hippopotamus (Franks *et al.*, 1958). Steppe indicators, such as horses, are absent from the Ipswichian faunas (Moore, 1986).

Coope (1974a) described a fossil insect assemblage from the type locality at Bobbitshole, Ipswich (Fig. 1, No. 16). The fauna was deposited in a shallow, eutrophic pond or lake, with abundant marsh vegetation. The composition of the assemblage closely matches that from Trafalgar Square.

The fossil record shows that the Ipswichian biota enjoyed a relatively long, stable climatic regime that was sufficiently protracted to allow establishment of mature deciduous forests including poplar, birch, and maple, accompanied by their particular bark beetles (Coope, 1990). Along with the subtropical mammalian megafauna, a rich dung beetle fauna also occupied Britain during this interglacial. Several species were found, which today inhabit only the Mediterranean region. Lemdahl and Coope (2007) estimate that actual mean July temperatures were about 21°C, or 3°C warmer than present-day southern England, and that mean January/February temperatures may have been about 4°C, which is about the same as in modern day southern England.

The termination of the Ipswichian is difficult to investigate because these sediments have often been eroded away, but a borehole at Elsing, Norfolk, contained Ipswichian deposits immediately overlain by sediments with insect faunas indicative of rapidly oscillating climates, ranging from temperate to arctic conditions (Taylor and Coope, 1985).

The next warm interval, the Chelford Interstadial in Britain (Fig. 2), has been correlated with MIS 5c (Lemdahl and Coope, 2007). It took place at about 100,000 yr BP. Insect faunas from this interstadial, including assemblages from Chelford, Four Ashes, and Wretton (Fig. 1, Nos. 25, 48, and 125, respectively), are comprised of species with modern distributions overlapping in southern Finland (between 60° and 65°N latitude). Beetle faunas characteristic of mixed conifer and birch woodland are known from at least two sites of this age in the English Midlands. The Chelford Interstadial is thus interpreted as being warmer than prior and subsequent stadial episodes. Lemdahl and Coope (2007) estimated that the summer temperatures were similar to those in central England today, but the winters were substantially colder. McCabe *et al.* (1987) described an interstadial flora and fauna from Aghnadarragh, Northern Ireland (Fig. 1, No. 2), that is possibly also of Chelford age. However, as with the other pre-middle Devensian deposits, there is some controversy about the age of the Aghnadarragh deposits (Worsley, 1991).

There are at least two separate interstadial beetle faunas associated with MIS 4 in central and southern England, separated from one another by a severely cold interval. Both of the temperate episodes postdate MIS 5 (Coope, 2000a) and the associated pollen spectra indicate treeless landscapes. British beetle assemblages that date from the earlier of these interstadials come from Isleworth (Fig. 1, No. 62) (Coope and Angus, 1975), Cassington (Maddy *et al.*, 1998), and Latton (Lewis *et al.*, 2006). These are rich, diverse assemblages composed of temperate/oceanic species, and yield MCR estimates of both T_{max} and T_{min} that are very close to modern parameters (Lemdahl and Coope, 2007). At Cassington, the temperate interstadial fauna is overlain by a deposit that yielded an arctic fauna. This highly characteristic assemblage was also found at the South Kensington site in London (Fig. 1, No. 97), providing a faunal link between the two sites (Coope *et al.*, 1997). The arctic fauna gave way to another temperate interstadial fauna at Cassington. The picture that is emerging from these discoveries is that there were multiple, intense warm intervals in the Early Devensian, though perhaps they were short-lived. As new studies are done, they will likely add evidence of additional interstadial events, and refined dating techniques may help resolve the timing of these events.

The Upton Warren Interstadial in Britain (Fig. 3), is named for deposits at the Upton Warren site (Fig. 1, No. 114), near Birmingham. This interstadial is correlated with MIS 3. However, the timing of the earlier parts of this interstadial remains in question, because it is on the fringe of reliable radiocarbon dating. Coope (1975, 1977a) summarized the Upton Warren faunas, dating between 48,000 and 30,000 cal yr BP. In his original interpretation, the Upton Warren interval was a complex of environmental changes that began with a rather abrupt warming that lasted roughly 1,000–2,000 years, followed by a lengthy period of conditions cooler than modern, but not fully glacial. The Isleworth fauna was originally thought to represent this warm spike that began this MIS 3 interstadial, but it now appears that the Isleworth fauna represents an earlier interstadial event (see above). The beetle assemblages from many deposits that can be reliably dated by radiocarbon (i.e., from about 44,000 to 24,000 cal yr BP) are relatively homogeneous in content, including several species with exclusively arctic and Asiatic distributions today (Coope, 1995). Several assemblages from the Four Ashes site in the Midlands (Fig. 1, No. 48) (Morgan, 1973) reflect these cooler conditions. They contain a mixture of temperate and northern species, and lack the extreme thermophiles seen from warmer interstadial assemblages. Later in the interstadial complex, further deterioration is marked in assemblages from the Upton Warren site (Coope *et al.*, 1961). These assemblages are also noteworthy because they include Asiatic elements, such as the Tibetan dung beetle, *Aphodius holdereri*, suggesting a shift to continental climate. MCR estimates have been made on over 20 of these faunas (Coope, 1987b). Average values for T_{max} are 9–11°C, and average values or T_{min} are −25 to −10°C.

Coope (1987a) summarized the insect faunal sequences between 47,000 and 13,000 cal yr BP (Fig. 4). As ice sheets approached the English midlands from the north, the remaining insect faunas south of the ice became increasingly impoverished, with just a few species identified from deposits dating to the Late Devensian stadial maximum, beginning about 31,000 cal yr BP (Fig. 4). Between 21,500 and 18,000 cal yr BP, even southern England experienced extremely cold, harsh climates. Conditions were probably similar to modern polar desert. There are no fossil insect records from this interval in Britain, and Coope (1987a) speculated that the beetle fauna may have been extirpated.

I discussed the British lateglacial insect faunas in some detail in Chapter 5. As in the Upton Warren Interstadial Complex, the lateglacial insect faunal sequence shows remarkably rapid and dramatic changes from arctic faunas prior to 13,000 yr BP at Hawks Tor, Cornwall (Fig. 1, No. 57) (Coope, 1977a), and Glanllynnau, northern Wales (Fig. 1, No. 50) (Coope and Brophy, 1972), to temperate faunas during the brief Windermere interstadial. Insects from a deposit at Low Wray Bay, Windermere (Fig. 1, No. 69), document the rapid warming following 15,800 cal yr BP (Coope, 1977a), as do insects from the Roberthill site, in southern Scotland (Fig. 1, No. 87) (Bishop and Coope, 1977). Again, the Windermere interstadial faunas reflect summer conditions

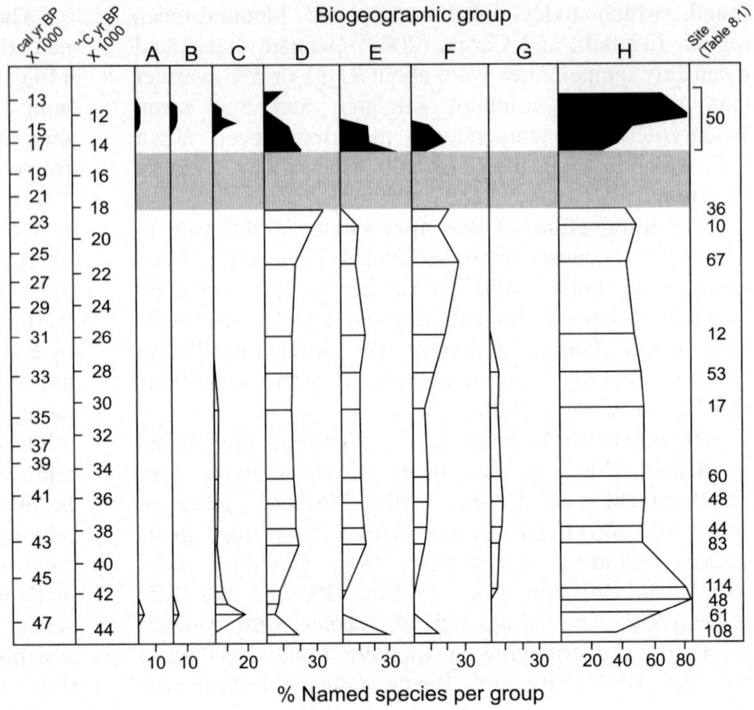

Fig. 4. Percent composition of British insect fossil assemblages during the middle and Late Devensian Glaciations, based on modern geographic distribution of species. Gray zone represents interval devoid of insect fossils. (A) Southern European species; (B) southern European species whose ranges just fail to reach Britain; (C) southern species whose ranges are south of central Britain; (D) widespread species whose ranges are north of central Britain; (E) boreal and montane species whose ranges extend down into the upper part of the coniferous forest zone; (F) boreal and montane species whose ranges are above treeline; (G) eastern Asiatic species, some of which range into North America; (H) cosmopolitan species (wide geographic ranges). See Table 1 for site names (modified from Coope, 1987a).

as warm as modern. Insect faunas dating between 14,200 and 12,900 cal yr BP from a number of sites suggest a cool, temperate climate.

The Windermere interstadial was followed by a striking deterioration between 12,900 and 11,000 cal yr BP. This is called the Loch Lomond Stadial in Britain, the latter part of which correlates with the Younger Dryas pollen zone in continental Europe. Temperate faunas were replaced once more with arctic species, including the return of several Asian species (Coope, 1977a). This rapid, large-scale cooling event may well have been linked to changes in North Atlantic circulation (Duplessy *et al.*, 1996). The subsequent amelioration to the Flandrian, or Holocene, was equally rapid and extreme, bringing summer temperatures as warm as or warmer than modern by 10,800 cal yr BP (Coope, 1987a). Most of the studied British Holocene insect assemblages come from archeological sites (Chapter 7) or from sites that contain a mixture of archaeological and natural deposits (e.g., Smith *et al.*, 2005b).

2. Research in the North Atlantic Region

Fossil arthropod research has been underway in Iceland and Greenland in recent decades. Haarløv (1967) published the first results of Quaternary insect fossils from Greenland. He studied Late Holocene peats from Sermermiut, near Jacobshavn, western Greenland (Fig. 5). The peats contained oribatid mites and the remains of leafminer flies (Diptera: Agromyzidae). The fauna comprises species associated with bogs and humid heaths in arctic and subarctic regions, though three of the mites identified from the assemblages have not been collected in modern Greenland.

Böcher and Bennike (1991) described **Eemian interglacial** (Isotope Stage 5e) insect assemblages from Hesteelv and Langelandselv, Jameson Land, east

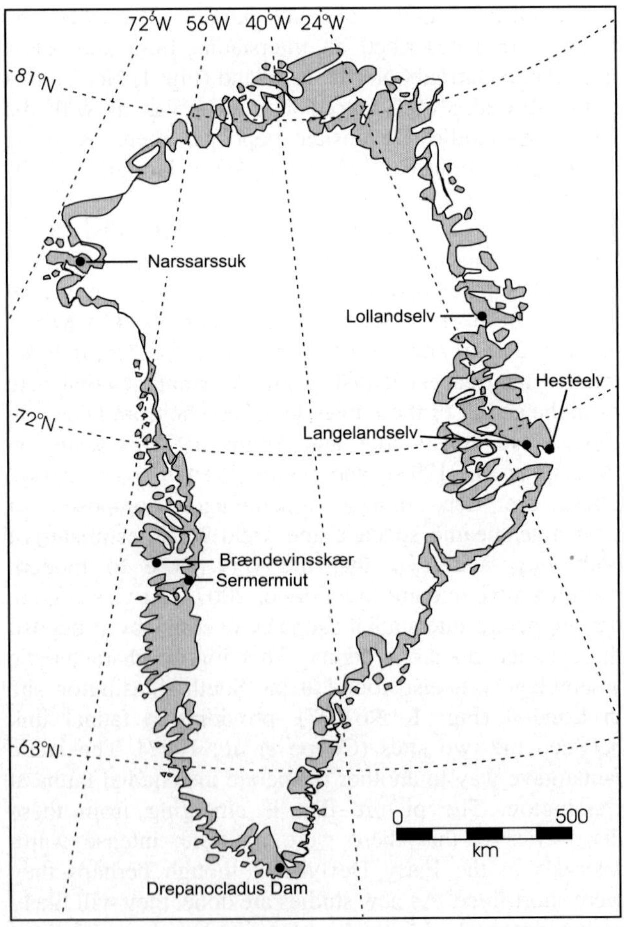

Fig. 5. Map of Greenland, showing Quaternary insect fossil localities discussed in text.

Greenland (Fig. 5). The assemblages contained only eight insect taxa, but the modern ranges of the identified beetle species are considerably south of the fossil site. Böcher and Bennike concluded that the fossil assemblages reflect climatic conditions markedly warmer than modern. Another MIS 5e site has been described from the Daugaard Jensen Land region by Bennike (2000). This organic deposit yielded only the remains of the seed bug, *Nysius groenlandicus*. Experimental studies on the environmental tolerances of this species (Böcher and Nachman, 2001) indicate that Greenland populations are adapted to warm, dry microclimates.

Böcher (1989b) also reported on the discovery of *Amara alpina* (Carabidae) fossils in an organic deposit at Narssarssuk in northwestern Greenland (Fig. 5). The deposit contained mollusc shells which yielded amino acid racemization ratios equivalent to those found in regional shells tentatively assigned to Isotope Stage 5a. In Greenland, this interval is called the Qarmat interstadial. *Amara alpina* is absent from Greenland today, and the fossil locality is also slightly farther north than the farthest north populations of that species in nearby arctic Canada. Bennike and Böcher (1992) also described a small insect fauna thought to be of Early Weichselian interstadial age from Thule in northwest Greenland. This faunal assemblage contained only two beetle species: *Amara alpina* and the arctic weevil *Isochnus arcticus*.

Fredskild *et al.* (1975), and Böcher and Bennike (1996) reported on very small assemblages of insect remains recovered from Holocene peat samples at Drepanocladus Dam in southern Greenland, and Lollandselv in Jameson Land, southeast Greenland. Bennike (1992) also reported on a few fossil insect remains from the site of Brædenvinskær, near Disco Island in western Greenland. The Lollandselv fauna dates to the Early Holocene. It contained *Nysius groenlandicus* and the leaf beetle, *Phratora polaris*. This species feeds on dwarf willow, and is found today in **subalpine** and alpine regions of central and northern Europe, and Iceland. There are no known Chrysomelidae living in Greenland today (Böcher, 1988a).

Buckland *et al.* (1986a,b) have investigated a number of sites in Iceland, comparing insect faunas before and after the arrival of Viking settlers. The Viking landings conveniently coincided with the deposition of the Landnám tephra in Iceland. The tephra thus serves as a marker horizon, separating the assemblages into pre- and post-settlement faunas. I have previously discussed the post-Landnám (archeological) faunas in Chapter 7. Pre-Landnám faunas have been described by Buckland *et al.* (1986b) from Einhyrningur, Hofsá, Holt, Kópavogur, Merkigil, Ósabakki, Skálafelsjökull, and Thjórsárbrú (Fig. 6, Nos. 11, 30, 31, 37, 59, 72, 88, and 92), and from Ketilsstadir (Fig. 6, No. 34) (Buckland *et al.*, 1986a). The pre-Landnám fauna was considerably less diverse than the post-Landnám, because of the successful establishment of Viking-introduced species. The pre-Landnám faunas, all of Holocene age, are adapted to cool-temperate rather than arctic climates. As such, Buckland *et al.* (1986b) argue that the Icelandic insect fauna did

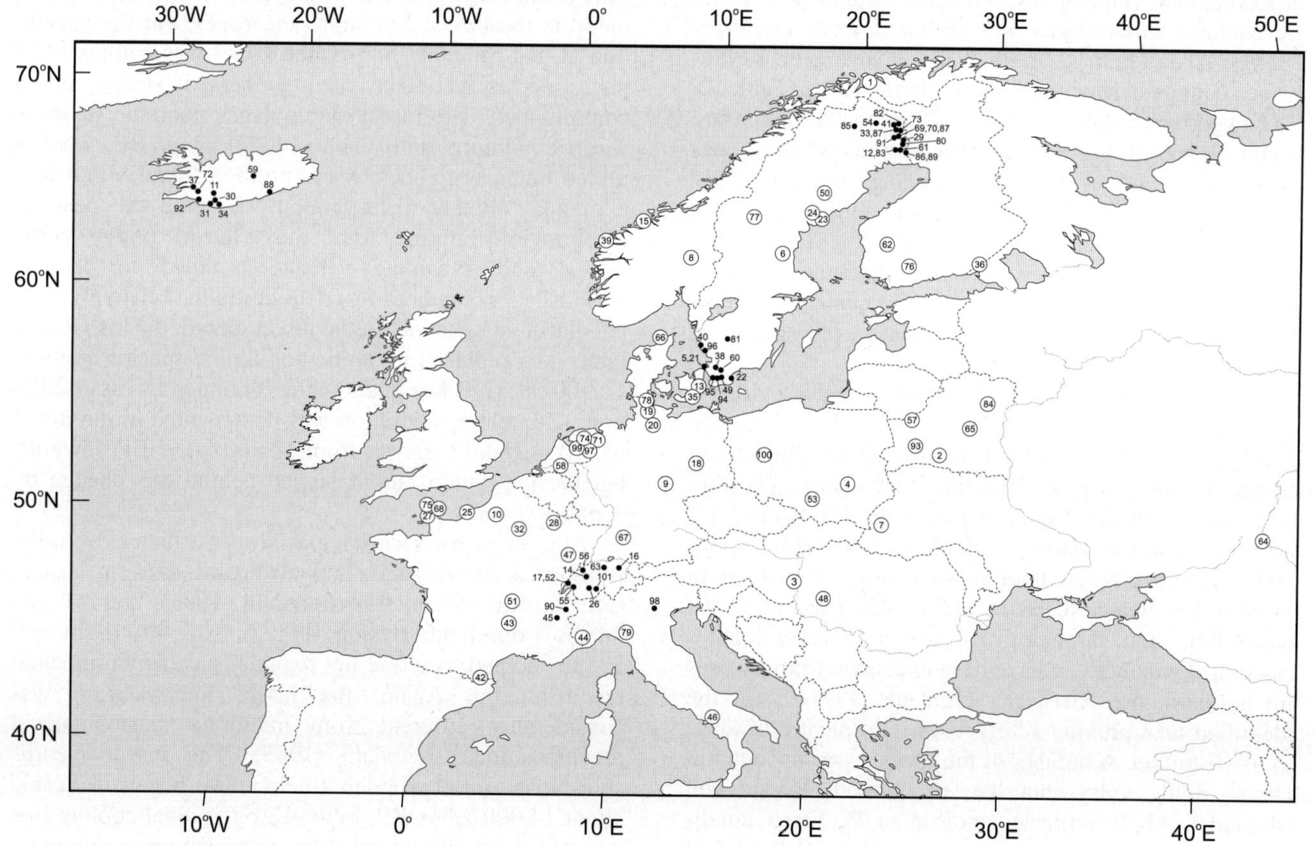

Fig. 6. Map of Europe, showing Quaternary insect fossil localities. See Table 2 for site names.

not survive the Weichselian glaciation *in situ*, but rather migrated to Iceland following deglaciation. Fully winged species may have been carried to Iceland from northwest Europe by winds, but many of the pre-Landnám beetle species are flightless. Buckland *et al.* (1986b) proposed a transport mechanism for these beetles in which sediment-covered ice floes from Scandinavia and northern Britain drifted to the northwest to Iceland, beaching their passengers on Iceland's newly deglaciated shores. This intriguing hypothesis is in need of additional tests.

3. Western European Research

While the number of studies in continental Europe is substantially smaller than in Britain, good progress has been made in many regions during the last 20 years (Table 2). The number of published Pleistocene beetle faunas has doubled in this interval. Sweden, France, and Switzerland have been centers of Quaternary insect research in recent years.

3.1. Sweden

As discussed in the first chapter, Scandinavian paleo-entomological research began in earnest with Carl Lindroth's (1948) work on interglacial insect fossils from Bollnäs, Härnön, Längsele, and Pilgrimstad in central Sweden (Fig. 6, Nos. 6, 23, 50, and 77). Later, Lindroth and Coope (1971) reported their conclusions about insect fossils from deposits thought to be of interglacial age, at Leveäniemi (Fig. 6, No. 54). The composition of the Leveäniemi assemblages was different from the other "interglacial" faunas, in that the Leveäniemi assemblages contain substantially more thermophiles (Table 3). The majority of the identified species in this fauna are found today in the oak and coniferous forest zones of southern and central Fennoscandia, but the study site is in Swedish Lapland, near the northern treeline. The Leveäniemi fauna is thus indicative of climatic conditions considerably warmer than present, and suggests continental climate. The Längsele fauna is the most thermophilous of the interstadial faunas, but it reflects conditions colder than at present.

Based on these differences and stratigraphic evidence developed by Lundqvist (1967), Lindroth and Coope ascribed the Bollnäs, Harnön, Längsele, and Pilgrimstad faunas to an Early Weichselian interstadial, possibly correlative with the Brørup of Denmark. The Brørup has been radiocarbon dated at 59,430 yr BP (Lundqvist, 1986), but more recent thinking places this warm interval in MIS 5c, greater than 100,000 yr BP (see Fig. 2 for correlations and chronology). Fossil insect assemblages associated with MIS 5c have been examined from 10 sites in northernmost Sweden (Lemdahl, 1997). All the identified taxa provide a very consistent picture of xeric to mesic tundra. A number of the species are not found in Fennoscandia today, but live in northern Russia and Mongolia. MCR estimates indicate cold, harsh conditions, with mean July temperatures of 9–11°C (Lemdahl and Coope, 2007). Lemdahl (1997) studied insect assemblages dating to MIS 5a, the Odderade Interstadial, from four Swedish sites. These faunas are characterized by arctic tundra species, and low faunal diversity. The climatic conditions seem to have been even more severe than those of the Brørup.

Early Weichselian interstadial faunas dated ca. 100,000 yr BP were recovered from three sites in the Norrbotten district of northernmost Sweden (Aronsson *et al.*, 1993). The faunas are all indicative of colder-than-modern climate, yielding an MCR estimate for T_{max} of 9–11°C, compared with modern T_{max} in this region of 15°C. Thus the authors concluded that these faunas represent a cool interstadial environment, with a more continental climate than is found there today.

Scandinavia was covered by ice during the Late Weichselian glaciation which began about 47,000 yr BP. Southern Sweden was deglaciated by about 14,000 yr BP, and northern Sweden by about 8,500 yr BP (Lundqvist, 1986). Two lateglacial beetle assemblages have been identified from Kullaberg, Sweden (Fig. 6, No. 40) (Lemdahl, 1988a, cited in Coope *et al.*, 1998). Although these include only a few species, they resemble lateglacial faunas from Britain, including an initial warming, followed by the Younger Dryas oscillation.

Lemdahl has examined a number of lateglacial sites in Scania, southern Sweden. Insect faunas from the Håkulls mosse site (Fig. 6, No. 21) span the interval 15,900–11,600 cal yr BP. Prior to 14,800 cal yr BP southern Sweden was a periglacial landscape with stagnant ice, sparse vegetation cover, and soils made unstable by **solifluction** (Berglund *et al.*, 1984). The earliest deposit containing insects dates from ca. 15,800 cal yr BP. This assemblage contains thermophilous insects that live today in the boreal and subalpine zones, but the vegetation at that time (Bølling pollen zone) was composed of plant species that today are members of steppe–tundra communities. The insect assemblages continue to show warm conditions until about 14,800 cal yr BP, when a mixed fauna of temperate and northern species indicates a gradual climatic deterioration following the peak of amelioration inferred from the **Allerød pollen zone**. Boreal and subalpine vegetation failed to become established in southern Sweden during the relatively brief lateglacial interstadial. In the insect record, the interstadial fauna was replaced by arctic and alpine species at about 12,900 cal yr BP, signaling the Younger Dryas oscillation. Climatic amelioration was documented in the insect record at Håkulls mosse from 12,400 cal yr BP onwards, but the vegetation record lagged behind this change by about 300 years.

Additional work on lateglacial insect faunas from the Björkeröds mosse and Toppeladugård sites in Scania (Fig. 6, Nos. 5 and 94) (Lemdahl, 1985), and Allerød faunas from Lake Bysjön (Fig. 6, No. 49) (Lemdahl, 1988b) helped to refine the regional paleoenvironmental reconstruction. Again, the insect interpretation was considerably different from traditional paleobotanical reconstructions (Lemdahl, 1985). The insect records show that a brief, sudden amelioration began in Scania about 14,900 cal yr BP, followed by gradual cooling that began by 14,500 cal yr BP, culminating in Younger Dryas cooling following 13,200 cal yr BP which saw the

Table 2. Continental European Quaternary insect fossil sites and references.

Site	Reference(s)
(1) Andøya, Norway	Fjellberg (1978)
(2) Artuki, Byelorussia	Nazarov (1984, 1991)
(3) Bad Tatzmannsdorf, Austria	Schweiger (1967)
(4) Belchatów, Poland	Morgan *et al.* (1982a), Kasse *et al.* (1998)
(5) Björkeröds Mosse, Sweden	Lemdahl (1985, 1988a)
(6) Bollnäs, Sweden	Lindroth (1948)
(7) Borislav, Ukraine	Angus (1973)
(8) Brumunddal, Norway	Helle *et al.* (1981)
(9) Burgtonna, Germany	Von Knorre (1978)
(10) Conty, France	Antoine *et al.* (2003), Ponel *et al.* (2005)
(11) Einhyrningur, Iceland	Buckland *et al.* (1986b)
(12) Empohultet, Sweden	Lemdahl (1997)
(13) Fakse Bugt, Denmark	Bennike and Jensen (1995)
(14) Gerzensee, Switzerland	Lemdahl (2000a)
(15) Godøy, Norway	Birks *et al.* (1993)
(16) Gossau	Jost-Stauffer *et al.* (2005)
(17) Grand Marais	Gaillard and Lemdahl (1994b)
(18) Gröbern, Germany	Walkling and Coope (1996)
(19) Grossensee, Germany	Günther (1983)
(20) Gross Todtshorn, Germany	Walkling (1997)
(21) Håkulls Mosse, Sweden	Berglund *et al.* (1984), Lemdahl (1985, 1988a)
(22) Hanobukten, Sweden	Gaillard and Lemdahl (1994a)
(23) Härnön, Sweden	Lindroth (1948)
(24) Härnosänd, Sweden	Lindroth (1948), García Ambrosiani and Robertsson (1992)
(25) Havre, France	Ters *et al.* (1971)
(26) Hérémence, Switzerland	Lemdahl (2000a)
(27) Herquemoulin, France	Coope *et al.* (1987)
(28) High Ardennes, France	Damblon *et al.* (1977)
(29) Hirvijärvi, Sweden	Lemdahl (1997)
(30) Hofsá, Iceland	Buckland *et al.* (1986b)
(31) Holt, Iceland	Sveinbjarnardottir (1983), Buckland *et al.* (1986b)
(32) Houdancourt, France	Ponel *et al.* (2005)
(33) Kainulasjärvi, Sweden	Lemdahl (1997)
(34) Ketilsstadir, Iceland	Buckland *et al.* (1986a)
(35) Kobbelgård, Denmark	Bennike *et al.* (1994)
(36) Koivusilta, Finland	Bondestam *et al.* (1994)
(37) Kópavogur, Iceland	Buckland *et al.* (1986b)
(38) Körslättamossen, Sweden	Hammarlund and Lemdahl (1994)
(39) Kråkenes, Norway	Birks *et al.* (2000), Lemdahl (2000b)
(40) Kullaberg, Sweden	Lemdahl (1988a)
(41) Kurujärvi, Sweden	Aronsson *et al.* (1993)
(42) La Borde, France	Ponel *et al.* (1999)
(43) Lac d'Issarlés, France	Ponel and Gadbin (1988)
(44) Lac Long Inférior	Ponel *et al.* (2001)
(45) La Côte, France	Field *et al.* (2000)
(46) Lago di Monterosi, Italy	Roback (1970)
(47) La Grand Pile, France	Ponel (1994, 1995), Ponel *et al.* (2003)
(48) Lake Balaton, Hungary	Dévai and Moldován (1983)
(49) Lake Bysjön, Sweden	Lemdahl (1988b), Gedda *et al.* (1999)
(50) Längsele, Sweden	Lindroth (1948)
(51) La Taphanel, France	Ponel and Coope (1990)
(52) Lausanne, Switzerland	Gabus *et al.* (1987)
(53) Lednica, Poland	Lemdahl (1991f)
(54) Leveäniemi, Sweden	Lindroth and Coope (1971)
(55) Leysin, Switzerland	Lemdahl (2000a)
(56) Lobsigensee, Switzerland	Elias and Wilkinson (1983)

(*Continued*)

Table 2. (*Continued*)

Site	Reference(s)
(57) Logoza, Byelorussia	Nazarov (1991)
(58) Mark Valley, The Netherlands	Bohncke *et al.* (1987)
(59) Merkigil, Iceland	Buckland *et al.* (1986b)
(60) Mickelsmossen, Sweden	Lemdahl and Persson (1989)
(61) Missesberget, Sweden	Lemdahl (1997)
(62) Mustalampi, Finland	Bondestam *et al.* (1994)
(63) Niederwenigen, Switzerland	Coope (2007a)
(64) Nikol'skoye, Russia	Bidashko and Proskurin (1988)
(65) Nizhninsky Rov Ravine, Byelorussia	Nazarov (1991)
(66) Nørre Lyngby, Denmark	Coope and Böcher (2000)
(67) Oerel, Germany	Behre *et al.* (2005)
(68) Omonville-la-Rouge, France	Coope *et al.* (1987)
(69) Onttoharjut, Sweden	Lemdahl (1997)
(70) Onttovaara, Sweden	Lemdahl (1997)
(71) Orvette, The Netherlands	Cappers *et al.* (1993)
(72) Ósabakki, Iceland	Buckland *et al.* (1986b)
(73) Outojärvi, Sweden	Lemdahl (1997)
(74) Peelo, The Netherlands	Coope (1969)
(75) Petit Beaumont, France	Coope *et al.* (1987)
(76) Piilonsuo, Finland	Koponen and Nuorteva (1973), Karppinen and Koponen (1974)
(77) Pilgrimstad, Sweden	Lindroth (1948)
(78) Poolsee, Germany	Hofmann (1983)
(79) Prato Spilla, Italy	Ponel and Lowe (1992), Ponel (1997b)
(80) Purasharju, Sweden	Lemdahl (1997)
(81) Ranstad, Sweden	Wastegård *et al.* (1995)
(82) Riipiharju, Sweden	Lemdahl (1997)
(83) Rotheden, Sweden	Lemdahl (1997)
(84) Rubezhnitsa, Byelorussia	Nazarov (1979)
(85) Särkasvaara, Sweden	Lemdahl (1997)
(86) Särkivuoma, Sweden	Lemdahl (1997)
(87) Sivakkojärvi, Sweden	Lemdahl (1997)
(88) Skálafelsjökull, Iceland	Buckland *et al.* (1986b)
(89) Snårberget, Sweden	Lemdahl (1997)
(90) Taillefer Massif, France	Ponel *et al.* (1992)
(91) Takkanenmännikö, Sweden	Lemdahl (1997)
(92) Thjórsárbrú, Iceland	Buckland *et al.* (1986b)
(93) Timoshkovichi, Byelorussia	Nazarov (1991)
(94) Toppeladugård, Sweden	Lemdahl (1985, 1991a)
(95) Torreberga, Sweden	Berglund and Digerfeldt (1970)
(96) Tvååker, Sweden	Lemdahl and Gustavson, 1997
(97) Usselo, The Netherlands	Van Geel *et al.* (1989)
(98) Verona, Italy	Foddai and Minelli (1994)
(99) Voorthuizen, The Netherlands	Angus (1975)
(100) Zabinko, Poland	Lemdahl (1991b)
(101) Zeneggen, Switzerland	Lemdahl (2000a)

return of arctic conditions and fauna. The additional sites confirmed the Håkulls mosse data showing rapid amelioration beginning at 12,400 cal yr BP.

Lemdahl (1991a) applied the MCR method to the Toppeladugård assemblages to reconstruct summer and winter mean temperatures from 13,700 to 11,500 cal yr BP. The MCR reconstruction shows mean summer temperatures of 15–18°C from 13,700 to 13,500 cal yr BP, and mean summer temperatures of 10–13°C by 10,800 yr BP. Following the Younger Dryas, summer temperatures rose to 14–16°C by 12,800 cal yr BP.

3.2. Norway

The first Norwegian study was by Coope (in Helle *et al.*, 1981). This study concerned Early Weichselian interstadial deposits at Brumunddal, southeastern Norway (Fig. 6, No. 8). Both pollen and insect data indicate climatic conditions with summer temperatures 2–3°C cooler than present. The lower of two insect assemblages is comprised of species found today in central Scandinavia (i.e., slightly north of the fossil locality). The upper assemblage is distinctly different; it contains only

Table 3. Modern habitat and Fennoscandian distributions of species found in the interglacial beetle fauna from Leveäniemi, Sweden.

Species	Moisture requirements*	Geographic regions**	Habitat zones in Fennoscandia***					
			Alpine	Subalpine	Montane	Birch forest	Coniferous forest	Oak forest
Bembidion doris (Carabidae)	H	S					●	
Patrobus septentrionis (Carabidae)	H	N	●	●	●	●	●	
Gyrinus marinus (Gyrinidae)	A	G					●	●
Hydraena riparia (Hydraenidae)	A	S					●	●
Chaetarthria seminulum (Hydrophilidae)	H	S					●	●
Olophrum consimile (Staphylinidae)	H	N			●	●	●	○
Acidota crenata (Staphylinidae)	M	G	○	○	●	●	●	●
Riolus nitens (Dryopidae)	A	S					●	●
Phalacrus caricis (Phalacridae)	H	S					●	●
Otiorrhynchus dubius (Curculionidae)	M	N	●	●	●	●	●	○
Notaris aethiops (Curculionidae)	H	G			●	●	●	●
Eubrychius velutus (Curculionidae)	A	S					●	●

Source: Data from Lindroth and Coope, 1971).
*Moisture requirement abbreviations: A, aquatic; H, hygrophilous; M, mesic
**Geographic abbreviations: G, generally distributed in Fennoscandia; N, northern Fennoscandia; S, southern Fennoscandia
***Habitat zone symbols: black circles, commonly found in this region; white circles, accidental or relict occurrences

cold-adapted, arctic species. Taken together, the upper fauna indicates summer temperatures at or just below 10°C. Coope correlated the Brumunddal insect faunal sequence with the Jämtland interstadial faunas of central Sweden. Lundqvist (1986) correlated the Brumunddal deposits with the Brørup interstadial of Denmark.

A brief treatment was published by Fjellberg (1978), concerning Mid- and Late Weichselian insects from lake sediments at Andøya, northern Norway (Fig. 6, No. 1). The lake sediments show continuous deposition from >21,500 cal yr BP to present. The record from 21,500 to 14,000 cal yr BP shows arctic tundra vegetation and arctic chironomids, and oribatid mites.

Lateglacial insect faunas have been studied from two sites in Norway: Kråkenes (Fig. 6, No. 39) and Godøy (Fig. 6, No. 15). The Kråkenes site (Lemdahl, 2000b) yielded a small faunal assemblage, dated about 14,000 cal yr BP. This fauna has arctic/subarctic affinities, yielding an MCR reconstruction showing mean July temperatures to be 10–13°C, similar to modern summer temperatures at the site in the subalpine zone of western Norway. During the subsequent cooling associated with the Younger Dryas chronozone, beetle remains became scarce in the sediments, making environmental reconstruction difficult. The fossil beetle assemblage from the island of Godøy (Birks *et al.*, 1993) is dated to the Allerød–Younger Dryas transition. It shows climatic cooling, with temperatures falling to 2.5–7.5°C. The subsequent climatic amelioration took place after 12,100 cal yr BP, as reflected in the replacement of arctic and alpine species with temperate species at the Tvååker site in southwestern Sweden (Fig. 6, No. 96) (Lemdahl and Gustavson, 1997).

3.3. Finland

Koponen and Nuorteva (1973) published the results of their study of a peat bog at Piilonsuo (Fig. 6, No. 76). The faunal assemblages range in age from 11,500 to 600 cal yr BP, and include abundant conifer-associated species (many specimens were taken from under preserved bark of buried tree stumps). In total, 259 taxa from more than 70 families in 11 orders of insects and arachnids were recovered. All of the identified species occur in Finland today. The fossil mite faunas were reported separately by Karppinen and Koponen (1974). The faunal assemblages provide a wealth of paleoecological data, reconstructing the history of forest and bog development in southern Finland. Bondestam *et al.* (1994) reported on insect faunas from the Younger Dryas and Early Holocene intervals at Koivusilta (Fig. 6, No. 36) and Mustalampi (Fig. 6, No. 62), in Finnish Karelia. They found that faunas associated with the Younger Dryas interval are composed either dominantly or exclusively of tundra-dwelling species. These faunas come from horizons dated 11,900–11,200. After 11,000 yr BP, the faunal composition of regional faunal assemblages changes suddenly, as arctic/subarctic species give way to temperate species. MCR analyses of these two sets of assemblages show a warming of as much as 10°C during the rapid transition from Younger Dryas to Holocene (Preboreal) intervals.

3.4. Denmark

In spite of the pioneering work by Henriksen (1933), surprisingly little more recent work on Pleistocene insect fossils has been done in Denmark. Henriksen's publication presented almost 200 pages of research on Danish and southern Swedish insect fossil assemblages, but it was necessarily weak on faunal chronologies, since radiocarbon dating had not yet been invented. Even the lateglacial interval has not been studied in recent decades, ironically in a country that coined the names used for the lateglacial pollen zones of Allerød, Bølling, and Younger Dryas! Bennike *et al.* (1994) described a fossil insect assemblage from the Kobbelgård site on the Island of Møn (Fig. 6, No. 35). The fauna is of Middle Weichselian age, and represents a steppe–tundra environment. It includes the cold-adapted ground beetles, *Nebria rufescens* and *Pterostichus (Cryobius) pinguedineus*. The former species is widespread in arctic and subarctic, and alpine regions throughout much of the Northern Hemisphere. The latter species is found today throughout arctic North America, and in western Siberia, as far west as the Yenisei River (Lindroth, 1966).

3.5. Germany

A considerable amount of paleoentomological work has gone on in Germany in recent years. Much of this work has focused on MIS 5-age faunas, revealing the complexities of the series of environmental changes within this interval. Behre *et al.* (2005) studied MIS 5 beetle faunas from the Oerel site, in northwest Germany (Fig. 6, No. 67). These were divided in two faunal units. The older of these comprises a species-rich fauna, indicative of temperate conditions during the Brørup interstadial (MIS 5c). The fauna indicates that the study region was a mosaic of varied habitats during this interval, including wetlands, woods, and open ground. The younger fauna contains only 20 species, and lacks thermophiles. This assemblage is thought to represent the Rederstall stadial (MIS 5b). The fauna represents boreal environments.

Other insect assemblages from the Oerel and Gross Todtshorn (Fig. 6, No. 20) sites in Germany (Walkling, 1997) are attributed MIS 5d, the Herning Stadial. These are rather poor faunas, made up of only a few taxa. At Oerel, a few species confined to a marshy environment were identified. At Gross Todtshorn typical heath taxa were recovered. MIS 5a interstadial beetle assemblages from Oerel and Gross Todtshorn contain only a few species, which are indicative of shrub tundra and subarctic climate. At the Oerel site, two insect assemblages

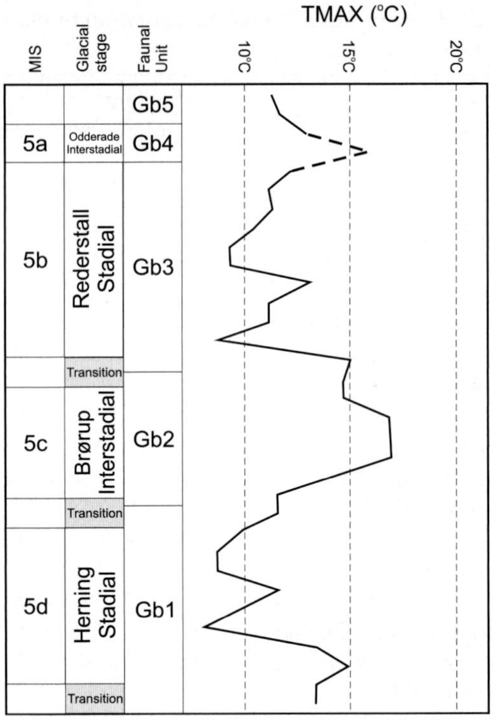

Fig. 7. MCR reconstruction of mean July temperatures (T_{max}) from MIS 5 faunal assemblages at Gröbern, Germany (after Walkling and Coope, 1996).

reflect two different climate regimes during this interstadial. The older fauna, probably representing the first part of the interstadial, is characterized by a thermophilous marsh fauna. The younger fauna contains cold stenotherm species, signaling climatic deterioration. At the Gröbern site (Fig. 6, No. 18) in eastern Germany, the beetle succession showed large-scale post-Eemian climatic oscillations associated with MIS 5d to 5a (Fig. 7). These involved changes of at least 7°C in mean July temperatures (Walkling and Coope, 1996).

Insect remains have also been described in Germany from deposits associated with the Oerel interstadial (MIS 4). These come from both the Oerel and Gross Todtshorn sites. The beetle assemblages are composed of northern Palearctic species, indicating a treeless environment and cold climate with average summer temperatures 10–11°C colder than today (Lemdahl and Coope, 2007). The MIS 4 fauna from Oerel contains a number of arctic and alpine species, such as the rove beetles, *Pycnoglypta lurida*, *Eucnecosum brachypterum*, *Boreaphilus henningianus*, and *Tachinus elongatus*. These are all open ground inhabitants, and the accompanying paleobotanical evidence suggests that the tallest vegetation on the local landscape was probably dwarf birch (*Betula nana*) (Behre *et al.*, 2005).

3.6. Poland

Polish research includes studies of several Late Weichselian sites. Morgan *et al.* (1982) published an abstract about a small, mid-Weichselian fauna from

Belchatów, in southeastern Poland (Fig. 6, No. 4). The fauna, including *Holoboreaphilus nordenskioeldi* and *Diacheila polita*, is indicative of open ground conditions under a cold, continental climate, prior to 30,000 cal yr BP. This initial study was followed by a more detailed study by Kasse *et al.* (1998), in which attention was given to the climate reconstruction of the early phase of the last glaciation (31,000–29,000 cal yr BP). The climate during the early Late Pleniglacial was extremely harsh, and permafrost was present, indicating a mean annual air temperature less than −4°C. The landscape was essentially analogous to modern arctic tundra. The beetle assemblage from Bechatów is one of the most cold-adapted faunas so far recorded from Europe. MCR estimates indicate that T_{max} was about 8°C and T_{min} was about −27°C.

Lemdahl (1991b,f) published the results of studies on lateglacial insects from Lednica and Zabinko (Fig. 6, Nos. 53 and 100). The Zabinko samples cover the brief interval of 14,900–14,200 cal yr BP. There are no significant faunal changes in the sequence. Rather, the faunas indicate climatic amelioration from the oldest assemblage onward, with summer temperatures estimated at 14–15°C. Lemdahl contrasted this early warming with the fossil records from southern Sweden, in which assemblages of this age show summer temperatures of only 10–12°C. In fact, lateglacial warming in Scandinavia began about 500 years after warming in Poland. Deglaciation took place much earlier in Poland (ca. 21,000 cal yr BP) than in Scandinavia. Late lying ice apparently kept southern Scandinavia colder than ice-free regions at similar latitudes. The timing of amelioration recorded from Zabinko is essentially synchronous with that recorded from Britain, France, and Switzerland. However, the British warming was more dramatic than elsewhere in Europe, probably because of the overwhelming effects of sea surface temperature increases on these islands.

3.7. Byelorussia

Nazarov (1984 and written communication, 1991) has done extensive work on insect fossil assemblages dating back to the Late Cretaceous in Byelorussia and adjacent Russia. I review here only his Quaternary studies. I provide more details for these studies in this section than for other European studies, because most of this literature is in Russian, and not easily obtained outside the former Soviet Union. The Quaternary assemblages are from sites at Artuki, Logoza, the Nizhninsky Rov Ravine, Rubezhnitsa, and Timoshkovichi (Fig. 6, Nos. 2, 57, 65, 84, and 93).

Cold intervals in the Early Pleistocene brought the first boreal species to Byelorussia. Beginning about 730,000 years ago, at least five Pleistocene glaciations occurred; during three stadials the region was almost completely ice covered. Regional landscapes south of the ice during these glaciations were dominated by periglacial environments with low summer and winter temperatures, and reduced precipitation. During Pleistocene glaciations, fossil insect data suggest that summer temperatures were about 10–13°C, and winter temperatures were as low as −32°C. The beetle fauna associated with periglacial environments includes species associated with tundra, forest-tundra, and montane steppe habitats. The tundra associated species include *Diacheila arctica*, *Pterostichus tundrae*, and *Helophorus obscurellus*.

During the Early and Middle Pleistocene interglaciations, warm climates allowed the return of forests and thermophilous beetles. Glacial stadial environments were extremely cold and continental, and cold-adapted Asiatic beetle species were found in these deposits. Middle Pleistocene interglacial faunas are among the richest of the Quaternary, comprised of species still living today in Byelorussia.

The long Berezina Glaciation began about 440,000 years ago, as evidenced by the return of tundra species, including *Diacheila polita*, *D. arctica*, and *Bembidion dauricum*. Even after glacial retreat, periglacial conditions persisted for a long time, and during this interval the tundra ground beetles in the *Pterostichus* (*Cryobius*) group first appeared in Byelorussian deposits.

The Dnieper Glacier covered almost all of Byelorussia, but the glacial advance was preceded by a lengthy cold period in which the arctic rove beetle *Tachinus arcticus* first appeared in the fossil record of this region. This species is an important element in cold-stage faunas from Beringia. The faunal assemblages from the Dnieper Glaciation are a mixture of arctic and steppe species; the latter live today in Central Asia.

The Eemian interglaciation, known in Byelorussia as the Murava interglaciation (see Table 9.1 for Late Quaternary correlations and chronology), fostered the most thermophilous beetle fauna of the Quaternary. This fauna was enriched by southern European species. In addition, some species from the Soviet Far East were able to penetrate central Europe.

The first cold stage of the last glaciation began about 115,000 years ago, when beetles such as *Elaphrus splendidus*, *Pterostichus pinguedineus*, *P. tundrae*, *Bembidion dauricum*, *Patrobus septentrionis*, and *Amara torrida* arrived from northern Europe and Siberia, replacing thermophilous species. The timing of this faunal change in Byelorussia was synchronous with climatic cooling signaled by insect faunas in Britain and France. This stage was followed by the Early Poozerje interstadial complex. Thermophilous beetles returned to Byelorussia during the warmest part of the interstadial, in which the climate was similar to today.

During the stadials of the last glaciation several cold-tolerant species inhabited Byelorussia, including *Tachinus arcticus*, *Chrysolina septentrionalis*, *Chrysomela tajmyrensis*, and *Trichalophus korotyaevi*. The Middle Poozerje interstadial complex, which was colder than the previous interstadial, allowed the penetration of new tundra species into the region.

The interval 25,000–20,000 cal yr BP was characterized by a beetle fauna enriched by Central Asian steppe species such as *Stephanocleonus eruditus*, *Pterostichus (Derus) majus*, and some taiga zone species. The last (Late Weichselian) cold stage was more prolonged; the Poozerje glacier advanced to cover the whole of northern Byelorussia by about 20,000 years ago. Shrub-associated and arctic tundra beetle species dominated the fossil assemblages from the beginning of this glacial interval onwards. The tundra ground beetles, *Pterostichus* (*Cryobius*), *Pterostichus vermiculosus*, *P. sublaevis*, *P. tundrae*, *Stereocerus haematopus*, and *Amara alpina* characterize the fauna of the last-glacial stadial.

During the lateglacial, the Allerød interstadial saw the brief return of a boreal fauna to Byelorussia. During the Younger Dryas interval, some tundra species returned, including *Amara alpina* and *Patrobus septentrionis*.

In the Holocene, the open ground fauna was replaced by coniferous forest species. Thermophilous species became established by 8,800 cal yr BP, indicating warm, moist climates. At 6,300 cal yr BP, the climate became warm and dry, and steppe species appeared in the fossil assemblages. Cooler climate in the Late Holocene brought the return of boreal species.

3.8. European Russia

Bidashko and Proskurin (1988) studied Middle Pleistocene insect fossils from the Nikol'skoye site on the Volga River in western Russia (Fig. 6, No. 64). The faunas represent a forest–steppe environment, similar to that found today in forest–steppe regions of western and central Siberia. They indicate that the study region was a mosaic of meadows, meadow–steppe, and steppe habitats, interspersed with coniferous forests.

More extensive Quaternary insect studies have been performed in Siberia, the topic of the next chapter.

3.9. The Netherlands

Angus (1975) studied Early Weichselian stadial faunas (stratigraphically positioned between the Brørup and Hengelo interstadials) from Voorthuizen (Fig. 6, No. 99). A lower fossil bed yielded an arctic fauna, reflecting summer temperatures below 10°C. A younger assemblage was less climatically diagnostic, but was thought to indicate a slight warming.

Coope (1969) studied mid-Weichselian beetles from Peelo (Fig. 6, No. 74). This fauna was indicative of glacial stadial conditions, and included the tundra beetle *Pterostichus vermiculosus*, which lives today only in arctic regions of North America and Siberia. The fauna also included a number of species whose modern distributions are exclusively arctic/Asiatic, reinforcing the reconstruction of a cold, dry climatic regime.

Coope (in Bohncke *et al.*, 1987) also studied lateglacial insect assemblages from deposits in the Mark Valley of the Netherlands (Fig. 6, No. 58). The assemblages range in age from about 14,900 to 12,100 cal yr BP. The insect faunas from a peat unit (14,900–12,900 cal yr BP) showed little change. Using the MCR method, Coope estimated mean summer temperatures of 15–18°C for this interval. The beetle fauna from an overlying sand unit (12,900–12,100 cal yr BP) documents the marked cooling associated with the Younger Dryas chronozone. The temperate fauna of the peat unit was replaced by a fauna of arctic affinities. MCR estimates of summer temperatures associated with this fauna were 10–11°C.

Van Geel and Coope (in van Geel *et al.*, 1989) examined lateglacial mites and insects from Usselo, The Netherlands (Fig. 6, No. 97). Beginning at 15,900 cal yr BP, the arthropod data suggest warm conditions (mean summer temperatures of 15–20°C). A slight cooling was recorded by the fauna of the Allerød zone, followed by substantial cooling during the Younger Dryas (mean summer temperatures 10–11°C). As in British and Swedish lateglacial studies, the arthropod results from Usselo differed somewhat from paleobotanical interpretations.

3.10. France

More extensive research has been carried out in France, by both English and French researchers. At Herquemoulin on the Cotentin Peninsula (Fig. 6, No. 27), Coope *et al.* (1987) studied insect assemblages from coastal deposits of Eemian to Early Weichselian age. The oldest studied faunas reflect interglacial conditions that correspond with cool-temperate mixed forest in the paleobotanical record. Above this unit, the flora and insect fauna are indicative of arctic/subarctic conditions. Finite radiocarbon ages of various regional deposits were considered to be far too young. Based on stratigraphic correlation with Uranium-series dated sites from nearby Jersey, the cold-adapted fauna may be 115,000 years old, signaling rapid climatic deterioration following the Eemian interglacial.

Ponel (1995) identified insect remains from 10 assemblages associated with the Eemian interglacial at La Grande Pile (Fig. 6, No. 47). The beetle assemblages were divided into an early and a late faunal unit. The early unit contains many beetle species that depend on deciduous trees, as well as thermophilous predators and scavengers. The younger faunal unit probably relates to the later part of the interglacial. It includes species associated with both deciduous and coniferous trees. A number of species indicate a mild and humid climate. One Grande Pile sample is associated with MIS 5d. It includes a number of species that live on open ground, such as open grassland. An MCR estimate from this fauna indicates that summer temperatures were about 5–8°C cooler than modern, so the fauna does not represent full-glacial environments. Beetle assemblages dating to the Rederstall stadial (MIS 5b) from La Grande Pile show a fall in the number of tree-dependent taxa and the appearance of cold-adapted species. During the Odderade (MIS 5a), the number of tree-dependent beetles increased again. The climate was probably slightly cooler than during the previous Brørup interstadial (MIS 5c) (Lemdahl and Coope, 2007). Middle Weichselian (MIS 4) beetle assemblages from La Grande Pile are characterized by arctic, continental species. Obligate tree-dependent species were not found in these assemblages. The inferred climate was generally cold throughout this interval, with mean July temperatures depressed by about 10°C (Lemdahl and Coope, 2007). LGM beetle assemblages from La Grande Pile indicate very cold and harsh conditions with mean summer temperatures around

10°C and mean winter temperatures well below −10°C (Lemdahl and Coope, 2007).

Ponel and Coope (1990) collaborated on an important study of lateglacial and Early Holocene insect faunas in lake sediments from La Taphanel in the Massif Central region (Fig. 6, No. 51). The paleoclimatic sequence from La Taphanel was described in Chapter 6. The beetle fauna documents a rapid climatic warming beginning at 15,900 cal yr BP. The paleobotanical signal was substantially out of synchroneity with the insect signal until the Allerød zone. Unlike assemblages from elsewhere in Europe, the Taphanel faunas show a marked cooling between temperate Bølling zone assemblages and cooler Allerød faunas. This cooling corresponds to the Older Dryas pollen zone. Marked cooling occurred in the Younger Dryas, followed by rapid amelioration after 12,100 cal yr BP.

A lateglacial sequence from Lac Long Inférior in the French Alps (Fig. 6, No. 44) has also yielded a series of insect faunas documenting regional high-altitude environments during this interval (Ponel *et al.*, 2001). A total of 55 samples were investigated, ranging in age from about 17,000 to 6,400 cal yr BP. An assemblage correlated with the Allerød–Younger Dryas transition, is indicative of alpine grassland with stands of birch and pine growing downslope from the site. Bark beetles in these assemblages were undoubtedly transported up slope to the fossil site, most likely by winds. This phenomenon has been seen in numerous high-altitude fossil beetle assemblages from the Rocky Mountains (Elias, 1985). Early Holocene assemblages show a marked decrease in species associated with snow patches, although several cold-adapted species, such as the water-scavenger beetle *Helophorus glacialis* and the rove beetle *Olophrum consimile* persisted through the Holocene sequence. This is to be expected in a high-altitude site, where climates have never been fully temperate. Ponel *et al.* (1999) also studied lateglacial insect fossils from a high-altitude site in the eastern Pyrenees, at La Borde. This sequence contained a large faunal assemblage that dates to the Allerød–Younger Dryas transition, about 12,900 cal yr BP. The fauna contains a mixture of high-altitude forest species and alpine grassland species, marking the climatic deterioration into the Younger Dryas interval.

Studies of Holocene environments in France include insect fossil analyses from peat deposits in the High Ardennes region of northeast France (Fig. 6, No. 28) (Damblon *et al.*, 1977), and Ponel and Gadbin's (1988) study of lake sediments from Lac d'Issarlés in the Ardèche region of southeastern France (Fig. 6, No. 43). The Ardennes study chronicles the effects of human activity on the ecology of regional peat bogs, and the Lac d'Issarlés study verified paleobotanical reconstructions of the regional establishment of hazel (*Corylus avellana*) forests.

Ponel *et al.* (1992) studied insect fossil assemblages from the Taillefer Massif region of the French Alps (Fig. 6, No. 90). They reconstructed treeline fluctuations for the Holocene, and documented the faunal and floral succession from open ground tundra through mature forest. After 2,000 yr BP, the forest shifted downslope. The beetle evidence helped to establish that the cause of this movement was climatic cooling, rather than human influence.

Shotton and Osborne (in Ters *et al.*, 1971) studied fossil insects from Early Holocene estuarine deposits at Le Havre (Fig. 6, No. 25). The lower deposit (9,400 cal yr BP) contained a fauna associated with a reed swamp environment. Middle and upper deposits (older than 8,650 cal yr BP) yielded more diverse faunas, indicative of a variety of aquatic and semiaquatic habitats. While marshland deposits such as these might be taken to represent salt marsh environments, the beetle fauna lacks halobionts (salt tolerant species), and Shotton and Osborne argue that the marshes grew in freshwater. All of the identified beetle species live in northern France today, so the climate was probably very similar to modern parameters.

3.11. Switzerland and Austria

Studies in these two alpine countries have focused mainly on lateglacial and Holocene-age faunas, with two exceptions. Jost-Stauffer *et al.* (2005) published the results of a study of two insect assemblages of Middle Weichselian (early MIS 3) age from the Gossau site (Fig. 6, No. 16). The fossils were recovered from compressed peat (lignite). They include several exclusively arctic and Asiatic species that do not occur anywhere in central Europe today. Among these are the arctic ground beetles *Diacheila arctica*, *D. polita*, and *Chlaenius costulatus*. MCR estimates indicate that summer temperatures were depressed by at least 8–10°C from the modern level. Interestingly, the Gossau faunas lack some of the extreme-continental, cold-adapted species found in British faunal assemblages from this interval. This suggests that the climate of Switzerland may have been less cold and continental than that of northwest Europe at that time (Jost-Stauffer *et al.*, 2005).

Coope (2007a) published the results of his study of insects associated with mammoth remains at the Niederwenigen site, near Zurich (Fig. 6, No. 63). The fossils are from Weichselian-age sediments, dated at about 45,000 ^{14}C yr BP (Hajdas *et al.*, 2007). The assemblages include species with modern distributions limited to arctic and alpine regions of Scandinavia, Siberia, and the Altai Mountains of Central Asia. The local environment was open ground tundra with alder, birch, and willow shrubs. Paleoclimate reconstructions based on insect assemblages suggest average summer temperatures ranging from 15°C for the basal sample to 10°C for the uppermost sample. Mean January temperature estimates range from −10°C for the basal assemblage to −14.5°C for the youngest assemblage. Precise dates are unavailable for these faunas, but they appear to document the transition from an Early Weichselian interstadial to a stadial. The Niederwenigan assemblages may be more-or-less contemporaneous with those from Gossau. Both indicate similar vegetation cover and climatic regime. The Niederwenigen faunas contain many of the same cold-adapted, arctic species found in Gossau, in addition to other well-known indicators of cold climates, such as the ground beetles *Amara alpina* and *A. torrida*, the water-scavenger beetle *Helophorus glacialis*, and the rove beetle, *Holoboreaphilus nordenskioeldi*.

Studies on lateglacial insect faunas in Switzerland began with the investigation of Lobsigensee, on the Swiss Plateau (Fig. 6, No. 56) (Elias and Wilkinson, 1983), in which a group of cold-adapted insects at the base of the sequence was rapidly replaced by a thermophilous group. The date of the climatic amelioration was about 15,600 cal yr BP, but it should be emphasized that this is an interpolated age, not fixed by radiocarbon dates at the climatic boundary. Based on beetle and caddisfly data, estimated mean July temperatures at the site remained at 14–16°C throughout the Bølling and Allerød intervals. Insect preservation in Younger Dryas-age sediments was poor, and inferences drawn from insects in this interval are inconclusive, but offer little positive evidence of a Younger Dryas cooling.

As mentioned in Chapter 7, Russell Coope and I investigated a second lateglacial site in Switzerland, the Champreveyres site at Neuchâtel (Fig. 7 of Chapter 7, No. 11). While this is an archeological site, the lateglacial insect faunas reflect essentially nonanthropogenic environments, and as such provide useful paleoenvironmental evidence. The Champreveyres insect data show the intensity of lateglacial climatic amelioration (Coope and Elias, 2000). This site contains stenothermic thermophiles (e.g., the ground beetle *Calosoma inquisitor*) very close to the beginning of the lateglacial period. The degree of increase in summer temperatures (about 7°C) was of the same order in northwest Europe. This sudden and intense warming episode was synchronous over much of Western Europe. Following the initial warming, an ecologically diverse beetle fauna became established at Champreveyres. Unfortunately, the Younger Dryas zone is represented by a coarse gravel deposit with little organic preservation, so the Champreveyres insect sequence is uninformative about this period.

Lemdahl (in Gabus *et al.*, 1987) described a lateglacial insect assemblage from a terrace above Lake Geneva at Lausanne (Fig. 6, No. 52). The terrace deposits date to 14,900 cal yr BP. The insect fauna is indicative of open ground environments, but lacks arctic and alpine species. It is essentially a temperate zone fauna, in contrast to the fossil flora, which suggests arctic tundra.

Gaillard and Lemdahl (1994b) published the results of a study of lateglacial pollen, plant macrofossils, and insects from the Grand Marais de Boussens site (Fig. 6, No. 17). As at Lobsigensee and Champreveyres, the initial warming following the last glaciation is clearly marked by the replacement of cold-adapted species by temperate species, at about 15,900 cal yr BP. The faunal assemblages indicate that temperatures rose dramatically during this transition, especially winter temperatures, which rose by about 10°C. Unfortunately, the Grand Marais de Boussens assemblages lacked the necessary stenothermic species to provide a reconstruction of climatic oscillations between 15,900 and 10,800 cal yr BP.

Lemdahl (2000a) subsequently published the results of studies from four lateglacial sequences at Gerzensee (Fig. 6, No. 14), Leysin (Fig. 6, No. 55), Hérémence (Fig. 6, No. 26), and Zeneggen (Fig. 6, No. 101). Unlike previous studies of lateglacial insect faunas from Switzerland, the Gerzensee and Zeneggen sites yielded useful faunas dating to the Younger Dryas interval. These faunas show that summer temperatures had reached essentially modern levels during the Allerød interstadial, followed by a rapid depression in temperatures during the Younger Dryas, as discussed in Chapter 5. The transition from this cold interval to the subsequent Early Holocene (Preboreal) interval was extremely abrupt. Cold-adapted faunas were replaced by temperate faunas between two contiguous samples, separated by only 1 cm of sediment.

In eastern Austria, Schweiger (1967) investigated lateglacial insects from a borehole at Bad Tatzmannsdorf (Fig. 6, No. 3). Although the samples contained mixed horizons, Schweiger discerned two distinct faunas. The majority of specimens were thought to be associated with the **Allerød pollen zone**. This suite comprises thermophilous beetles that are found today in southern Europe, including arid regions of the eastern Mediterranean. The other faunal group has been assigned to an earlier, cold climatic interval. It includes boreo-montane beetles, and was correlated with the Older Dryas pollen zone, based on the **palynology** of the sediments.

3.12. Italy

Only a few studies of Pleistocene insects have been done in Italy. Fossil arthropods in an assemblage dating around 22,500 cal yr BP, were recovered from a site close to Verona, Italy (Fig. 6, No. 98) (Foddai and Minelli, 1994). They indicate climate conditions colder and wetter than today. Average summer temperatures were depressed by as much as 8–9°C from the modern level. This faunal assemblage contains the arctic and alpine ground beetle, *Amara alpina*. In Europe today, this cold-adapted species is found only on alpine tundra in Scandinavia, and in the Arctic. During glacial intervals such as the LGM, the regions in which this species dwells today were covered by ice. It was recovered from a full-glacial age fauna north of the Alps at Niederwenigen, Switzerland (Coope, 2007a), but otherwise has not been found in glacial stadial-age faunas in central or southern Europe. Nilsson *et al.* (1993) considered the most probable refugium during the last glaciation for the present arctic-alpine beetles such as *Amara alpina* was the region south of the **Würmian ice sheet**. The LGM specimens from Verona support this conclusion.

Ponel and Lowe (1992), and Ponel (1997b) reported on a sequence of lateglacial insect faunas from a high-altitude site in the Northern Apennines, called Prato Spilla (Fig. 6, No. 79). The faunas document a progressive climatic cooling during the lateglacial interval, culminating in the Younger Dryas. About 175 taxa were identified from these assemblages, sampled from an exposure along the banks of a small river. Unlike some of the cold-adapted species found in the samples from Verona, all of the identified Prato Spillo species are still living in Italy today, with the probable exception of the cold-adapted rove beetle, *Pycnoglypta lurida*. As with other mountain sites, it was difficult for Ponel to determine whether the presence of a given species was indicative of its presence on the local landscape, or whether it had been transported up slope to the site by mountain winds. Elias (1985) attempted to circumvent the problem of interpreting such high-altitude faunas by restricting his paleoenvironmental reconstructions to flightless species. Nearly all of the species identified from the Prato Spilla faunas are able to fly, making faunal interpretations more difficult. Even with this caveat, the climatic cooling associated with the Younger Dryas is particularly well marked at this high-elevation site, as it was in high-elevation sites of Gerzensee and Zeneggen in Switzerland, and at La Taphanel in France.

9

Siberian Studies

The differences in thermal regime between north- and south-facing slopes are equivalent to differences in climate on a latitudinal scale between the steppes of Central Asia and the arctic tundra of Siberia.

– Alfimov and Berman (2001)

During the last 35 years, Russian scientists have devoted a great deal of effort to the study of Quaternary environments of Siberia. This vast, remote region, the size of a continent, has grudgingly yielded its secrets to a few dedicated scientists. Most fossil studies have been carried out along the banks of major rivers (Figs. 1 and 2). Riverbank exposures have attracted researchers for two reasons: (1) unlike other areas, they can be reached by boat during the summer months and (2) the banks cut by the rivers continuously offer access to extensive, fresh exposures of Pleistocene organic-rich silts and sands. Some of these exposures contain unconsolidated sedimentary sequences which span millions of years. The sediments are in permafrost, and the fossils they contain (from pollen grains to whole mammoth carcasses) are extremely well preserved. Since the Early Holocene, these exposures have thawed and eroded little by little through successive summers, gradually yielding their frozen treasure trove of fossils.

When I reported on Russian research in ''Quaternary Insects and Their Environments'' in 1994, I said that most writings about Siberian paleoentomology had appeared in Russian publications not easily obtained outside the former Soviet Union, but a few review articles and book chapters have been published in English (Kiselyov, 1973; Kiselyov and Nazarov, 1984). Some Russian colleagues kindly provided translated summaries of their Russian papers to aid my 1994 book, but the lack of papers published in English was a real stumbling block for non-Russian readers. I am happy to report now that much more of this research has been published in English in recent years. The study sites are listed in Table 1, and the geographic positions of the localities are shown in Figs. 1 and 2. Excellent summaries of eastern Siberian research are found in Sher *et al.* (2005), Kuzmina and Sher (2006), and Sher and Kuzmina (2007). Useful summaries of western Siberian research have been published by Zinovjev (2006, 2008). Correlations between the Siberian Pleistocene sequence and MISs can be found in Table 2.

The Siberian research is fascinating, revealing major exchanges of insect species between the western sector of Beringia and central Asia, Europe, and eastern Beringia. Siberia appears to have been the principal region in which steppe–tundra developed, and small patches still exist there.

1. Western Siberian Research

The southernmost study published in English (Kiselyov, 1973) is based on a site from the banks of the Tura River (Fig. 1, No. 90). A small insect fauna, dating from the middle of the last glaciation (ca. 60,000–40,000 yr BP) was sampled from a riverbank exposure. The assemblage was a mixture of steppe, tundra, or forest–tundra, and boreal forest (taiga) species. The tundra/forest–tundra element includes three ground beetle species: *Nebria nivalis*, *Pelophila borealis*, and *Diacheila polita*. Approximately one-third of the identified species are indicative of dry steppe environments. There is no modern analog for either the beetle assemblage or the environmental conditions it reflects. Kiselyov concluded that the fauna represents a time of rapid climatic change, in which elements of various biological communities shifted distributions across the landscape. I discuss the phenomenon of steppe and tundra insect communities further in the next chapter, because this combination appears repeatedly in Siberian (and Alaskan) Pleistocene assemblages.

On the Jamal Peninsula in northwestern Siberia, Erochin and Zinovjev (1991) have studied lateglacial insect faunas from the Ljabtosjo and Ngojun sites (Fig. 1, Nos. 56 and 65), and Kiselyov (1988) examined insects from Ust-Yuribey (Fig. 1, No. 96). The lateglacial assemblages reflect arctic tundra conditions, similar to those found today on the Taimyr Peninsula. The assemblages included *Carabus truncaticollis*, *Amara glacialis*, *Pterostichus tundrae*, and *Tachinus arcticus*.

Kiselyov (1988) also studied a series of Pleistocene assemblages from exposures along the Ob River (Fig. 1, Nos. 14, 22, 31, 32, 42, 58, and 65) and the Gyda (Fig. 1, Nos. 15, 26, and 99), and Taimyr Peninsulas (Fig. 1, Nos. 4, 78, and 82) in western Siberia. During interglacial and interstadial intervals of the Middle to Late Pleistocene, insect assemblages suggest summer temperatures averaging 2–3°C warmer than present. Summer temperatures during glacial stadials were about 10°C cooler than present, and an arctic tundra biota dominated the region. Both mesic tundra (e.g., *Pterostichus* (*Cryobius*) species) and xeric tundra beetles have been identified from stadial assemblages, in addition to some weevils found in steppe habitats.

DEVELOPMENTS IN QUATERNARY SCIENCES
VOLUME 12 ISSN 1571-0866

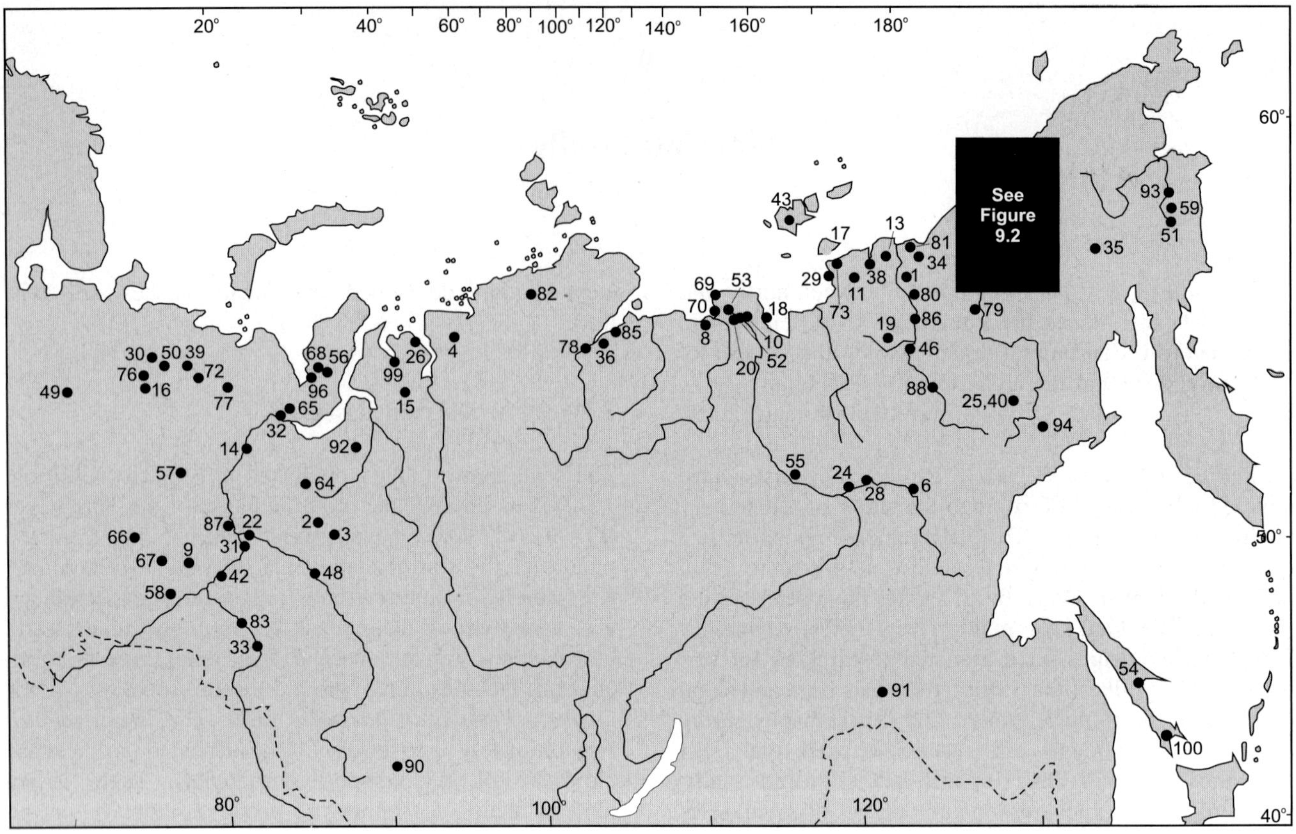

Fig. 1. Map of central and northern Asia, showing location of insect fossil localities in Siberia. Inset box is shown in larger scale map in Fig. 2.

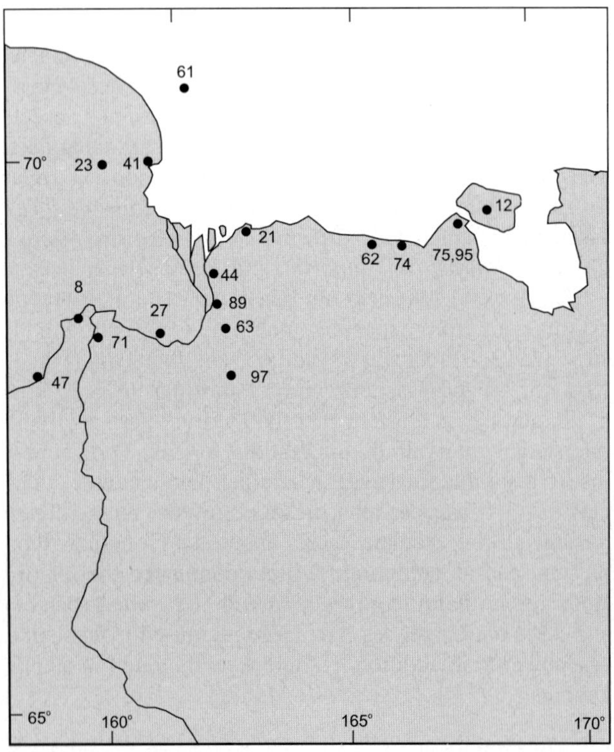

Fig. 2. Map of northeastern Siberia, showing location of insect fossil sites.

Zinovjev (2006, 2008) presented the results of numerous studies of Middle and Late Pleistocene insect fossil assemblages from the Ural Mountains and western Siberia (Table 3). Environmental reconstructions based on these assemblages have been made largely by comparison with modern regional faunas, although knowledge of the modern ecology and distribution of most regional species remains far from complete. The fossil faunas tend to fall within defined ecological groups (Table 4). These include arctic and subarctic tundra species, boreal species, and steppe species. As in many other regions, there are no exact modern geographic analogs for many of the fossil assemblages from western Siberia. The oldest site studied thus far is at Tchembaktchinskiy Yar (Fig. 1, No. 87). This site contains faunal assemblages that have been dated by thermoluminescence to 650,000 ± 100,000 and 313,000 ± 80,000 yr BP. These faunas are dominated by cold-adapted species, associated with forest or forest–steppe environments. At the Kul'egan site (Fig. 1, No. 48), a suite of fossil assemblages spanning 360,000–290,000 yr BP shows a series of environmental changes from subarctic to boreal environments, followed by a climatic cooling in which arctic species replaced the boreal species. Subsequent climatic amelioration was reflected by a rise in subarctic species, followed by the return of a boreal fauna. A fauna dating to the end of MIS 3 is dominated by subarctic species.

Table 1. Locations and references of Siberian Quaternary insect fossil sites.

Site	Reference(s)
(1) Achchagyy-Allaikha	Kiselyov and Nazarov (1984)
(2) Agan River	Zinovjev (2006)
(3) Aganskiy uval	Zinovjev (2006)
(4) Agapa	Kiselyov and Nazarov (1984), Kiselyov (1988)
(5) Alazeya River	Kuzmina and Sher (2006)
(6) Aldan River	Kuzmina and Sher (2006)
(7) Alyoshkina Zaimka	Kiselyov and Nazarov (1984)
(8) Anabar River	Golosova *et al.* (1985)
(9) Andryushino	Zinovjev (2006)
(10) Arga-Bilir-Aryta and Samoylov Islands	Kuzmina and Sher (2006)
(11) Ary-Mas River	Golosova *et al.* (1985)
(12) Ayon Island	Kiselyov and Nazarov (1984), Golosova *et al.* (1985)
(13) Berelekh	Kiselyov and Nazarov (1984)
(14) Berezovo	Kiselyov (1988)
(15) Bol'shiye Kheta	Kiselyov (1988)
(16) Bol'shoy Aranets	Kiselyov and Nazarov (1984)
(17) Bol'shoy Lyjakhovsky Island	Golosova *et al.* (1985)
(18) Bykovsky Peninsula	Kuzmina and Sher (2006)
(19) Bylgnyr	Kuzmina and Sher, (2006)
(20) Byor-Khaya	Sher and Kuzmina (2007)
(21) Cape Letyatkin	Golosova *et al.* (1985)
(22) Chembakcheno	Kiselyov (1988)
(23) Chukoch'ya River	Matthews (1974a), Kiselyov and Nazarov (1984)
(24) Chukskoye Exposure	Kiselyov and Nazarov (1984)
(25) Dima Mammoth Site	Kiselyov *et al.* (1982), Kuzmina and Sher (2006)
(26) Dorofeyevskaya	Kiselyov (1988)
(27) Duvanny Yar	Kiselyov and Nazarov (1984)
(28) Dygdal	Kiselyov and Nazarov (1984)
(29) Entrykayskiy Ravine	Kiselyov and Nazarov (1984)
(30) Garevo	Kiselyov and Nazarov (1984)
(31) Gornopravdensk	Kiselyov (1988)
(32) Gorny Kazymsk	Kiselyov (1988)
(33) Kazakovka	Zinovjev (2006)
(34) Keremsit River	Kiselyov and Nazarov (1984), Golosova *et al.* (1985)
(35) Kergoli River	Golosova *et al.* (1985)
(36) Khatanga	Kiselyov and Nazarov (1984)
(37) Khomus-Yuryakh	Sher and Kuzmina (2007)
(38) Khroma River	Kiselyov and Nazarov (1984), Krivolutsky and Druk (1990)
(39) Kipiyevo	Kiselyov and Nazarov (1984)
(40) Kirgiljakh River	Kiselyov and Nazarov (1984), Krivolutsky and Druk (1986)
(41) Kon'kovaya	Kiselyov and Nazarov (1984), Golosova *et al.* (1985)
(42) Koshelevo	Kiselyov (1988)
(43) Kotelny Island	Kuzmina and Sher (2006)
(44) Kray Lesa	Kiselyov and Nazarov (1984)
(45) Krazivoye	Sher and Kuzmina (2007)
(46) Krest-Mayor	Kuzmina and Sher (2006)
(47) Krestovka River	Kiselyov and Nazarov (1984), Sher *et al.* (1977)
(48) Kul'egan	Zinovjev (2006)
(49) Kur'yador	Kiselyov and Nazarov (1984)
(50) Kushshor	Kiselyov and Nazarov (1984)
(51) Ledovy Obryv	Kiselyov and Nazarov (1984)
(52) Lena Delta	Kuzmina and Sher (2006)
(53) Lena River	Golosova *et al.* (1985)
(54) Leonidovka	Kiselyov and Nazarov (1984)
(55) Lepiske	Kiselyov and Nazarov (1984)
(56) Ljabtosjo	Erochin and Zinovjev (1991)
(57) Loz'va	Zinovjev (2006)

(*Continued*)

Table 1. (*Continued*)

Site	Reference(s)
(58) Malkovo	Kiselyov and Nazarov (1984)
(59) Mamontov Obryv	Kiselyov and Nazarov (1984)
(60) Mamontovyy Khayata Cliff	Sher *et al.* (2005)
(61) Medvezhyi Islands	Kuzmina and Kolesnikov (2000)
(62) Milkera River	Kiselyov and Nazarov (1984), Golosova *et al.* (1985)
(63) Molotkovskiy Kamen'	Kiselyov and Nazarov (1984)
(64) Nadym	Kiselyov (1988)
(65) Ngojun	Erochin and Zinovjev (1991)
(66) Nikitino	Zinovjev (2006)
(67) Nizhnjaya Tavda	Zinovjev (2006)
(68) Nyamu	Kiselyov (1988)
(69) Nykolay Lake	Kuzmina and Sher (2006)
(70) Olenyok Channel, Nagym	Sher and Kuzmina (2007)
(71) Omolon River	Kiselyov and Nazarov (1984), Golosova *et al.* (1985)
(72) Os'van'	Kiselyov and Nazarov (1984)
(73) Oyogossikiy Ravine	Kiselyov and Nazarov (1984)
(74) Primorskiy	Kiselyov and Nazarov (1984)
(75) Rauchua River	Golosova *et al.* (1985)
(76) Rodionovo	Kiselyov and Nazarov (1984)
(77) Rogovaya	Kiselyov and Nazarov (1984)
(78) Romanikha	Kiselyov and Nazarov (1984)
(79) Sededema	Kiselyov and Nazarov (1984)
(80) Shamanovo	Kiselyov and Nazarov (1984)
(81) Shandrin River	Grunin (1973), Kiselyov and Nazarov (1984)
(82) Shrenk	Kiselyov and Nazarov (1984)
(83) Skorodum	Zinovjev (2006)
(84) Stanchikovsky Yar	Sher and Kuzmina (2007)
(85) Syndassko Bay	Golosova *et al.* (1985)
(86) Sypnoy Ravine	Kiselyov and Nazarov (1984)
(87) Tchembakchinskiy yar	Zinovjev (2006)
(88) Tirekhtyakh	Kiselyov and Nazarov (1984)
(89) Tretiy Ruchey	Kiselyov and Nazarov (1984)
(90) Tura River	Kiselyov (1973)
(91) Tynda	Kiselyov and Nazarov (1984)
(92) Urgenoi	Kiselyov (1988)
(93) Ust'-Alganskiy	Kiselyov and Nazarov (1984)
(94) Ust'-Omchug	Kiselyov (1994); Kuzmina and Sher (2006)
(95) Ust'-Rauchua	Kiselyov and Nazarov (1984)
(96) Ust-Yuribey	Kiselyov (1988)
(97) Utlinskiy Kamen'	Kiselyov and Nazarov (1984)
(98) Yakutskoe Lake	Sher and Kuzmina (2007)
(99) Yuribey River	Kiselyov and Nazarov (1984), Krivolutsky and Druk (1986)
(100) Yuzhno-Sakhalinsk	Golosova *et al.* (1985), Krivolutsky and Druk (1986)

The other studies discussed in Zinovjev's (2006, 2008) papers concern assemblages from the last glaciation. Faunas from the last interglacial (MIS 5e) reflect boreal forest conditions as warm as or warmer than those at the study sites today. Faunas from the Mal'kovo and Andryushino sites that are associated with the interval 70–60 ka are dominated by arctic or boreo-arctic species, reflecting climatic cooling associated with MIS 4 (Zinovjev, 2008). Several faunas from western Siberia are associated with the interstadial conditions of MIS 3. Several of these faunas contain forest–steppe species. Interstadial faunas from two sites along the lower reaches of the Irtysh River, dated about 31,000 cal yr BP, present seemingly conflicting faunas. Even though these sites are relatively close to each other, the Kazakova site fauna (Fig. 1, No. 33) contained species indicative of steppe–tundra environments, while the Skodorum site fauna (Fig. 1, No. 83) contained a mixture of forest–steppe and subarctic species. A fauna of similar age from the Nikitino site (Fig. 1, No. 66), further west (Transuralia region), likewise indicated forest–steppe environments. Forest–steppe covers extensive regions of western Siberia today. The northern edge of this ecosystem stretches from the Ural Mountains along the southern end of the West Siberian Plain, and south to the foothills of the Altai and Sayan Mountains. The southern reach stretches along the

Table 2. Correlation of eastern European and Siberian Pleistocene sequences with oxygen isotope stages.

Isotope stage	Approximate age (cal yr BP)*	Byelorussian sequence**	European Russian sequence***	Siberian sequence****
1	0–11 K	Holocene	Holocene	Holocene
2	11–26 K	Late Poozerje glaciation	Late Valdai glaciation	Upper Zyrianka/ Sartanian stadial
3	26–55 K	Mid-Poozerje interstadial complex	Bryansk/Dunaevo interstadial	Karginsky interstadial
4–5a	55–110 K	Early Poozerje stadial (began ca. 115 K)	Early Valdai stadial	Lower Zyranka stadial
5e	110–130 K	Murava interglaciation	Mikulino interglacial	Kazantsevo interglacial
6	130–180 K		Dnieper glaciation	Taz glaciation

*Ages after Arkhipov *et al.* (1986)
**Byelorussian correlations after Nazarov (1991)
***Eastern European corrrelations after Velichko and Faustova (1986)
****Siberian correlations after Arkhipov *et al.* (1986)

Table 3. Summary of Quaternary insect fossil assemblages from western Siberia and the Ural Mountains (data from Zinovjev, 2006, 2008).

Site	Region	Age (cal yr BP)	Inferred environment
Tchembaktchinskiy Ravine	Lower reaches of Irtysh river, W. Siberia	650 ± 110 ka 313 ± 80 ka	Cold-adapted fauna associated with forest or forest–steppe vegetation
Kul'egan 2251	Ural Mountains	360–290 ka	Subarctic → Boreal → Arctic → Subarctic → Boreal
Karymkary	Middle reaches of Ob River, W. Siberia	ca. 130 ka	Interglacial – boreal
Loz'va-2	Loz'va River, Urals	ca. 125 ka	Interglacial – boreal
Multiple	Lower reaches of Ob River, W. Siberia	110–105 ka	Arctic fauna
Mal'kovo	Tura river, near Tyumen, Siberia	70–60 ka	Arctic/boreo-arctic fauna
Andryushino	Tavda river, SW Siberia	70–60 ka	Arctic/boreo-arctic fauna
Kazakovka	Lower reaches of Irtysh river, W. Siberia	ca. 31 ka	Steppe–tundra fauna
Skorodum	Lower reaches of Irtysh river, W. Siberia	31 ka	Forest–steppe and subarctic fauna
Nikitino	Middle Transuralia	29.2 ka	Forest–steppe
Kul'egan 2247	Ural Mountains	ca. 25 ka	Subarctic fauna
Agan-1082/1	Agan River, W. Siberia	ca. 20 ka	Steppe–tundra fauna
Agan-1082/2	Agan River, W. Siberia	ca. 15 ka	Boreal fauna
Agan-4068/2	Agan River, W. Siberia	13 ka	Arctic fauna associated with boreal vegetation
Kul'egan 2241	Ural Mountains	12.7 ka	Boreal fauna associated with open-ground tundra vegetation
Lugovskoye	Lower reaches of Ob river	10 ka	Subarctic
Vansevat	Middle reaches of Ob river	9.3 ka	Warmer-than-modern fauna
Loz'va	Northern Urals	6 ka	Boreal forest

Russia–Kazakhstan border; it crosses the reaches of the Middle Irtysh River in East Kazakhstan (Gvozdetskiy and Mikhailov, 1987). The biota of this vast region is extremely diverse, containing mixtures of forest and steppe species. Patches of coniferous woodland are interspersed with steppe grasslands in this semiarid ecosystem.

Lateglacial assemblages from the Agan River in western Siberia (Fig. 1, No. 2) and from the Kul'egan site in the Urals contain faunas with ecological requirements seemingly at odds with the coincident vegetation. The former assemblage, dated ca. 13,000 cal yr BP, contains an arctic fauna associated with boreal vegetation. The latter fauna, dated ca. 12,700 cal yr BP, contains a boreal fauna associated with open-ground tundra vegetation. Zinovjev (2006) discusses such biotic anomalies in terms of vegetation lagging behind rapid climatic change. The Agan River fauna may thus indicate rapid climatic

Table 4. Ecological groups of beetle taxa found in Late Pleistocene fossil assemblages of western Siberia.

Ecological group	Description	Principle species
Arctic and subarctic tundra insects	Arctic tundra: up to 70% of individuals in predaceous groups Subarctic tundra: similar in composition to modern fauna of shrub tundra regions of Yamal Peninsula	Arctic: *Pterostichus costatus*, *P. sublaevis*, *Blethisa catenaria*, *Elaphrus lapponicus*, *Tachinus arcticus*, *Chrysolina cavigera*, *C. instabilis*, *C. subsulcata*, *Lepyrus nordenskioeldi*; detritus feeders include *Hypnoidus* and *Oedostetus* (Elateridae) Subarctic: *Pterostichus vermiculosus*, *P. adstrictus*, *Amara alpina*, *A. brunnea*, *Lepyrus nordenskioeldi*
Subarctic insects Boreal (taiga) insects	Boreal forest inhabitants; increased taxonomic diversity over tundra zone faunas; tree-associated families include Scolytidae, Anobiidae, Erotylidae, Cucujidae, Cisidae, Curculionidae	*Tachyta nana*, *Dromius* spp., *Xantholinus*, *Nudobius*, *Hylobius*, *Pissodes*, *Magdalis*
Steppe zone insects	Steppe-region species; large numbers of detritivores, especially in Tenebrionidae, Meloidae, Scarabaeidae (dung beetles)	*Harpalus*, *Zabrus*, *Polystichus connexus*, *Poecilus nitens*, *Amara (Curtonotus) fodinae*, weevil subfamily Cleoninae

Source: After Zinovjev (2006).

deterioration during the Younger Dryas interval, as reflected in the arctic beetle fauna. The Kul'egan fauna may represent a subsequent rapid climatic amelioration, in which boreal-zone insects invaded the region well before the arrival of boreal vegetation.

Postglacial faunas from western Siberia show warmer-than-modern conditions in the Early Holocene, such as a fauna from the Vansevat site on the Ob River, dated 9,300 cal yr BP. By 6,000 cal yr BP, regional climates had cooled in the northern Urals (Zinovjev, 2008).

2. Research in Northeastern Siberia

The most thoroughly studied Asian region is northeastern Siberia, especially the Indigirka lowland (Fig. 1, Nos. 1, 13, 19, 34, 46, 80, 86, and 88) and the Kolyma lowland (Fig. 2). Regional Quaternary faunas are composed of tundra, forest–tundra, and steppe species. The steppe beetles are mostly weevils in the subfamily Cleoninae. The steppe species in the fossil assemblages live today in regions further south in Siberia and in Mongolia. However, some species still persist in arid habitats in northern Siberia, from the Taimyr Peninsula eastward. Most of northeastern Siberia was unglaciated during the Pleistocene, but this region experienced cold, dry, continental climatic conditions during glacial stadials. Regional insect assemblages from these intervals were dominated by steppe species, in combination with beetles associated with xeric tundra habitats, including the ground beetle species *Amara alpina* and *A. glacialis*, the leaf beetle species *Chrysolina septentrionalis* and *C. subsulcata*, and the pill beetle (Byrrhidae) *Morychus viridus* (Kiselyov and Nazarov, 1984; Berman, 2001). Sher and Kuzmina (2007) developed a fuller treatment of the regional faunas, dividing the species found in Pleistocene assemblages into several ecological groups (Table 5). The xeric group is made up of species adapted to dry environments, including steppe, meadow-steppe, and cold-steppe habitats. Steppe faunas include weevils in the genus *Stephanocleonus*, whereas meadow-steppe faunas typically contain weevils in the genus *Coniocleonus*. The cold-steppe habitat is characterized by the pill beetle *Morychus viridis* (see discussion of this species, below). The dry tundra group consists of species found today on well-drained sites with diverse herbaceous vegetation. These beetles seek out the warmest patches of ground, and many have populations today in open-ground habitats within the boreal forest (taiga) zone of northeast Asia. This group includes a suite of ground beetle species (see Table 4), as well as the leaf beetle *Chrysolina marginata borealis* and the weevils *Sitona borealis*, *Hypera ornata*, and *Vitavitus thulius*, a species also found in steppe–tundra faunal assemblages in Alaska and the Yukon. The arctic tundra group is characterized by cold-adapted beetle species found in various arctic habitats today. These include three species of the leaf beetle *Chrysolina*, and the weevil, *Isochnus arcticus*. The mesic tundra group includes species that live in tundra environments where the moisture conditions range from medium to wet, including bogs in the boreal zone. This faunal group shares most of its species with the mesic tundra faunas of the Alaskan Pleistocene, including ground beetles in the *Cryobius* group of the genus *Pterostichus*, *Blethisa catenaria*, and *Diacheila polita*. Rove beetles in this group include *Tachinus arcticus* and *T. brevipennis*, and the leaf beetle *Chrysolina septentrionalis*. Finally, the forest–tundra group includes species found in boreal forest regions today, such as bark beetles and ants. It also includes species that live in boreal meadows, and species associated with shrub vegetation, but not with dwarf arctic shrubs.

Table 5. Ecological groups of beetle taxa found in Late Pleistocene fossil assemblages of northeastern Siberia.

Ecological group	Description	Principle species
Xeric (dry-adapted) insects	Includes inhabitants of steppe, meadow-steppe, cold-steppe, and other xeric habitats; many steppe species found today in steppe regions to the south; meadow-steppe and cold-steppe species found today in patches of relict steppe in the north, or in dry patches within boreal zone.	Steppe: *Harpalus pusillus*, *Cymindis arctica*, *Chrysolina perforate*, *C. brunnicornis bermani*, *Galeruca interrupts circumdata*, *Stephanocleonus eruditus*, *S. fossulatus*, *S. incertus* Meadow-steppe: *Harpalus vittatus kiselevi*, *Troglocollops arcticus*, *Chrysolina arctica*, *Phyllobius kolymensis*, *Coniocleonus cinerascens*, *C. ferrugineus*, *C. astragali* Cold-steppe: *Morychus viridis* Other xeric habitats: *Notiophilus aquaticus*, *Aphodius* spp.
Dry tundra insects	Found on warmest sites in the tundra zone: well-drained sites with diverse herbaceous cover. Many found in taiga zone today, in suitable biotopes. Some species also found on relict steppe patches.	*Bembidion dauricum*, *Pterostichus (Derus) nearcticus*, *Pterostichus abnormis*, *P. sublaevis*, *Stereocerus haematopus*, *Amara (Curtonotus) alpina*, *A. interstitialis*, *A. glacialis*, *Dicheirotrichus (Trichocellus) mannerheimi*, *Chrysolina marginata borealis*, *Mesotrichapion wrangelianum*, *Hemitrichapion tschemovi*, *Sitona borealis*, *Hypera ornate*, *H. diversipunctata*, *Vitavitus thulius*
Arctic tundra insects	Cold-resistant species from various Arctic habitats.	*Chrysolina tolli*, *C. subsulcata*, *C. bungei*, *Isochnus arcticus*
Mesic tundra insects	Found on mesic to moist tundra in cold regions; some found today in bogs within boreal zone.	*Carabus truncaticollis*, *Blethisa catenaria*, *Diacheila polita*, *Pterostichus (Cryobius) ventricosus*, *P. (Cryobius) pinguedineus*, *P. (Cryobius) brevicomis*, *P. vermiculosus*, *P. costatus*, *P. agonus*, *Cholevinus sibiricus*, *Olophrum consimile*, *Tachinus arcticus*, *T. brevipennis*, *Chrysolina septentrionalis*
Forest–tundra insects	Includes inhabitants of boreal forest, boreal meadow, and shrub habitats. Shrub species feed mainly on willows, but not arctic dwarf-shrub willows.	Boreal forest beetles: *Pterostichus magus*, *Denticollis varians*, *Hylobius piceus*, *Pissodes insignatus*, *P. irroratus*, *Ips cembrae*, *Polygraphus* sp., ants: *Leptothorax acervorum*, *Formica gagatoides*, *Camponotus herculeanus* Boreal meadows: *Byrrhus fasciatus*, *Cytilus sericeus*, *Hypnoidus hyperboreus*, *Bromius obscurus*, *Phaedon concinnus*, *P. armoraciae*, *Phyllobius virideaeris*, *Sitona lineellus* Shrubs: *Chrysomela blaisdelli*, *Phratora polaris*, *P. vulgatissima*, *Lepyrus nordenskioeldi*, *L. gemellus*, *Dorytomus rufulus amplipennis*, *Isochnus flagellum*.

Source: After Sher and Kuzmina (2007).

Fossil assemblages from northeastern Siberia generally contain a restricted number of species. These are dominated by species associated with tundra, steppe, and other xerophilic habitats. This makes the species composition of Pleistocene fossil assemblages from northeastern Siberia rather monotonous. They differ from each other mostly in the relative abundance of fossils of certain taxa (Sher and Kuzmina, 2007). This situation is unlike that of many other regions, such as Western Europe or temperate North America, where different climatic intervals are marked in the fossil assemblages by wholesale changes in species composition. Siberian researchers, starting with Kiselyov (1981), have therefore used changes in the relative proportion of the various ecological groups through time to track the history of environmental changes in this region.

Eastern Siberia formed the western part of the larger unglaciated arctic landmass called Beringia (Fig. 3). Beringia was a refuge for cold-adapted biota during the Pleistocene, when most of the Arctic was covered with ice. Beringia comprised unglaciated regions of eastern Siberia, Alaska, the western part of the Yukon Territory, and the exposed continental shelf between the continents.

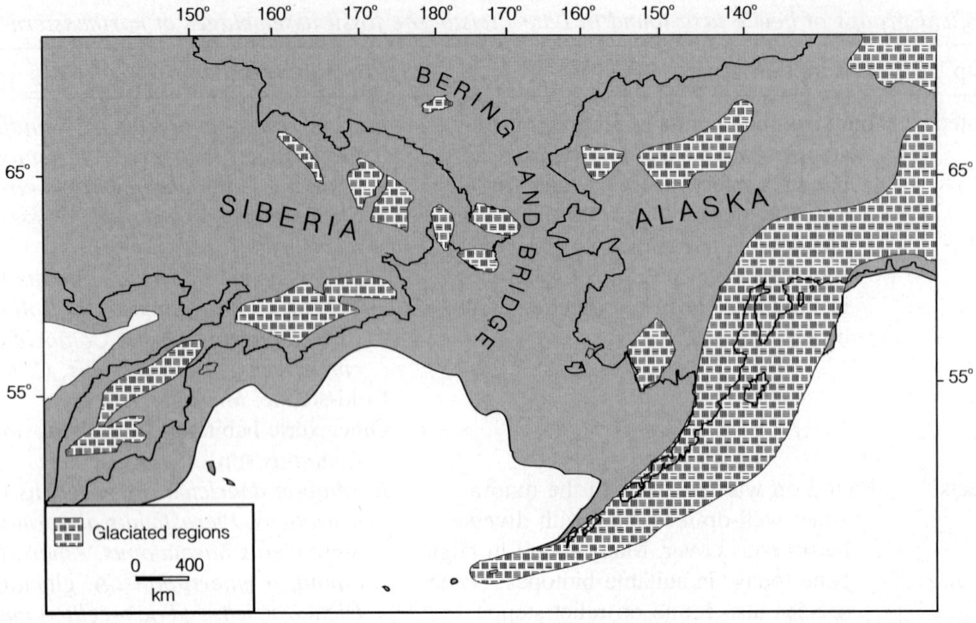

Fig. 3. Map of Beringian environment at about 18,000 yr BP, showing exposed regions of continental shelf and position of glacial ice (after Barry, 1982).

2.1. Modern Comparative Studies

Berman (1990) studied Pleistocene records of the pill beetle, *Morychus*, from Siberian deposits. Pill beetle (Byrrhidae) fossils play an important part in many Quaternary assemblages throughout Beringia, comprising as much as 60% of some fossil assemblages. Because of their importance to paleoecological studies, Berman conducted modern ecological research on pill beetles in eastern Siberia. Although many pill beetles are moss feeders, Berman found that one of the important species in Quaternary assemblages, *M. viridis*, does not live today in mesic and wet or semiaquatic moss habitats (such as bogs, swamped woodlands, and shoreline localities). Rather, it is found in xeric plant communities, such as relict steppes, mountain tundra (mostly on south-facing slopes), and various lichen-dominated communities in tundra–steppe, in places where the sedge, *Carex argunensis*, is present. This sedge species grows in well-drained sites, which are confined regionally to the axial parts of mountains and foothills, on steep slopes, plateaus, and other similar localities (Fig. 4). It is a member of various plant community types. The larvae of *M. viridis* are generally confined to dense patches of the moss, *Polytrichum piliferum*, on which they live and feed.

The habitat preferences of *Morychus aeneus* differ sharply from those of *M. viridis*. *M. aeneus* is found on Kolyma River terraces in meadow localities with sandy loam soils and abundant *Ceratodon purpureus* moss. Observations on the ecology of *Morychus viridis* lead to the following environmental correlations: (1) scarcity of snow and strong winds in winter; (2) dry summer conditions, ranging in temperature from cold to locally warm (microclimatic steppe); (3) dominance of a mosaic vegetation cover of various groups, including sedges and mosses; and (4) variations in the ratio of steppe to tundra plants, depending on the availability of heat and permafrost conditions in microhabitats.

How do steppe-adapted species survive today in northeastern Siberia? Alfimov and Berman (1991, written communication) have studied modern communities of relict steppe in this region. Isolated biotic communities associated with warm, dry environments occupy microhabitats on south-facing slopes on the arctic tundra. The principal ranges of these taxa are far to the south of the tundra biome in Asia and southeastern Europe, and their presence in Siberia makes an interesting problem in modern as well as paleobiology, because the fossil record demonstrates their presence in northeastern Siberia throughout much of the Pleistocene.

Two main hypotheses have been proposed for the existence of relict steppe in northern Siberia. Species of the steppe biota either persist in the north because they are adapted to cold climates or they are confined to specific microclimates formed within steppe habitats that differ sharply from those of the adjacent tundra regions. The second hypothesis postulates that either the thermal conditions in the relict steppe communities are similar to those found in the southern steppes of Central Asia or that thermal conditions are not an important factor determining the distribution of these beetles.

Alfimov and Berman (2001) studied the biota, microclimate, and soil conditions on both north- and south-facing slopes in the Kolyma lowland. The north-facing slopes are consistently colder, and the active layer (the layer which thaws seasonally) is considerably thinner than on the south-facing slopes, where steppe vegetation persists in patches. The soils at the steppe sites thaw deeply in summer, and are well enough drained to create xeric habitats. The soils on north-facing slopes remain moist, and support mesic- to moist-tundra vegetation. The study showed that by June or early July, soil temperatures at 20 cm depth on south-facing slopes average 25–26°C and soil surface temperatures reach 58–62°C. On north-facing slopes, soil temperatures at 20 cm depth reach only 1–2°C, and the water content of the soil approaches

(a)

(b)

Fig. 4. (A) Photograph of Morychus viridis *habitat on dry, south-facing slope exhibiting steppe vegetation, Kolyma lowland (photo by Daniel Berman, used with permission). (B) Photograph of relict steppe-tundra vegetation on south facing slope in eastern Alaska (photo by Scott Armbruster, used with permission).*

saturation. By mid-September, the soils on the north-facing slope are frozen, whereas the soils on the south-facing slope remain thawed as late as mid-October. The steppe microclimate is very hot and dry in comparison with surrounding tundra environments, and the relict steppe communities survive in the north because of the stability of the microclimate established on south-facing slopes.

Alfimov and Berman suggested that the faunal similarities between Early Pleistocene thermophilous insect assemblages and modern steppe faunas indicate that throughout at least the last 2.5 million years, some steppe habitats have persisted in northeastern Siberia. Highly continental climate apparently dominated most of the Pleistocene in western Beringia. Based on the thermal requirements of such steppe-adapted taxa as the weevil genus *Stephanocleonus*, mean summer temperatures must have been at least 10–11°C, even during glacial stadial intervals.

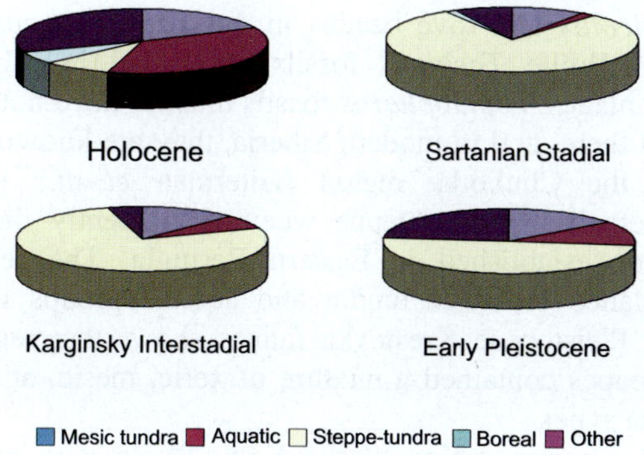

Fig. 5. Pie charts of percentage composition of ecological groups in fossil insect assemblages from Holocene (Duvanny Yar Sample 13–16), Sartanian glacial stadial (Duvanny Yar Sample 7–8), Karginsky interstadial (Duvanny Yar Sample 3), and Early Pleistocene (Krestovka River Sample 67) deposits in northeastern Siberia (data from Giterman et al., 1982).

2.2. Early Pleistocene Faunas

Conditions suggestive of steppe–tundra began to take shape in the Kolyma lowlands before 700,000 yr BP (Giterman *et al.*, 1982). Small patches of dry tundra expanded to large steppe-like landscapes by the last glaciation, probably under the influence of very cold, dry, continental climates. This continentality was fostered by the closing of oceanic circulation between the Pacific and Arctic oceans when sea level fell, exposing the broad continental shelf regions between the continents (Barry, 1982). This created the Bering Land Bridge, which was in existence for much of the Late Pleistocene (Matthews, 1982).

During interglacial and interstadial warm periods, the steppe–tundra assemblages were replaced by boreal forest and mesic tundra assemblages, the latter including species in the *Cryobius* group of the ground beetle genus, *Pterostichus*. Sher *et al.* (1977) studied Early to Mid-Pleistocene insect faunas from an exposure on the Krestovka River in the Kolyma lowland (Fig. 2, No. 47). No radiocarbon dates are available from the Krestovka section, and much of it is undoubtedly well beyond the range of ^{14}C dating (Giterman *et al.*, 1982). Beetles identified from Early Pleistocene assemblages include both mesic and xeric tundra species, but with a large numbers of *Morychus aeneus*, in addition to some steppe weevils. Fig. 5 shows the percentage of ecological groups in this Early Pleistocene assemblage. The composition of Pleistocene beetle faunas from eastern and western Beringia differs significantly. Western Beringian faunas reflecting cold, dry conditions are dominated by xeric-adapted byrrhids (e.g., *Morychus*) and steppe weevils now found in southern and central Asia, whereas Eastern Beringian faunas indicative of similar conditions include large numbers of the weevil species *Lepidophorus lineaticollis* and rove beetles in the *Tachinus* (*apterus*) group. While *Tachinus* fossils are found in Siberian assemblages, *Lepidophorus* fossils thus far have not been found there, and in modern Siberia, they are known only from the Chukotka region (Giterman *et al.*, 1982). Conversely, Asiatic steppe weevils apparently did not become established in Eastern Beringia. The relative abundance of mesic tundra and aquatic groups in the Early Pleistocene Krestovka faunas shows that regional landscapes contained a mixture of xeric, mesic, and wet habitat types.

The upper section of the Early Pleistocene reflects more mesic conditions. The beetle faunas include water beetles, boreal carabids, few *Morychus*, and no steppe weevils.

Mid-Pleistocene environments have been described as the Olyor Suite in northeastern Siberia. An exposure on the Chukoch'ya River (Fig. 2, No. 23) yielded peats studied by Matthews (1974a). He identified a small fauna (30 taxa), including *Pterostichus* (*Cryobius*) species, *Amara alpina*, *Trichocellus puncatellus*, and abundant *Morychus* specimens. Again, these elements combine in a steppe–tundra environment. The Eastern Beringian fauna with the greatest affinity to Olyor Suite is the Late Pleistocene fauna from Cape Deceit, Alaska (Chapter 10). This fauna also suggests steppe–tundra, but roughly contemporaneous (Mid-Pleistocene) assemblages from Cape Deceit indicate mesic tundra, not steppe–tundra.

At the Krestovka River site, Olyor Suite faunas and floras once again indicate steppe–tundra environments. The beetle assemblages include steppe weevils, the rove beetle *Tachinus apterus*, and the ground beetle *Bembidion dauricum*, and also the carpenter ant *Camponotus herculeanus*, which needs trees to make nests. Younger Olyor Suite assemblages from Krestovka show increased numbers of boreal forest insects, including the boreal ground beetle species, *Trachypachus zetterstedti*, and *Camponotus herculeanus*. These results appear to differ from Matthews' Olyor Suite faunas; the species composition is of a very different character. It is likely that the sites are not precisely contemporaneous. After all, the term "Mid-Pleistocene" covers many tens of thousands of years, and except for extinct mammalian faunas, few chronostratigraphic markers establish the age of these Siberian deposits.

2.3. MIS 5e Interglacial Faunas

Interglacial deposits are not easily found in northeastern Siberia. In Alaska and the Yukon (i.e., in Eastern Beringia), interglacial deposits are often marked by the presence of "forest beds." These are layers of woody forest litter that clearly mark interglacial deposits exposed in cut banks of major rivers. In Western Beringia, the sedimentology of interglacial deposits is generally quite similar to that of stadial and interstadial intervals. Nevertheless, a number of insect fossil assemblages inferred to be of this age, based on stratigraphic position, have been found in northeastern Siberia. The most notable characteristic of these assemblages is the extreme heterogeneity of their ecological composition (Sher and Kuzmina, 2007). Many of the species found in these assemblages do not occur together today. Their modern ranges are often separated by many hundreds of kilometers. The most extreme examples of this come from the northernmost assemblages, such as Bolshoy Lyakhovsky Island (Fig. 1, No. 17), where about 80% of identified insect species currently live much further south. This fauna indicates a substantially warmer climate than today, with higher summer and probably somewhat higher winter temperatures. This is in agreement with the paleobotanical data (Andreev *et al.*, 2004), suggesting summer temperatures that were 4–5°C higher than present. The fossil insect data suggest that the climate of Western Beringia during MIS 5e was more continental than it is today, but less continental than during the subsequent Weichselian interval (Sher and Kuzmina, 2007). There is good evidence for the presence of tall shrubs in the arctic latitudes at that time. Thick winter snow cover would have allowed the growth of these shrubs at high latitudes.

2.4. Early Weichselian Insect Assemblages

Because of dating problems, it is difficult to confidently assign any fossil insect assemblage to the Early Weichselian (Lower Zyranka) interval. However, there are a number of studied deposits that may belong to this time interval. For example, at the Duvannyy Yar site (Fig. 1, No. 27), organic remains from the lower part of the section yielded infinite radiocarbon dates. It is thought that these deposits may come from the Lower Zyranka (Sher, 1991). The same is possible for the Oyagosskiy Yar assemblage (Fig. 1, No. 73) and the sections on the southern coast of Bolshoi Lyakhovsky Island (Fig. 1, No.17). In the Kolyma lowland, insect assemblages thought to come from this interval are dominated by the pill beetle *Morychus viridis*, with small numbers of other xerophiles (usually below 10%) (Sher *et al.*, 1979). However, fossil sites further north, such as Oyagosskiy Yar, Bolshoi Lyakhovsky Island, and the Lena Delta (Fig. 1, No. 52), contain probable Lower Zyranka faunas that are dominated by tundra species, sometimes with a high percentage of arctic tundra or mesic tundra species. The xerophile group (mostly *Morychus viridis*) usually constitutes less than 15% of the fauna, while true steppe species are either extremely rare or absent (Sher and Kuzmina, 2007).

2.5. MIS 3 Insect Assemblages

Many organic deposits that were once thought to date to MIS 3 (Karginsky Interstadial) have subsequently been shown to be much older, including some deposits that possibly date to MIS 5e (Sher and Kuzmina, 2007). However, where more or less continuous sampling of Late Pleistocene sections has been possible, MIS 3 deposits have been more confidently identified (see Fig. 6). Sher *et al.* (2005) developed a highly detailed record of Late Pleistocene environments in the delta region of the Lena River, Bykovsky Peninsula,

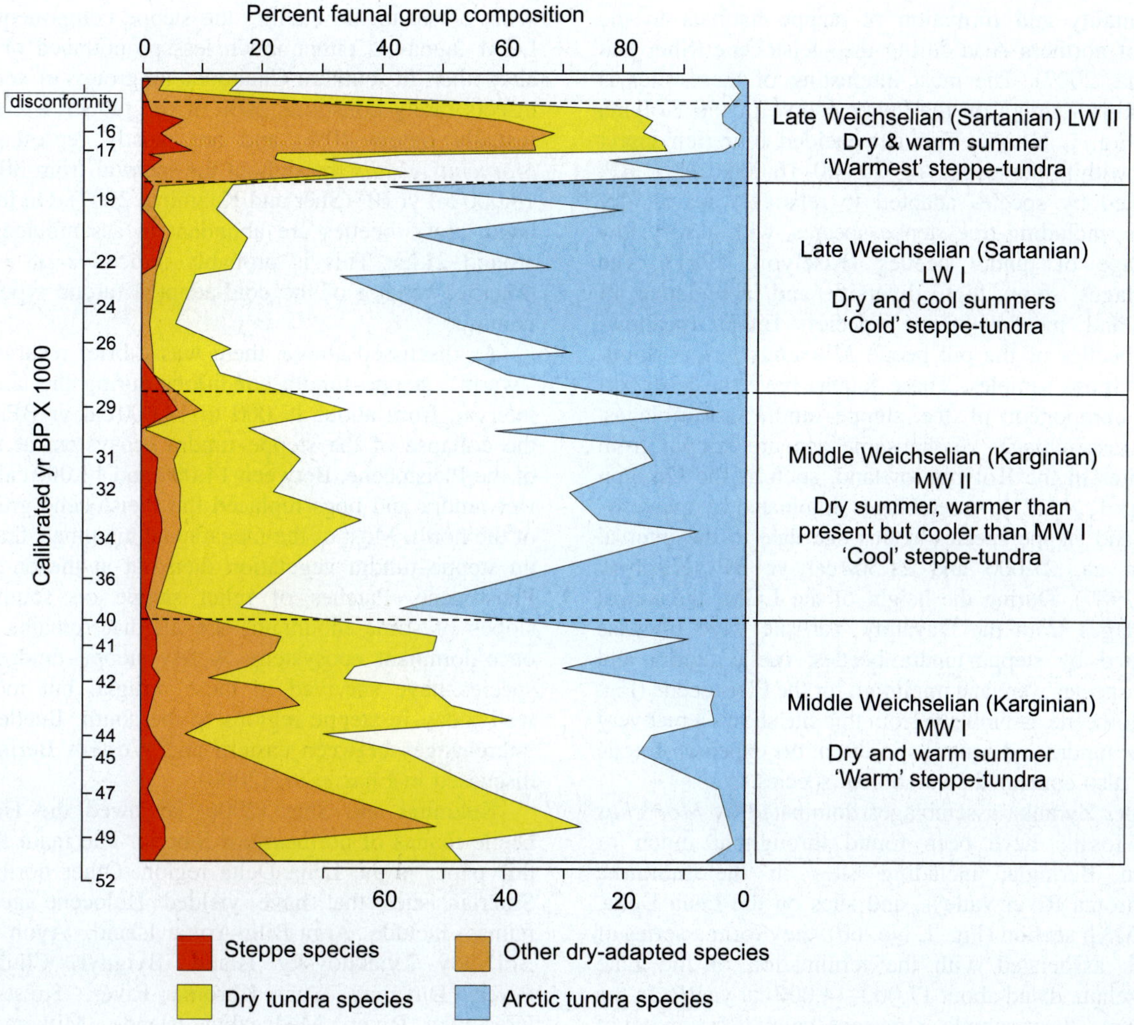

Fig. 6. Reconstruction of fossil insect communities and major climatic intervals at the Lena Delta, northeastern Siberia, based on percentage composition of four ecological groups (after Sher et al., 2005).

based on insect fossil and other studies from MKh Cliff (Fig. 1, No. 60). A total of 70 insect fossil assemblages were recovered from this site, spanning the interval from ca. 50,000–14,000 cal yr BP. Steppe–tundra vegetation dominated the region throughout this interval, but there were distinct changes in the insect faunas through time. Tundra-dominated insect assemblages persisted in the Lena Delta throughout MIS 3 combined with a more or less consistent presence of xerophilic (dry-adapted) species. The contribution of xerophilic species was more noticeable in the earlier part of MIS 3, especially in the assemblages radiocarbon dated to about 48,000–50,000 cal yr BP. Although modern tree line (sparse larch forest) is not far south of the site, no fossil assemblages of the forest or even forest–tundra type have been found in the Lena Delta region. By about 40,000 cal yr BP (34,000 ^{14}C yr BP), the numbers of arctic tundra beetles increased substantially, marking a cooler phase of the interstadial (MIS 3).

A series of assemblages thought to be associated with MIS 3 were found in the Primorsky site in western Chukotka (Fig. 1, No. 74). The sampled unit contained tree birch wood, dated >27,870 yr BP (Kiselyov, 1994). The fossil beetle assemblages include notable percentages of forest–tundra species (7%), including forest-dwelling rove beetles in the genera *Acidota*, *Arpedium*, *Xantholinus*, *Anthobium*, and *Bolitobius*, as well as the birch-feeding weevil *Betulapion simile*. These assemblages also contain high percentages of the pill beetle *Morychus viridis* and a few true steppe beetles. In contrast to this, in southern Chukotka, insect assemblages from the same age interval (39,000–29,000 cal yr BP) do not contain any forest insects (Kiselyov, 1994). The percentage of xerophilic species is high (38%), and true steppe and meadow-steppe species are extremely rare.

2.6. Upper Zyranka Insect Assemblages

The cooling trend that was noted from the Lena Delta faunas during the latter phases of MIS 3 continued through about 20,000 cal yr BP, when arctic tundra species became dominant. This corresponds with the Upper Zyranka (Sartanian) glacial interval. The fossil insect assemblages dating from about 29,000 to 14,000 cal yr BP were first studied in the Kolyma lowland, which later proved to be almost the centre of continentality and formation of steppe habitats in the whole of northern Asia during the Pleistocene (Sher and Kuzmina, 2007). The most interesting of these sites is Alyoshkina Zaimka on the lower course of the Kolyma River (Fig. 1, No. 7). This site yielded four rich insect faunas within the range of 20,000–16,000 cal yr BP, dominated by species adapted to relatively warm, dry climates, including true steppe species, with a very low percentage of tundra beetles (Kiselyov, 1981). The assemblages show high diversity and abundance of steppe and meadow-steppe species. Either meadow-steppe beetles or the pill beetle *Morychus viridis* dominate all these samples. These faunas have the strongest steppe component of the steppe–tundra assemblages. Similar assemblages of the same age are known from other sites in the Kolyma lowland, such as the Omolon site (Fig. 1, No. 71). Assemblages dominated by meadow-steppe and steppe species at this site date to the interval between ca. 27,000 and 21,500 cal yr BP (Kiselyov *et al.*, 1987). During the height of the LGM, the faunal assemblages from the Duvanny Yar site were likewise dominated by steppe–tundra beetles; mesic tundra and aquatic species were at a minimum for the Pleistocene (Fig. 6). A Holocene assemblage from this site shows a recovery of mesic tundra and aquatic species, at the expense of xeric taxa. It also contained some boreal species.

Upper Zyranka assemblages dominated by *Morychus viridis* fossils have been found throughout much of Western Beringia, including sites in the Indigirka and Khroma River valleys, and sites on the Lena Delta. In the MKh section (Fig. 1, No. 60), they form a series of samples associated with the termination of the Late Weichselian, dated about 17,000–14,000 cal yr BP. In the Lena Delta, these samples represent the highest content of *Morychus*, steppe, and other xerophilic species during the Last glaciation so this stage has been recognized as the LGE, a brief interval of higher summer temperatures and drier environments in the Siberian Arctic (Sher *et al.*, 2005). The LGM insect assemblages in the Lena Delta are strikingly different from those in the Kolyma lowland. Most of these assemblages are dominated by arctic tundra insects, mainly the weevil, *Isochnus arcticus,* which feeds on arctic willows. Xerophilic insects constitute only a small percentage of the fauna, in fact the lowest values in from MIS 4-2. These unusual assemblages, dominated by arctic species, have been interpreted as evidence of a very cold climate with dry summers (Sher *et al.*, 2005). Outside the Lena Delta, this type of arctic tundra–steppe fauna has only been found in a single, small fauna of Late Weichselian age from the northern Yana-Indigirka Lowland (Sher and Kuzmina, 2007). Further south, LGM beetle faunas included increasing numbers of species indicating relatively warm, dry, and steppe environments. For example, at the Krazivoye site (Fig. 1, No. 45), an assemblage dated about 23,000 cal yr BP was dominated by dry tundra and xerophilic beetles, combined with sedge steppe, meadow-steppe, and true steppe species. Assemblages with an even more pronounced xerophilic component are common in Late Weichselian deposits throughout Western Beringia. Many exposures in the Kolyma lowland have yielded faunas with 65–90% xerophiles. In contrast to this, at the eastern edge of Asia, closer to the Pacific, the steppe component in the LGM faunas is either much less pronounced or absent altogether. In southern Chukotka, all groups of xerophilic insects rarely constitute 30% of the faunal assemblages (usually below 10%) and are mostly represented by *Morychus viridis* throughout the interval from 30,000 to 18,000 cal yr BP (Sher and Kuzmina, 2007). On the other hand, arctic beetles are abundant in assemblages dated around 21 ka. This is probably indicative of a milder (Pacific) version of the cold-adapted (arctic type) LGM faunas.

As discussed above, there was a brief resurgence of ''warm'' steppe–tundra conditions during the lateglacial interval, from about 17,000 to 14,000 cal yr BP, before the collapse of the steppe–tundra ecosystem at the end of the Pleistocene. Between 14,000 and 11,000 cal yr BP, wet tundra and bogs replaced the Pleistocene grasslands of the north. Most of the megafaunal mammals that relied on steppe–tundra vegetation died out at the end of the Pleistocene. Patches of relict steppe on south-facing slopes of some mountains are all that remains of this once-dominant ecosystem. A few steppe–tundra beetle species have survived in these refugia, but most live well today in steppe regions to the south. Beetle faunal interchanges between eastern and Western Beringia are discussed in Elias *et al.* (2000).

Kuzmina and Sher (2006) reviewed the Holocene beetle faunas of northeastern Siberia. The main focus of this paper is the Lena Delta region. Other northeastern Siberian sites that have yielded Holocene-age insect faunas include Arga-Bilir-Aryta Island, Ayon Island, Bol'shoy Lyjakhovsky Island, Bylgnyr, Chukoch'ya River, Duvanny Yar, Khroma River, Krest-Mayor, Krestovka River, Medvezhyi Islands, Milkera River, Omolon River, Primorsky, and Sededema River (Fig. 1, Nos. 10, 12, 17, 19, 23, 27, 38, 46, 47, 61, 62, 71, 74, and 79). Regional faunas of Early Holocene age represent some of the most diverse fossil assemblages ever seen from this region. The faunas are dominated by mesic tundra species, such as ground beetles in the *Cryobius* group of the genus *Pterostichus*. However, these Early Holocene faunas also include specimens associated with steppe–tundra environments, such as the weevil *Stephanocleonus eruditus* and the pill beetle *Morychus viridis*. Based on studies of recent sediments in this region, Kuzmina and Sher (2006) have concluded that it is likely that at least some of these steppe–tundra species may be reworked from local Pleistocene deposits, in which their remains were ubiquitous (especially *Morychus viridis*). However, the possibility remains that patches of steppe–tundra habitat persisted in the Lena Delta region during the Early Holocene. The persistence of such habitats is supported by another line of evidence. Several species of relatively thermophilous beetles are found in Early Holocene fossil assemblages from the Lena Delta and elsewhere. These species, including the ground beetle *Loricera pilicornis* and the bark beetle *Polygraphus*, are currently restricted to the boreal forest (taiga) zone, indicating that Early Holocene climates were markedly warmer than modern climates in the Arctic. Faunas from the Lena Delta region that date from 9,200 to 8,300 cal yr BP include boreal-zone insects such as the carrion beetles

Phosphuga atrata and *Aclypea opaca*, as well as the carpenter ant *Camponotus herculeanus*.

Regional climates cooled in the latter half of the Holocene. Trees, tall shrubs, and associated insects retreated southwards, toward their modern limits. Arctic species became dominant in these assemblages, including the arctic leaf beetles *Chrysolina wollosowiczi*, *C. tolli*, and *Camponotus bungei*, as well as the arctic rove beetle *Tachinus arcticus* (Kuzmina and Sher, 2006).

Other fossil arthropod studies have also been performed in northeastern Siberia. Golosova *et al.* (1985), Krivolutsky and Druk (1986), and Krivolutsky *et al.* (1990) discussed the Quaternary oribatid mite fauna of Siberia. The oribatids are classified in five adaptive types: inhabitants of the soil surface, deep soil, small soil holes, nonspecialized forms, and aquatic oribatids. Siberian studies include fossil faunas from peat samples at the Yuzhno-Sakhalinsk site on Sakhalin Island (Fig. 1, No. 100) and northern Siberia, including the mammoth carcass localities on the Kirghilyakh and Yuribey Rivers (Fig. 1, Nos. 40 and 99). Only one oribatid species was found in the fossil assemblages which is not a part of the modern Siberian fauna. The mite fauna of the Olyor Suite (Mid-Pleistocene) was indicative of rather severe climates and tundra environments similar to modern eastern Siberia. Holocene mite faunas were the most diverse and abundant of the Quaternary.

Grunin (1973) described an extinct species of botfly, associated with a mammoth carcass found on the Shandrin River in northeastern Siberia (Fig. 1, No. 81). Larval exuviae of *Cobboldia russanovi* were recovered from the mammoth carcass. Some species of botflies are known to parasitize modern elephants, but these fossils do not match any known extant species, so perhaps they became extinct along with their host.

3. Regional Paleoenvironmental Syntheses

As we have seen, Russian research in Siberia offers a unique synthesis of fossil and modern biological studies. Kiselyov (1991, written communication) has synthesized the paleoenvironmental interpretations drawn from the Siberian insect fossil studies. In Siberia, as elsewhere, ecosystem structure depends on the dynamics of environmental change. The fossil record reveals a high level of structural stability in the biotic communities of northeastern Siberia. During the Late Pleistocene, latitudinal differences in response to environmental changes over this vast region were slight. The dominant climatic factor affecting biotic communities in the north was probably precipitation, rather than temperature. In spite of several decades of research, there are huge spatial and temporal gaps in the fossil insect record of this enormous region. Little is known of sites south of the Arctic Circle in northeastern Siberia, and the arctic regions west of the Lena River delta are also poorly studied.

Based on the steppe elements in the faunas, Kiselyov estimated that mean July temperatures in the Late Pleistocene of northeastern Siberia were 14–15°C. Climate changed only slightly until postglacial times. The climate approached modern parameters by 13,000–11,000 cal yr BP. In the modern arctic coast of western Chukotka, mean summer temperatures reached 12°C (they are now 6°C). This warmer-than-present event has also been registered by contemporaneous insect fossil assemblages from the exposed Chukchi Sea shelf (Elias *et al.*, 1992a) and the Arctic Coastal Plain of Alaska (Nelson and Carter, 1987). Insect fossils from lateglacial assemblages in western Siberia indicate that the climate was essentially modern there by 13,000 cal yr BP; there is no evidence of a Younger Dryas cooling in these insect fossil records.

In the Early Holocene, paleoenvironmental reconstructions based on insect data show that climate was the main factor affecting Siberian biotic communities. Evidence for latitudinal differences in climatic regime and biotic provinces is slight, in marked contrast to today. Kiselyov interpreted maximum Holocene warming from 9,000 to 8,000 yr BP, based on range expansions of such thermophilous beetle taxa as the chrysomelids, *Donacia* and *Plateumaris*.

There are few records of Middle Holocene insect faunas from northeastern Siberia. Paleobotanical evidence suggests that tall shrubs began to retreat from the arctic regions by about 8,300 cal yr BP. Sea level rose throughout the Early Holocene, eventually reaching its modern position by about 5,400 cal yr BP. There must have been an appreciable cooling effect of the approaching sea upon arctic regions that had formerly been well removed from the coast. Modern vegetation (and presumably climate) was established in the arctic regions between 5,200 and 3,800 cal yr BP (MacDonald *et al.*, 2000; Anderson and Lozhkin, 2002).

The next chapter reviews insect fossils and Quaternary environments from across the Bering straits, in Eastern Beringia.

10

Eastern Beringian Studies

How far into the past must we probe to understand the natural periodicities of world climate? I believe we must look back at least to the late Tertiary.

– John Matthews (1989)

In this chapter, I pick up the threads of the Beringian story, as they have developed on the eastern side of the arctic refugium. The origins of the modern arctic ecosystems date back to Late Tertiary environments in this region. To put the Beringian story in its proper context, I begin here with a summary of arctic environmental history during that period.

1. Late Tertiary/Early Quaternary Faunas

During the past 30 years, Quaternary entomologists working in the Arctic and the subarctic have had the opportunity to study extremely rare deposits of fossil insects that date back millions of years. In some cases, these fossil assemblages represent Late Tertiary environments that preceded the earliest glaciations of the Quaternary. Other assemblages represent Early Quaternary environments, when glacial–interglacial cycles were beginning. These fossil assemblages afford us rare glimpses into the history of the north, during periods when regional environments were startlingly different from today. Only a handful of such sites have been discovered thus far. They are mostly in very remote regions of the Arctic (Fig. 1). In most cases, there are no modern analogues for the fossil assemblages. It has been challenging to make ecological sense of these fossil assemblages, and to reconstruct the environments they represent. What makes these Late Tertiary and Early Quaternary fossils unique is that they are millions of years old, yet they are essentially the same kind of exoskeletal remains found in much younger (Pleistocene and Holocene) sediments (Fig. 2). This is because of their preservation in permafrost. Once in a frozen sediment matrix, the beetles' exoskeletal remains have been protected from the kind of abrasion associated with sands and silts that are not frozen. Bacterial decomposition is likewise almost negligible in permafrost.

One of the most difficult aspects of this research has been establishing the age of the fossil assemblages. While uranium-series dating has been used to establish the age of a few samples (notably the Lava Camp assemblages from Alaska), most of the assemblages discussed below have only been dated on the basis of site stratigraphy. These assemblages have only approximate age estimates, and the chronologies are very likely to change in future, as more geologic research is done in the study regions (Elias, 2007a).

As might be expected, fossil sites containing insect remains more than 1,000,000 years old are extremely rare. Permafrost environments occur only in the high latitudes, and most high latitude regions were repeatedly glaciated in the Quaternary. Glaciers and ice sheets have obliterated underlying (Late Tertiary and Quaternary) deposits, first stripping the land surface down to bedrock in many arctic regions, then leaving a mantle of reworked debris as their margins retreated. However, there were some regions in the Arctic that remained unglaciated through much, if not all, of the Pleistocene. Chief among these was Beringia, which included the unglaciated lowlands of northeastern Russia, Alaska, and the Yukon Territory, linked together by the Bering Land Bridge.

The fossil insect faunas of Late Tertiary and Early Quaternary age from North America and Greenland comprise 437 taxa of insects and arachnids (Elias *et al.*, 2006). Of these, 180 have been identified to the species level. Kiselyov (1981) and Kuzmina (1989) have published beetle faunas of this age range from northeastern Siberia. These include 139 beetle taxa, of which 114 species have been identified. Surprisingly, there are only 10 species in common between the Siberian and North American/Greenland faunal lists for these ages of fossil assemblages (Elias, 2007a). Late Tertiary Canadian arctic faunas have been discussed by Matthews (1979a,b, 1980a, 1981, 1989) and Matthews *et al.* (1986).

1.1. Alaskan and Canadian Faunas

Eleven faunal assemblages dating from about 5.7 to 1.8 mya have been published (Table 1). Geographically, the sites range from just south of the Arctic Circle at the Lost Chicken and Palisades sites, to north of 80° latitude at the Meighen Island and Wolf Valley sites (Fig. 1). The faunal assemblages from these sites document very warm, generally maritime Late Tertiary climates. The faunas are quite mixed, in terms of their modern ranges. Some of these species live today only in Asia; others live today in the temperate regions of eastern North America. As might be expected for such warm-adapted insect faunas, there is considerable taxonomic diversity in these assemblages.

By about 2.5–2 mya, both the insect and plant fossil records indicate climatic cooling. Nevertheless, temperatures remained well above their modern levels in the high

DEVELOPMENTS IN QUATERNARY SCIENCES
VOLUME 12 ISSN 1571-0866

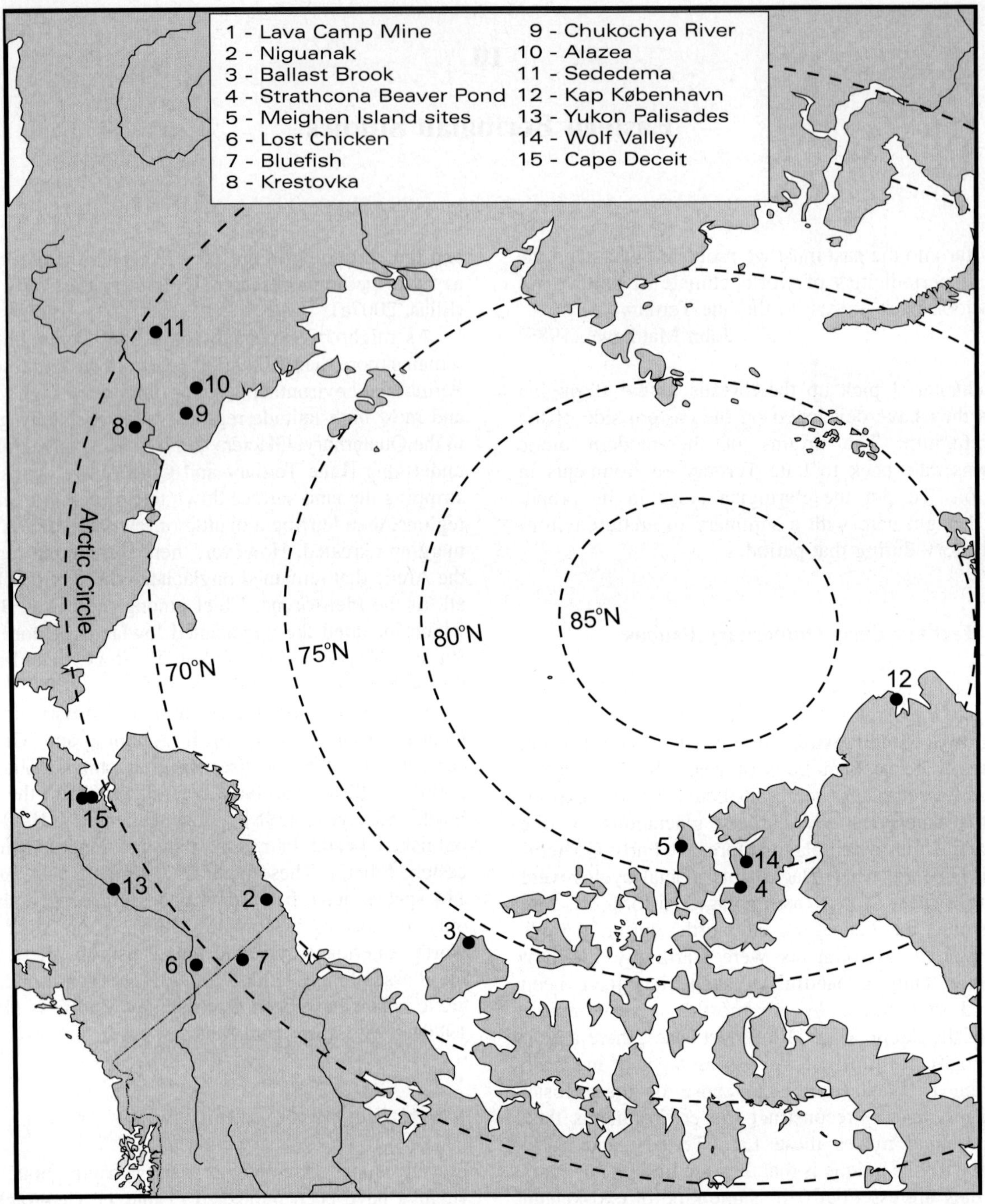

Fig. 1. Map of the Arctic, showing location of Late Tertiary and Early Quaternary insect fossil sites (from Elias, 2007a, © Elsevier).

arctic. Sites such as Wolf Valley, Ellesmere Island (2.1–1.7 mya), and Cape Deceit, Alaska (1.8 mya), contain mixtures of arctic tundra and coniferous forest insect fauna and flora. The Cape Deceit fauna includes some elements that typify the mesic tundra environments that existed throughout much of the Late Pleistocene, including a wide variety of ground beetle species in the *Cryobius* group of the genus *Pterostichus*.

Two faunal assemblages from the Palisades site on the Yukon River were deposited 25 cm below the Eva Creek tephra, which has been dated at 2.02 ± 0.14 mya (Westgate *et al.*, 2003). These assemblages are dominated by dry tundra and mesic tundra taxa, with minor representation of plant litter, and riparian and aquatic species. The faunas demonstrate that as early as the Late Pliocene, dry and mesic tundra habitats were established in interior Alaska. This view contrasts with that of Matthews and Telka (1997) that the earliest lowland tundra existed only on or near the Bering Land Bridge or the arctic coast of Alaska – not in interior regions.

1.2. Siberian Faunas

The Siberian faunal assemblages of Late Tertiary and Early Quaternary age have been described from four sites

Fig. 2. Light microscope photographs of fossil beetle specimens from Olyorian faunal assemblages at the Alazea site, northeastern Siberia (A–D) and the Kap København formation in Greenland (E–H). (A) Pronotum of Thanatophilus dispar, *(B) pronotum of* Upis ceramboides, *(C) elytron of* Carabus odoratus, *(D) elytron of* Chrysolina brunnicornis, *(E) elytron of* Elaphrus tuberculatus, *(F) elytron of* Scolytus piceae, *(G) elytron of* Grypus equiseti, *(H) pronotum of* Diacheila polita *from Kap København, and (I) pronotum of* Diacheila matthewsi. *Scale bars equal 1 mm (from Elias, 2007a, © Elsevier).*

Table 1. Late Tertiary and Early Quaternary insect fossil sites.

Site	Age (millions of years ago)	Insect fauna	Environmental reconstruction	Reference(s)
(1) Lava Camp Mine, Alaska	5.7 ± 0.2 mya (Late Miocene)	83 insect and arachnid taxa, including several extinct species; fauna indicative of coniferous forest	Warm, maritime climate; fossil flora includes species found today in the forests of the Pacific Northwest region	Hopkins *et al.* (1971), White *et al.* (1997, 1999), Elias and Matthews (2002)
(2) Niguanak, Alaska	Early Pliocene?	47 insect and arachnid taxa, including extinct species, *Diacheila matthewsi*; fauna not yet fully analyzed, but has some Asian affinities	Plant fossils indicate coniferous forest, including hemlocks; climate was far warmer and less continental than today	Matthews and Telka (1997), Elias and Matthews (2002)
(3) Ballast Brook, Banks Island, NWT, Canada	5–3 mya	39 insect and arachnid taxa, indicative of boreal environment	Coniferous forest, similar to modern forests in the subarctic regions of Canada; there are indications that the site was near northern tree line at the time of fossil deposition; climate substantially warmer than today	Fyles *et al.* (1994), Elias and Matthews (2002)
(4) Strathcona Beaver Peat, Ellesmere Island, NWT, Canada	> 3.3 mya	86 insect and arachnid taxa, including both arctic and boreal elements; some species with Asian affinities	Fauna and flora indicative of northern tree line setting; climate substantially warmer than today	Matthews and Telka (1997), Elias and Matthews (2002)
(5) Meighen Island sites, NWT, Canada	3 mya	198 insect and arachnid taxa, including both boreal and arctic tundra species; faunal affinities with eastern North America and Asia	Flora and fauna indicative of northern tree line environment; climate substantially warmer than today	Matthews (1974d, 1976b, 1977a, 1979b), Matthews and Telka (1997), Elias and Matthews (2002)
(6) Lost Chicken, Alaska	3 mya	83 insect and arachnid taxa, including blind weevil, *Otibazo*, found today in Japan; boreal fauna with several extinct species	Flora and fauna indicative of coniferous forest; several warmth-loving plant species; climate with warmer winters than today	Matthews and Telka (1997), Elias and Matthews (2002)
(7) Bluefish, Yukon Territory	< 3 mya (Late Pliocene)	26 insect and arachnid taxa, including *Notiophilus aeneus*, now confined to eastern North America, and extinct species, *Helophorus meighensis*	Bluefish plant fossil assemblages include species indicative of substantially warmer, less continental climate than exists in the Yukon today; MCR results indicate T_{max} up to 3°C warmer than today	Matthews and Ovenden (1990), Matthews and Telka (1997), Elias and Matthews (2002)
(8) Krestovka, NE Siberia, Russia	Late Pliocene (Kutuyakh Fm.)	22 beetle taxa; mixture of boreal and arctic fauna; some elements of steppe–tundra fauna already present	Larch forest tundra landscape with steppe patches	Kiselyov (1981)

(*Continued*)

	Late Pliocene to Early Quaternary (Olyorian Suite)	51 beetle taxa; most found today in arctic tundra regions; some taxa with steppe affinities	Steppe–tundra landscape with patches of shrub tundra and coniferous woodlands	Kiselyov (1981), Kuzmina (1989)
(9) Chukochya River, NE Siberia, Russia	Late Pliocene to Early Quaternary (Olyorian Suite)	80 beetle species; most found today in arctic tundra regions; some taxa with steppe affinities	Steppe–tundra landscape with patches of shrub tundra and coniferous woodlands	Kiselyov (1981), Kuzmina (1989)
(10) Alazea, NE Siberia, Russia	Late Pliocene to Early Quaternary (Olyorian Suite)	56 beetle species; most found today in arctic tundra regions; some taxa with steppe affinities	Steppe–tundra landscape with patches of shrub tundra and coniferous woodlands	Kiselyov (1981), Kuzmina (1989)
(11) Sededema, NE Siberia, Russia	Late Pliocene to Early Quaternary (Olyorian Suite)	55 beetle species; most found today in arctic tundra regions; some taxa with steppe affinities	Steppe–tundra landscape with patches of shrub tundra and coniferous woodlands	Kiselyov (1981), Kuzmina (1989)
(12) Kap København, Greenland	2.5–2 mya (Early Quaternary)	154 insect and arachnid taxa, indicative of northern tree line environments	Probably deposited after first glaciation; less faunal diversity than Meighen Island faunas, indicating climatic cooling; climates still substantially warmer than today; T_{max} about 8°C warmer than today; T_{min} about 10°C warmer than today	Böcher (1995), Elias and Matthews (2002)
(13) Palisades of the Yukon	2 mya	24 beetle species, indicative of dry tundra and mesic tundra environments	No evidence of forest at site; T_{max} estimate between 6 and 7°C cooler than modern, and a T_{min} estimate 3–5°C cooler than modern	
(14) Wolf Valley, Ellesmere Island, NWT, Canada	2–1.7 mya	21 insect and arachnid taxa, indicative of northern tree line environments	Plant and insect fossils suggest northern tree line near site at time of deposition; climate cooler than that of Beaufort Formation faunas; T_{max} 5–7°C warmer than today	Matthews and Fyles (2000), Elias and Matthews (2002)
(15) Cape Deceit, Alaska	1.8 mya	86 insect and arachnid taxa, indicative of northern treeline environments; some elements of tundra fauna seen for first time in Eastern Berngian region	The vegetation record suggests that temperatures were slightly warmer than they are today; MCR results indicate T_{max} about 2°C warmer than today; T_{min} similar to modern winter temperatures	Matthews (1974b), McDougall (1995), Matthews and Telka (1997), Elias and Matthews (2002)

Source: From Elias (2007a), with permission © Elsevier.

(Fig. 1, Table 1). In contrast to the contemporaneous fossil record of the North American arctic, the Late Pliocene assemblages of northeastern Siberia already contain cold-adapted insects. These assemblages include species of *Pterostichus* (*Cryobius*) and *Blethisa catenaria* that live on mesic to wet tundra and forest–tundra, as well as woodland species, especially weevils associated with conifers (Kiselyov, 1981). Some steppe-associated beetles made their first appearance in northern Siberian faunal assemblages of the Late Pliocene. The combined faunal and floral evidence suggests a larch forest–tundra landscape with patches of steppe.

The Olyor Suite of fossil assemblages is now thought to represent the Tertiary–Quaternary transition. Deposits of this age are very widespread in northeastern Siberia, especially in the Kolyma lowland region. These beetle assemblages are virtually identical in composition to regional faunas from the Middle and Late Pleistocene. The only difference recognized thus far (Kuzmina, 1989) is that Olyor Suite faunas have a smaller steppe component. The typical Olyorian insect faunas consist mostly of tundra species, with dry tundra insects dominating over mesic and wet tundra species.

1.3. Greenland Fauna

Only one Greenland site has been discovered that contains insect fossil assemblages in the Late Tertiary–Early Pleistocene age range. Most of Greenland has been covered by ice during the many glaciations of the Quaternary, making the fossil site all the more remarkable. Kap København (Fig. 3) is the northernmost site yet studied for insect fossils of this age. It is situated near the northern tip of Greenland (Fig. 1). Böcher (1995) estimated the age of the deposits between 2.5 and 2 mya, approximately at the Pliocene–Pleistocene boundary. Although the site is well north of 80° latitude, the insect fauna is indicative of northern treeline environments. It appears that the Kab København fossil beds were deposited soon after the first glacial period, near the end of the Pliocene. The Meighen Island fauna from northern Canada is more diverse, and contains more warm-adapted species than the Kab København fauna. The former fauna is thought to be 500,000–700,000 years older than the latter (Matthews and Telka, 1997).

2. Importance of the Late Tertiary Faunas

The fossil insect research described above has demonstrated the morphological constancy of exoskeletal features in beetles and some other insects, over several million years for some species. This, alone, would seem a strong indicator that the species in question have remained constant throughout this time interval. However, insect exoskeletons provide little or no data on the internal physiology or ecological requirements of these organisms. In order to test these aspects of species' constancy, other indirect methods must be employed. Fortunately, we do have a method of examining the ecological requirements of apparently extant species in fossil assemblages – through the analysis of their modern counterparts. Based on the modern climate envelopes developed for the species in question, Elias and Matthews (2002) analyzed the climatic compatibility of the Late Tertiary and Early Pleistocene fossil assemblages from Greenland, Canada, and Alaska, and in all cases the species in the fossil assemblages were found to be climatically compatible. Taking this process a step further, the climate envelopes of the species in the fossil assemblages overlapped to yield a MCR, a set of climatic parameters within which all the species in the assemblage can live. The MCR study indicates that the species in question have not evolved new sets of climatic tolerances since the Late Tertiary. Dozens of species were used in the MCR reconstructions. If any significant physiological evolution had taken place, then it would be extremely unlikely that the climate envelopes based on the modern distribution of these beetles would overlap to produce sensible MCR estimates.

This is not to say that there has been no evolution within the Coleoptera in the last several million years. Matthews and Telka (1997) discussed several Late Tertiary beetle species that have apparently become extinct, and these are mainly primitive forms for which we can trace descendent species with more advanced morphological traits. One such case involves the ground beetle genus *Asaphidion*. Matthews and Telka (1997) described an extinct species closely related to the contemporary subarctic species *Asaphidion yukonense*. A fossil elytron from the Lost Chicken site (Fig. 4C) clearly shows bare (impunctate) patches that characterize the modern species, but the patches are less well developed and elytral microsculpture is better developed than in the extant species. Fossils that are similar to *Asaphidion yukonense* have been found in other Late Tertiary assemblages from the Arctic, but some of those have even more primitive elytral sculpture than the Lost Chicken specimens. These fossils could represent various stages in the evolution of the modern species. Matthews (1976a) also described Late Tertiary fossil specimens of the water-scavenger beetle genus *Helophorus* that appear to be precursors to the modern species, *Helophorus tuberculatus*. The fossils from Lava Camp Mine and Meighen Island exhibit more primitive elytral sculpture. Matthews traced the possible evolution of these fossil types into the extant species. Matthews (1970) also discussed the possible evolution of the more primitive rove beetles, *Micropeplus hoogendorni* and *Micropeplus hopkinsi*, leading to the modern species, *Micropeplus cribratus* and *Micropeplus punctatus*. Matthews assumed that the two former species had become extinct in the Quaternary, but *M. hoogendorni* has subsequently been found to be extant, as discussed in Chapter 4.

3. Eastern Beringian Quaternary Studies

> To recreate a landscape, green with life or windswept and barren, and then repopulate it with animals and men is a formidable task (Schweger *et al.*, 1982).

Fig. 3. Kap København. Upper: view of the modern polar desert landscape; lower: view of the Early Quaternary fossil beds. Photographs by Jens Böcher (from Elias, 2007a, © Elsevier).

Beringia has captured the imagination of many researchers, because it was essentially a world unto itself for much of the Pleistocene. A unique refuge for arctic biota, it spanned the margins of two continents, and was nearly surrounded by ice (sea ice to the north; continental ice sheets to the east, west, and south in some regions). As reported in Chapter 9, Western Beringia experienced cold, dry environments during much of the Late Pleistocene, with steppe elements more important than mesic tundra elements in a mosaic of biological communities.

The nature of Pleistocene environments in Eastern Beringia is a topic of considerable debate. Paleobotanical research has indicated that steppe–tundra was important, especially in the interior basins of Alaska and the Yukon Territory (Matthews, 1982). Insect fossil studies add a new dimension to this scenario. The most recent review of the data is presented in Elias and Crocker (2008). Pleistocene fossil sites from Eastern Beringia are listed in Table 2. The site numbers are coordinated with the localities shown in Fig. 5.

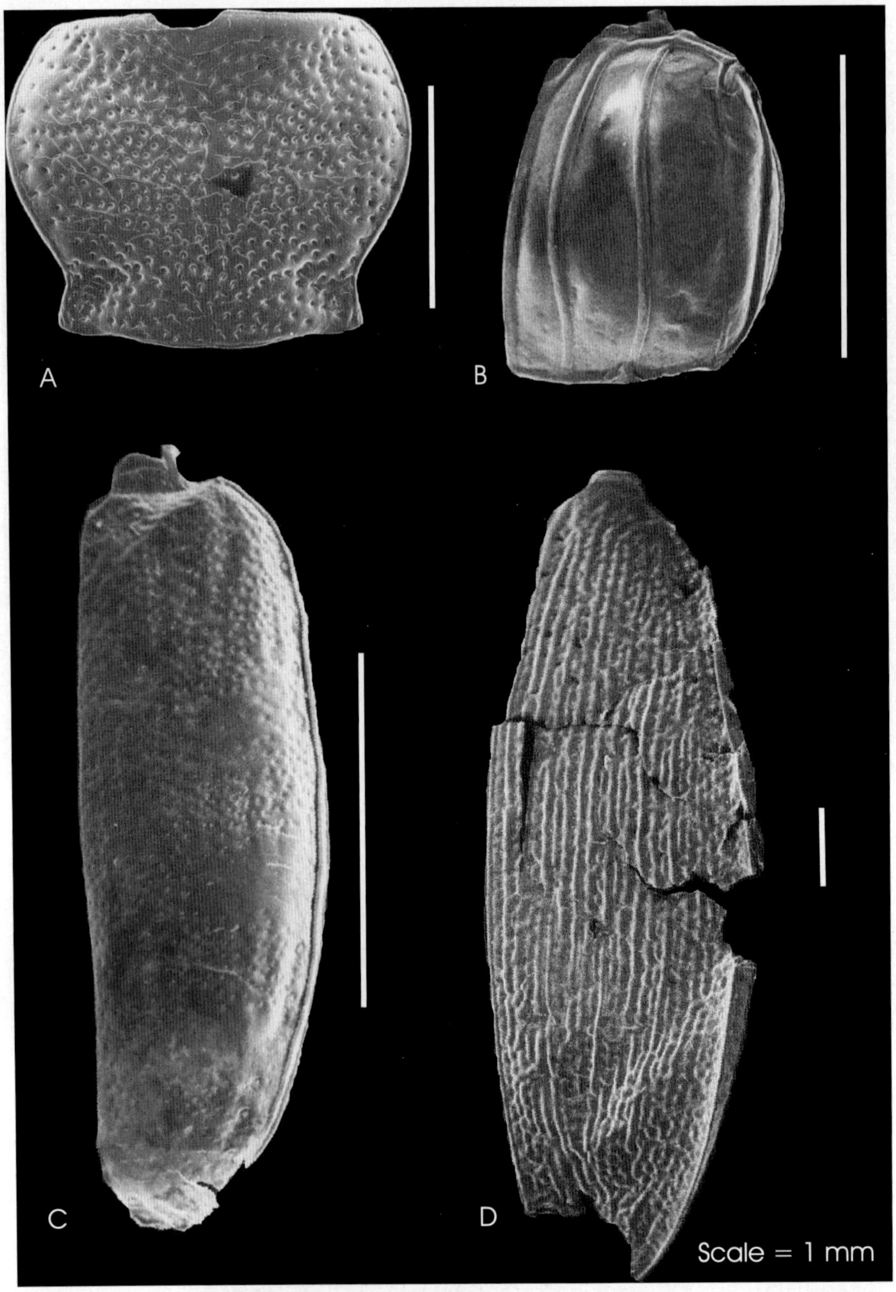

Fig. 4. Scanning electron micrographs of fossil beetle specimens from Canada and Alaska. (A) Pronotum of Diacheila matthewsi *from the Beaufort Formation, Meighen Island, (B) elytron of extinct species of* Kalissus *from the Beaufort Formation, Meighen Island, (C) elytron of extinct species of* Asaphidion *from the Lost Chicken site, Alaska, and (D) elytron of extinct species of* Carabus *from the Beaufort Formation, Meighen Island. Photos by Alice Telka and John Matthews, Geological Survey of Canada (from Elias, 2007a, © Elsevier).*

Macrofossil data indicate that Eastern Beringia was not a uniform steppe–tundra ecosystem during the Pleistocene, but rather a mosaic of different biological communities, many of which no longer exist. For instance, research in western Alaska and the Bering Land Bridge indicates the persistence of mesic- and moist-tundra habitats, even during the height of the last glaciation (Elias *et al.*, 1996c, 1997; Elias, 2007b).

While about 40 publications describe Quaternary insect fossil studies from Eastern Beringia, an enormous amount of work is yet to be done. Many regions remain untouched, and there are many temporal gaps. As in eastern Siberia, much of Alaska and the Yukon remain nearly inaccessible wilderness. Likewise, most Eastern Beringian investigations have been of deposits from riverbank exposures. Many of these exposures, like their Siberian counterparts, contain organic deposits in permafrost, spanning as much as several million years. I will examine the fossil insect record of Eastern Beringia chronologically, starting with the early Middle Pleistocene.

3.1. Middle Pleistocene Faunas

Middle Pleistocene interglacial faunas have been described from interior Alaskan sites, and these reflect a

Table 2. Pleistocene insect fossil assemblages from Eastern Beringia.

Site	Reference(s)
(1) Baldwin Peninsula	Hopkins *et al.* (1976)
(2) Barter Island	Wilson and Elias (1986)
(3) Bering Land Bridge Park	Elias (2001), Kuzmina *et al.* (2008)
(4) Bering Shelf	Elias *et al.* (1996c, 1997)
(5) Birch Creek	Edwards *et al.* (in press)
(6) Blow River	Matthews and Telka (1997)
(7) Blue Babe Site	Guthrie (1990)
(8) Bluefish	Matthews and Telka (1997)
(9) Cape Deceit	Matthews (1974b)
(10) Chester Bluff	
(11) Ch'ijee's Bluff	Matthews *et al.* (1990a)
(12) Chukchi Shelf	Elias *et al.* (1996c, 1997)
(13) Clarence Lagoon	Matthews (1975a)
(14) Coffee Point	Lea *et al.* (1991)
(15) Colorado Creek	Elias (1992b)
(16) Colville River	Morgan *et al.* (1979)
(17) Cutler River	Elias (1997)
(18) Dalton Highway	
(19) Eagle River	Matthews & Telka (1997)
(20) Ekuk Bluffs	Lea *et al.* (1991)
(21) Etolin Point	Lea *et al.* (1991)
(22) Eva Creek	Matthews (1968)
(23) Flounder Flat	Elias (1992b)
(24) Foraker River	Waythomas *et al.* (1993)
(25) Fox Permafrost Tunnel	Hamilton *et al.* (1988)
(26) Goldbottom Creek	Zazula *et al.* (2006)
(27) Holukuk Mountain	Short *et al.* (1992)
(28) Hungry Creek	Hughes *et al.* (1981)
(29) Igushik	Lea *et al.* (1991)
(30) Ikpikpuk River	Nelson (1986), Nelson and Carter (1987)
(31) Isabella Basin	Matthews (1974c)
(32) Koyukuk KY-11	Edwards *et al.* (in press)
(33) Kruzof Island	Klinger *et al.* (1990)
(34) Kulukbuk Bluffs	Elias (1992a)
(35) Kuskokwim/Big River	Elias (1992a)
(36) Kvichak Peninsula	Lea *et al.* (1991)
(37) Landslide Bluff	Waythomas *et al.* (1989)
(38) Last Chance Creek	Zazula *et al.* (2002, 2007)
(39) Mayo Village	Matthews *et al.* (1990b)
(40) Niguanak Uplands	Wilson and Elias (1986)
(41) Noatak River	Elias *et al.* (1999b), Edwards *et al.* (2003)
(42) Nuyakuk River	Elias and Short (1992)
(43) Old Crow	Matthews (1975b), Morlan and Matthews (1983)
(44) Palisades of the Yukon	
(45) Quartz Creek	Zazula *et al.* (2007)
(46) Revenue Creek	Matthews and Telka (1997)
(47) Rock River	Matthews and Telka (1997)
(48) Stevens Village	
(49) Titaluk River	Nelson (1986), Nelson and Carter (1987)
(50) Toklat River	Churchill and Nelson (1992), Elias *et al.* (1996b)
(51) Upper Porcupine River	Matthews and Telka (1997)

wide variety of environments. The two earliest of these faunal assemblages come from the Palisades site (Fig. 5, No. 44), representing early Middle Pleistocene interglacial environments. The older of these assemblages contains an almost even mixture of dry tundra and mesic tundra taxa, as well as several species associated with boreal forest environments. The younger assemblage contains more dry tundra indicators, as well as the steppe-associated species, *Connatichela artemisiae* (Fig. 6). Another faunal assemblage from early in the Middle

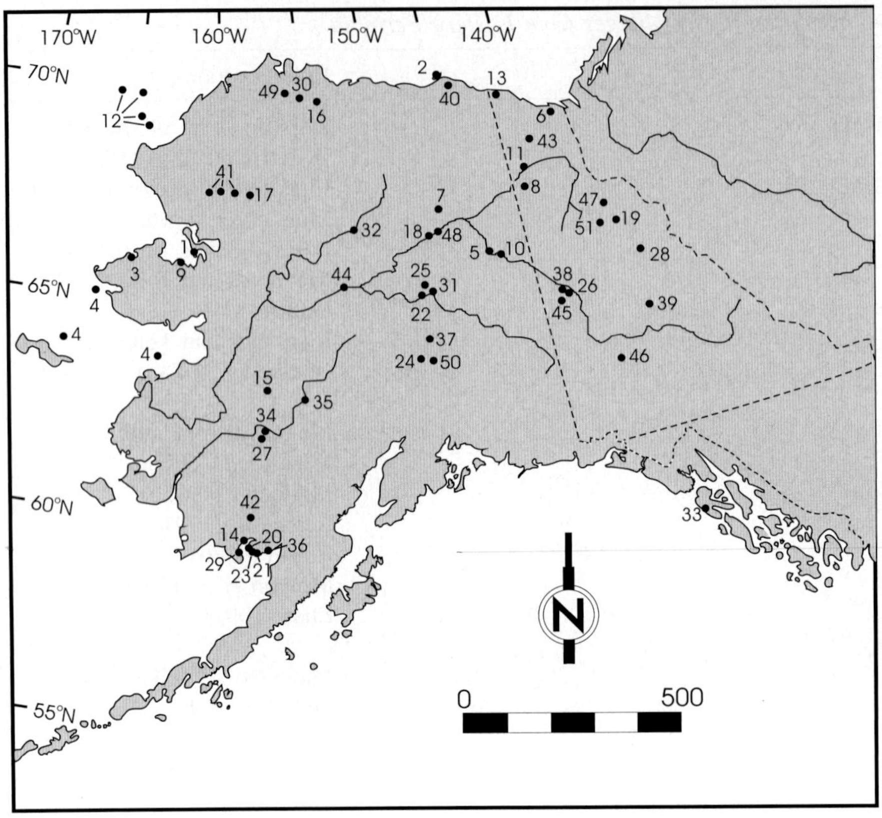

Fig. 5. Map of Alaska and the Yukon Territory, showing location of Quaternary insect fossil sites, listed in Table 2.

Pleistocene interglacial sequence comes from Chester Bluff (Fig. 5, No. 10). This fauna is almost equally divided into dry tundra, mesic tundra, and forest-dwelling species. These three assemblages are all indicative of colder-than-modern climates. The two Middle Pleistocene assemblages from Palisades both yielded T_{max} estimates of about 8–10°C, which is about 6–8°C colder-than-modern T_{max} at the site. As usual, the T_{min} estimates based on these faunal assemblages are much broader, but both faunal assemblages yielded T_{min} estimates that are as much as 7–8°C colder than modern. The lower of the two Palisades beetle faunas from this interval contained a mixture of boreal and arctic tundra species, perhaps indicating that this part of Eastern Beringia was at or near treeline at that time. The younger interglacial fauna from Chester Bluff yielded a more temperate climatic reconstruction. The T_{max} estimate from this assemblage was about 2–2.5°C cooler than modern conditions. However, this fauna likewise contained no definitive forest indicators, such as bark beetles or ants.

It is difficult to know how to interpret these Middle Pleistocene faunas, because we are looking at ''snapshots'' that represent intervals of good organic preservation. These intervals may or may not represent full interglacial conditions, as might be supposed from the use of the term ''forest bed.'' These organic-rich deposits are separated from each other by many meters of loess deposits, most commonly associated with glacial environments (see discussion of the deposits at these sites in Froese *et al.*, 2003, and Matheus *et al.*, 2003).

3.2. MIS 11–7 Faunas

Two faunal assemblages from the Dalton Highway site (Fig. 5, No. 18) are associated with MIS 11. Both assemblages are dominated by dry tundra beetles, and steppe-associated beetles are present in both assemblages, but both also contain substantial numbers of forest-dwelling beetles. The older of the two assemblages yielded a T_{max} estimate of 15–15.5°C, which is within about 1.0–1.5°C of the modern T_{max} in this region. The T_{min} estimate for this sample is 2.5–4.5°C warmer-than-modern mean winter temperatures. The younger sample (B2) yielded a T_{max} estimate of 12–15.5°C, suggesting that mean summer temperatures were as much as 4.8°C cooler than modern values. The T_{min} estimate for this assemblage was too broad to be of use. All that can be said with any certainty is that summer temperatures were ''less than or equal to'' those of the older MIS 11 assemblage. These faunal assemblages contain weevils, bark beetles, and ants that rely on the presence of conifers in their habitats, either for food (the weevils and bark beetles) or for nesting (ants). The insect faunal evidence suggests that MIS 11 was the warmest Pleistocene interglacial prior to MIS 5e in interior Eastern Beringia.

By MIS 13–11, the fossil beetle evidence indicates that steppe-tundra habitats were persisting, even through interglacial intervals. This persistence may have been fostered by generally arid climates. South-facing slopes, such as the Dalton Highway fossil site, were probably the warmest and driest patches of interior Alaska and the Yukon, and thus could have supported steppe–tundra

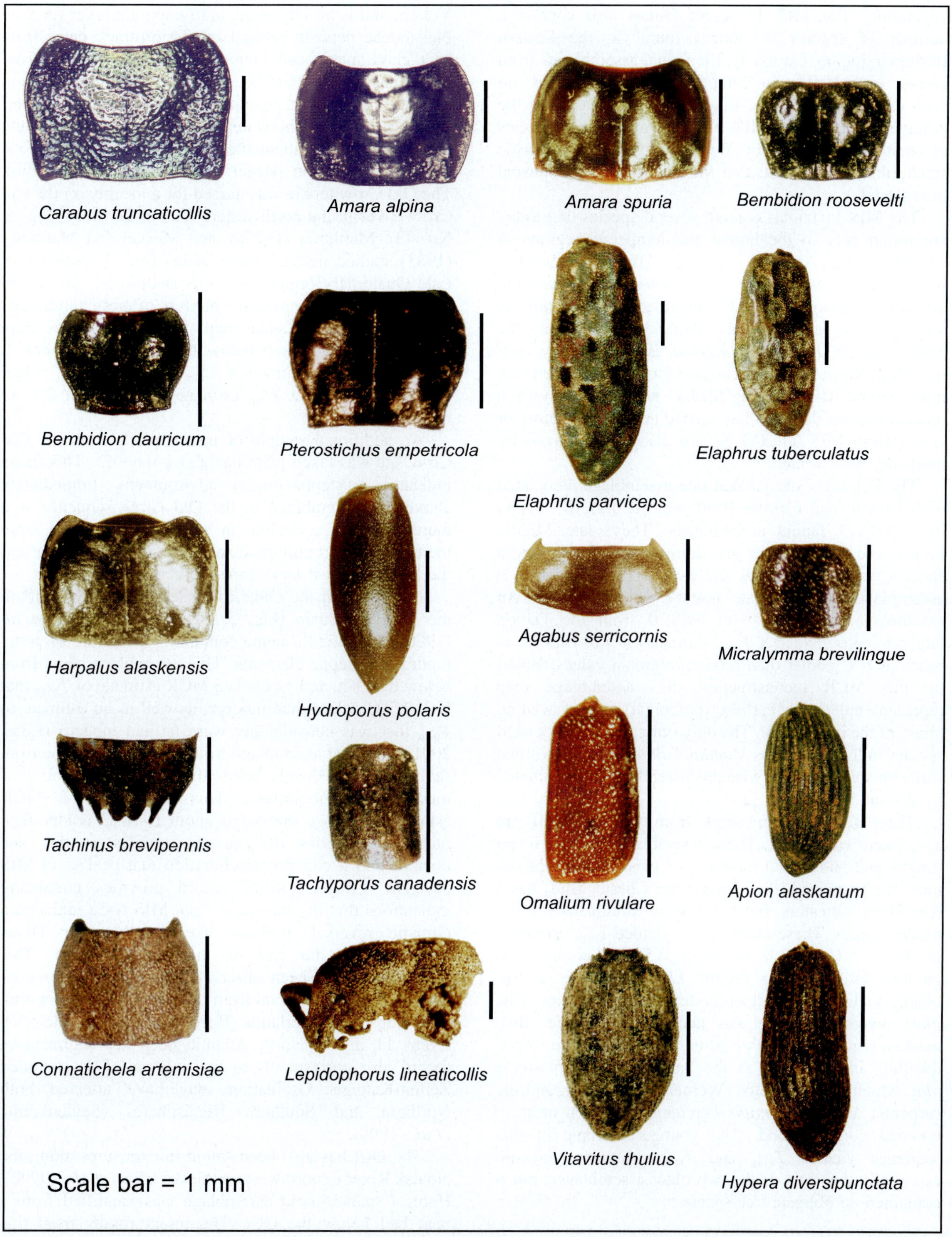

Fig. 6. Late Pleistocene insect fossil specimens from sites in Eastern Beringia (photos by Svetlana Kuzmina; used with permission).

vegetation. The MIS 11 insect faunas also contain a number of species no longer found in the Eastern Beringian region. Curiously, the faunal assemblages from earlier interglacials in interior Alaska contained no such extralimital species. Likewise, the faunas from the younger interglacials, MIS 9–5, contained only species associated with Eastern Beringia. From a faunistic standpoint, this suggests that MIS 11 was a very unusual interglacial.

The MIS 11 faunas contain several species that today are found only in the boreal and temperate regions of eastern North America, including the rove beetles, *Arpedium cribratum*, *Acidota subcarinata*, and *Olophrum obtectum*. Campbell (1983) described *O. obtectum* as 'endemic to eastern North America.' Likewise, the ground beetle, *Bembidion nigrum*, is found today only in eastern North America. The presence of these beetles in north-central Alaska during MIS 11 argues strongly for a moister climate than that has existed here, either before or since. Both MIS 11 and 5e are also characterized by markedly mild winters.

The Palisades site yielded one assemblage associated with MIS 9, and Chester Bluff yielded one MIS 9 and three MIS 7 faunal assemblages. These late Middle Pleistocene assemblages are dominated by dry tundra species, especially MIS 9 and early MIS 7. The MIS 9 assemblages lack obligate forest-dwelling species. An assemblage associated with MIS 9 from the Dalton Highway site yielded a T_{max} estimate of 7.5–10.5°C, as much as 9°C cooler than modern regional values. Based on this MCR reconstruction, this assemblage may represent either the earliest (preboreal) or the waning phase of the interglacial. The only obligate tree-associated species in this fauna was *Magdalis alutacea*, a weevil that feeds on spruces. Otherwise this assemblage is dominated by dry tundra species.

Three faunal assemblages from Chester Bluff are associated with MIS 7. These assemblages contain forest insects and substantial numbers of mesic tundra indicators. The younger MIS 7 fauna from Chester Bluff has a few forest indicators, but is otherwise dominated by dry tundra beetles. These assemblages yielded T_{max} estimates of 13–13.5°C, which is about 2.5–3°C cooler than modern values in this region. Likewise, these assemblages yielded colder-than-modern T_{min} estimates. The most precise of these was about 4–5°C cooler than modern regional T_{min}. Two of the samples contain forest-dwelling insects, such as the bark beetle, *Phloeosinus pini*, which attacks spruce (Werner *et al.*, 2006), and the carpenter ant, *Camponotus herculeanus*, which nests in downed conifer wood. The youngest sample in this sequence yielded T_{max} and T_{min} estimates that are essentially the same as the two older assemblages, but it contained no obligate tree species.

3.3. MIS 6 Faunas

A few fossil insect assemblages from Eastern Beringia most likely date to the late stages of the MIS 6 stadial, based on their close association with the Old Crow tephra (OCt). This tephra is widespread in Alaska and the Yukon, and is an important stratigraphic marker for Late Pleistocene deposits. Based on fission-track dates from OCt glass shards and luminescence dates on silts from immediately above and below the tephra, in combination with its stratigraphic position at eight Alaskan localities, the OCt must have been deposited between 140,000 and 130,000 yr BP, at about the time of the Marine Isotope Stage 6/5 transition (Begét, 1996; Elias *et al.*, 1999b). The Old Crow tephra was named for a locality on the Old Crow River in the northwestern Yukon Territory (Fig. 5, No. 43). Matthews (1975a), and Morlan and Matthews (1983) studied insect fossil remains from a sequence of Late Quaternary organic deposits at this site. The fauna relating to MIS 6 contained a mixture of mesic tundra and xeric open-ground faunas with few hygrophilous taxa. This sample was taken below the Old Crow tephra. It yielded an MCR estimate of T_{max} that was 6–7°C colder than modern, and a T_{min} estimate within about 2°C of modern values.

An additional sample of insect fossils from the Old Crow site was taken from just above the OCt. This fauna indicates a steppe–tundra environment. Immediately above this assemblage in the Old Crow sequence is a fauna showing a decline in xeric species and a large increase in hygrophilous taxa. This fauna may represent the MIS 5e, but it lacks forest insects.

Not far from the Old Crow site at Ch'ijee's Bluff in the Bluefish Basin (Fig. 5, No. 11), Matthews *et al.* (1990b) identified a fauna containing a mixture of arctic tundra and steppe elements. This assemblage came from below the OCt, and yielded an MCR estimate of T_{max} that is 3–3.5°C colder than modern, as well as an estimate of T_{min} that was considerably warmer than modern (Elias, 2001). A faunal assemblage from this site that came from sediments immediately below the OCt was indicative of much colder conditions. This fauna yielded MCR estimates of T_{max} that were about 6–10°C colder than modern. A series of younger samples at this site documented a climatic amelioration that peaked in MIS 5e. The Ch'ijee's Bluff record provides tantalizing indications that the transition from MIS 6–5e included a climatic reversal, perhaps like the Younger Dryas oscillation at the end of the last glaciation. This phenomenon has been observed in paleoclimatic reconstructions of this interval from other regions of the world, including the Netherlands (Beets *et al.*, 2006). Heinrich Event 11 in the North Atlantic deep sea sedimentary record also corresponds to this interval. This so-called Zeifen-Kattegat Oscillation may have affected both Northern and Southern Hemispheres (Seidenkrantz *et al.*, 1996).

The OCt has also been found in exposures along the Noatak River in northwestern Alaska (Elias *et al.*, 1999b). Here, a sparse insect assemblage was identified from a peat bed below the OCt. The insect fossils from this assemblage are typical of arctic tundra inhabitants today. These include the ground beetle *Pterostichus mandibularoides*, the rove beetle *Micralymma brevilingue* (Fig. 6), and the predaceous diving beetle *Agabus arcticus*. This fossil assemblage yielded an MCR reconstruction with T_{max} values from 7.5 to 10.5°C. Modern T_{max} at the site is about 11°C, so the MIS 6 reconstruction is at most 2.5°C colder than modern. Thus the climatic signal of Noatak faunas matches that of Ch'ijee's Bluff in indicating that climatic amelioration at the end of MIS 6 was well underway *before* the deposition of the OCt. Two Noatak faunal assemblages in close stratigraphic proximity to the OCt indicate that T_{max} fell by as much as 7°C about the time the tephra was being deposited.

3.4. *MIS 5 Faunas*

Two of the Palisades faunal assemblages are thought to represent MIS 5. The older of these is dominated by forest and riparian/aquatic beetles, with smaller numbers of dry tundra and mesic tundra species. The younger assemblage contains a greater proportion of riparian/aquatic species, and far fewer forest species. Its tundra-dwelling species are essentially the same as those found in the older of the two assemblages. These two assemblages contain the greatest percentages of aquatic and riparian taxa of all the Pleistocene assemblages from the Palisades site. This could reflect increased moisture, but it might equally well be a taphonomic phenomenon, based on the close proximity of the fossil site to the ancient Yukon River. The older assemblage yielded a T_{max} estimate of 16–16.5°C, which essentially matches the modern mean July temperature at the site. It also yielded a T_{min} estimate that is 4–5°C warmer-than-modern mean January temperature. This fauna has a clear boreal forest signature, and is the earliest Pleistocene sample from Eastern Beringia that reflects interglacial climates as warm as modern, with especially mild winter temperatures. A second sample from MIS 5 may represent either the end of MIS 5e or a later substage of MIS 5, based on its stratigraphic position. This sample yielded a T_{max} estimate about 6–7°C cooler than modern. However, this sample contains two obligate tree-associated species, the weevil *Hylobius* cf. *pinicola*, which feeds on the roots of pine, larch, and spruce (Anderson, 1997), and the carpenter ant, *C. herculeanus*, which nests in downed conifers (Sanders, 1964).

Full interglacial warming peaked at levels that varied from region to region in Eastern Beringia. In southwestern Alaska, the height of the interglacial saw temperatures as much as 3.5°C warmer than modern. At one of the Noatak River sites in northwestern Alaska, T_{max} climbed to as much as 4.5°C above modern levels (Edwards *et al.*, 2003). Further east, summer temperatures were probably closer to modern levels. The best constrained estimate of average winter temperatures during MIS 5e comes from Ch'ijee's Bluff. This estimate suggests that T_{min} was approximately 4–7°C warmer than today, even though average summer temperatures at this site may have been slightly cooler than modern (Elias, 2007c). Boreal insects and paleobotanical evidence combine to document the presence of coniferous forests in much of Eastern Beringia during MIS 5e. For instance, at Old Crow and Ch'ijee's Bluff, the MIS 5e faunas were very diverse, including bark beetles, boreal carabids, and abundant aquatic and hygrophilous species. The boreal ground beetle fauna included species living far to the south of the site today in British Columbia and the southern Yukon. This fauna is indicative of boreal forest and climatic conditions warmer than present. The Ch'ijee's Bluff Sangamon fauna is indicative of conditions in the heart of the boreal forest. Beetle faunas associated with younger intervals of MIS 5 have yielded MCR estimates showing climatic cooling, with T_{max} levels declining 2–3°C below modern parameters.

3.5. *MIS 4 Faunas*

Two faunas thought to derive from MIS 4 indicate that summer temperatures continued to cool during this interval. A faunal assemblage from the Igushik site in southwestern Alaska (Fig. 5, No. 29) yielded an MCR estimate of T_{max} that is 3–4°C colder than modern. An MIS 4 faunal assemblage from Hungry Creek, Yukon (Fig. 5, No. 28), yielded a T_{max} estimate 6–9°C cooler than modern (Elias, 2007c). Otherwise, few faunal assemblages have been found that can confidently be placed in this stadial interval, although numerous assemblages associated with infinite or near-infinite radiocarbon ages may in fact come from this interval.

3.6. *MIS 3 Faunas*

The long MIS 3 interstadial complex is represented by 25 faunal assemblages from 18 different sites. Warming intervals are indicated by faunas from the Titaluk River in northern Alaska (Fig. 5, No. 49). These faunas are dated at 46,400, 36,000, and 33,600 cal yr BP (Elias, 2007c). Likewise, a fauna from Cape Deceit on the Seward Peninsula (Fig. 5, No. 9) yielded a fauna indicating interstadial conditions, dated to 42,200 cal yr BP. A fauna from Kulukbuk Bluffs in southwestern Alaska also indicated interstadial warming at 35,000 cal yr BP. The strongest indication of interstadial warming comes from the Titaluk River fauna dated 33,600 cal yr BP. This fauna yielded a T_{max} estimate 0.5–2°C warmer than modern. The other faunas discussed previously yielded T_{max} estimates that were 0.5–2°C cooler than modern. Interestingly, a fauna dated 31,500 cal yr BP from Mayo Village, Yukon (Fig. 5, No. 39), indicates that regional T_{max} had fallen to 5–6°C colder-than-modern levels. Likewise, a fauna dated 35,200 cal yr BP from Eva Creek, interior Alaska (Fig. 5, No. 22), indicated T_{max} levels 7–8°C colder than modern (Elias, 2001). Thus, within a span of 2,000 years, temperatures appear to have oscillated dramatically in Eastern Beringia (Elias, 2007c). The beetle faunas that yielded the indications of interstadial warming are composed of species found today in open ground habitats within the boreal forest. They do not rely on the presence of trees. The paleobotanical evidence suggests that most, if not all, of Eastern Beringia remained open-ground tundra or steppe–tundra throughout MIS 3. Steppe tundra appears to have dominated interior Alaska and the Yukon during much of MIS 3 and all of MIS 2, while mesic tundra persisted in northwestern and southwestern Alaska. However, the question remains – Did coniferous forests become established in parts of Eastern Beringia during MIS 3? There are now several regions in Eastern Beringia that provide evidence of boreal forest establishment during the mid-Wisconsin interstadial, including sites in the Yukon Territory and the Alaskan interior. The distribution of sites with boreal assemblages suggests a loose geographic trend, with forest establishment in interior sites and persistence of tundra elsewhere. Brubaker *et al.* (2005) mapped pollen percentage data for the LGM and postglacial intervals in Beringia, and concluded that both spruce and pine survived in refugia through the LGM. Since it is highly unlikely that these conifers entered Beringia from more southerly regions during the LGM, they would have had to be there in the previous interval (MIS 3).

Fossil beetle and paleobotanical evidence indicate that the climatic regimes of the mid-Wisconsin interstadial were apparently quite varied in southwestern Alaska (Lea *et al.*, 1991), as indicated by the succession of fossil beetle faunas. During the early (and perhaps warmest) part of the interval, the composition of the beetle faunas (Fig. 7) was similar to that of the Sangamon and Holocene faunas, with the exception that bark beetle remains and other evidence for coniferous forest are lacking. The hygrophilous element is also not as pronounced as in the full interglacial faunas, and the xeric element is slightly more prominent. Nevertheless, abundant, diverse insect faunas, rich in mesic tundra species, were preserved in deposits of mid-Wisconsin age. Unfortunately, the mid-Wisconsin interstadial is not dated accurately because much of it is beyond the reliable range of radiocarbon dating. Therefore, the rates of environmental change at the onset and within the interstadial cannot be postulated clearly. Younger (stratigraphically higher) sediments of mid-Wisconsin age yield faunas subtly different from earlier interstadial faunas. The only remarkable difference is that the hygrophilous and riparian species are diminished, compared to earlier mid-Wisconsin faunas. Climatic deterioration is suggested by the presence of the cold-adapted ground beetles, *Dyschirius frigidus* and *Patrobus septentrionis*, and the rove beetle, *Tachinus frigidus*, which are found only in late mid-Wisconsin assemblages.

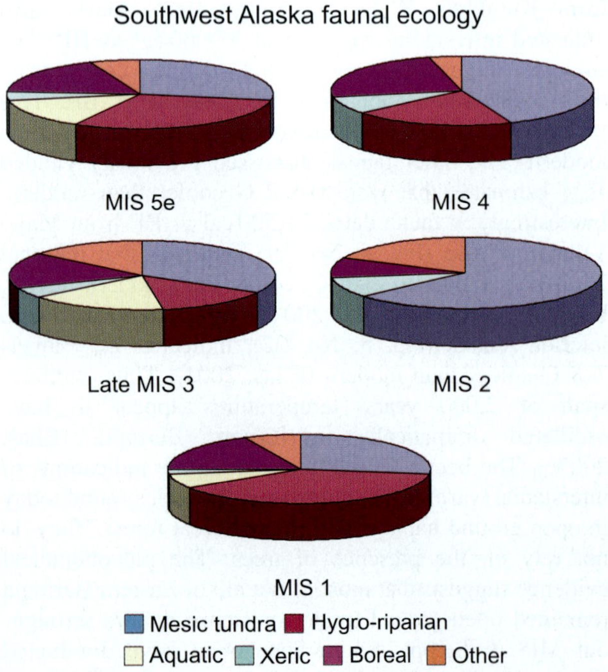

Fig. 7. Pie charts of percentage composition of ecological groups in Late Quaternary fossil insect assemblages from southwestern Alaskan sites (after Elias, 1992c).

3.7. MIS 2 Faunas

In general, it appears that aridity increased in most regions of Eastern Beringia during the LGM. One site that may serve as a boundary marker between Late Pleistocene ecosystems in Eastern Beringia is the Colorado Creek mammoth site, near McGrath in the Alaskan interior (Fig. 5, No. 15). Insect fossils were extracted from organic-rich silts surrounding mammoth remains, and from a mammoth dung bolus (Elias, 1992a). The fauna dates to the late Wisconsin stadial (ca. 18,000 cal yr BP), with two components to the fossil insect assemblage: one associated with the mammoth remains, and another from the surrounding upland environment. Not surprisingly, the mammoth dung contained abundant *Aphodius* dung beetles (*Aphodius congregatus*), and the mammoth remains yielded the carrion beetle species (Silphidae), *Silpha coloradensis*. In addition, abundant blowfly pupae were found in the nasal cavities of the mammoth skull.

The small fauna representing the upland environment included the pill beetle genus, *Morychus*, the leaf beetle genus, *Chrysolina*, as well as a few *Cryobius* specimens, in combination with *Lepidophorus lineaticollis* (Fig. 6). This small assemblage may represent a xeric habitat developed within a broader region of mesic tundra, or it may have been part of the larger ''steppe–tundra'' biome of the Alaskan interior. If this is so, then the geographic location of the site is critical, because all the studied sites south of Colorado Creek have yielded mesic and hygrophilous faunas that lack xeric-adapted taxa. The insect fossil fauna, although too small to be definitive, offers at least a suggestion that the Colorado Creek site was near the southwestern boundary of steppe–tundra environments in the Late Wisconsin. To the south, mesic tundra environments persisted, even through the LGM (Fig. 7). To the north and east, more xeric, steppe–tundra environments dominated (Fig. 8).

During the LGM, the beetle evidence indicates that T_{max} was depressed in Eastern Beringia, but T_{min} was within 1–2°C of modern levels in most regions (Elias, 2007c). Elias *et al.* (1999a) discussed difficulties in reconstructing T_{min} from faunal assemblages sampled from coastal sites. First, the ancient coastline of Beringia was removed from its current position by more than 1,000 km in some regions. Therefore, the modern T_{min} of a coastal site, greatly influenced by maritime climate, bears little relation to the ancient climate of the same site when it was hundreds of kilometers inland because of lowered sea level. Also, even the modern T_{min} of coastal sites is a poor reflection of winter temperatures at these sites because of the incursion of arctic air masses during the winter season.

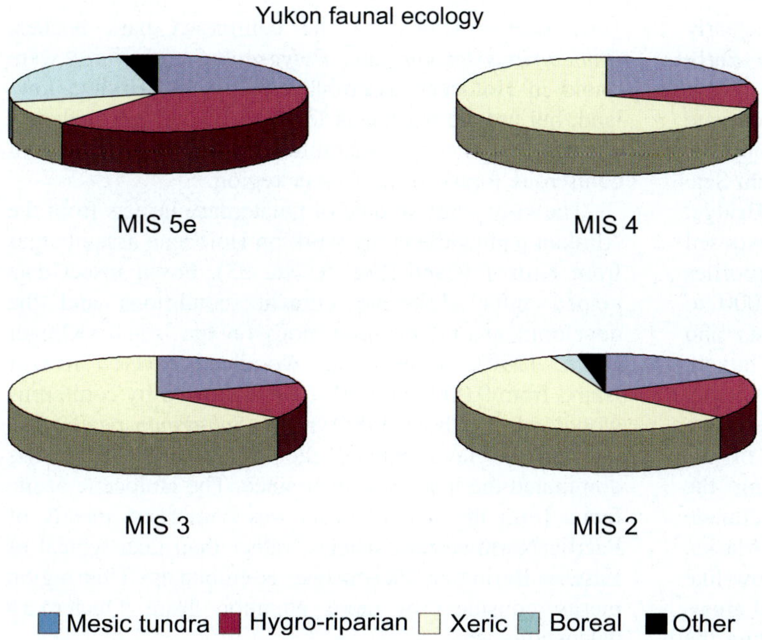

Fig. 8. Pie charts of percentage composition of ecological groups in Late Quaternary fossil insect assemblages from northern Yukon sites. Sangamon fauna from Ch'ijee's Bluff, Unit 4; Early Wisconsin fauna from Old Crow site, Sample 77-51; mid-Wisconsin fauna from Old Crow site, Sample CRH-32; Late Wisconsin fauna from Ch'ijee's Bluff, Sample HH-75-9 (data from Matthews (1983) and Matthews et al. (1990b)).

These periods of very low temperature are sufficiently long, and sufficiently frequent, to virtually eliminate the less cold-resistant beetle fauna of coastal Alaska.

The last glacial episode brought periglacial environments to the lowlands of southwestern Alaska (Lea, 1989; Waythomas, 1990). Organic deposits were limited to thin layers of detritus dispersed in aeolian silt. Regional beetle faunas were once again reduced to a few species, with most assemblages dominated by *Tachinus brevipennis* (Fig. 6). Hygrophilous and riparian taxa were reduced to their lowest levels in the faunal sequence, and aquatic species disappeared from regional records (Fig. 7). Xeric taxa increased to their highest levels, even though mesic tundra species continued to be regionally important.

3.8. Lateglacial and Holocene Faunas

Evidence for lateglacial and Holocene environments comes from a number of sites across Eastern Beringia. No fossil beetle assemblages from Eastern Beringia date to the period between 18,000 and 14,500 cal yr BP, but the fossil beetle record from western Alaska indicates that mean summer temperatures rose to modern levels by at least 14,500 cal yr BP, and beetle assemblage data from northern Alaska indicate that T_{max} values rose as much as 7°C and T_{min} values rose as much as 5°C above modern values just before the inundation of the Bering Land Bridge at 13,000 cal yr BP (Elias, 2000). This large-scale amelioration of temperatures was probably brought about by a combination of two main factors. First, T_{max} values rose because summer insolation values were very high in the high northern latitudes during the lateglacial interval (Berger, 1978). Second, T_{max} and T_{min} values rose because the progressive flooding of land bridge regions brought a more moderate, maritime climate progressively north into arctic Beringia, culminating in the reestablishment of circulation between North Pacific and Arctic Ocean waters, probably a few centuries before 13,000 cal yr BP. The reentry of relatively warm Pacific waters into the previously closed Arctic Basin probably reduced the arctic sea ice and snow cover, which may have led to warmer winter temperatures (Broccoli and Manabe, 1987).

Between 13,000 and 11,500 cal yr BP, beetle assemblages from northern Alaska record a rapid, large-scale cooling of both summer and winter temperatures. The record of T_{max} cooling from the Noatak River assemblages coincides with the timing of the Younger Dryas cooling marked in the North Atlantic region (Atkinson *et al.*, 1987; Dansgaard, 1987; Dansgaard *et al.*, 1989). The cooling shown in the Noatak record was similar in scale to the cooling estimated from the North Atlantic region: a decline in T_{max} of 7°C over a period of at most 350 years. A climatic cooling during the Younger Dryas interval has also been noted in pollen records from Kodiak Island (Peteet and Mann, 1994) and central Alaska (Bigelow and Edwards, 2001). The pollen and beetle evidence both strongly indicate that the lateglacial period in Eastern Beringia saw a series of large-scale, rapid environmental changes, especially between about 14,500 and 11,500 cal yr BP.

Evidence for large-scale early postglacial warming comes from the arctic coastal plain site on the Ikpikpuk River (Fig. 5, No. 30). Nelson and Carter (1987) discussed an insect fauna dating to 10,600 cal yr BP. The deposit also contained extralimital *Populus* macrofossils and pollen, suggestive of climatic conditions warmer than present. The insect fauna contained more than 100 beetle taxa and 60 identified species, including abundant tundra carabids, with 10 species of *Pterostichus (Cryobius)*, *Diacheila polita*, and *Carabus chamissonis*. It also contained extralimital carabids, such as *Harpalus amputatus*, *Blethisa multipunctata*, *Agonum quinquepunctatum*, *Agonum quadripunctatum*, *Amara erratica*, *Harpalus fulvilabrus*, and the byrrhid, *Cytilus alternatus*. All of these are found today in the boreal zone of north-central Alaska and adjacent Canada. While the extralimital *Populus* data might be taken to indicate only increased moisture at the site, the beetle data clearly indicate North Slope climatic conditions in the Early Holocene 3–5°C warmer than present.

Perhaps the most intriguing evidence for rapid climate change in the Arctic comes from terrestrial peat deposits, which are now beneath 50 m of water in the Chukchi Sea. These deposits formed on the Bering Land Bridge, when the shallow continental shelf regions were exposed during the Late Pleistocene. Cores have yielded profiles of terrestrial peats, ranging in age from >40,000 to 11,000 yr BP (Elias *et al.*, 1992a, 1996c). Peats and organic-rich silts from a transect of sites on the Chukchi shelf (Fig. 5, No. 12) yielded a small insect fauna, dating from 13,200 to 13,000 cal yr BP. The insect fossils are indicative of arctic coastal habitats like those of the Mackenzie Delta region, suggesting that during the lateglacial, the exposed Chukchi shelf had a climate 6–10°C warmer than the modern arctic coast of Alaska. The insect and pollen data suggest a meadow-like graminoid tundra with shrubs growing in sheltered areas. These peats were overlain by marine sands, marking the marine transgression that took place around 13,000 cal yr BP. The timing of this transgression coincided with the marked warming seen in the arctic insect faunas. As long as the Bering Land Bridge was in existence, it fostered increased climatic continentality in Eastern Beringia. However, once the land bridge was flooded, that continental climatic regime probably broke down. The Chukchi shelf evidence suggests that this effect was both rapid and intense, not unlike the amelioration seen in British insect faunas following the return of warm Gulf Stream waters after the Younger Dryas oscillation.

The climatic amelioration that followed the last glacial interval began by at least 14,500 cal yr BP in southwestern Alaska (Short *et al.*, 1992). Lateglacial faunas document this environmental change in several ways. Xeric taxa declined as hygrophilous and riparian species increased in numbers. Furthermore, several thermophilous species were recorded in lateglacial deposits, including the ground beetle *Patrobus stygicus* and the rove beetle *Tachyporus rulomus*. These species are associated with open-ground habitats within the boreal forest. They indicate a boreal climatic regime, even though the boreal forest was still absent from southwestern Alaska.

Holocene assemblages in southwestern Alaska indicate a return of the hygrophilous and riparian species to very high percentages, concurrent with a sharp decline in xeric beetles (Fig. 7). The coniferous bark beetles, *Phloeotribus lecontei* and *Polygraphus rufipennis*, were found in Holocene assemblages from the Holitna lowland, but not the Nushagak lowland. The insect evidence, combined with conifer macrofossils, indicate the return of coniferous forest to the former region.

The only study to date of Quaternary insects from the Alaskan panhandle is my work on Holocene assemblages from Kruzof Island (Fig. 5, No. 33). Fossil insect data helped infer Holocene climatic conditions and the development of *Sphagnum* bogs on the island (Klinger *et al.*, 1990). A deciduous woodland covered Kruzof Island from 9,000 to 6,000 yr BP, followed by coniferous woodland. At about 3,000 yr BP, *Sphagnum* peatland is inferred to have choked the conifer forest and has dominated the landscape ever since. The Holocene beetle fauna from the Kruzof peats was comprised mostly of Pacific Northwestern species, rather than taxa typical of Eastern Beringian Pleistocene assemblages. This region merits considerably more attention from Quaternary paleoecologists.

This review has shown that Eastern Beringia was not a uniform, coherent steppe–tundra ecosystem during the Late Quaternary. Given the continental size of Beringia, and the variety of environmental conditions impinging on the various landscapes, this should surprise no one. Early and Middle Pleistocene "forest bed" faunas mostly indicate colder-than-modern climates, but these faunas represent snapshots of paleoenvironments, rather than long sequences of environmental change. As such it is impossible to determine whether these snapshots represent the height of interglacial warming, or intervals of good organic preservation that either preceded or followed such warm intervals. So it is too soon to say whether the faunas thought to represent Middle Pleistocene interglacial intervals are giving reliable reconstructions of full interglacial conditions. MIS 11 stands out as highly unusual, because these faunas include species found today only in eastern North America. We now know enough about the climatic regimes of MIS 5e in Eastern Beringia to be able to say that the levels and styles of warming (i.e., winter vs. summer temperature ameliorations) differed across Eastern Beringia. One would expect considerable climatic diversity in a region as large as this. As I stated in my 1994 book, the data in hand represent the tip of a very large iceberg, offering tantalizing glimpses into this vanished realm.

11

Other Studies in the New World

> The state of our knowledge of the effects of the Pleistocene on the present patterns of insect distribution is, at best, inadequate.
>
> – Henry F. Howden (1969)

In this chapter, I review Quaternary fossil insect studies from the conterminous United States, southern Canada, and South America. Some of these regions have received considerable attention from Quaternary entomologists, while many other regions have not been studied at all. The environmental reconstructions based on insect data are therefore concentrated in a few study regions.

1. North American Studies

Quaternary entomology has come a long way in North America since Henry Howden made his observation. Up to the 1960s, the only published Quaternary insect studies were those by Scudder and Pierce, which were fraught with identification errors. During the 1970s, several paleoentomologists turned their attention to North America. By 1980, Morgan and Morgan cited 23 publications treating 33 sites in North America. The number of publications concerning Quaternary entomology more than tripled between 1980 and 1990, but has slowed since then. In *Quaternary Insects and Their Environments*, I reported 119 published insect fossil sites for North American regions, south and east of Alaska, and the Yukon Territory. Since 1994, that list has grown to 139 sites (Table 1). Sadly, a number of researchers who were formerly active in North American research have either retired (such as John Matthews) or their studies of new North American sites have slowed to a trickle (such as Allan Ashworth, Clarke Gary, Alan and Anne Morgan, Randy Miller, Bob Nelson, and Don Schwert). Nevertheless, a lot of interesting work has been done, in spite of the fact that the regional and chronological coverage is spotty at best. In the following review, I provide regional summaries of the work thus far published.

1.1. Eastern and Central United States and Canada

More than 50 insect fossil studies have been published from sites south of the Laurentide ice sheet in eastern and central North America (Table 1, Fig. 1). The fossil assemblages range in age from pre-MIS 5e interglacial to Late Holocene (see Table 2 for North American stratigraphic nomenclature), but most are Late Wisconsin to Early Holocene-age. Most MIS 5e and Holocene faunas reflect similar environmental conditions, but the composition of regional faunas has changed appreciably between these interglaciations. Many of the beetles in the MIS 5e faunas are found only in western North America today.

Unlike in Europe, very few insect fossil assemblages that predate the last interglacial interval have been described from southern Canada and the lower 48 states of the United States. Garry *et al.* (1997) described two faunal assemblages from the Loraff Farm site in northeastern South Dakota (Fig. 1, No. 71) and the Nordick Farm site in western Minnesota (Fig. 1, No. 82) that are associated with MIS 6 (Illinoian glaciation). These small faunal assemblages are distinctly arctic/subarctic in character, containing such cold-adapted species as the ground beetle *Diacheila polita*, the water-scavenger beetle *Helophorus arcticus*, and the rove beetle *Olophrum latum*. The Loraff Farm fauna is sufficiently diagnostic to yield an MCR paleotemperature reconstruction. I estimate that T_{max} at the site was 10.5–11.5°C, which is 10–11°C colder than the modern T_{max}. I estimate that T_{min} at the site was −30 to −23°C, which is 11–18°C colder-than-modern T_{min} (Fig. 2). The fauna thus indicates full-glacial levels of cooling, similar to those found during the LGM in the Midwestern United States.

Mott (1990) and Mott and Matthews (1990) reviewed paleoenvironmental evidence from the MIS 5e interglacial in Canada. Because MIS 5e is beyond the range of radiocarbon dating, most assemblages cannot be confidently assigned to strict chronostratigraphic units. MIS 5e sites in the Maritime provinces include a few studied insect assemblages. These are the East Milford site, Nova Scotia (Fig. 1, No. 33) (Mott *et al.*, 1982), and the Baie du Bassin and Portage du Cap sites (Fig. 1, Nos. 7 and 89) on Magdalen Island, Quebec (Prest *et al.*, 1976). The faunas indicate warmer-than-modern conditions, including some species with modern distributions ranging north only to southernmost Canada. The fauna from Portage du Cap is sufficiently diverse and contains enough specimens identified to the species level to allow an MCR reconstruction (Fig. 2). The T_{max} estimate falls between 3.6°C colder than modern and 1°C warmer than modern. The T_{min} estimate likewise straddles the modern T_{min}.

Anderson *et al.* (1990a) described an MIS 5e flora and fauna from the Pointe-Fortune site in southern Quebec (Fig. 1, No. 88). These MIS 5e assemblages were indicative of warmer-than-modern climates. The MIS 5e

DEVELOPMENTS IN QUATERNARY SCIENCES
VOLUME 12 ISSN 1571-0866

Table 1. Pleistocene insect fossil sites in North America (excluding Alaska and the Yukon Territory).

Site	Reference(s)
(1) Adams Mill, Indiana	Morgan *et al.* (1983b)
(2) Ajo Mountains, Arizona	Hall *et al.* (1989)
(3) Amaguadees, Nova Scotia	Miller and Elias (2000)
(4) Athens, Illinois	Morgan (1987)
(5) Au Sable River, Michigan	Morgan *et al.* (1985)
(6) Baby Vulture Den, Texas	Elias and Van Devender (1992)
(7) Baie du Bassin, Quebec	Prest *et al.* (1976)
(8) Barehead Creek, Ontario	Warner *et al.* (1987)
(9) Beales, Ohio	Morgan (1987)
(10) Bechan Cave, Utah	Elias *et al.* (1992b)
(11) Benacadie, Nova Scotia	Miller and Elias (2000)
(12) Bennett Ranch, Texas	Elias and Van Devender (1992)
(13) Bida Cave, Arizona	Elias *et al.* (1992b)
(14) Biggsville, Illinois	Carter (1985)
(15) Bishop's Cap, New Mexico	Elias and Van Devender (1992)
(16) Bonfils Quarry, Missouri	Schwert *et al.* (1997)
(17) Bongards, Minnesota	Schwert and Ashworth (1985)
(18) Boniface River, Quebec	Lavoie *et al.* (1997b)
(19) Bonneville Estates, Utah	Rhode (2000), Elias (2007c)
(20) Brampton, Ontario	Morgan and Freitag (1982), Morgan (1987)
(21) Brookside, Nova Scotia	Mott *et al.* (1986)
(22) Campbell, Nova Scotia	Miller and Elias (2000)
(23) Cañon de la Fragua, Coahuila	Elias *et al.* (1995)
(24) Cataviña, Baja California	Clark and Sankey (1999)
(25) Charlevoix, Quebec	Lavoie (2001)
(26) Chaudiere Valley, Quebec	Matthews *et al.* (1987)
(27) Cincinnati, Ohio	Lowell *et al.* (1990)
(28) Clarksburg, Ontario	Warner *et al.* (1988)
(29) Clinton, Illinois	Morgan (1987)
(30) Collins Pond, Nova Scotia	Miller and Elias (2000)
(31) Conklin Quarry, Iowa	Baker *et al.* (1986)
(32) Discovery Park, Washington	Nelson and Coope (1982)
(33) East Milford, Nova Scotia	Mott *et al.* (1982)
(34) Eighteen Mile River, Ontario	Ashworth (1977)
(35) Elkader, Iowa	Schwert (1992), Schwert *et al.* (1997)
(36) Ennadai Lake, Kivalliq Region, Nunavut	Elias (1982a)
(37) Ernst Tinaja, Texas	Elias and Van Devender (1990)
(38) Escalante River Caves, Utah	Elias *et al.* (1992b)
(39) False Cougar Cave, Montana	Elias (1990b, 1996)
(40) Fort Dodge, Iowa	Schwert (1992)
(41) Gage Streeet, Ontario	Schwert *et al.* (1985)
(42) Gardena, Illinois	Morgan (1987)
(43) Garfield Heights, Ohio	Coope (1968d)
(44) Gervais Formation, Minnesota	Ashworth (1980)
(45) Gods River, Manitoba	Dredge *et al.* (1990)
(46) Goulais River, Ontario	Warner *et al.* (1987)
(47) Guadalupe Mountains, New Mexico	Elias and Van Devender (1992)
(48) Guadalupe Mountains, Texas	Elias and Van Devender (1992)
(49) Henday, Manitoba	Nielsen *et al.* (1986)
(50) Hornaday Mountains, Sonora	Hall *et al.* (1988)
(51) Hueco Mountains, Texas	Elias and Van Devender (1992)
(52) Huntington Canyon, Utah	Elias (1996)
(53) Innerkip, Ontario	Pilny and Morgan (1987)
(54) Johns Lake, North Dakota	Ashworth and Schwert (1992)
(55) Kaetan Cave, Arizona	Elias *et al.* (1992b)
(56) Kalaloch, Washington	Cong and Ashworth (1996)

(*Continued*)

Table 1. (*Continued*)

Site	Reference(s)
(57) Kewaunee, Wisconsin	Garry *et al.* (1990a)
(58) Klondike Bog, Northwest Territories	Matthews (1980b)
(59) Lake Emma, Colorado	Elias *et al.* (1991)
(60) Lake Isabelle Delta, Colorado	Elias (1985, 1996)
(61) Lake Isabelle Fen, Colorado	Elias (1985, 1996)
(62) Lamb Spring, Colorado	Elias and Toolin (1989), Elias (1996)
(63) La Poudre Pass, Colorado	Elias (1982b, 1996)
(64) Last Chance Canyon, New Mexico	Elias and Van Devender (1992)
(65) Leamington, Ontario	Karrow *et al.* (2007)
(66) Lefthand Reservoir, Colorado	Elias (1985, 1996)
(67) Limestone River, Manitoba	Dredge *et al.* (1990)
(68) Lockport Gulf, New York	Miller and Morgan (1982)
(69) Longs Peak Inn, Colorado	Elias (1996)
(70) Longswamp, Pennsylvania	Morgan *et al.* (1982b)
(71) Loraff Farm, South Dakota	Garry *et al.* (1997)
(72) Makinson Inlet, Ellesmere Isl., Nunavut	Blake and Matthews (1979)
(73) Maravillas Canyon Cave, Texas	Elias and Van Devender (1990)
(74) Marias Pass, Montana	Elias (1988a, 1996)
(75) Mary Hill, British Columbia	Miller *et al.* (1985a)
(76) Mary Jane, Colorado	Short and Elias (1987), Elias (1996)
(77) McKittrick, California	Miller (1983)
(78) Mosbeck, Minnesota	Ashworth *et al.* (1972)
(79) Mount Ida Bog, Colorado	Elias (1985, 1996)
(80) Newton, Pennsylvania	Barnowsky *et al.* (1988)
(81) Nichols Brook, New York	Fritz *et al.* (1987)
(82) Nordick Farm, Minnesota	Garry *et al.* (1997)
(83) Norwood, Minnesota	Ashworth *et al.* (1981)
(84) Oregon jack Creek, British Columbia	Hebda *et al.* (1990)
(85) Owl Creek, Ontario	Miller (1990), Mott and DiLabio (1990)
(86) Owl Roost, Arizona	Elias *et al.* (1992b)
(87) Pasley River, Nunavut	Dyke and Matthews (1987)
(88) Pointe-Fortune, Quebec	Anderson *et al.* (1990a)
(89) Portage du Cap, Quebec	Prest *et al.* (1976)
(90) Portland, Maine	Anderson *et al.* (1990b)
(91) Port Moody, British Columbia	Miller *et al.* (1985a)
(92) Powers, Michigan	Morgan (1987)
(93) Puerto Blanco Mountains, Arizona	Hall *et al.* (1990)
(94) Puerto de Ventanillas, Coahuila	Elias *et al.* (1995)
(95) Quillin, Ohio	Morgan (1987)
(96) Quitman Mountains, Texas	Elias and Van Devender (1992)
(97) Radisson, Quebec	Lavoie and Arseneault (2001)
(98) Rainy River, Ontario	Schwert and Bajc (1989), Bajc *et al.* 2000
(99) Rancho La Brea, California	Doyen and Miller (1980), Miller (1983)
(100) Rat's Nest Cave, Alberta	Bain *et al.* (1997)
(101) Roaring River, Colorado	Elias *et al.* (1986)
(102) Roberts Creek, Iowa	Schwert (1996)
(103) Rocky Arroyo, New Mexico	Elias and Van Devender (1992)
(104) Rostock, Ontario	Morgan *et al.* (1983a), Pilny *et al.* (1987)
(105) Rous Lake, Ontario	Morgan (1987)
(106) Russellville, Indiana	Morgan *et al.* (1983b), Morgan (1987)
(107) Sacramento Mtns, New Mexico	Elias and Van Devender (1992)
(108) San Andres Mtns, New Mexico	Elias and Van Devender (1992)
(109) Sandy River, Maine	Nelson (1987)
(110) St. Charles, Iowa	Baker *et al.* (1991)
(111) St. Eugene, Quebec	Mott *et al.* (1981)
(112) St. Flavien, Quebec	Lavoie *et al.* (1997a)

(*Continued*)

Table 1. (*Continued*)

Site	Reference(s)
(113) St. Hilaire, Quebec	Mott *et al.* (1981)
(114) Salt Creek Canyon, Utah	Elias *et al.* (1992b)
(115) Salt River, Missouri	Schwert *et al.* (1997)
(116) Salt Spring Hollow, Illinois	Schwert *et al.* (1997)
(117) Saylorville, Iowa	Schwert (1992)
(118) Scarborough Bluffs, Ontario	Williams and Morgan (1977), Williams *et al.* (1981)
(119) Seibold, North Dakota	Ashworth and Brophy (1972)
(120) Shafter, Texas	Elias and Van Devender (1992)
(121) Shelter Cave, New Mexico	Elias and Van Devender (1992)
(122) Sierra de la Misericordia, Coahuila	Elias *et al.* (1995)
(123) South Organ Mtns, New Mexico	Elias and Van Devender (1992)
(124) Streeruwitz Hills, Texas	Elias and Van Devender (1992)
(125) Terlingua, Texas	Elias and Van Devender (1990)
(126) Tinajas Atlas Mountains, Arizona	Hall *et al.* (1988)
(127) Titusville, Pennsylvania	Cong *et al.* (1996)
(128) Todd Mountain, New Brunswick	Miller and Elias (2000)
(129) Tonica, Illinois	Schwert (1992)
(130) Trollwood Park, North Dakota	Yansa and Ashworth (2005)
(131) Tunnel View, Texas	Elias and Van Devender (1990)
(132) Two Creeks, Wisconsin	Morgan and Morgan (1979)
(133) Umiakoviarusek, Labrador	Elias (1982c)
(134) Wales, Ontario	Miller *et al.* (1987)
(135) Weaver Drain, Michigan	Morgan *et al.* (1981)
(136) Wedron, Illinois	Garry *et al.* (1990b), Schwert *et al.* (1997)
(137) West Mabou, Nova Scotia	Miller and Elias (2000)
(138) Winter Gulf, New York	Schwert and Morgan (1980)
(139) Woodbridge, Ontario	Karrow *et al.* (2001)

deposit was overlain by younger organic horizons which contained a beetle fauna suggesting climatic cooling, and hence a waning of interglacial warmth. My MCR reconstruction for faunal unit 4A (post-MIS 5e in age) is 15–18°C, which represents summer temperatures 2–5°C colder than modern.

Three regions in Ontario have yielded MIS 5e insect faunas. These are the Scarborough Bluffs site, Toronto (Fig. 1, No. 118) (Williams and Morgan, 1977; Williams *et al.*, 1981); the Innerkip site, southwest of Toronto (Fig. 1, No. 53) (Pilny and Morgan, 1987), and the Owl Creek beds, northeastern Ontario (Fig. 1, No. 85) (Miller, 1990; Mott and DiLabio, 1990). The Scarborough insects are indicative of climate like that in region of the lower Great Lakes today, which implies MIS 5e temperatures warmer than modern. The Innerkip assemblages are similar in composition to the Scarborough Bluffs assemblages; however, MCR analysis of this fauna yielded T_{max} and T_{min} estimates slightly cooler than modern levels at the site (Fig. 2).

The Owl Creek fauna comes from the upper parts of cores from the Timmins region. The fauna is comprised of species found today in the boreal and subarctic regions of Canada. The fauna is indicative of northern tree line conditions, which suggests conditions similar to present (Miller, 1990). Older sediments in the Owl Creek beds lack insect fossils, but have produced pollen spectra suggesting interglacial conditions warmer than present (Mott and DiLabio, 1990).

Insect fossils from the Missanaibi Formation in northern Manitoba also indicate interglacial environments. Lists of taxa have been published for five sites, as summarized in Nielsen *et al.* (1986) and Dredge *et al.* (1990). Assemblages from the Henday site (Fig. 1, No. 49) are indicative of a northern tree line environment, similar to that found today at nearby Churchill, Manitoba. Fossil beetles from the Limestone River site (Fig. 1, No. 67) are indicative of open-ground environments at or near tree line. These colder than present assemblages may represent the later stages of the MIS 5 (i.e., MIS 5c or 5a). Insects from the Echoing River, Flamborough, and Gods River interglacial deposits (Fig. 1, No. 45) reflect boreal-zone climate, similar to modern regional conditions. The Gods River fauna included several species of bark beetles associated today with boreal forests in Canada. These fossils were probably deposited during the height of interglacial warming (MIS 5e).

Early Wisconsin-age organic sediments, assigned to the Massawippi Formation, have been studied from the Chaudière Valley in southern Quebec (Fig. 1, No. 26) (Matthews *et al.*, 1987). Macrofossils from this formation yield infinite radiocarbon ages, but are thought to date from a cold stage during late MIS 5 or the Early Wisconsin (Isotope Stages 5a or 4), based on their stratigraphic position. The biotic evidence from these deposits suggests the type of arctic environment found today in northernmost Quebec. My MCR estimate from this assemblage indicates that T_{max} was about 8°C colder

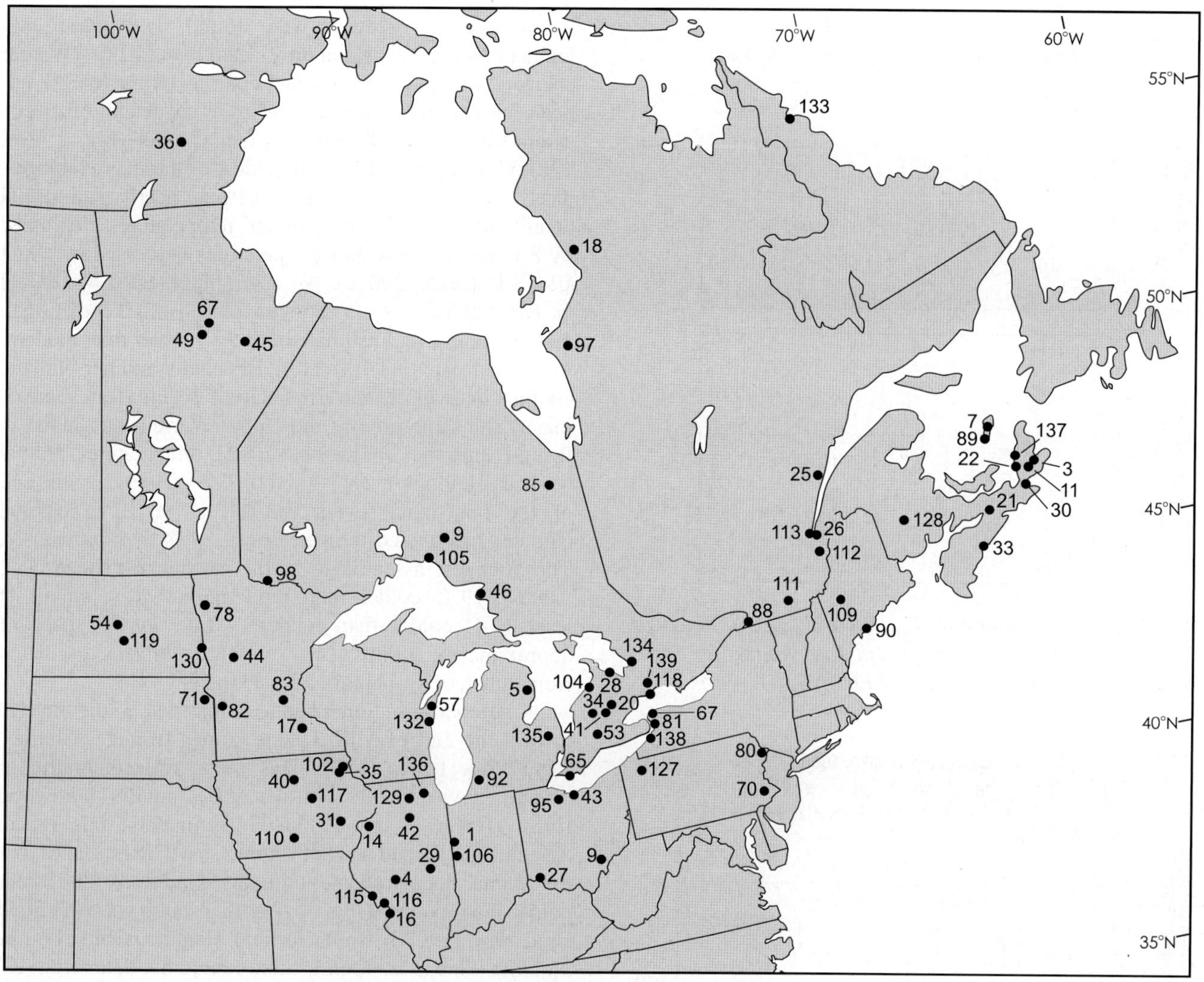

Fig. 1. Map of eastern and central North America, showing location of Quaternary insect fossil sites, listed in Table 1.

Table 2. Correlation of North American Pleistocene sequences with oxygen isotope stages.

Isotope stage	Approximate age (cal yr BP)	Eastern and central North America American sequence*	Western North America American sequence†
1	0–11 K	Holocene	Holocene
2	11–26 K	Late Wisconsin glaciation	Pinedale glaciation
3	26–55 K	Mid-Wisconsin interstadial	Mid-Wisconsin interstadial
4–5a	55–110 K	Early Wisconsin glaciation	Early Wisconsin glaciation
5e	110–130 K	Sangamon interglacial	Sangamon interglacial
6	130–180 K	Illinoian glaciation	Bull Lake glaciation

*After Gibbard *et al.* (2005)
†After Pierce (2004)

than today, and T_{min} was at least 10°C colder than today at the site (Fig. 2). The insect assemblages include several arctic species which were subsequently extirpated from regions east of Hudson Bay, and have not become reestablished in northern Quebec during the Holocene.

Karrow *et al.* (2001) described a small beetle faunal assemblage of probable MIS 4 age from Woodbridge, near Toronto, Ontario (Fig. 1, No. 139). This fauna came from the local equivalent to the Scarborough Formation, described from Toronto. It contained a number of cold-adapted species, including the ground beetles *Diacheila polita* and *Elaphrus lapponicus*, as well as the arctic water-scavenger beetle *Helophorus arcticus*. The combined fossil evidence from insects, molluscs, and plant remains indicate a T_{max} of 10–12°C, placing the site near northern tree line. However, there is no definitive age classification for this deposit. It could represent one or more intervals from late MIS 5 to late MIS 4 (Karrow *et al.*, 2000).

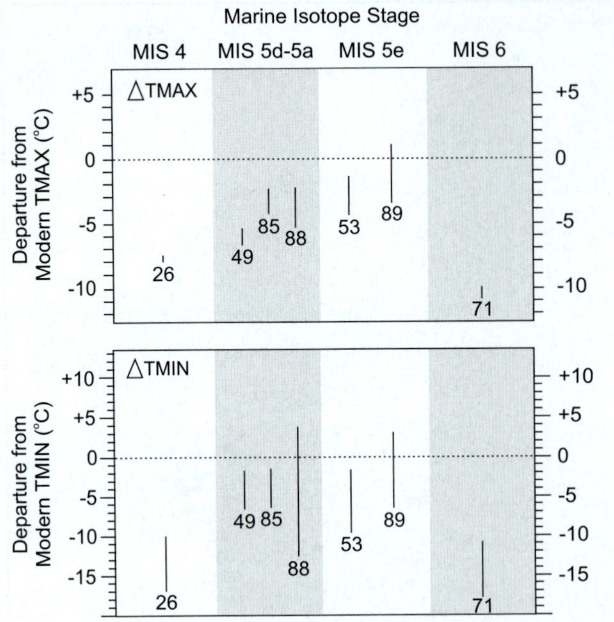

Fig. 2. MCR estimates of T_{max} and T_{min}, displayed as departures from modern T_{max} and T_{min} values, for fossil beetle assemblages associated with MIS 6-4.

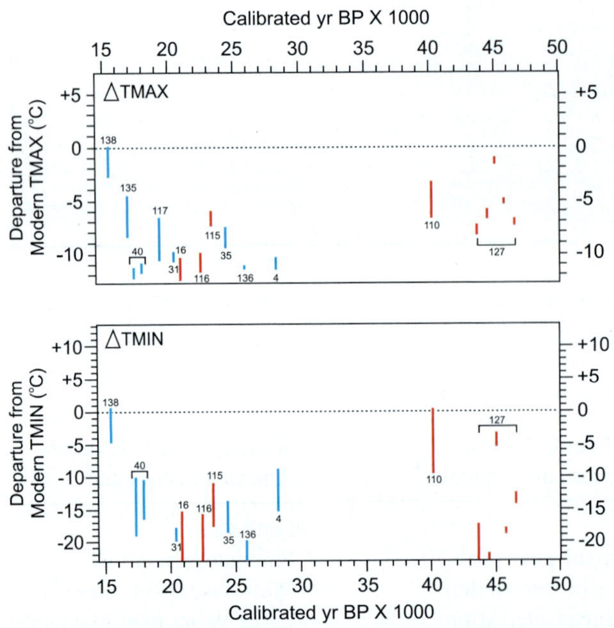

Fig. 3. MCR estimates of T_{max} and T_{min}, displayed as departures from modern T_{max} and T_{min} values, for fossil beetle assemblages associated with MIS 3-2. Red bars indicate estimates from sites well removed from the margins of the Laurentide ice sheet. Blue bars indicate estimates from ice-proximal sites.

MIS 3 has been shown to have been an interval of rapid, extreme climatic changes, rather than an interval marked by a gradual interstadial warming signal (Fig. 3). Arctic conditions in central Ontario have been shown by an insect assemblage of Early- or Mid-Wisconsin age from Clarksburg, Ontario (Fig. 1, No. 28). The fauna includes the *Pterostichus* (*Cryobius*) species such as *P. pinguedineus*, *P. ventricosus*, and *P. brevicornis*. No forest-associated species were found, and mean July temperatures were apparently less than 10°C (Warner *et al.*, 1988). A climatic amelioration during MIS 3 has been clearly demonstrated from a sequence of beetle assemblages from Titusville, Pennsylvania (Fig. 1, No. 127) (Cong *et al.*, 1996). The oldest of these assemblages, dated ca. 46,500 cal yr BP, reflects nearly full-glacial conditions, with mean summer temperatures depressed by 8°C and mean winter temperatures depressed by about 12–13°C (Elias, 2007c). An assemblage dated 45,000 cal yr BP yielded a T_{max} estimate only 1–2°C cooler than modern, and a T_{min} estimate of 3–5°C cooler than modern. The fauna dated 43,700 cal yr BP marks another climatic cooling interval. The warm interval occurred at approximately the same time as the Upper Warren interstadial in Britain and the warm sea-surface temperature interval between Heinrich events 4 and 5 (Elias, 1999). By the end of the interstadial event (43,200 cal yr BP), regional climates had cooled to even lower levels than previously. The youngest assemblage in this sequence yielded T_{max} values 8–9°C colder than modern, and T_{min} values at least 17°C colder than modern. Thus, in the space of approximately 3,000 years, regional climates oscillated from subarctic to boreal and back again.

A number of regional beetle faunas date to the interval toward the end of MIS 3. One comes from St. Charles, Iowa (Fig. 1, No. 110). This fauna reflects prairie or savanna environments with patches of conifer–hardwood forest (Baker *et al.*, 1991). Unfortunately, this faunal assemblage yielded a set of imprecise MCR estimates, but the fauna does indicate climatic conditions that were colder than modern (Fig. 3). Another fauna was described by Coope (1968d) from Garfield Heights, Ohio (Fig. 1, No. 43). This is an open-ground fauna that dates from ca. 32,500 to 29,000 cal yr BP. No strict tundra dwellers were identified, although Coope (April, 2009, written communication) identified a ground beetle in the *Cryobius* group of the genus *Pterostichus*, and most of the species in this group are tundra dwellers. Isolated conifer macrofossils in the deposit suggest small stands of trees. The identified beetles live today in open-ground habitats in eastern Canada. Coope estimated mean July temperatures at the site to have been less than 15°C.

A suite of sites from south of the Great Lakes has also yielded insect faunas from this interval. These faunas are suggestive of boreal-style climate and coniferous woodlands. At Athens and Gardena, Illinois (Fig. 1, Nos. 4 and 42), Morgan (1987) described insect assemblages ranging in age from 31,700 to 27,000 cal yr BP. These are boreal faunas, with species found today in central Canada, where modern July temperatures are 15–17°C. My MCR analysis of the Athens fauna yielded one of the greatest departures from modern T_{max}: 9.6–11.6°C colder than modern (Elias, 2007c). The Athens fauna thus clearly marks the trend in climatic cooling that culminated in the LGM. The sketchy picture that emerges for the Early and Mid-Wisconsin intervals is one of spatial and temporal heterogeneity in which populations of plants and insects were responding more dynamically than at any time during the Holocene (Ashworth, 2004).

By approximately 30 kya, the beetle evidence suggests that central North American climates had cooled to full-glacial levels (Fig. 3). Regional faunas contained mixtures of boreal and arctic species, depending on latitude. Southerly sites, such as Salt Spring Hollow, Illinois (Fig. 1, No. 116), contained boreal beetle assemblages just before and during the LGM. More northerly sites, such as Conklin Quarry, Iowa (Fig. 1, No. 31), contained beetles with arctic and subarctic affinities. The Adams Mill and Russellville sites in Indiana (Fig. 1, Nos. 1 and 106) date to about 26,500 and 25,000 cal yr BP, respectively (Morgan *et al.*, 1983b). These sites yielded small insect faunas, including aquatic beetles, caddis larvae, and some scolytids and conifer macrofossils, suggesting the presence of some trees. Mean July temperatures during this interval in Indiana were estimated at 14–16°C, indicating that T_{max} was 8–10°C cooler than modern.

From 25,800 to 17,700 cal yr BP, arctic and subarctic species inhabited a discontinuous tundra zone along the margin of the ice sheet. Fossil assemblages typical of this time have been reported from Iowa to New York (Ashworth, 2004). At the end of the glaciation, these cold-adapted faunas were forced to migrate in order to survive. Some species found refuge in high mountains, such as the Rocky Mountains in the west and the tops of the Appalachian Mountains in the east. Other species died out in the middle latitudes of the continent, and only their Beringian populations in ice-free regions of northwestern North America survived beyond the end of the Pleistocene.

In the Midwestern region, arctic beetles replaced the boreal forest fauna along the southern margin of the Laurentide ice sheet at approximately 25,800 cal yr BP. Colonization was probably from populations that dispersed southward in front of the growing ice sheet and westward and eastward from montane refugia in the Appalachian and Rocky Mountains, respectively (Schwert and Ashworth, 1988). This narrow band of periglacial habitat supported insects associated with arctic tundra or with open-ground habitats of the subarctic and boreal regions, in mixtures of species not seen in any one region today. During the last (Wisconsin) glacial advance, the coniferous woodland ecosystem remained intact only further to the south. Insect fossil sites dating to the last (Wisconsin) glaciation have been described from Clinton, Illinois (Fig. 1, No. 29), and from an upper horizon at Gardena. The deposits are dated about 24,500 and 23,300 cal yr BP. The older fauna contains a mixture of arctic and boreo-arctic beetles. The younger sample includes only arctic and subarctic species. This assemblage reflects mean July temperatures of only 11–12°C (Morgan, 1987).

Another full-glacial fauna has been described by Morgan and Pilny (in Lowell *et al.*, 1990) from Cincinnati, Ohio (Fig. 1, No. 27). This deposit has been dated at 23,300 cal yr BP. Laurentide ice advanced to within 5 km of the site at 23,400 cal yr BP. The beetle fauna suggests mean July temperatures of 10–12°C. These conditions are found today at northern tree line in Canada, and they represent a departure from modern T_{max} of 12–14°C.

Regional insect faunas indicate that climatic warming began soon after 18,000 cal yr BP, but local environmental histories were complicated by waxing and waning of ice lobes and **proglacial lakes** (see Morgan, 1987; Elias *et al.*, 1996a). Accordingly, regional lateglacial insect faunas reflect a variety of conditions, ranging from arctic conditions at ice-proximal sites to temperate conditions in more southerly locations. The earliest of the regional lateglacial insect fossil records comes from Longswamp, Pennsylvania (Fig. 1, No. 70). These date from ca. 18,250 to 17,250 cal yr BP. Paleobotanical evidence (Watts, 1979) suggests tundra conditions at this time, but even the earliest insect assemblages include boreal species from open-ground habitats. By 14,800 cal yr BP, conifers and bark beetles arrived. This represents a lag of as much as 3,450 years between the establishment of boreal-zone climate and the arrival of conifers, even though conifer forests were not far south of the site in the Late Wisconsin (Watts, 1979).

By 17,600 cal yr BP, a boreal climatic regime was established once again in regions immediately south of the Laurentide ice sheet. This is demonstrated by an insect fauna from Quillin, Ohio (Fig. 1, No. 95), which included boreal insect taxa and conifer macrofossils. Morgan (1987) estimated mean July temperatures of about 15°C for this assemblage. As the ice retreated northward, newly exposed ground supported a mixture of arctic and northern boreal insects, as shown from an assemblage at Weaver Drain, Michigan (Fig. 1, No. 135), dated at 16,900 cal yr BP. The Weaver Drain site was ice proximal at 17,000 cal yr BP, but MCR analysis (Fig. 2) shows that T_{max} had started to rise above the levels reconstructed from a 17,300 cal yr BP assemblage at Fort Dodge, Iowa (Fig. 1, No. 40). The Weaver Drain fauna suggests conditions similar to modern tree line in Canada. While this is considerably cooler than the conditions suggested by the Quillin and Longswamp faunas, it probably reflects persistent regional cooling due to the proximity of the waning Laurentide ice sheet (Morgan *et al.*, 1981). By comparison, a contemporaneous fauna that lived far from the ice margin at Beales, Ohio (Fig. 2, No. 8), indicates summer temperatures of 16–17°C (Morgan, 1987).

As discussed in Chapter 5, lateglacial climates in eastern and central North America have been somewhat difficult to reconstruct from fossil beetle assemblages, because of the complicating factors of some sites being in close proximity to the retreating Laurentide ice sheet, or to large proglacial lakes. The lateglacial insect faunas from southern Ontario trace the retreat of Laurentide ice, followed by rapid warming. Several centuries later, the mixed forest compatible with this warming finally became established. Lateglacial assemblages from basal organic sediments in southern Ontario reflect cold, ice-proximal environments. These include faunas from Leamington (Fig. 1, No. 65) (Karrow *et al.*, 2007), dated 16,150 cal yr BP; the Rostock site (Fig. 1, No. 104) (Pilny *et al.*, 1987), dated 15,800 cal yr BP; and the Brampton site (Fig. 1, No. 20) (Morgan and Freitag, 1982; Morgan, 1987), dated >14,800 cal yr BP at the base. All these assemblages contained open-ground faunas with no conifer-associated taxa. Following regional deglaciation, younger assemblages from Rostock (14,000 cal yr BP) reflect warming of summer temperatures, and by 11,000 cal yr BP, the

Brampton assemblages suggest mean July temperatures rising to near modern levels.

The most complete lateglacial insect sequence is from the Gage Street site in Kitchener, Ontario (Fig. 1, No. 41) (Schwert *et al.*, 1985). The base of the organic sequence was dated at 15,300 cal yr BP, and contained a mixture of open-ground and boreal species, including bark beetles. Mean July temperatures reconstructed from this assemblage were 15–17°C at 15,300 cal yr BP, rising to nearly 21°C by 11,000 cal yr BP. The Holocene fauna (9,600–8,800 cal yr BP) indicates summer temperatures of 22–23°C, followed by an increase to 25°C by 7,750 cal yr BP, at the top of the sedimentary sequence. However, sites north of the Great Lakes in Ontario indicate summer temperatures only in the 17–18°C range at this time. These sites include Barehead Creek, north of Lake Superior (Fig. 1, No. 8), and the Goulais River site near Sault St. Marie (Fig. 1, No. 46) (Warner *et al.*, 1987).

Lateglacial sites in upstate New York include the Winter Gulf site (Fig. 1, No. 138), dating to 15,000–14,800 cal yr BP. The beetle fauna implies an open mire surrounded by trees (Schwert and Morgan, 1980). A lateglacial site at Nichols Brook, New York (Fig. 1, No. 81), reported by Fritz *et al.* (1987), is contemporaneous with Winter Gulf. The basal insect fauna from Nichols Brook is indicative of conditions at the center of the boreal zone. Younger assemblages show that summer temperatures rose as high as 18°C by 11,000 cal yr BP.

Ashworth and Schwert (in Barnowsky *et al.*, 1988) published the results of a study of lateglacial insects associated with a mammoth skeleton at Newton, Pennsylvania (Fig. 1, No. 80). The insect assemblages were from a horizon dated at 14,100 cal yr BP, and from below this horizon. The insects represent open-ground conditions and climate similar to modern tree line. The assemblages include *Pterostichus pinguedineus* and *Diacheila polita*; the latter species is thought to have been extirpated from the American midwest by about 15,000 yr BP.

Another lateglacial fauna was described from Two Creeks, Wisconsin (Fig. 1, No. 132) (Morgan and Morgan, 1979). Assemblages dating from the interval 14,000–13,700 cal yr BP comprise an open boreal fauna reflecting July temperatures of 12.5–13.5°C.

Garry *et al.* (1990a) described a contemporaneous beetle fauna from Kewaunee, Wisconsin (Fig. 1, No. 57). While the Kewaunee fauna has many species in common with Two Creeks, it also contains some cold stenotherms indicative of slightly colder conditions. The difference in temperature reconstruction may reflect microclimatic differences between the sites.

Anderson *et al.* (1990b) studied lateglacial sediments from Portland, Maine (Fig. 1, No. 90). Insect fossils dating to 13,400 cal yr BP included boreal-zone beetles and ants, but some of the beetles are indicative of open-ground situations within the boreal forest.

The Champlain Sea was a large proglacial lake, occupying much of southeastern Ontario, southwestern Quebec, and adjacent regions of New England at the end of the last glaciation. Insect fossil assemblages have been described from sites along the Champlain Sea at Ste. Eugene and Ste. Hilaire, Quebec (Mott *et al.*, 1981). The Ste. Eugene site (Fig. 1, No. 111), dated at 12,950 cal yr BP, includes a northern boreal and tree line beetle assemblage. The MCR estimate of T_{max} from this assemblage (10–12.5°C) reflects the cooling effects of the site's proximity to the Champlain Sea (Elias *et al.*, 1996a).

Lateglacial boreal forest environments have also been documented in Michigan and New York. At the Powers site, in southern Michigan (Fig. 1, No. 92), a boreal fauna was found in association with mastodon remains, dated at 13,000 cal yr BP. This fauna reflects mean July temperatures of 17°C (Morgan, 1987). Miller and Morgan (1982) studied insect fossils from contemporaneous assemblages at Lockport Gulf, New York (Fig. 1, No. 68). Organic remains from a pond deposit spanning 12,900–10,300 cal yr BP yielded abundant insect assemblages. The faunas are indicative of boreal forest habitats from 12,900 to 11,200 cal yr BP, followed by a transition to a mixed forest fauna from 11,000 to 10,300 cal yr BP. As inferred from many other insect faunas in the lateglacial to Holocene transition, the Lockport Gulf faunas indicate that lateglacial climatic conditions were already suitable for the establishment of the plant communities that would finally become stable in the Holocene. The vegetation succession from boreal to mixed forest therefore reflects ecological succession, rather than climatic warming in the Early Holocene.

At Eighteen Mile River, Ontario (Fig. 1, No. 34), a site dated at 12,400 cal yr BP documented conditions on the margin of glacial lake Algonquin (Ashworth, 1977). This fauna contained a mixture of elements (mostly boreal) with no modern analogue. As at the Kewaunee site in Wisconsin, the Eighteen Mile River assemblages suggest a cold microenvironment, surrounded by a warm macroclimate.

1.2. Midwestern Studies

The oldest Quaternary insect fauna in the Midwestern region is beyond the range of radiocarbon dating, but is probably Early Wisconsin in age. It is an assemblage from the Gervais Formation in northwestern Minnesota (Fig. 1, No. 44) (Ashworth, 1980). The fauna is a mixture of boreal and arctic taxa, reflecting conditions similar to modern tree line in Canada. Some of the species are known today only from northwestern North America, including the weevil *Vitavitus thulius*, which is common in interstadial assemblages from Eastern Beringia. The known modern range of this species includes the Yukon Territory and adjacent regions of the Northwest Territories in Canada. It inhabits dry tundra and steppe habitats, along with other species associated with steppe–tundra environments in the Beringian Pleistocene. This fauna yields an MCR estimate of T_{max} (15.5–16°C) that is about 6°C colder than modern, and an estimate of T_{min} (−19 to −17°C) that is about 4–6°C colder than modern.

The other insect faunas from this region range in age from 40,000 to 13,600 cal yr BP. They have been studied from sites in Illinois, Iowa, Minnesota, North Dakota, Wisconsin, and northwestern Ontario. The fossil faunas indicate regional responses to a series of marked environmental changes from before the last glaciation through present. The earliest of these faunas is from the

St. Charles site in Iowa (Fig. 1, No. 110) (Baker *et al.*, 1991). Insect assemblages dated at 40,000 cal yr BP are indicative of open-ground environments, but all the species live today in the upper Great Lakes region, which Baker *et al.* (1991) interpreted to mean that the climate was only slightly cooler than modern. As discussed above, the MCR estimate from this fauna is imprecise, but it indicates that summer temperatures were at least 3.3°C colder than modern, which agrees well with the interpretation of Baker *et al.*

The Midwestern faunal sequence in the Late Wisconsin is similar to that discussed above for regions farther east. Midwestern regional sites that predate the last glaciation span the interval 31,700–25,000 cal yr BP. These are indicative of closed spruce forest. Boreal faunas were replaced by tundra and forest–tundra faunas during the Late Wisconsin glaciation. By 17,000 cal yr BP, the cold-adapted species were extirpated by climatic warming, and replaced by a fauna of boreal and western montane species. This paleoenvironmental reconstruction has been summarized by Schwert (1992). The results of individual studies are summarized in Table 3.

Sites predating the last glaciation include Biggsville, Illinois (Fig. 1, No. 14); Gardena, Illinois (Fig. 1, No. 42); Athens, Illinois (Fig. 1, No. 4); and Wedron, Illinois (Fig. 1, No. 136). These assemblages are characterized by

Table 3. Summary of Quaternary insect fossil sites in the Midwestern United States.

Site	Age range (cal yr BP × 1,000)	Environmental interpretations	Reference(s)
Pre-last glacial faunas			
Biggsville, Illinois	32.4–21.4	Diverse forest assemblage, similar to modern fauna of southern boreal forest in central Canada	Carter (1985)
Gardena, Illinois	30.3	Diverse forest assemblage, similar to modern fauna of southern boreal forest in central Canada	Morgan and Morgan (1986)
Wedron, Illinois	25.7	Diverse forest assemblage, similar to modern fauna of southern boreal forest in central Canada	Garry *et al.* (1990b)
Last glacial faunas			
Gardena, Illinois	23.5	Subarctic assemblage with modern analogues near tree line in Mackenzie Delta region of Canadian Northwest Territories; T_{max} = 10–12°C	Morgan and Morgan (1986)
Conklin Quarry, Iowa	21.8–19.9	Subarctic assemblage with modern analogues near tree line in Mackenzie Delta region; T_{max} = 10–12°C	Baker *et al.* (1986)
Saylorville, Iowa	20.3–17.2	Subarctic assemblage with modern analogues near tree line in Mackenzie Delta region; T_{max} = 10–12°C	Schwert (1992)
Fort Dodge, Iowa	18.5–18.3	Subarctic assemblage with modern analogues near tree line in Mackenzie Delta region; T_{max} = 10–12°C	Schwert (1992)
Lateglacial sites			
Fort Dodge, Iowa	17.5–16.3	Subarctic assemblages with open-ground species; no modern analogue	Schwert (1992)
Tonica, Illinois	17.5	Open-ground faunas replaced by spruce forest faunas	Schwert (1992)
Saylorville, Iowa	17.3–16	Open-ground faunas replaced by spruce forest faunas	Schwert (1992)
Norwood, Minnesota	16–13.1	Subarctic assemblages replaced by spruce forest faunas after 12,400 yr BP	Ashworth *et al.* (1981)
Fort Frances, Ontario	13.1–11.6	Fauna from Lake Agassiz shore indicative of boreal forest found today in central and southern Canada	Schwert and Bajc (1989)
Holocene sites			
Seibold, North Dakota	11.2–?	Documents transition from boreal forest to tall grass prairie between 9,000 and 8,500 yr BP	Ashworth and Brophy (1972)
Bongards, Minnesota	3.8–recent	Assemblages reflect modern conditions throughout sequence	Schwert and Ashworth (1985)

diverse forest faunas, dominated by scolytids and weevils, and swamp-dwelling carabids and staphylinids. Close modern analogues for these assemblages are in the southern boreal forest regions of central Canada. By 27,500 cal yr BP, the forest was replaced by open-ground as glaciers spread southward.

Assemblages from the last glaciation have been described from the upper section of the Gardena site, Conklin Quarry, Iowa (Fig. 1, No. 31); Saylorville, Iowa; and Fort Dodge, Iowa (Fig. 1, No. 40). These faunas are characterized by subarctic species, with modern analogues in tree line situations west of the Mackenzie Delta region of northwestern Canada. Mean July temperatures in this region are 10–12°C. My MCR estimates for these sites indicate that T_{max} was depressed by 10–12°C from modern levels at these sites, and that T_{min} was 18–20°C colder than modern (Fig. 3).

Three faunal assemblages from sites well south of the Laurentide ice sheet have been described from southern Illinois and Missouri (Schwert *et al.*, 1997). The Salt River site in Missouri (Fig. 1, No. 115) yielded a fauna dated 23,300 cal yr BP that is indicative of climatic conditions that are colder than modern, but not at the full-glacial level of cooling. My MCR estimate of T_{max} or this fauna was 7.5–6°C colder than modern, and T_{min} was 11–18°C colder than modern (Elias, 2007c). A fauna dated 800 years younger than this from nearby Salt Spring Hollow, Illinois (Fig. 1, No. 116), yielded a T_{max} estimate 10–12°C colder than modern, and a T_{min} estimate 16–24°C colder than modern. The fauna from Bonfils Quarry, Missouri, just south of Salt Spring Hollow on the west bank of the Mississippi, dates to 20,800 cal yr BP, and yielded essentially the same MCR estimates as the Salt Spring Hollow fauna. These sites were more than 100 km southwest of the LGM ice limit, and together with the other regional sites of LGM age, they document the boundaries of three regional ecosystems that existed at that time. The Bonfils Quarry and Salt Spring Hollow faunas are indicative of boreal forest environments. These faunas include the ground beetle *Chlaenius alternatus*, and numerous conifer-feeding bark beetles. The Salt River and Saylorville faunas are indicative of forest–tundra environments. These faunas include the ground beetles *Diacheila arctica* and *Pelophila borealis.* The Conklin Quarry fauna is indicative of arctic tundra, as shown by the presence of tundra-dwelling species such as *Diacheila polita* and *Pterostichus vermiculosus* (Schwert *et al.*, 1997). The compression of these three different ecosystems into relatively narrow bands may have been promulgated by the proximity of the Laurentide ice sheet, bringing cold northwesterly winds to the unglaciated regions to the south and west (Schwert *et al.*, 1997) (Fig. 4). The temperature regime was likewise compressed. The range of T_{max} over a latitudinal span of about 300 km was 11°C in the north (Conklin Quarry) to 16°C in the south

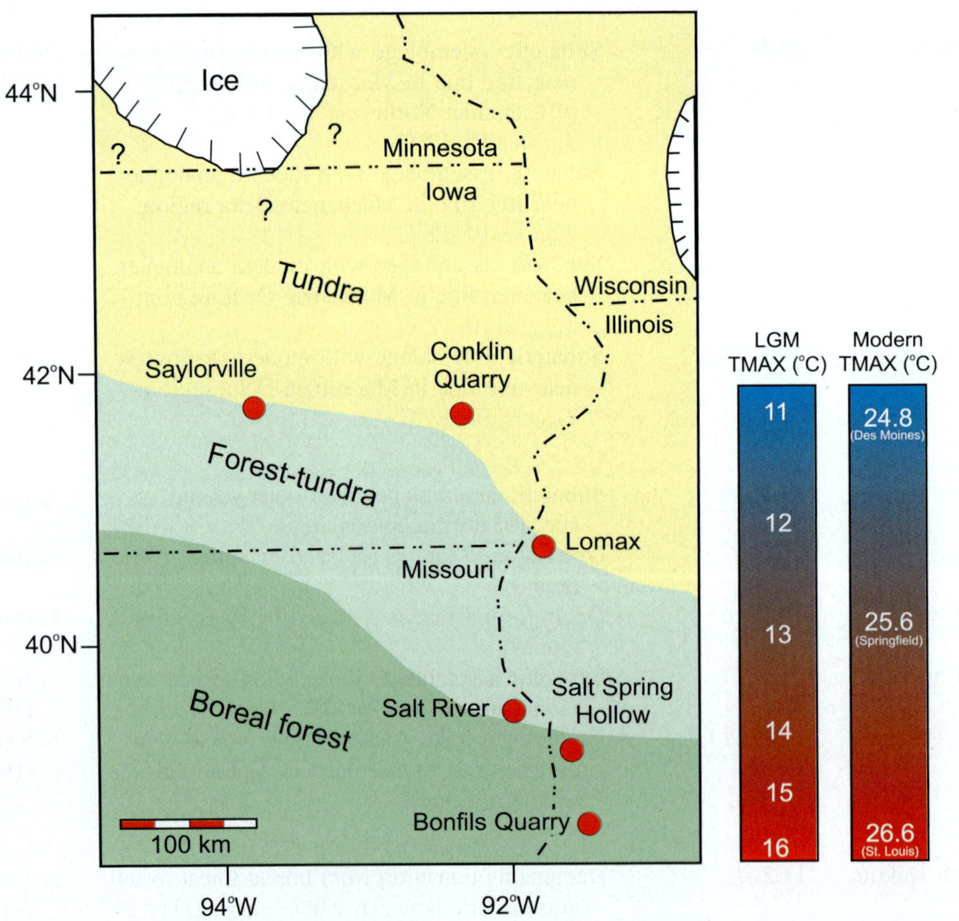

Fig. 4. Reconstruction of major ecosystem boundaries (left) and T_{max} estimates based on LGM-age faunas from the Midwestern region of the United States (after Schwert et al., 1997).

(Bonfils Quarry). Today this same latitudinal span has T_{max} values ranging from 24.8°C in the north to 26.6°C in the south, less than half of the temperature range estimated for this span of sites during the LGM (Fig. 4).

Lateglacial sites in the Midwestern United States are also summarized in Table 3. These include the upper part of the Ft. Dodge sequence, Tonica, Illinois (Fig. 1, No. 129); upper horizons of the site at Saylorville, Iowa; and Norwood, Minnesota (Fig. 1, No. 83). Lateglacial faunas are dominated by subarctic species until 17,250 cal yr BP at Saylorville, but contain a mixture of open-ground species with no exact modern analogues. By 17,500 cal yr BP at Tonica, and by 14,600 cal yr BP at Norwood, the open-ground fauna was replaced by closed spruce forest species, even though the sites bordered the ice margin. As ice retreated, newly exposed ground was colonized by thermophilous, open-ground species; these were rapidly replaced by boreal forest species. Holocene sites in the Midwestern United States are also summarized in Table 3. These include Seibold, North Dakota (Fig. 1, No. 119); Trollwood Park, North Dakota (Fig. 1, No. 130); and Bongards, Minnesota (Fig. 1, No. 17).

One of the major trends in the Midwestern scenario was the regional extinction of arctic species at the end of the last glaciation. These beetles had nowhere to go in the midwest, because their migration northward was blocked by stagnant ice, and local conditions were becoming increasingly too warm for these cold stenotherms to tolerate. Some species managed to migrate into the Appalachian Mountains, where they are described as Pleistocene relicts today. Most others were extirpated south of the waning ice sheet; Beringian populations of these species have recolonized western Canada during the Holocene. In eastern Canada, late lying ice made northern migration almost impossible, because it persisted well into the Mid-Holocene (see Chapter 6 for details).

1.3. Studies in Arctic and Subarctic Canada

A few fossil insect investigations have been undertaken in northern Canada, split between sites representing the last interglacial and the Holocene. The Laurentide ice sheet persisted on the center Labrador-Ungava peninsula of eastern Canada until about 6,800 cal yr BP (Andrews and Dyke, 2007). This late lying ice effectively blocked the immigration of insects until the Mid-Holocene. I studied a Late Holocene insect fauna from the Umiakoviarusek site in northeastern Labrador (Fig. 5, No. 115) (Elias, 1982b). The peat profile from which the insects were extracted was dated between 2,770 cal yr BP and recent. Assemblages from the interval 2,770–950 cal yr BP were indicative of boreal woodland environment. Younger assemblages suggest a climatic cooling, as arctic tundra species became dominant.

I also worked on Holocene insect fossils from two peat bank profiles at Ennadai Lake, Nunavut (Fig. 5, No. 36) (Elias, 1982a). The lake is situated at northern tree line today. The Ennadai region was deglaciated by about 8,850 cal yr BP (Dyke and Prest, 1986). Peat growth probably commenced after the draining of glacial Lake Kazan, which occupied the region as the ice sheet melted. The Ennadai faunas span the interval 7,200–600 cal yr BP. The earliest peat layers contained evidence of spruce establishment, but the insects from the basal peat reflect open-ground environments, and no bark beetles were in the basal peats (ca. 7,200–6,800 cal yr BP). Between 6,800 and 2,900 cal yr BP, insect assemblages typical of northern boreal forest were dominant. A decline in conifer pollen from 5,500 to 5,200 cal yr BP had previously been interpreted as a climatic cooling, forcing a shift in tree line to south of Ennadai Lake (Nichols, 1975). However, coniferous bark beetles persisted at Ennadai through this interval, documenting the continued presence of trees.

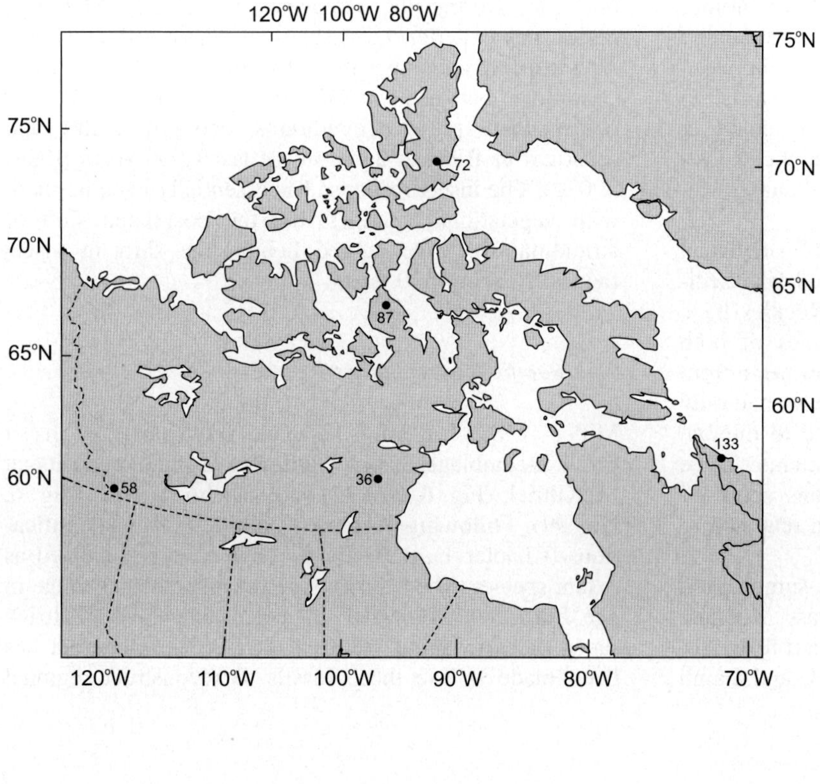

Fig. 5. Map of northern North America, showing location of regional Quaternary insect fossil sites, listed in Table 1.

I inferred from this that regional stands of spruce underwent some type of stress which caused them to reduce or stop pollen production for a few centuries. Both insect and paleobotanical data indicated a climatic cooling and forest retreat between 2,200 and 1,500 cal yr BP, followed by a return of woodland by about 1,000 yr BP.

Dyke and Matthews (1987) studied organic deposits exposed on the Pasley River, Boothia Peninsula, Nunavut (Fig. 5, No. 87). These deposits contained plant macrofossils and insects indicative of warmer-than-present conditions, and have been tentatively correlated with the MIS 5e interglacial. The provenance of some of the organic horizons is confounded by the inclusion of reworked organic materials, some of which may have come from Beaufort Formation (Late Tertiary) deposits.

Blake and Matthews (1979) studied another peat deposit from a site well beyond modern tree line in the Canadian arctic. Peat from Makinson Inlet, Ellesmere Island (Fig. 5, No. 72), yielded three infinite radiocarbon ages. A small insect fauna extracted from the peat included only species with modern distributions south of the study site. Blake and Matthews concluded that the deposit was deposited during an interval with warmer-than-modern climate, probably the last interglacial.

Matthews (1980b) studied insect fossils from Klondike bog in southwestern Mackenzie, northwest Territories (Fig. 5, No. 58). The samples were taken from a core, rather than from an exposure. The few species in the sample, initially thought to be of Holocene age, indicated little departure from present climatic conditions. The assemblage is now thought to represent a Wisconsin interstadial environment (Matthews, 1992, written communication).

1.4. Studies in the Rocky Mountain Region

The vertical relief of the Rocky Mountains has provided the regional biota with a wide variety of physical environments, expressed in the development of biomes ranging from grasslands in the east through montane and subalpine forests and alpine tundra. Ecotones between these biomes have undergone measurable shifts in elevation in the Late Quaternary, facilitating study of environmental changes. Fossil insects from the Rocky Mountains have left records of environmental change not found in adjacent lowlands.

I have studied a north–south transect of lateglacial assemblages from peat bogs, sedge fens, and lake sediments (Elias, 1988b, 1990b, 1995a). The Rockies have been a refuge for many cold-adapted species of both plants and animals. Species that shifted southward in front of advancing glacial ice in the American interior had only two options when the ice receded. They had to migrate either northward or up the slopes of mountains. Thus, cold-adapted beetles that lived on the plains south of Denver toward the end of the last glaciation retreated to the alpine tundra in Colorado.

The transect of Rocky Mountain sites is summarized in Table 4. Lateglacial sites include Marias Pass, Montana (Fig. 6, No. 74); False Cougar Cave, Montana (Fig. 6, No. 39); Mary Jane, Colorado (Fig. 6, No. 76); and Lamb Spring, Colorado (Fig. 6, No. 62). Based on the modern distributions and thermal tolerances of insect species in the fossil assemblages from these sites, summer temperatures were depressed by about 8–10°C from 21,500 to 17,000 cal yr BP (Elias, 1991). During this interval, regional insect communities comprised mixtures of species no longer found coexisting in any one region. This level of cooling has also been postulated by glacial geologists as the necessary depression of summer temperature to allow the growth of mountain glaciers to their maximum Late Wisconsin (Pinedale) size. Leonard (2007) has estimated that, given no change in precipitation between modern levels and the LGM, a cooling of summer temperatures of 6–8°C would be necessary to maintain LGM glacial ice in Colorado and Wyoming. If precipitation during the LGM was only half of what it is today, the necessary summer temperature depression would be 8–10°C.

Beginning at about 17,000 cal yr BP, regional climates began warming, based on the replacement of cold-adapted beetle species with thermophilous taxa. Fossil insect data suggest that this amelioration was gradual from 17,000 to approximately 13,400 cal yr BP. During this interval, insect fossil evidence suggests that summer temperatures rose from about 10°C cooler than present to about 5°C cooler than present. After 13,400 cal yr BP, insect fossil data suggest that regional climates warmed extremely rapidly, with summer temperatures reaching modern conditions within a few centuries. Insect faunas indicative of higher-than-present tree line suggest that Early Holocene climates may have been warmer than present.

Holocene sites in the Rocky Mountain region are summarized in Table 4. These include La Poudre Pass, Colorado (Fig. 6, No. 63); Lake Emma, Colorado (Fig. 6, No. 59); Roaring River, Colorado (Fig. 6, No. 101); Mount Ida Bog, Colorado (Fig. 6, No. 79); Lake Isabelle, Colorado (Fig. 6, No. 60); Lefthand Reservoir, Colorado (Fig. 6, No. 66); and Huntington Canyon, Utah (Fig. 6, No. 52). During the Holocene, the Rocky Mountain region has experienced a series of climatic fluctuations, with insect assemblages indicative of warmer than present conditions between 10,800 and 7,800 cal yr BP, and colder than present conditions between 5,200 and 3,200 cal yr BP, and again in the last 1,000 years (Elias, 2007d). The insect response has essentially been in phase with vegetational changes, but the postulated shift of altitudinal tree line lagged behind the shift in insect faunas by about 500 years.

1.5. Pacific Coast Studies

Miller (1983) listed the taxa and provenance of insect fossil assemblages from California asphalt deposits at McKittrick (Fig. 6, No. 77) and Rancho La Brea (Fig. 6, No. 99). Following Pierce's (1946–1957) misidentifications (Chapter 1), these fossils have been reevaluated as extant species. The Rancho La Brea assemblages range in age from >40,000 yr BP to recent, and the McKittrick fauna dates to about 7,800 cal yr BP. Little attempt has been made to use these fossils to reconstruct regional

Table 4. Summary of Quaternary insect fossil sites in the Rocky Mountain region of the United States.

Site	Age range (cal yr BP × 1,000)	Environmental interpretations	Reference(s)
Last glacial faunas			
Lamb Spring, Colorado (1,731 m elevation)	21.7	Cold-adapted grassland species indicate cold, dry conditions similar to modern grassland regions in western Canada; T_{max} depressed by about 8–10°C	Elias and Nelson (1989), Elias and Toolin (1989)
Lateglacial sites			
Lamb Spring, Colorado (1,731 m elevation)	17.6	Alpine tundra species indicate cold, moist conditions and open-ground environment; T_{max} depressed by about 8–10°C	Elias and Nelson (1989), Elias and Toolin (1989)
Mary Jane, Colorado (2,882 m elevation)	16.8–14.6	Tree line depressed to below site. Alpine tundra fauna suggests summer temperatures depressed by 7°C at 13,700 BP; after 12,800 yr BP, tree line was near site, and T_{max} depressed by about 5–6°C	Short and Elias (1987)
Marias Pass, Montana (1,548 m elevation)	14.0	Arctic and alpine fauna indicates tree line depression of at least 650 m below modern level; this equates with T_{max} depression of at least 5°C	Elias (1988a)
Holocene sites			
False Cougar Cave, Montana (2,590 m elevation)	11.3	Insect fauna suggests essentially modern climatic conditions	Elias (1990b)
La Poudre Pass, Colorado (3,103 m elevation)	11.3–recent	Basal assemblages indicate conditions as warm as modern, although conifer forest arrived 500 years later. Faunas show climatic cooling from about 6.8–6.3 ka, 5.2–2.8 ka. Ameliorations peaked at 8.8, 5.8, and 2.0 ka	Elias (1983), Elias *et al.* (1986)
Huntington Canyon, Utah (2,730 m elevation)	10.7	Insect fauna suggests essentially modern climatic conditions	Elias (1990b), Gillette and Madsen (1992)
Lake Isabelle, Colorado (3,324 m elevation)	10.0–recent	Oldest assemblages indicate climatic conditions at least as warm as modern; climatic cooling suggested by faunas dating 4,500 yr BP. Younger faunas indicate close proximity of tree line from 3,800–500 yr BP	Elias (1985)
Lake Emma, Colorado (3,740 m elevation)	10–9.2	Insect assemblages show tree line and summer temperatures above modern levels	Elias *et al.* (1991)
Mount Ida Pond, Colorado (3,520 m elevation)	10.1–5.3	Insects suggest modern climatic conditions from 10.1 to 9.3 ka; climatic optimum reached just after 9.3 ka, then a gradual cooling; youngest fauna showing coolest conditions	Elias (1985)
Lefthand Reservoir, Colorado (3,224 m elevation)	6.1–recent	Insects suggest cooler than modern climate from 5.5 to 3.9 ka, then rapid warming at 3.2–2.9 ka, a cool interval from 2.0 to 1.3 ka, and a moderate warming about 0.8 ka	Elias (1985)
Roaring River, Colorado (2,800 m elevation)	2.4	Insect fauna indicative of modern conditions	Elias *et al.* (1986)

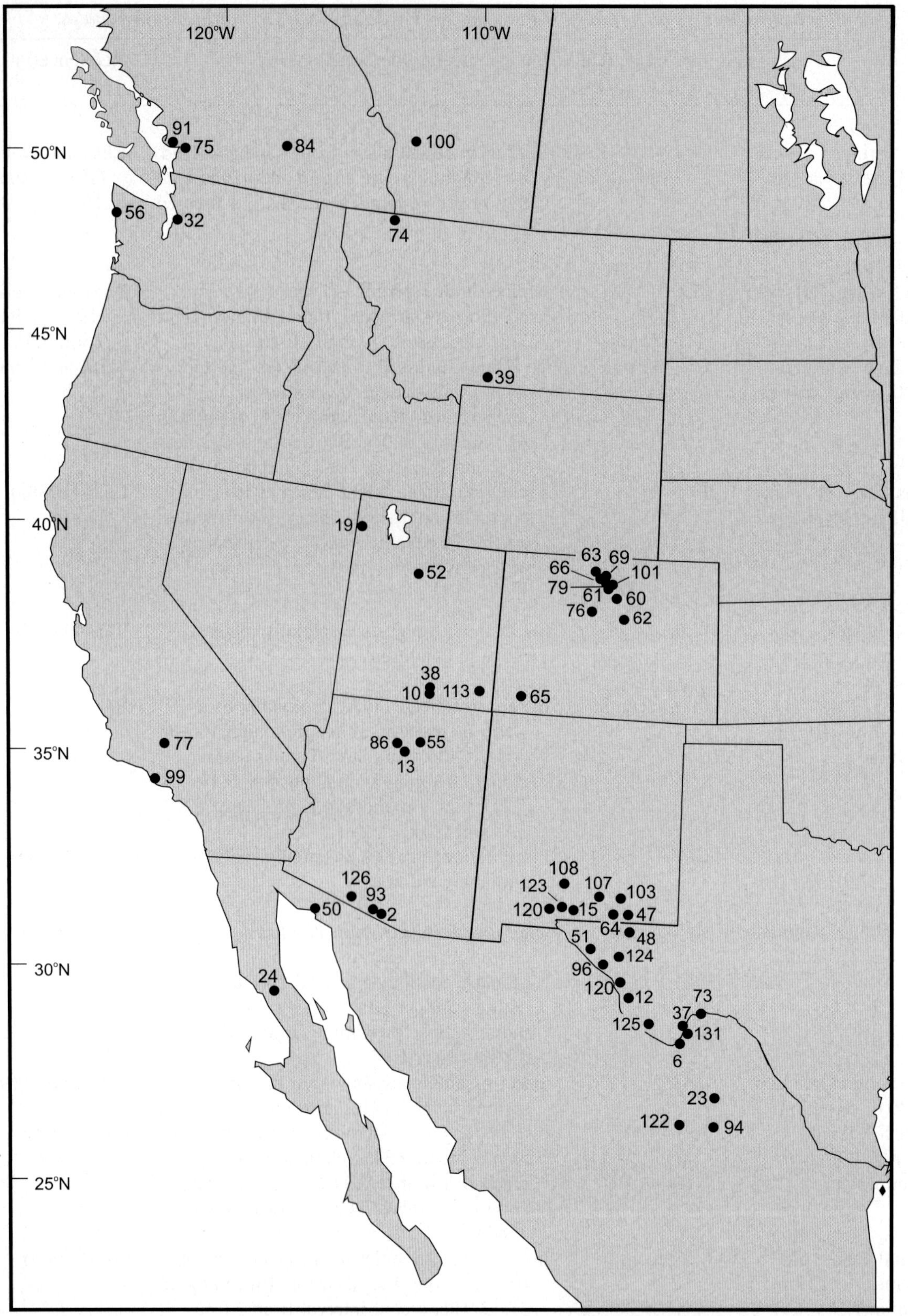

Fig. 6. Map of western North America, showing location of regional Quaternary insect fossil sites, listed in Table 1.

paleoenvironmental conditions, although some assemblages have been shown to reflect conditions similar to those in southern California today.

Two groups of faunal assemblages have been described from a coastal bluff at Kalaloch, Washington, by Cong and Ashworth (1996) (Fig. 6, No. 56). These assemblages range in age from about 52,000 to 17,000 cal yr BP. The older of the two faunal units (ca. 52,000–50,000 cal yr BP) contains a mixture of species found in the study region today, and in northern and montane forest regions. A number of bark beetles recovered from this sample feed on Pacific northwest tree species, such as

Douglas-fir (*Pseudotsuga menziesii*), Sitka spruce (*Picea sitchensis*), and western hemlock (*Tsuga heterophylla*). The authors estimated that mean summer temperatures were only about 1°C cooler than modern, based on the mixture of species. The younger faunal unit is characterized by the lack of bark beetles and other tree indicators. This fauna represents open-ground habitats, and the beetle evidence suggests a climatic cooling. My MCR estimate for the fauna dated ca. 44,000 cal yr BP indicates that T_{max} was about 4–5°C cooler than modern at that time. The younger faunal assemblages contain too few species to allow paleotemperature reconstructions. If this level of cooling was maintained, it would not have been sufficient to cause the replacement of coniferous forest by open-ground (predominantly herbaceous) vegetation. Based on regional pollen evidence, Heusser *et al.* (1980) interpreted that regional LGM climates had only about one-third of the moisture that falls in this region today.

Nelson and Coope (1982) described a diverse insect fauna dating from 19,800 cal yr BP from Discovery Park, Seattle, Washington (Fig. 6, No. 32). Although the pollen spectra associated with this assemblage suggest conditions substantially colder than present, the insects are characteristic of the modern Puget lowland. Nelson and Coope suggest that the discrepancy between the flora and insect fauna may be due to increased climatic continentality just before the last (Vashon) glacial advance.

In southern British Columbia, Miller *et al.* (1985a) studied insect fossils from 22,400 to 22,000 cal yr BP at Mary Hill and Port Moody (Fig. 6, Nos. 75 and 91). These assemblages represent an open coniferous forest floor community, developed in cool, dry climatic conditions during an interval between two advances of the Cordilleran ice sheet.

Hebda *et al.* (1990) examined insect fossils from two Late Holocene packrat middens in the arid interior of British Columbia. The Oregon Jack Creek site (Fig. 6, No. 84) yielded a small fauna dated at 1,000 cal yr BP. This is the northernmost packrat midden locality to produce identifiable insect fossil remains. The Oregon Jack Creek fauna was comprised of silken fungus beetles (Cryptophagidae), minute brown scavenger beetles (Lathridiidae), and click beetles (Elateridae). These beetles were probably scavengers in the packrat's nest. While all these families have been found in packrat midden assemblages from the desert southwest, none of these are important constituents there. Conversely, beetle families that dominate desert midden assemblages, such as darkling beetles (Tenebrionidae), spider beetles (Ptinidae), and checkered beetles (Cleridae), were absent from the Oregon Jack Creek assemblages.

1.6. Studies in the Arid Southwest

Elias (1987, 1992b), Elias and Van Devender (1990, 1992), and Elias *et al.* (1995) studied insect fossils from a north–south transect of packrat midden sites in the Chihuahuan Desert. The Chihuahuan Desert is an interior continental desert, stretching from southeastern Arizona to northern Zacatecas, Mexico. Three sites in the Bolson de Mapimi region, in the southern Chihuahuan Desert of Mexico (Fig. 6, Nos. 23, 94, and 122), yielded 45 fossil packrat midden assemblages ranging in age from 16,400 cal yr BP to recent. These middens contained insects, arachnids, and millipedes; most of the identified taxa are beetles. Modern studies have shown that the great majority of arthropods found in packrat nests are facultative inquilines. This means that they are species which normally dwell in the open, by occasionally entering the rockshelter housing, a packrat nest, in order to take advantage of the moderated microclimate. Once there, many of these arthropods prey on other species, or scavenge food from the piles of plant detritus accumulated by the rat.

The fossil assemblages from the Bolson de Mapimi region are unique among those studied thus far from the Chihuahuan Desert, because they contain mixtures of desert- and temperate-zone species in almost every chronozone from the lateglacial through the Late Holocene (Elias *et al.*, 1995). Midden assemblages from locations farther north in the Chihuahuan desert are generally separated into glacial-age faunas with temperate-zone affinities and postglacial faunas with desert-zone affinities. The ''no modern analogue'' faunal assemblages indicate that the Late Quaternary environments in this part of the desert were unlike any that exist today (Elias *et al.*, 1995). This conclusion is also borne out in the paleobotanical record from the Bolson de Mapimi, which likewise shows unique combinations of floristic elements. Elias and Van Devender (1990) also studied fossil insects from 50 assemblages at 5 localities in and around Big Bend National Park, Texas (Fig. 6, Nos. 6, 37, 73, 125, and 131), the samples ranging in age from >41,000 cal yr BP to recent. Plant macrofossil analyses have been conducted and radiocarbon chronologies have been developed for these and the other Chihuahuan Desert sites.

The lateglacial assemblages from the southern Chihuahuan Desert, dated between 16,500 and 14,400 cal yr BP, comprise a mixture of temperate and desert taxa. However, most of the desert-associated species are phytophages (plant feeders), more indicative of certain desert floristic elements than of xeric climate *per se*. The lateglacial faunas have species with tropical to subtropical affinities. The ground beetle *Amara chalcea* is the only species in the lateglacial assemblages that today is confined the temperate zone of the United States. This beetle lives in xeric habitats within these regions.

A temporal hiatus in samples from 14,000 to 10,000 cal yr BP is followed by a loss of most of the lateglacial insect species from the southern Chihuahuan records. Beetles associated with desertscrub communities dominated Early and Middle Holocene assemblages. In the Late Holocene, several insect species were recorded, including both temperate and desert species. Paleobotanical and arthropod records indicate that this region served as a refugium for desert biota during the Late Pleistocene. Its southerly latitude appears to have dampened the climatic effects of the Wisconsin glaciation, allowing the survival of desert species during the lateglacial interval and presumably earlier. A mosaic of temperate and xeric habitats was available to the regional biota, even during the last 1,000 years, when other regions of the Chihuahuan Desert have experienced extremes of aridity.

The Big Bend regional faunas indicate that there was greater effective moisture from 34,000 to 14,000 cal yr BP. An especially diverse, mesic fauna characterized the earlier half of this interval (34,000–24,000 cal yr BP). In the Late Wisconsin, many grassland species, now confined to higher elevations and cooler, moister regions to the north, lived in the Big Bend region. After 14,000 cal yr BP, most of these species were replaced by either desert species or by more cosmopolitan taxa. Few stenothermic species were preserved in the Late Wisconsin records from Big Bend, but the faunal change suggested a climatic shift from cool, moist conditions of Wisconsin glacial times to hotter, drier conditions of latest Wisconsin and Early Holocene.

A third part of my study included a transect of midden sites in the northern Chihuahuan Desert of western Texas and southern New Mexico (Fig. 6, Nos. 12, 49, 50, 53, 108, 124, 125, and 126). These samples range in age from >46,900 cal yr BP to recent, yielding a wide variety of insects, arachnids and millipedes (Fig. 7). In the northern Chihuahuan Desert, full-glacial (26,500–21,600 cal yr BP) arthropod records suggest widespread coniferous woodland at elevations as low as 1,200–1,400 m asl. From about 26,000 to 13,000 cal yr BP, woodland environments persisted, but the insect data suggest considerable open ground, with grasses at least locally important at the midden sites. The grassland nature of the arthropod fauna was also suggested in the regional vertebrate record. The transition from the temperate Wisconsin fauna to the more xeric postglacial fauna started by 14,800 cal yr BP, the timing of this faunal change being essentially synchronous throughout the northern Chihuahuan Desert. The Big Bend assemblages suggested that the transition took place after 14,000 cal yr BP, but that chronology was based on far fewer data.

A major difference between the Big Bend and northern Chihuahuan Desert scenarios is in the nature of this temperate-to-xeric faunal change. In the Big Bend region, the transition was characterized by the disappearance from the record of all but one of the temperate insect species at about 14,000 cal yr BP. However, the xeric-adapted fauna did not appear in the Big Bend records until about 8300 cal yr BP. In the northern Chihuahuan Desert assemblages, the xeric species first appeared at 14,800 cal yr BP, and several of the temperate grassland species from the Wisconsin interval persisted well into the Holocene. This mixture of xeric and temperate elements makes sense from an ecological standpoint, given that these northern faunas were living close to the edge of the Chihuahuan Desert. The gradual shifting of northern desert boundaries in the Holocene probably created many marginal habitats for temperate species in ecotones between grassland and desertscrub.

By about 8,300 cal yr BP, the appearance of more xeric species indicates establishment of desert environments, including desert grasslands, throughout the Chihuahuan Desert region. After 2,600 cal yr BP, the last of the temperate species were replaced by species associated with desertscrub communities.

On the whole, there is good agreement between regional reconstructions based on paleobotanical evidence and the fossil arthropod evidence. The two sets of records agree on the nature of Wisconsin-age and Holocene environments, including temperature and moisture conditions. This was also true of the previous study of insect fossils compared with paleobotanical data from the nearby Fra Cristobal and San Andres Mountains in south central New Mexico (Elias, 1987). The only serious disparity between insect and plant macrofossil records comes during intervals of rapid climate change, especially the transition from Wisconsin glacial to Holocene (postglacial) environments. In this case, the arthropod evidence shows that temperate Wisconsin environments gave way to postglacial environments with increasing temperatures and shifting precipitation patterns by 14,800 cal yr BP. The principal botanical change that marks this transition, the disappearance of pinyon pines, occurred between about 13,000 and 12,000 cal yr BP, depending on the region. However, detailed analyses of the Hueco Mountains flora revealed shifts in plant abundance and diversity, including pinyon pine, at about 14,800 cal yr BP. Xeric desertscrub species such as lechugilla, creosotebush, and ocotillo appeared in the Chihuahuan Desert paleobotanical records about 5,200 cal yr BP. Xeric-adapted insects also began to dominate regional faunal assemblages during the Mid-Holocene, with a transition to a completely xerophilous fauna occurring by 2,600 cal yr BP. It is interesting to note that the transition from temperate-to-xeric faunas took such a long time to complete. Since there is good agreement between different sources of proxy data, one can only surmise that the climatic transition was indeed gradual in this region.

I have also done research on Late Quaternary insect faunas from the Colorado Plateau region of Utah and Arizona (Fig. 6, Nos. 10, 13, 57, 87, and 115), in collaboration with colleagues at Northern Arizona University. Studies of samples ranging in age from 34,800 to 1,500 cal yr BP suggest that Late Wisconsin climatic conditions were cooler and moister than present, and that the plateau supported a mosaic of grassland and shrub communities without modern analogue (Elias *et al.*, 1992b).

One of the most interesting aspects of the Colorado Plateau research was the opportunity to study fossil mammoth dung from Bechan Cave in southeastern Utah. This sandstone cave apparently served as the toilet area for generations of Columbian mammoths (*Mammuthus columbianus*) during the Late Pleistocene. Over 300 m^2 of the cave floor was covered with dried dung boluses (Mead *et al.*, 1986). I was anxious to test samples of this dung for the presence of dung beetles, because I anticipated one of two possible scenarios. First, perhaps there was a dung beetle in the American southwest during the Pleistocene that was adapted to feeding either exclusively or principally on mammoth dung (much as there are modern African dung beetles that feed exclusively on elephant dung), and that species of beetle must now be extinct, since its food sources has disappeared. The second scenario would be that an extant species of dung beetle formerly fed on mammoth dung

Fig. 7. SEMs of insect fossils from packrat midden assemblages from the northern Chihuahuan Desert in Texas and New Mexico. Top row: Right elytron of Pachybrachis mitis *(Chrysomelidae); right elytron of* Onthophagus lecontei *(Scarabaeidae); left elytron of* Lebia *cf.* lecontei *(Carabidae); middle row: exoskeleton of* Ptinus *sp. (Ptinidae); head of* Onthophagus cochisus *(Scarabaeidae); head of* Piscatopus griseus *(Curculionidae); bottom row: left elytron of* Discoderus parallelus *(Carabidae); left elytron of* Hypocaccus estriatus *(Histeridae); head capsule of* Rhagodera costata *(Zopheridae).*

Fig. 8. Modern specimen of the dung beetle, A. fossor, feeding on cow dung (photo by Guy Hanley, used with permission).

(not exclusively), and has since switched food sources, feeding on other kinds of herbivore dung in the American southwest. In the end, the results of my study did not agree with either of these two scenarios. The dung beetle I found associated with the mammoth dung from Bechan Cave was *Aphodius fossor* (Fig. 8). This beetle is considered by taxonomists to be an Old World species that has been introduced into North America in recent times (Lindroth, 1957). Today it feeds mainly on cow dung (Floate and Gill, 1998). It may or may not have been extirpated from North America at the end of the Pleistocene. Along with other dung beetles, this species could easily have become established in eastern North America as an accidental immigrant from Europe. Cows transported by boat to North America would have provided the necessary dung that would have attracted dung beetles on board ships docked in European harbors, and beetle-infested dung would have been shoveled out of the holds of ships when they arrived in the New World, giving the dung beetles an opportunity to become established, and spread across the continent. This is just one of the ways in which Old World beetles have been accidentally introduced into North America (Buckland *et al.*, 1996b).

As discussed in Chapter 5, I developed MCR estimates for packrat midden fossil assemblages from the American southwest. MCR analysis of these arid region faunas (Elias, 1998) suggests that effective moisture exerts greater control on desert beetle distributions than does temperature. One of the more interesting results of this study concerned a reconstruction of temperature and precipitation for a site near Bonneville, Utah, in the eastern Great Basin (Fig. 6, No. 19). The Bonneville Estates site yielded a beetle fauna dated 13,600 cal yr BP. Several of the species in this fauna live today in the Pacific northwest region. MCR analysis of this fauna indicated that T_{max} was 5–6°C cooler than today, and mean annual precipitation (MAP) was 525–675 mm, compared to modern MAP at the site of approximately 190–210 mm. These results agree well with paleoclimate reconstructions based on ancient pluvial lake levels (Benson, 2004) and regional paleobotany (Thompson *et al.*, 2004). MCR analysis of Late Pleistocene beetle assemblages from the Mojave Desert (Elias, 2007c) indicates that T_{max} was 7–8°C cooler than it is today from 38,000 to 36,000 cal yr BP, and MAP was 3.5–4.4 times greater than it is today.

Hall *et al.* (1988, 1989, 1990) have studied insect faunas from several sites in the Sonoran Desert of southwestern Arizona and northwestern Sonora, Mexico (Fig. 6, Nos. 2, 25, and 94). Unlike the records from the Chihuahuan Desert insect fauna, the Sonoran Desert fauna indicates little change, in that all taxa recovered from Sonoran middens probably live within a few kilometers of the midden sites today. In their study of fossils from middens in Organ Pipe Cactus National Monument, Hall *et al.* (1990) noted that the fauna was only moderately diverse in the Middle Holocene, despite increased precipitation during that interval. The record of the arthropod fauna showed a marked increase in diversity during the Late Holocene, as more subtropical plants and warmer climates became established about 4,500 cal yr BP. The increase in diversity in fossil arthropod assemblages has been correlated with the frequency of winter freezes, and secondarily with the amount of summer precipitation. Many warm-stenothermic insects probably dispersed into the study region from Sonora, Mexico, during the last 4,500 cal yr. This Late Holocene peak in species richness is in sharp contrast to the Chihuahuan Desert insect record, which showed the least number of species in Late Holocene samples. The Chihuahuan Desert fauna may have been an important source of immigrant species into the Sonoran Desert in the Late Holocene.

The fossil arthropod record from northwestern Sonora is similar to that of the Puerto Blanco Mountains of Organ Pipe Cactus National Monument in that it shows an increase in species richness in the Late Holocene (Hall, *et al.*, 1988). A major difference is that the northwestern Sonoran record shows the establishment of a relatively modern fauna by the Early Holocene. However, this conclusion is primarily based on identifications made only to the generic or family level. Additional specific identifications may help to clarify the issue of stable versus changing faunas in the Holocene. If the Sonoran Desert insect fauna was truly stable through the Late Quaternary, then this represents a significant difference between the Sonoran and Chihuahuan Desert insect faunal histories. Since climatic factors appear to have been the most important element in bringing about the large-scale distributional changes seen in the Chihuahuan Desert faunas during the last 40,000 years, this might suggest that the late Quaternary climatic regimes in the Sonoran Desert region were more stable than those of the Chihuahuan Desert.

2. Overview of North American Studies

Quaternary entomology has developed markedly in North America during the last 30 years, but the pace has slowed markedly in the last decade. Paleoentomologists have learned a great deal about insect response to climate change during the Wisconsin glaciation and the Holocene, especially in the central and northeastern

United States, southeastern Canada, and the Chihuahuan Desert. However, much remains to be done. Extensive regions, including the prairies of central Canada, most of the Northwest Territories, the southern United States, the Great Basin, and the Pacific coast states, are virtually unstudied. Also, very little is known about faunas predating the last interglacial.

Clearly, insects have responded to Late Quaternary environmental changes in North America as much as they have elsewhere, namely, by making marked distributional shifts. As in Europe, certain suites of species appear to have maintained some form of association in past communities. Some of these associations had great longevity during the Wisconsin interval, and have only dissolved in the Holocene (Schwert, 1992). In a Quaternary perspective (i.e., last 2,600,000 years), our present biological communities are ephemeral associations of species, lasting only a small fraction of the life spans of the species involved. It seems that the environmental forces that shaped North American ecosystems during most of the last 100,000 years have held sway for most of the Quaternary. Cold environments have persisted through long glacial and interstadial episodes. Warm interglacials are short-lived. Moreover, the fossil record indicates that North American insects have responded rapidly and with great sensitivity to these shifting environmental patterns. Their almost constant geographic shifts may have prevented speciation in the Quaternary; evidence of extinction or speciation in the North American record of the last million years is very limited.

These are the same inferences that emerged from the study of the British fossil insect record. North American studies show that the British story was not exceptional or unique.

3. South American Studies

> In recent years, the view that Pleistocene climatic events played a major role in the evolution of the biotas of southern, primarily tropical continents has begun to displace the previously held conviction that these areas were relatively stable during the Quaternary (Beryl Vuilleumier, 1971).

The fossil record of South America received the attention of Quaternary entomologists in the 1980s and 1990s. Vuilleumier's seminal paper, while it did not discuss insect fossils (at that time, very little had been done!), set the tone for much that has come afterward. Recent studies in South America have amply demonstrated that the biota of this continent underwent a series of dynamic changes in response to Pleistocene climates. In the tropical regions, the changes were mostly in the amount and seasonality of precipitation. In the higher latitudes of southern South America, changes in temperature and precipitation brought mountain glaciers and subantarctic conditions. Most Quaternary insect research in South America has been conducted in the latter region (Fig. 9 and Table 5).

Only two studies have dealt with fossil insect faunas from tropical latitudes in South America. Binford (1982) investigated the paleolimnology of Lake Valencia, in Venezuela (Fig. 9, No. 1), including chironomid larval remains in sediment cores. Basal sediments were radiocarbon dated at about 12,500 yr BP. The lake was relatively shallow and subject to drying out until about 10,500 yr BP, when it began to fill rapidly. Between 10,000 and 7,000 yr BP, sedimentation decreased; lake waters became less turbid; and chironomid species characteristic of deep, clear water were dominant.

From 7,000 to 2,200 yr BP, the water level of the lake dropped below the outflowing streams; the lake increased in salinity, and most invertebrate populations declined. Lake levels and water quality have undergone a series of fluctuations in the last 1,200 years.

A second study in equatorial South America was an investigation of insect remains from tar-seeps at Talara, Peru (Fig. 9, No. 7), by Churcher (1966). The mode of deposition of these tar-seeps is similar to that of the asphalt deposits at Rancho La Brea, California, and the ecological composition of the insect faunas is likewise similar at the family level. The Talara faunas are dominated by aquatic beetles that lived in water just above the tar, scavengers that fed on carcasses of animals trapped in the tar, and terrestrial insects that strayed onto the tar surface and were trapped. The age range of the studied assemblages is between 14,400 and 13,600 yr BP. Unfortunately, no specific determinations were made, in spite of remarkably good preservation, so Churcher was not able to make a paleoclimatic interpretation from the data.

Ashworth and Hoganson have studied a number of lateglacial and Holocene insect faunas from the central and southern coastal regions of Chile (Fig. 9). As I discussed in Chapter 7, they examined insect fossils from essentially natural deposits at the Monte Verde archeological site (Fig. 9, No. 2). At Monte Verde, insect faunas were extracted from a 13,000 yr BP peat horizon (Ashworth *et al.*, 1989). The Monte Verde peats yielded abundant, diverse insect assemblages. The environmental reconstruction based on the insect fauna is a shallow creek with sparsely vegetated sand bars and some bogs, flowing through Valdivian (southern South American) rain forest. The lateglacial paleoclimate was interpreted as very similar to modern; this interpretation is in agreement with all of the other lines of evidence except for the pollen interpretation, as discussed in Chapter 7.

To develop an adequate knowledge of the modern distribution and ecological requirements of the beetle fauna of central and southern Chile, Ashworth and Hoganson (1987) carried out an extensive collecting program for several field seasons, much of it in the Puyehue National Park, which contains a range of biological communities from lowland deciduous forest near sea level to alpine tundra above 1,200 m elevation. Ashworth and Hoganson identified 462 species of beetles from 48 families in 41 locations in and around the park. This intensive study allowed them to develop meaningful interpretations of Late Quaternary fossil assemblages.

Hoganson and Ashworth (1992) and Ashworth (2007) have summarized their work on Late Pleistocene insect faunas from south central Chile. The oldest fossil beetle assemblages to have been examined from this region are

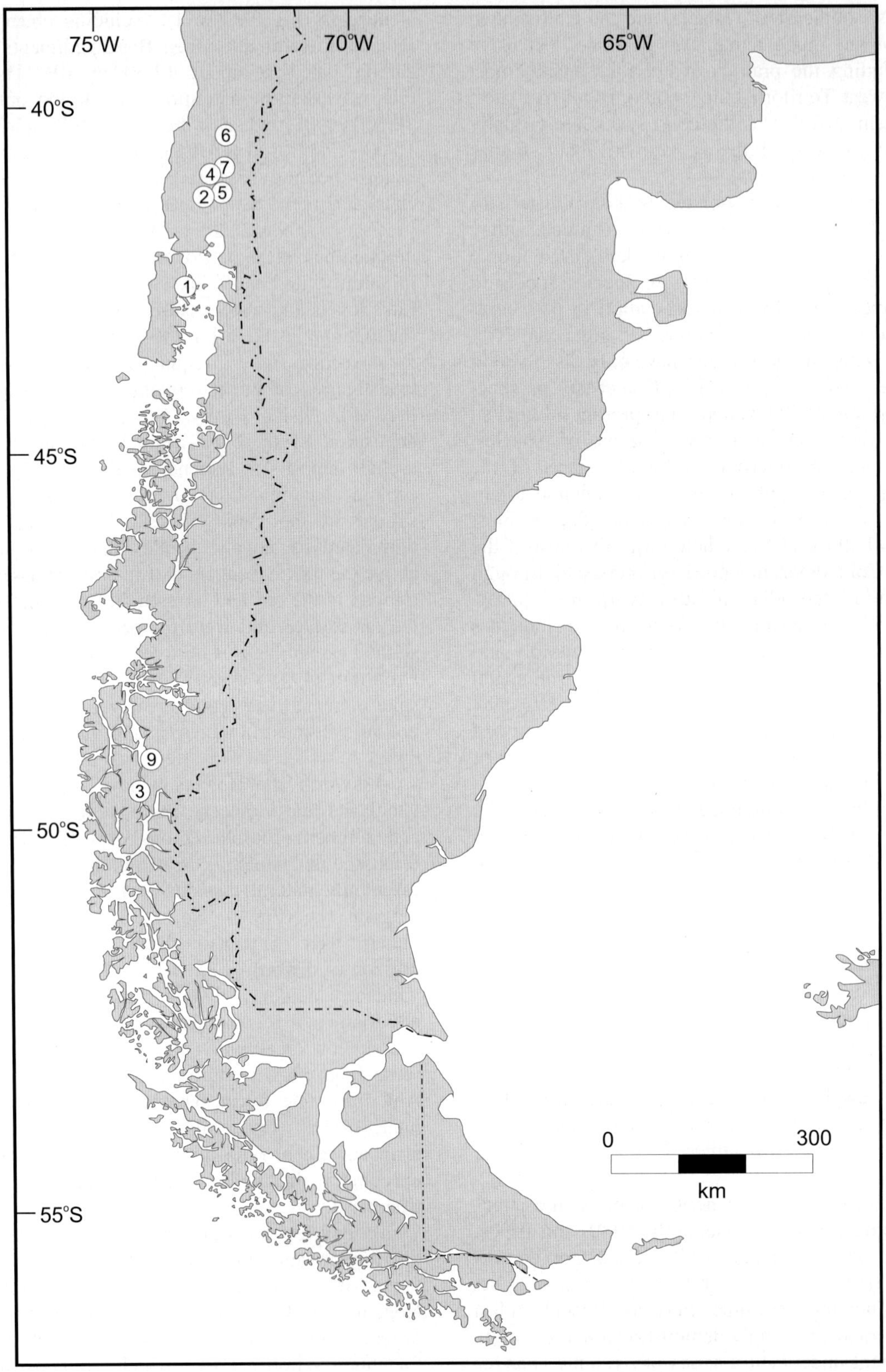

Fig. 9. Map of South America, showing location of fossil insect sites listed in Table 2.

from peat deposits at Lago Rupanco (Fig. 9, No. 7). Fossils were extracted from an *in situ* peat that is overlain by gravel, till, and volcanic deposits (Ashworth, 2007). The assemblages range in age from 23,300 to 28,800 cal yr BP; thus, they represent the environment of the lake region immediately prior to the LGM. The assemblages are the least diverse of those studied from the Chilean Lake Region, containing only 9–10 species. Fossils of the weevil *Germainiellus dentipennis* make up 90% of the fossils. Younger assemblages in the Chilean Lake Region are also dominated by this species until after 17,500 cal yr BP. The youngest of these assemblages (17,500 cal yr BP) comes from a site near Dalcahue on La Isla Grande de Chiloé (Fig. 9, No. 1). This small fauna includes two

Table 5. Quaternary insect fossil sites in South America.

Site	Reference(s)
(1) Dalcahue	Ashworth (2007)
(2) Monte Verde	Ashworth *et al.* (1989)
(3) Puerto Edén	Ashworth *et al.* (1991)
(4) Puerto Octay	Hoganson and Ashworth (1992)
(5) Puerto Varas	Hoganson and Ashworth (1992), Ashworth (2007)
(6) Rio Caunahue	Ashworth and Hoganson (1987), Hoganson and Ashworth (1992)
(7) Rupanco	Ashworth (2007)
(8) Talara Tar Seeps	Churcher (1966)
(9) Témpano Sur	Ashworth and Markgraf (1989)

species that today inhabit higher elevations in the Andes: the weevil *Paulsenius carinicollis* and the ground beetle *Cascellius septentrionalis*. Based on the composition of fossil beetle assemblages, Ashworth and Hoganson (1993) estimated that regional T_{max} during the LGM was 4–5°C lower than today. This estimate is more conservative than that of Denton *et al.* (1999), who estimated that snow and tree line were lowered by 1,000 m between 33,700 and 17,500 cal yr BP, which would correspond with a 6–8°C low of T_{max} compared with today.

Based on fossil assemblages from sites at Puerto Octay (Fig. 9, No. 4), Puerto Varas (Fig. 9, No. 5), and Rio Caunahue (Fig. 9, No. 6), Hoganson and Ashworth (1992) postulated that cold-adapted beetles colonized regional lowlands following glacial retreat, about 21,700 cal yr BP. Depauperate moorland faunas dominated the interval 21,700–17,200 cal yr BP; these were replaced by more thermophilous taxa, including arboreal beetles. This amelioration continued until 14,800 cal yr BP, by which time the beetle fauna was comprised solely of rain forest species.

Ashworth (2007) examined the long-term faunal trends in the Chilean fossil beetle assemblages, plotting both the total taxa per assemblage and the ratio of arboreal taxa to total taxa for 1,000-year intervals between 24,000 and 9,000 ^{14}C yr BP (28,000–10,000 cal yr BP) (Fig. 10). As discussed above, he found that the beetle fauna from 28,000 cal yr BP to the time of the last glacial advance (about 17,000 cal yr BP) was composed of just a few species associated with wet moorland habitats. Between 17,000 and 15,500 cal yr BP, diversity remained low but the number of percentage of arboreal taxa began to rise dramatically. Between 15,500 and 14,000 cal yr BP, assemblage diversity rose to more than five times higher than those of the glacial interval. The faunal evidence is clear that by 15,500 cal yr BP, the moorland ecosystem, which had dominated the lowlands of this region for more than 10,000 years, had been replaced by essentially modern forest ecosystems.

This climatic amelioration was of the order of 4–5°C in mean annual temperature, and it brought a fivefold increase in beetle species diversity over the previous (glacial) fauna. This transition to postglacial conditions was the last major regional climatic change shown in the insect faunal sequence. As in the Monte Verde assemblages, Hoganson and Ashworth saw no evidence of a Younger Dryas cooling in this region.

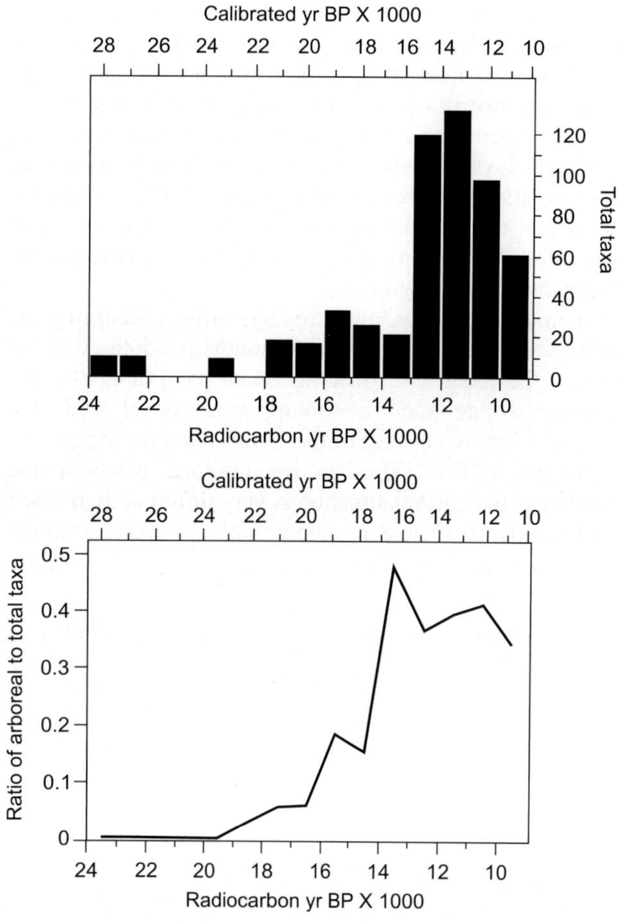

Fig. 10. Long-term faunal trends for the interval 28,000–10,000 cal yr BP in southern South America, represented by plots of the total number of taxa (bar graph) and the ratio of obligate arboreal taxa to other taxa (line graph) (after Ashworth, 2007).

To the south, Ashworth and Markgraf (1989) and Ashworth *et al.* (1991) studied lateglacial and Holocene insect faunas at two sites in the Chilean Channel region. This region was heavily glaciated during the Late Pleistocene, but these studies have shown that there must have been ice-free regions in which cold-adapted biota survived. At Puerto Edén (Fig. 9, No. 3), Ashworth described assemblages ranging in age from 15,900 cal yr BP to recent. Insects apparently invaded newly deglaciated terrain from regional refugia. The fossil data indicate that this colonization occurred within a few decades of deglaciation. The earliest beetle assemblages were comprised of species associated with barren, open-ground environments. The pioneering vegetation was a heath. However, both the paleobotanical and insect data suggest that the pioneer heathland also supported shrubs and trees.

A southern beech woodland became established at Puerto Edén after 15,900, persisting until about 10,800 cal yr BP; this supported a mixture of open-ground and woodland beetles. This same environment was reflected by insect and paleobotanical assemblages dating from the lateglacial interval at the nearby Témpano del Sur site (Fig. 9, No. 8) (Ashworth and Markgraf, 1989). The basal peat from the Témpano del Sur site is dated between 13,100 and 11,750 cal yr BP. Fossil beetle and pollen assemblages from the peat indicate a fauna and flora very similar to that inhabiting stable substrates within the area today (Ashworth, 2007). The implication is that during the Younger Dryas interval, the South Patagonian ice field was no larger than it is today and could even have been smaller (Ashworth and Markgraf, 1989). In fact, no evidence was found in the beetles, pollen, or plant macrofossils for a proposed Younger Dryas cooling in the Chilean Channel region.

During the 1990s, a controversy arose concerning the existence of a cooling event in southern South America during the lateglacial interval. Some palynologists were convinced that such a cooling was coeval with the Younger Dryas cooling in the North Atlantic region. As Ashworth (2007) notes in his regional paleoclimatic summary, the current situation is very different than when disagreements between interpretations were initially stated. Currently, even those who argue for this regional lateglacial climatic oscillation admit that it was subtle phenomenon, and that it began 500–1,200 years before the Younger Dryas chronozone. Moreno (1997), Moreno *et al.* (1999), and Hajdas *et al.* (2003) still consider this cooling to be part of a globally synchronous event, but Ashworth and Hoganson (1993), Markgraf (1993), and Bennett *et al.* (2000) argue that there is no evidence to support climatic reversals at the time of the Younger Dryas (12,900–11,500 cal yr BP) or during the Antarctic Cold Reversal (17,600–15,600 cal yr BP).

The paleoenvironmental sequence in the Holocene included a mosaic of rainforest and moorland habitats from 10,800 to 6,300 cal yr BP, an increase in heathland importance from 6,300 to 3,200 cal yr BP, and a return to rainforest–moorland mosaic in the last 3,200 years. The Mid-Holocene heathland episode has been attributed to increasing aridity, which has also been found in pollen spectra from Tierra del Fuego.

Ashworth and Hoganson have made a significant beginning in southern South America, overcoming tremendous obstacles just to develop an understanding of the modern fauna, let alone Quaternary assemblages. Their scientific curiosity led them to explore the history of life in a region that many would consider "the end of the Earth." In order to identify and interpret fossil insect assemblages in this region, Ashworth and Hoganson spent years collecting modern specimens and tabulating their ecological requirements and the altitudinal zonation of their populations. This painstaking work paved the way for meaningful fossil studies. Sadly, this foundational work in the paleoentomology of southern South America has not been built upon in recent years.

12

Japanese Studies

> Japanese Quaternary entomology will continue to supply vital information on the evolution of Japanese beetle fauna. The framework of speciation in the islands differs from that in continental regions. The formation of land bridges between island and continent or island and island are the most important element of the regional zoogeography.
>
> – Masazaku Hayashi (2007)

Quaternary entomology began in Japan in 1978. Prior to the publication of *Quaternary Insects and Their Environments* in 1994, Japanese paleontologists and entomologists were not fully aware of the paleoentomological research going on in Europe and the Americas. Communication of the results of Japanese research in this field has likewise been hindered by the language barrier. Most of the literature on this topic has been written in Japanese. A major review of the Japanese research has recently appeared in English (Hayashi, 2007). Hopefully this will facilitate the exchange of ideas between Japanese researchers and those who live elsewhere.

The Japanese method of fossil extraction (discussed in Chapter 2) involves peeling back individual layers of peat in order to find fossil specimens. While this method allows articulated exoskeletal elements to be discovered *in situ*, it has almost certainly skewed their findings towards large and brightly colored fossil specimens. Small and dull-colored sclerites are easily overlooked when scanning peat surfaces.

1. Fossil Assemblages

Hayashi (2007) noted that Japanese faunal assemblage lists are dominated by ground beetles (Carabidae) and donaciine leaf beetles (Chrysomelidae: Donaciinae). As a result, the 25 sites discussed here (Table 1, Fig. 1) have yielded faunal assemblages comprising only about 130 species. Studied faunas range in age from the Pliocene through the Holocene.

1.1. Pliocene and Early Pleistocene Faunas

Japanese paleoentomological research extends back to Pliocene-age faunal assemblages, and faunas from the Plio-Pleistocene boundary. Pliocene faunas such as the assemblages from sites on the Yasu River (Fig. 1, No. 1) and the Ajimu Basin (Fig. 1, No. 2) are indicative of environmental conditions quite similar to those of today (Hayashi and Shiyake, 2002; Hayashi *et al.*, 2003b). The deposits yielding these faunas range in age from more than 6 mya to about 4.1 mya. The faunas are dominated by aquatic and riparian species.

The Tado Organizations Research Group (1998) studied faunal assemblages taken from the Tokai Group of formations in central Japan. The age estimate of the fauna (Pliocene/Pleistocene transition) is based on its stratigraphic position immediately above the Karegawa tephra. They recovered a small insect fossil assemblage (183 individuals from six families) from the Chikaro site (Fig. 1, No. 3). The climatic transition from the Pliocene to the Pleistocene is marked in this fauna, which included cold-adapted species such as the dytiscid *Ilybius* cf. *poppiusi*. Today this species ranges north into eastern Siberia. An aquatic leaf beetle, *Plateumaris constricticollis constricticollis*, was also identified. This species is endemic to Japan, and the *constricticollis* subspecies is found today in northern Japan, in regions that experience colder, snowier winters (Sota *et al.*, 2007). The beetle faunal evidence, in combination with the paleobotanical evidence, indicates climatic cooling about 1.75 mya.

Japanese paleoentomologists have devoted a great deal of effort to the study of fossil and modern aquatic leaf beetles (Chrysomelidae, subfamily Donaciinae) (Hayashi, 1998b, 1999a, 2002; Hayashi and Shiyake, 2002; Hayashi *et al.*, 2003a; Sota and Hayashi, 2007). Late Pliocene and Early Pleistocene fossil assemblages in Japan frequently contain specimens in this group of beetles. Hayashi (1999b) studied donaciine fossils from the Higashikubiki site, representing the Uonuma Formation in the Niigata Prefecture (Fig. 1, No. 4). Combining this information with other fossil donaciine records, he was able to piece together a history of several species in this group in Japan (Hayashi, 2004) (Fig. 2). He documented the presence of *Plateumaris constricticollis*, *Donacia japana*, *Donacia vulgaris*, and *Donacia ozensis* throughout at least the last 1.6 mya. He considered *Plateumaris constricticollis* and *Donacia ozensis* to be endemic Japanese species; *D. vulgaris* and *Donacia japana* were considered to be early colonizers from mainland Asia. Another donaciine species, *Plateumaris akiensis*, has been identified from Late Pliocene deposits in Japan. This species was apparently more widespread in Japan during at least part of the Quaternary, but it is now restricted to one district in the Hiroshima Prefecture (Hayashi and Shiyake, 2002).

Two other donaciines, *Plateumaris sericea* and *Donacia splendens*, are thought to be more recent colonizers, becoming established in Japan during the last

DEVELOPMENTS IN QUATERNARY SCIENCES
VOLUME 12 ISSN 1571-0866

Table 1. Pleistocene insect fossil sites in Japan.

Site	Age (cal yr BP × 1000)	Reference(s)
(1) Yasu River	Pliocene	Hayashi and Shiyake (2002)
(2) Ajimu Basin	Pliocene	Hayashi *et al.* (2003b)
(3) Chikarao	Plio/Pleistocene	Tado Organizations Research Group (1998)
(4) Higashikubiki	Plio/Pleistocene	Hayashi (1999b, 2001)
(5) Kitamimaki-mura	Early Pleistocene	Hayashi (1998b)
(6) Oyama-machi	Early Pleistocene	Hayashi *et al.* (2004a)
(7) Kitsuki	Middle Pleistocene	Hayashi *et al.* (2005)
(8) Mashiki-machi	Middle Pleistocene	Hayashi *et al.* (2004b)
(9) Kurijanami	ca. 100	Kashiwazaki Naumann's Elephant Research Group (1991)
(10) Ono	ca. 70–60	Hayashi (1998a)
(11) Nakamachi	70–11	FIRGNE (1993b, 1996b, 2000b)
(12) Ikejirigawa	50–11	FIRGNE (1987, 1990)
(13) Tategahana	50–11	FIRGNE (1980, 1984, 1987, 1990, 1993a, 1996a, 2000a, 2003)
(14) Taguro	>40	Nirei and Hayashi (1998)
(15) Garahagi	44–34	Hayashi and Miyatake (1996)
(16) Kotari	47–30	Hayashi *et al.* (2001a)
(17) Taira	30–29	Hayashi and Shimadu (2005)
(18) Nukui	30–29	Hayashi (1999a)
(19) Tarukuchi	30–29	Hayashi and Miyatake (1996)
(20) Owatasi II	30–14	Mori (1995)
(21) Tomizawa	29–23	Mori and Itoh (1992)
(22) Takasaki	29–16	Hayashi (2005), Nirei and Hayashi (2004)
(23) Sojya	17–16	Hayashi (1996)
(24) Miyanomae	17–14	Mori *et al.* (1997)
(25) Shimogose	?	Hayashi *et al.* (2002)

500,000 years. Hayashi also identified four donaciine species that are either extirpated from Japan today, or have possibly become extinct. He described *Plateumaris virens* (Fig. 3) as a fossil species that has become extinct, because no known modern populations exist. *Donacia versicolorea* was found as a fossil in Japan from sediments spanning the interval 1.6–1.1 mya, but subsequently it died out there. Today it is known only from Europe. *Donaciella nagokana* and *Donaciella uedana* are two species known only from Early to Middle Pleistocene assemblages in Japan.

Early Pleistocene insect faunas have been examined from a site at Kitamimaki-mura, in the Nagano region (Hayashi, 1998b) (Fig. 1, No. 5), and from the Oyama-machi site on the island of Kyushu (Hayashi *et al.*, 2004a) (Fig. 1, No. 6). None of the Oyama-machi material was identified to the species level. The Kitamimaki-mura fauna included the cold-adapted dytiscid, *Ilybius poppiusi*, and the first appearances of two other water beetles, *Rhantus erraticus* and *Hydrochara libera*, in the Japanese fossil record. These two species are only known from Japan today.

1.2. Middle Pleistocene Faunas

Middle Pleistocene beetle faunas have been identified from two sites on Kyushu: the Kitsuki site (Hayashi *et al.*, 2005) (Fig. 1, No. 7) and the Mashiki-machi site (Hayashi *et al.*, 2001b, 2004b) (Fig. 1, No. 8). Only two fossil taxa were identified to the species level from these faunas: the ground beetle *Elaphrus japonicus* and the click beetle (Elateridae) *Spheniscosomus cribricollis*. The latter species is currently Pan-Pacific in distribution, known from Japan and Korea, as well as many parts of North America. No other details of these Middle Pleistocene faunal assemblages are available in English-language publications.

1.3. Late Pleistocene Faunas

Sixteen Late Pleistocene fossil sites have been analyzed thus far (Table 1, Fig. 1). Most of these sites are on the main island of Honshu; one site is on the island of Kyushu. These Late Pleistocene fossil assemblages come from archeological excavations, terrace deposits, and lake sediments.

1.4. Last Interglacial Faunas

The oldest Late Pleistocene fossil assemblage thus far identified from Japan comes from the Kurijanami site (Hayashi, 2007) (Fig. 1, No. 9). The assemblage comes from the Yasuda Formation, in Kashiwazaki City, Niigata Prefecture, and is thought to date from the last interglacial (MIS 5e). The fossil assemblage included 24 identified taxa. This work was carried out as an adjunct to the excavation of the remains of a specimen of Naumann's

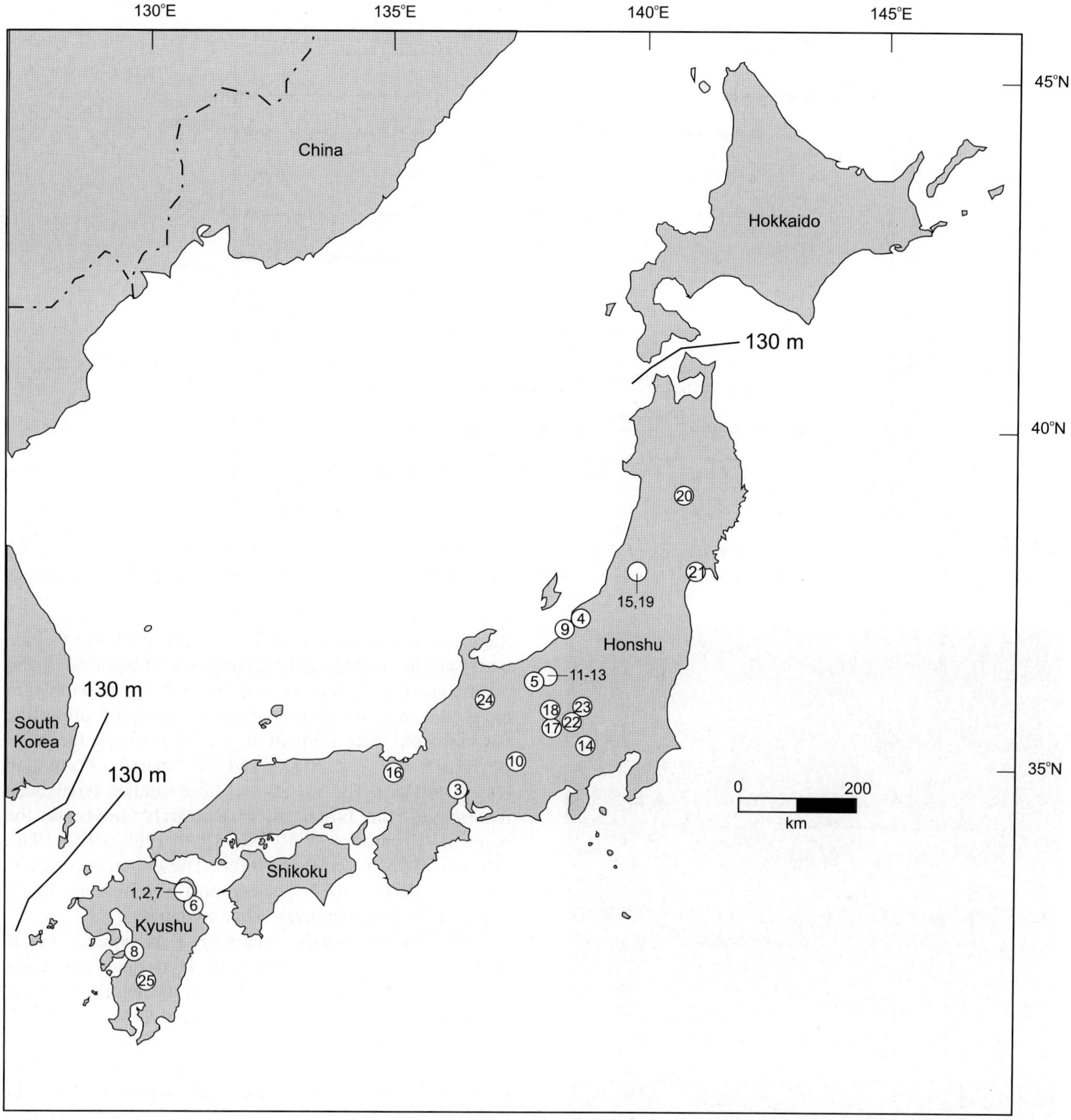

Fig. 1. Map of Japan, showing locations of fossil study sites (site numbers match those in Table 1). The heavy line drawn between the islands of Honshu and Hokkaido represents the Tsugaru Strait. The heavy lines between the island of Kyushu and South Korea represent the Tsushima and Korean straits. The numbers indicate waters depths in meters.

elephant (*Palaeoloxodon naumanni*) (Kashiwazaki Naumann's Elephant Research Group, 1991). The fauna is dominated by water beetles, ground beetles, phytophagous species, and dung beetles. Most of these species still live today in and around the fossil site, except the dung beetle *Copris tripartitus*, which no longer lives on the island of Honshu. The ground beetles include *Carabus insulicola*, *Scarites terricola pacificus*, *Lesticus magnus*, *Pterostichus planicollis*, *Epomis nigricans*, and *Chlaenius circumdatus*. The leaf beetles include four donaciine beetles: *Donacia ozensis*, *D. lenzi*, *D. provostii*, and *D. vulgaris*. Based on the fossil data, the MIS 5e paleoenvironment of the site has been reconstructed as including a range of local environments from forest edges to grassland, surrounding an area of still water with emergent and floating vegetation.

1.5. MIS 4–3 Faunas

The most concentrated studies of insect fossils from the last glacial interval come from the Nojiri-ko Site Group in

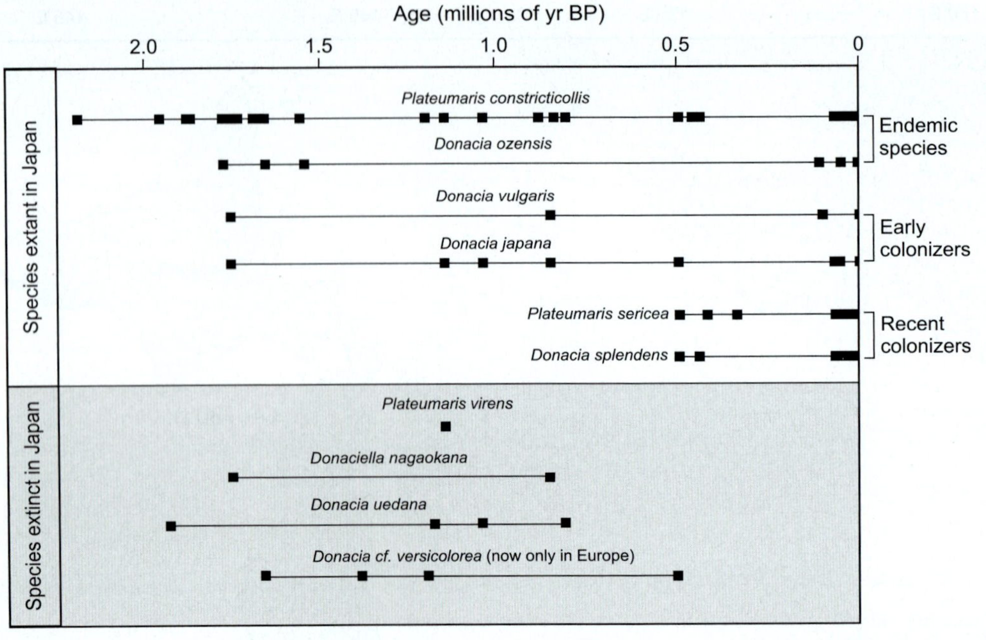

Fig. 2. Reconstructed history of several species of donaciine leaf beetles in Japan during the Quaternary (after Hayashi, 2004). Shaded section indicates species that no longer live in Japan. All but Donacia *cf.* versicolorea *are thought to be extinct.*

Fig. 3. Photograph of the type specimen of Plateumaris virens *(Chrysomelidae: Donaciinae) from the Higashikubiki site, Niigata Prefecture, Japan (from Hayashi, 1999b, used with permission).*

Shinano-machi, Nagano Prefecture. These studies include work at the Nakamachi, Ikejirigawa, Tategahana, and other sites (Fig. 1, Nos. 11–13). Fossil beetle assemblages have been recovered from three stratigraphic units: the Biwajima-oki Peat Formation (ca. 100–70 ka), the Kannoki Formation (ca. 60 ka), and the Nojiri-ko Formation (ca. 50–10 ka). The results of these studies have been reported by the Fossil Insect Research Group for the Nojiri-ko Excavation (FIRGNE, 1980, 1984, 1987, 1990, 1993a,b, 1996a,b, 2000a,b, 2003). These faunas comprise a total of 92 identified species, most of which are found today in the cool-temperate zone of Japan.

The faunas contain relatively rich aquatic beetle faunas, including five species of Dytiscidae and eight species of Hydrophilidae, as well as the whirligig beetle (Gyrinidae) *Dineutus orientalis*. Most of these taxa inhabit standing water, but the dytiscid *Platambus pictipennis* lives in streams. The terrestrial beetles in these faunas include Carabidae, Histeridae, and Silphidae. Most of them live in forest floor and/or forest edge habitats today. Five of the ground beetle species live on moist ground with vegetation, including *Carabus granulatus*, *Elaphrus japonicus*, *Pterostichus prolongatus*, *Epomis nigricans*, and *Chlaenius gebleri*. Two of the ground beetles are arboreal species: *Calosoma inquisitor* and *Calleida lepida*. The plant-feeding beetles include Lucanidae, Scarabaeidae (the genera *Mimela*, *Anomala*, *Rhomborrhina*, and *Eucetonia*), Cerambycidae, Chrysomelidae, Attelabidae, and Curculionidae. Most of these species live in forests and/or the forest edge. A variety of aquatic leaf beetles in the genera *Plateumaris*, *Donacia*, and *Limnobaris* feed on Cyperaceae, Potamogetonaceae, Sparganiaceae, Typhaceae, and Nymphaeaceae.

The modern ranges of the species found in Nojiri-ko faunas are shown in Table 2. Surprisingly, 55% of the 92 species are known today only from the Japanese islands. This is a remarkably high degree of endemism, but the

Table 2. Fossil beetle species identified from the Nojiri-ko fossil assemblages in Japan.

Species	Modern range
Carabidae	
Calleida lepida Redt.	Japan
Calosoma inquisitor cyanescens Mots.	Palearctic
Carabus albrechti Morawitz	Japan
Carabus granulatus L.	Palearctic
Carabus insulicola Chaud.	Japan
Carabus vanvolxemi Putz.	Japan
Chlaenius gebleri Gangl.	Japan
Chlaenius micans (Fab.)	Japan
Chlaenius pallipes Gebl.	Eastern Palearctic
Colpodes japonicus (Mots.)	Japan
Damaster blaptoides Kollar	Japan
Elaphrus japonicus S.Ueno	Japan
Pterostichus longinquus Bates	Eastern Palearctic
Pterostichus microcephalus (Mots.)	Eastern Palearctic
Pterostichus planicollis (Mots.)	Palearctic
Pterostichus prolongatus Morawitz	Japan
Dytiscidae	
Agabus conspicuous Sharp	Eastern Palearctic
Agabus japonicus Sharp	Japan
Agabus optatus Sharp	Japan
Cybister brevis Aube	Japan & Korea
Dytiscus czerskii Zaitzev	Eastern Palearctic
Platambus pictipennis (Sharp)	Japan and Sakhalin Island
Rhantus erraticus Sharp	Japan
Hydrophilidae	
Anacaena asahinai M. Sato	Japan
Cercyon ustus Sharp	Eastern Palearctic
Coelostoma orbiculare (Fab.)	Palearctic
Coelostoma stultum (Walker)	Japan
Enochrus japonicus (Sharp)	Japan
Hydrochara affinis (Sharp)	Japan
Hydrochara libera (Sharp)	Eastern Palearctic
Hydrophilus acuminatus Mots.	Eastern Palearctic
Pachysternum haemorrhoum Mots.	Japan
Sternolophus rufipes (Fab.)	Japan
Gyrinidae	
Dineutus orientalis (Modeer)	Eastern Palearctic
Histeridae	
Athous duodecimstriatus (Gyll.)	Palearctic (Europe to Japan)
Hister concolor Lewis	Japan and Kurile Islands
Hister simplicisternus Lewis	Japan
Margarinotus niponicus (Lewis)	Eastern Palearctic
Silphidae	
Eusilpha japonica (Mots.)	Japan
Nicrophorus vespilloides (Herbst)	Palearctic (Europe to Japan)
Oiceoptoma thoracicum (L.)	Palearctic (Europe to Japan)
Phosphuga atrata (L.)	Palearctic (Europe to Japan)
Silpha longicornis Portevin	Japan
Staphylinidae	
Paederus fuscipes (Curtis)	Palearctic (Europe to Japan)
Scaphidiidae	
Scaphidium rufopygum Lewis	Japan
Lucanidae	
Nipponodorcus rubrofemoratus (S van Voll.)	Eastern Palearctic
Platycerus acuticollis Y. Kurosawa	Japan

(*Continued*)

Table 2. (*Continued*)

Species	Modern range
Geotrupidae	
Geotrupes auratus Mots.	Japan
Scarabaeidae	
Anomala cuprea Hope	Japan
Anomala daimiana Harold	Japan
Aphodius brachysomus Solsky	Japan and Korea
Aphodius breviusculus (Mots.)	Japan
Aphodius elegans Allibert	Palearctic, India
Aphodius igai Nakane	Japan
Aphodius pusillus (Herbst)	Palearctic (Europe to Japan)
Aphodius quadratus Reiche	Eastern Palearctic
Aphodius rectus (Mots.)	Palearctic (Europe to Japan)
Aphodius rufipes (L.)	Palearctic (Europe to Japan)
Aphodius urostigma Harold	Malaysia, Southern Palearctic north to Japan
Aphodius yamato Nakane	Japan
Caccobius jessoensis Harold	Japan
Caccobius nikkoensis Lewis	Japan
Copris pecuarius Lewis	Japan
Eucetonia roelofsi (Harold)	Japan
Mimela testaceipes Mots.	Japan
Onthophagus lenzii Harold	Japan and Korea
Rhomborrhina unicolor Mots.	Japan
Byrrhidae	
Cytilus sericeus (Forst.)	Palearctic (Europe to Japan)
Elateridae	
Selatosomus puncticollis (Mots.)	Eastern Palearctic
Trogossitidae	
Leperina squamulosa (Gebl.)	Eastern Palearctic
Coccinellidae	
Harmonia axyridis (Pallas)	Japan, Korea, North America
Chrysomelidae	
Chrysolina aurichalcea (Mann.)	Palearctic (Europe to Japan)
Colasposoma dauricum Mann.	Japan
Donacia hiurai Kimoto	Japan
Donacia japana Chujo et Goecke	Japan
Donacia lenzi (Schonf.)	China and Japan
Donacia ozensis Nakane	Japan
Donacia sparganii Ahrens	Palearctic (Europe to Japan)
Donacia splendens Jacobson	Japan
Gastrolina depressa Baly	Japan
Gastrolina peltoidea (Gebl.)	Palearctic (Europe to Japan)
Gonioctena japonica Chujo et Kimoto	Japan
Linaeidea aenea (L.)	Palearctic (Europe to Japan)
Plateumaris constricticollis (Jacoby)	Japan
Plateumaris sericea (L.)	Palearctic (Europe to Japan)
Cerambycidae	
Pterolophia caudata Bates	Japan
Attelabidae	
Byctiscus puberulus (Mots.)	Japan
Byctiscus puberulus regalis (Roelofs)	Japan
Byctiscus venustus (Pascoe)	Japan
Euops punctatostriatus (Mots.)	Japan
Curculionidae	
Dyscerus orientalis (Mots.)	Japan
Limnobaris japonica Yoshihara et Morimoto	Japan

Source: Data from Fossil Insect Research Group for the Nojiri-ko Excavation publications.

modern biota of Japan exhibits consistently high levels of endemism. The Japanese flora contains 35% endemic species (Numata, 1974); the mammalian fauna is 49% endemic (Abe, 1994); the freshwater fishes are 24% endemic today (Masuda *et al.*, 1984). Based on molecular genetic evidence, the ancestry of the modern Japanese ground beetle fauna can be divided into two groups: (1) species whose ancestors inhabited ancient Japan at the time of its split from the Eurasian Continent (ca. 15 mya), and (2) species that invaded Hokkaido from the Eurasian Continent through land bridges from Sakhalin and/or the Kurile Islands, or invaded western Japan from the Korean Peninsula during the Pleistocene (Tominaga *et al.*, 2000). It would seem likely that the ancestry of the other Coleoptera families followed a similar dispersal and evolutionary pattern. Several species identified from the Nojiri-ko faunas are found today only in Japan and Korea, including the predaceous diving beetle *Cybister brevis* and the dung beetles *Aphodius brachysomus* and *Onthophagus lenzii*. The long-standing faunal connections between Japan and the Kurile and Sakhalin Islands are maintained today by several species found in the Nojiri-ko faunas, including the predaceous diving beetle *Platambus pictipennis*, and the histerid beetle *Hister concolor*.

The Ono Peat bed (Fig. 1, No. 10) dates to ca. 60,000 yr BP (Hayashi, 1998a). Only eight species of beetles were identified from this peat bed, including the dytiscid *Agabus japonicus* and the chrysomelid *Plateumaris sericea*. As with the Nojiri-ko faunas, the Ono fauna is comprised of species still found in Japan today. The same can be said of the faunas from the Garahagi and Tarukachi sites (Fig. 1, Nos. 15 and 19). The peat beds from Tarukachi date to the interval 30,000–29,000 cal yr BP. These deposits yielded fossils of the ground beetle *Carabus vanvolxemi*, the predaceous diving beetle *Agabus miyamotoi*, the carrion beetle *Silpha longicornis*, and the aquatic leaf beetle *Plateumaris sericea* (Hayashi and Miyatake, 1996). The Garahagi Site is somewhat older; organic deposits at this site yielded a radiocarbon age of >35,900 yr BP (Hayashi and Miyatake, 1996). Only nine beetle taxa were identified from this site, including *Elaphrus japonicus*, *Agabus* sp., *Coelostoma orbiculare*, *Donacia splendens*, and *Plateumaris constricticollis*.

1.6. MIS 2 and Lateglacial Faunas

Japanese beetle assemblages from the LGM are not dominated by cold-adapted immigrants from Siberia. The very limited fossil assemblages from the last glaciation from the Owatasi II, Sojya, and Miyanomae sites (Fig. 1, Nos. 20, 23, and 24) include a mixture of species found today in Japan and elsewhere in the Palearctic (Table 3).

Table 3. Fossil beetle species identified from LGM fossil assemblages in Japan.

Species	Modern range
Carabidae	
Elaphrus japonicus S. Uedo	Eastern Palearctic
Pterostichus leptis Bates	Japan and Sakhalin Island
Dytiscidae	
Ilybius poppiusi Zait.	Eastern Siberia, Mongolia, Japan
Ilybius weymarni Balfour-Browne	Japan and Sakhalin Island
Hydrophilidae	
Anacaena asahinai M. Sato	Japan
Coelostoma orbiculare (Fab.)	Palearctic (Europe to Japan)
Gyrinidae	
Dineutus orientalis (Modeer)	Eastern Palearctic
Lagriidae	
Luprops cribrifrons Marseul	Japan
Chrysomelidae	
Cneorane elegans Baly	Japan
Plateumaris constricticollis (Jacoby)	Japan
Plateumaris sericea (L.)	Palearctic (Europe to Japan)
Cerambycidae	
Spondylis buprestoides L.	Holarctic including Japan and Korea
Attelabidae	
Byctiscus puberulus (Mots.)	China, Eastern Russia, Japan
Curculionidae	
Hylobius pinastri (Gyll.)	Palearctic (Europe to Japan)

Source: Data from Mori (1995); FIRGNE (1980–2003).

However, there are several species in these assemblages whose modern ranges extend north into Sakhalin Island and Siberia, including the ground beetle *Pterostichus leptis*, and the predaceous diving beetles *Ilybius poppiusi* and *I. weymarni*. The pine-feeding weevil *Hylobius pinastri* was identified from the Owatari II peat beds, but these faunas range in age from the LGM interval to 14,000 cal yr BP, so this piece of evidence for the presence of pine woods may relate to a time of climatic amelioration following the LGM. There is good evidence from other proxies that Japan experienced dramatic cooling during the LGM. Nakagawa *et al.* (2002) reconstructed terrestrial climates for central Japan, based on pollen assemblage data. Their reconstruction indicates that both summer and winter temperatures reached a minimum around 24,000 cal yr BP, when they were about 10°C lower than present. The fossil beetle record of Japan has yet to document the climatic record of this interval, because very few LGM assemblages have been studied, and the few studied include only 14 species, few of which are stenotherms.

Could cold-adapted beetle species have migrated into Japan during the LGM from Siberia? At least one migration route from continental Asia into Japan was apparently available during the LGM. A land bridge linked southeastern Russia with Sakhalin Island, where the current water depth of the Mamiya Strait is only 12 m. Sakhalin Island was apparently connected to Hokkaido across a land bridge spanning the Soya Strait (water depth 55 m), and based on fossil mammalian evidence, there was also a land bridge linking Hokkaido with Honshu across the Tsugaru Strait (water depth 130 m) (Millien-Parra and Jaeger, 1999). It is thought that the horse *Equus* migrated along this series of land bridges and islands about 20,000 years ago (Kawamura, 1991). The deepest straits between Japan and mainland Asia are the Tsushima and Korean straits, separating Japan from Korea (Fig. 1). If sea level was lowered by 130 m during the LGM, this would have allowed a land bridge to form across these straits, but there is paleoceanographic evidence that suggests that there has been a connection between the Sea of Japan and the North Pacific throughout the Quaternary (Keigwin and Gorbarenko, 1992). This suggests that these straits remained open.

Few Holocene-age sites have been thus far analyzed for insect fossils in Japan. As interest in the discipline grows there will undoubtedly be opportunities for Japanese paleoentomologists to work with archeologists on post-glacial-age sites associated with human occupation.

13

Studies in Australia and New Zealand

> Perhaps the greatest limitation of Quaternary beetle research in Australia is the lack of suitable sites relative to other parts of the world where Quaternary entomology is established…
>
> – Porch and Elias (2000)

1. Australian Studies

I recently had the opportunity to fly over Australia, from the northwest corner of the continent to Sydney in the southeast. The sights below made a deep impression on me, as I had never seen so much desert before in my life. Much of what I saw out the window looked like the Great Basin desert regions of Nevada or western Utah, but the deserts of the Australian outback are orders of magnitude larger, with no ''wet patches'' such as the mountain ranges that break up the arid landscapes of the American West. It took about 6 h to cross the Australian deserts in a modern jet airplane – roughly the same amount of time it takes to fly across North America, with its multitude of different ecosystems. It is therefore not surprising that nearly all of the Australian human population lives within 100 km of a coast, where there is more moisture. All of the biodiversity ''hot spots'' of Australia are likewise in the less-arid zones of the tropical north, or the temperate coastal regions.

Paleoentomology has come late to Australia, for a number of reasons, discussed below. It began in earnest only in the 1990s, as Nick Porch began his doctoral dissertation research project on this topic at Monash University. Nick has been the lone pioneer of Australian paleoentomology thus far.

1.1. Nature of the Australian Biota

Australia has been biologically isolated from most of the rest of world since the breakup of the supercontinent of Gondwana was completed, about 99 mya (Veevers, 2006). Some of the biological peculiarities of Australia are well known, such as the near-absence of placental mammals, and the dominance of eucalypts and acacias in the arboreal vegetation. What is less well known is that Australia's beetle fauna is likewise unlike that of other continents of the world. Australia's beetle fauna is not small. With about 23,000 described species in 121 families (Lawrence and Britton, 1994), the level of beetle faunal diversity compares well with the diversity seen in other continents. However, different families of beetles are far more prominent in Australia than elsewhere, and families that are typically well represented on other continents are only poorly represented here. Likewise, within well-known families, unusual or Australian endemic taxa are very important. One example of this is the group of darkling beetles (Tenebrionidae) known as pie-dish beetles (genera *Helaeus* and *Pterohelaeus*). These form an important part of the Australian darkling beetle fauna of many arid regions, but are unknown elsewhere. Early on, Darlington (1943) noted the large number of endemic ground beetles, with almost a quarter of the known species being arboreal beetles associated with ''the shaggy trunks of Eucalyptus and other Australian trees.'' Certain subfamilies, tribes, and genera of beetles within widely distributed families are much more important here than in the Northern Hemisphere. Among the ground beetles, genera that exhibit a high degree of species diversity in the Northern Hemisphere are either poorly represented, or not represented at all in Australia. For instance, the riparian genus *Bembidion* that contains in excess of 400 species in North America (Arnett, 1973) is only represented by 12 species in Australia (Calder, 2002). In contrast, the genus *Clivina*, represented by only two species in Europe, contains 108 Australian species (Calder, 2002). Some endemic Australian carabid genera contain impressive numbers of species, such as *Carenum* (120 species), *Notonomus* (98 species), *Adelotopus* (123 species), and *Arthropterus* (79 species) (Calder, 2002).

1.2. Australian Paleoenvironments

Surprisingly little is known of Pleistocene environments in Australia. Most studies have focused on fossil pollen assemblages, but the vast majority of Australian pollen records are from the Holocene, with few extending far into the Pleistocene (Kershaw and van der Kaars, 2007). The major exception to this is that suitable fossil sites are preserved in the wetter regions of southeastern Australia, and in the tropical north. Vegetation changes that have taken place in the Late Pleistocene have been regional in nature. There is little evidence of long-distance migration of plant species in response to climate change. Temperature changes have dominated the climatic signal in southern Australia, whereas precipitation changes have dominated the climate signal in the northern tropics. Paleobotanical reconstructions of LGM temperatures for Australia suggest mean annual temperatures were 4.0–6.5°C cooler than present (Kershaw, 1995) and mean summer temperatures for Tasmania 1–7°C cooler during

DEVELOPMENTS IN QUATERNARY SCIENCES
VOLUME 12 ISSN 1571-0866

the LGM (Kirkpatrick and Fowler, 1998). Recent evidence from temperature-dependent racemization of amino acids in eggshell from the Lake Eyre basin suggest "millennial-scale average temperatures were at least 9°C lower between 45 and 16 kyr BP," and that the LGM may have been substantially cooler (Miller *et al.*, 1997).

There has been a general trend of increasing aridity in much of Australia throughout the Late Pleistocene, as evidenced by increased fire frequency and the increasing dominance of sclerophyllous woodlands in southeastern Australia. The timing of the increased fire frequency is thought to coincide broadly with the arrival of humans, about 45,000 yr BP (O'Connell and Allen, 2004). The lack of permanent water bodies in much of Australia has imposed severe limits on fossil preservation. Permanent lakes and bogs are found almost exclusively in the wettest parts of the continent. Paleoclimate reconstructions are necessarily limited by the inability of cold-adapted flora and fauna to enter Australia from other regions during glacial intervals. As discussed above, there have been no records of long-distance migration into or out of Australia during the Pleistocene, except for human beings. Similarly, there are no significant highland regions in Australia that would have provided a potential source for the internal migration of cold-adapted species during cold periods. The detection of cold climates in the prehistory of Australia is therefore difficult or impossible, based on biotic evidence.

1.3. Australian Entomology

Paleoentomology in Australia suffers from the incomplete state of knowledge of the modern beetle fauna. Australia's beetle fauna includes about 23,000 described species in 3,265 genera and 121 families, but the total species number is estimated to be in the range of 80,000–100,000 (Lawrence and Britton, 1994). If this is an accurate estimate, then the vast majority of the modern fauna remains undescribed. Furthermore, there are only limited data on the modern distributions and ecological requirements of most of the described beetle species. These are major roadblocks to paleoentomological research in any study region. Porch (2007a) has made the best of the situation by developing an alternative method of paleoenvironmental reconstruction that relies less heavily on complete knowledge of the modern range and environmental tolerances of species found in fossil assemblages (the BIOCLIM program discussed in Chapter 5).

The fossil beetle faunas thus far described from Australia range in age from the Late Pleistocene to the Holocene. There are no sites yet known (none have been tested) from the period of the Early Pleistocene and Late Tertiary, but several potential sites are known from this period that contain plant macrofossils or pollen. These sites may contain fossil insect faunas and are sites for future exploration. The Late Pleistocene faunas all come from deposits of water-lain sediments, associated with fluvial and lacustrine environments. There is also a small potential to develop insect fossil records from stick-rat (*Leporillus*) middens, but those that have been studied thus far are all of Late Holocene age (Pearson, 1999). Pleistocene fossil sites from Australia are listed in Table 1. Site numbers coordinate with the localities shown in Fig. 1. Beetle species identified in the various Australian fossil assemblages are shown in Table 2.

1.4. Yarra Creek, King Island, Tasmania – MIS 5

A fossil beetle assemblage associated with MIS 5 has been recovered from sediments exposed in coastal bluffs, near the outlet of Yarra Creek into the City of Melbourne Bay on King Island (Fig. 1, No. 1). The site has been dated by thermoluminescence, and the age within MIS 5 has not been precisely determined. The vegetation reconstruction associated with the beetle assemblage is that of a cool-temperate rain forest. These conditions do not occur regionally today, and almost certainly did not occur at any time during the Holocene (Porch, 2007a).

Table 1. Pleistocene insect fossil sites Australia and New Zealand.

Site	Age (cal yr BP × 1000)	Reference(s)
Australia		
(1) Yarra Creek	MIS 5	Porch (2007a)
(2) Spring Creek	>40	Porch (2007a)
(3) Caledonia Fen	50–30	Porch (2007a)
(4) Pulbeena Swamp	34–32.5	Porch (2007a)
(5) Pipe Clay Lagoon	30–25	Porch (2007a)
New Zealand		
(6) Auckland	MIS 7	Marra *et al.* (2006b)
(7) Banks Peninsula	MIS 6	Marra (2003b)
(8) Palliser Bay	MIS 5	Shulmeister *et al.* (2000), Marra (2003a)
(9) Westport	37–21	Burge and Shulmeister (2007)
(10) Howard Valley	37–21	Marra (2008)
(11) Waitotara Valley	34–33	Marra *et al.* (2007, 2009)
(12) Awatere Valley	24–22	Marra and Leschen (2004)
(13) Lyndon Stream	25–22	Marra *et al.* (2006a)

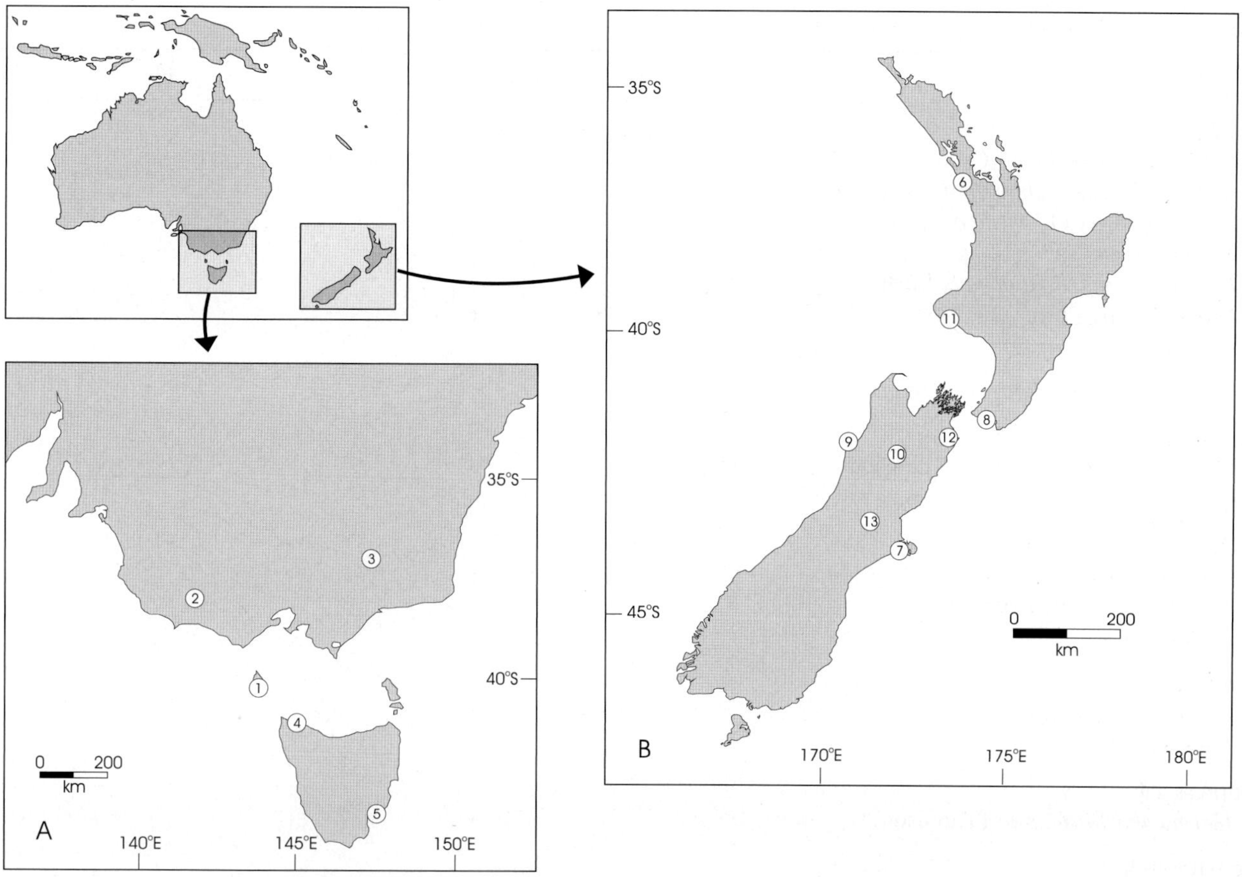

Fig. 1. Map of southern Australia and Tasmania (A) and New Zealand (B), showing fossil localities discussed in text (site numbers match those in Table 1).

The Yarra Creek insect assemblage accumulated in a fluvial deposit. The fauna is dominated by saproxylic (associated with decaying wood) beetles, and beetles found in the leaf litter of rain forests or wet sclerophyll forests. No beetles that feed on specific host trees were found in this assemblage, but it is clear that the assemblage has no modern analogues. It contains four faunistic elements. It is dominated by taxa that are widespread across mesic southeastern Australia, including Tasmania. The second element is species restricted to Tasmania but not reaching King Island, such as the stag beetle (Lucanidae) *Ceratognathus westwoodi*. The third element consists of taxa that are widespread in mainland southeastern Australia and just reach south to King Island today, including the prostomid beetle *Prostomis cornuta*, and the ground beetle *Mecyclothorax* "*cordicollis*." An additional group is composed of taxa that are currently restricted to the mainland and are not known to inhabit either King Island or Tasmania. This group includes the darkling beetle (Tenebrionidae) *Archaeoglenes australis*. Using the BIOCLIM method of paleoclimate reconstruction, Porch (2007a) estimated that the regional climate associated with the fossil assemblage was substantially different from modern conditions. Summer rainfall was greater than today, and the degree of seasonality was enhanced, resulting in slightly warmer summers and colder winters. Average winter temperatures are estimated to have been about 4°C colder than today. The paleobotanical interpretation agrees with this reconstruction.

1.5. Spring Creek – MIS 3?

The Spring Creek site in western Victoria (Fig. 1, No. 2) was brought to light through the study of megafaunal remains that were originally thought to date to the LGM (Flannery and Gott, 1984). A more recent reassessment of the age is based on four AMS radiocarbon determinations, obtained on bone (White and Flannery, 1995); the results indicated that the original age determination was too young, due to contamination of the bone samples. The authors proposed that the true age of the Spring Creek fauna was greater than 35 kyr. A series of AMS radiocarbon determinations from plant and insect remains yielded dates >40 kyr BP. Insect assemblages were taken from samples surrounding the megafaunal bones, and from a site downstream where the stratigraphic equivalent of the megafaunal bone bed was found. The insect faunas from all samples are essentially the same, suggesting rapid accumulation. The fauna is dominated by aquatic insects, especially caddisflies (Trichoptera) and aquatic beetles, indicative of a clean, rapidly flowing stream with a sandy or gravel bottom. These beetles include the elmids *Austrolimnius* and *Simsonia*, and ecnomid and hydropsychid caddisflies. The small terrestrial fauna is dominated by species indicative of open grassland or heathland environments. Only one beetle identified in these assemblages is associated with trees, the buprestid *Germarica lilliputana*, but this species is also found in shrubs. The dung beetle fauna includes three aphodiine

Table 2. Indicator beetle taxa identified from Australian Pleistocene assemblages.

Taxon	YC	SC	CF	PS	PCS
Carabidae					
Bembidion blackburni Csiki		+		+	
Cyphotrechodes gibbipennis Blackburn					+
Mecyclothorax cf. *cordicollis* Sloane	+				
Pseudoceneus sp.				+	
Promecoderus viridiaeneus Sloane				+	
Scopodes sp.		+			
Hydraenidae					
Tympanogaster (*Hygrotympanogaster*) sp.			+		
Lucanidae					
Ceratognathus westwoodi Thomson	+				
Scarabaeidae					
Ataenius basiceps Lea	+			+	
Proctophanes sculptus Hope		+			
Podotenus erosus Erichson					+
Podotenus sp.		+			
Elmidae					
Australolimnius sp.		+	+		
Simsonia sp.		+			
Buprestidae					
Germarica lilliputana (Thomson)		+			
Tenebrionidae					
Archaeoglenes australis Lawrence	+				
Prostomidae					
Prostomis cornuta	+				
Anthribidae					
Eucioides suturalis		+			

Source: Data from Porch (2007a).

species (*Ataenius basiceps*, *Proctophanes sculptus*, and *Podotenus* sp.). These beetles are commonly associated with marsupial dung. The environmental reconstruction based on the fossil insect fauna (and supported by the paleobotanical evidence) is of a small stream flowing through a treeless landscape.

The BIOCLIM paleoclimate reconstruction for the assemblages yielded temperature and precipitation estimates that overlap the modern site values, but there is a suggestion of slightly more continentality than today, with slightly increased summer precipitation. The reconstructed climatic conditions appear to be at odds with the ''glacial'' pollen record, which is dominated by grasses and other herbs (Poaceae and Asteraceae), and thought to represent cold and/or dry climates.

1.6. Caledonia Fen, Great Dividing Range, Victoria – MIS 3

Caledonia Fen is a montane fen at 1,280 m asl in the Great Dividing Range region of Victoria (Fig. 1, No. 3). The site exposure contains a continuous sequence of sediments, extending back to MIS 6 (Porch, 2007a). The pollen record indicates the presence of steppe grassland for most time intervals, except for the Holocene, MIS 5e, and a short period at the beginning of MIS 3 when a Eucalyptus forest grew there. A series of small beetle assemblages were recovered from core samples, and AMS radiocarbon ages from plant macrofossils and charcoal indicate that the insect faunas represent the interval from about 50,000 to 30,000 cal yr BP. Most of the samples yielded very few insect remains, but several samples contained larger faunas. These are indicative of subalpine environments. The older faunas (50,000–36,000 cal yr BP) include specimens of the hydraenid beetle *Tympanogaster* (*Hygrotympanogaster*) sp., and the riffle beetle (Elmidae) *Austrolimnius*. Species in the *Hygrotympanogaster* species group live today exclusively in rocky habitats that are continually sprayed with water, such as the spray zone of waterfalls or the splash zone of rocky streams. *Austrolimnius* species are restricted today to sandy and gravely substrates in flowing streams. The presence of these taxa indicates that through this period the fen was fed by a perennial stream (there is no inflowing stream presently), implying that effective precipitation was higher in early MIS 3 than it is today (Porch, 2007a).

1.7. Pulbeena Swamp, Northwestern Tasmania – MIS 3

Pulbeena Swamp is a large, drained swamp in northwest Tasmania (Fig. 1, No. 4). The stratigraphy of a 5-m section exposed during the late 1970s has been described by Colhoun *et al.* (1982), who also analyzed pollen spectra. The MIS 2 peat is well humified, and insect fossil preservation is poor. Older, less humified layers of peaty marl yielded better-preserved insect remains, dating to late MIS 3. The youngest assemblage of well-preserved beetles dates to 34,000–32,500 cal yr BP (Porch, 2007a). This faunal assemblage indicates riparian swamp forest. The forest/scrub leaf litter habitat is represented by several species of rove beetles and pselaphids. Closed canopy rain forest environments are represented by cryptorhynchine weevils and the ground beetle *Promecoderus viridiaeneus*. Open riparian habitats are indicated by the presence of ground beetles such as *Bembidion blackburni* and *Pseudoceneus*. The pollen record associated with these assemblages indicates that the dominant vegetation was likely *Melaleuca* or *Leptospermum* rather than Eucalyptus. The presence of *Promecoderus viridiaeneus* suggests that both annual precipitation and summer seasonal precipitation levels were at least as great as today, but that summer temperatures were as much as 6°C lower than today.

1.8. Pipe Clay Lagoon, Eastern Tasmania – MIS 3/2 Boundary

The Pipe Clay Lagoon site (Fig. 1, No. 5) was described by Colhoun (1977). The exposure is a coastal bluff on Pipe Clay Lagoon, a marine embayment in eastern Tasmania. Colhoun (1977) described the stratigraphy and pollen record of the site. The organic deposits from which insect fossils were extracted have yielded radiocarbon ages ranging from about 19,800 to 25,250 ^{14}C yr BP (23,700–30,200 cal yr BP). Thus the faunas represent the MIS 3–2 transition. The site is notable as it presents one of the first glacial-age pollen records from southern Australia. Organic preservation in this dry region is quite unusual (Porch, 2007a). The insect assemblages described by Porch (2007a) were sampled from the uppermost of two organic sand beds. The faunas are dominated by aquatic and riparian beetles. The fauna suggests deposition in shallow standing, peaty water. The riparian fauna includes a range of taxa common at the margins of standing water in southeastern Australia today. Among the riparian species is ground beetle *Cyphotrechodes gibbipennis*, which is restricted today to dense accumulations of litter beside water, especially *Sphagnum* and other mosses that grow in high rainfall areas where MAP approaches 1,000 mm. The terrestrial element of the fauna includes two aphodiine dung beetles: *Ataenius basiceps*, and *Podotenus erosus*. The former species ranges south into in the warm lowlands of northeastern Tasmania; the latter species is apparently restricted today to moist habitats in central and northern Tasmania. The BIOCLIM climatic reconstruction for the Pipe Clay Lagoon assemblage suggests that summer temperatures during the MIS 3–2 transition were quite similar to today, slightly cooler in winter, and substantially wetter (Porch, 2007a).

Probably the most intriguing aspect of the Australian story is the contrast between the beetle and pollen signals during the last glaciation. The beetle records from sites such as Pipe Clay Lagoon and Spring Creek indicate summer temperature regimes and precipitation levels very similar to modern site values. The modern potential vegetation at these sites is forest, but the fossil pollen spectra indicate treeless landscapes. Kershaw *et al.* (2004) have developed scenarios in which extreme winter temperature excursions, low CO_2 levels, or altered seasonality of precipitation are used to explain the dominance of herbaceous vegetation. The temperature and precipitation scenarios are pretty clearly contradicted by the beetle evidence, but both the beetle and pollen evidence indicates open-ground environments where trees are either very scarce or absent altogether. The problem remains unresolved at this point, but the beetle evidence suggests that changes in temperature and precipitation were not the driving force in the system.

1.9. Overview and Prospectus for the Future

Quaternary paleoentomology is off to a promising start in Australia. Porch has been working single-handed thus far, and it will be difficult for him to handle the enormous task of analyzing the fossil insect faunas of an entire continent, even if suitable organic preservation is restricted to the wetter regions of southeastern Australia and Tasmania. It seems that this research will always face certain limitations. The continent has few mountains sufficiently high to support alpine tundra ecosystems. This, combined with the geographic isolation of Australia means that truly cold-adapted insects cannot enter the country to provide a ''cold'' signal in fossil assemblages. Most other regions of the world where Quaternary entomology has been studied have been able to recruit cold-adapted species, either from highland regions or from higher latitudes.

It is clear that more modern data are needed to make Australian paleoentomology succeed. As discussed above, perhaps the majority of modern beetle species remain undescribed. Knowledge of the ecological requirements and modern ranges of the described species is likewise far from complete. These obstacles are bound to be overcome with time, but they remain formidable for the foreseeable future. Porch (2007a) has made the best of the situation through the use of the BIOCLIM program, which allows paleoclimate reconstructions to be made in spite of incomplete modern range data. However, in the long run, accurate, detailed paleoenvironmental reconstructions can only be made on the basis of sound modern data.

Only a handful of sites have thus far been investigated in Australia, although some work that has been done remains unpublished until now. In spite of the caveats discussed above, the work shows substantial promise. As in other parts of the world, the fossil insect story threatens to overturn the previous environmental reconstructions based mainly or solely on pollen analysis.

2. New Zealand

Like Australia, New Zealand formed part of the supercontinent of Gondwana in the Mesozoic Era. New Zealand separated from Gondwana about 85 mya, and has been isolated from other land masses since then (Suggate *et al.*, 1978). The two main islands of New Zealand (North Island and South Island) comprise about 268,000 km^2. This area is equivalent in size to Japan or the British Isles. The New Zealand islands stretch 1,600 km from the subtropical northeast to the cold-temperate southwest, and three-quarters of the land is above 200 m elevation. The geography of New Zealand imposes many barriers on the dispersal of insects and the tracking of ecosystems across the landscape in response to climate change (Marra, 2007). The topography of South Island is dominated by the Southern Alps, a series of mountain ranges that extend the length of the western side of the island, with peaks reaching over 3,000 m. During the last glaciation the Southern Alps were covered by an icecap that extended on the west to the coast in the southern half of the island. On the eastern side of the mountains the ice advanced into inland basins and coastal outwash fans. Although the North and South islands are separated by Cook Strait, a land bridge developed between the two islands during glacial intervals when sea level was lowered. The North Island topography is also quite varied, with a series of north-south trending mountain ranges, active volcanoes, and a high elevation central plateau.

The modern beetle fauna of New Zealand includes 5,223 native species known to science, and 356 introduced species (Klimaszewski and Watt, 1997). The actual number of species may be around 10,000–10,500 species (Klimaszewski and Watt, 1997), so only about half of the fauna has thus far been described. About 90% of the fauna is endemic (Marra, 2007).

Not surprisingly in such an isolated fauna, families that are relatively unimportant on other continents, such as Scydmaenidae and Colydiidae, are fairly prominent in the New Zealand fauna. Conversely, families that are relatively dominant on other continents, such as Carabidae and Chrysomelidae, are far less important in the New Zealand fauna (Table 3). The two most diverse families here are Curculionidae (over 1,500 known species) and Staphylinidae (over 1,000 known species).

One of the more peculiar aspects of New Zealand Pleistocene beetle assemblages is their lack of ground beetles, compared with fossil assemblages in other parts of the world. This is not due to small numbers of species in the New Zealand fauna. The modern beetle fauna contains about 650 species, of which 450 have been described (Larochelle and Lariviere, 2001). Marra (2003a) points to a taphonomic problem as one likely cause. There are no New Zealand ground beetles specifically associated with swamps or peat bogs, although such wetlands are often the source of water-lain fossil material in New Zealand. This situation is in stark contrast with North America and Europe, where many ground beetle species are associated with swampy habitats. This cannot be the only reason for the paucity of Carabidae in New Zealand assemblages, however. Northern Hemisphere Pleistocene assemblages often contain substantial numbers of upland ground beetle species, as well. Marra (2003a) also discussed the problem from the point of view of species' biology. Ground beetle collectors in New Zealand have observed that the fecundity and biomass of New Zealand forest Carabidae are very low. Pitfall trapping yields of ground beetles in the Northern Hemisphere are one to two orders of magnitude larger than in New Zealand, and some New Zealand species produce only a few eggs per year. Thus their diversity on the landscape may be relatively high, but their abundance remains low.

Because the modern fauna is so poorly known, Quaternary fossil work in New Zealand has been difficult to establish. One of the greatest difficulties to overcome has been the incomplete knowledge of ranges and ecological requirements of the modern fauna. Without these critical data, even well-identified fossil assemblages yield poorly defined reconstructions. In response to this situation, Marra *et al.* (2004) developed the MLE method for quantifying past climates from beetle fossils in New Zealand (see discussion of this method in Chapter 5).

2.1. New Zealand Environments: Past and Present

Much of New Zealand lies in the temperate and subantarctic latitudes. This, combined with the great topographic diversity of these islands, has meant that New Zealand has experienced a far greater range of climatic change during the Pleistocene than Australia.

Table 3. Dominant beetle families of New Zealand.

Family	Number of described species
Curculionidae (incl. Scolydidae)	1,542
Staphylinidae (incl. Pselaphidae)	1,021
Carabidae	445
Scydmaenidae	202
Colydiidae	196
Cerambycidae	188
Chrysomelidae	153
Scarabaeidae	144

Source: Data from Klimaszewski and Watt (1997).

Cold-adapted species presumably evolved in highland regions during the Tertiary. These species were able to spread downslope into the lowlands during cold intervals. The presence of alpine beetles in lowland fossil assemblages provides evidence for cold climates much more easily than in Australia, where virtually no alpine habitats have been available. The climate of New Zealand is far wetter, on average, than that of Australia. While only the tropical northern regions and wettest parts of Australia's coastal regions receive more than about 900 mm of precipitation per year, most meteorological stations in New Zealand report annual precipitation in excess of 1,000 mm, and Milford Sound on South Island receives an average of 6,749 mm of precipitation annually (NIWA, 2008). Many mountain regions receive even higher amounts of precipitation, either as rain in the summer or as snow in winter. Some high elevation regions receive 15,000 mm of precipitation per year (NIWA, 2008). Because so much of New Zealand is relatively wet, there are far more lakes, bogs, and other wetlands than in most of Australia, and therefore more water-lain organic deposits containing Pleistocene insect fossils.

Until the arrival of humans, more than three-quarters of New Zealand was forested. The highlands had subalpine grasslands and alpine tundra, dominated by tussock grasses. The principal forest types included temperate rain forest on North Island, dominated by podocarps (Southern Hemisphere conifers in the family Podocarpaceae) and giant kauri (*Agathis australis*, an endemic New Zealand conifer). The cooler climates of South Island supported mostly Southern Beech (*Nothofagus*) forests. Southern Beech expanded in range during some Late Pleistocene intervals, and the pollen record indicates that shrubland and steppe grassland dominated many regions during cooler and drier glacial periods (Heusser and van de Geer, 1994; Mildenhall, 2003), although the fossil beetle record contradicts this reconstruction in some cases (see below).

Lowland forest clearance began in AD 1200–1400 as the Maoris began to modify New Zealand landscapes (Wilmshurst *et al.*, 2004). Additional lowland forests were cleared by European colonists, when they began farming and ranching in the 19th century. It therefore seems likely that many beetle species that were associated with these forests have become extinct, so it may not be possible to identify some fossil specimens, because they have no modern counterparts. There may also be no specimens of these species in museum collection, because systematic insect faunal studies were only just getting underway in 19th century New Zealand, as lowland habitats were being radically altered.

In spite of these hindrances to fossil beetle research, much has been done in New Zealand in the last decade. Fossil insect assemblages ranging in age from MIS 7–2 have been studied, offering tantalizing glimpses into the ancient ecosystems and physical environments of these islands. This research has recently been summarized in an article by Marra (2007). New Zealand Pleistocene fossil sites are listed in Table 1. The site numbers coordinate with the localities shown in Fig. 1.

2.2. Manukau Harbor, Auckland (MIS 7)

The oldest faunal assemblage thus far studied from New Zealand comes from a coastal site at Manukau Harbor, near Auckland (Fig. 1, No. 6). The modern beetle fauna of the Auckland region is particularly well known, because of the Lynfield Survey (Kuschel, 1990), carried out in the 1980s. This survey allowed the ancient faunal assemblage to be interpreted much more fully. The assemblage was taken from organic-rich muds associated with an ancient forest that was buried by volcanic deposits traced to an explosive eruption of the nearby Maungataketake Volcano. Optically stimulated luminescence dating of the volcanic deposits, combined with paleoecological evidence, indicate that the faunal assemblage dates from late MIS 7.

The fauna and associated forest flora clearly represent interglacial environments. All but two of the fossil beetle species are found today in the Auckland region. The fauna includes the weevils *Heterotyles argentatus* (Fig. 2C), *Inosomus rufopiceus* (Fig. 2D), and *Psepholax sulcatus* (Fig. 2G). Fossil preservation at this site is exceptionally good, because of the highly unusual mode of fossil burial. The forest floor was almost instantaneously buried by the explosive volcanic eruption, sealing the sediments and preventing normal decomposition (Marra *et al.*, 2006b). Thus this fauna represents a sort of time capsule of the latter part of the MIS 7 interglacial in northern New Zealand.

About 40% of the fossil beetle assemblage is made up of taxa associated with trees, including beetles found on

Fig. 2. Light microscope photographs of fossil beetle specimens from New Zealand Pleistocene assemblages. (A) Head and right elytron of Bembidion rotundicolle *(Carabidae); (B) leg of* Ectopsis ferugalis *(Curculionidae); (C) head of* Heterotyles argentatus *(Curculionidae); (D) right elytron of* Inosomus rufopiceus *(Curculionidae); (E) right elytron of* Platypus caviceps *(Platypodidae); (F) right elytron of* Psepholax coronatus *(Curculionidae); (G) head of* Psepholax sulcatus *(Curculionidae). Scale bars equal 1 mm (photos by Maureen Marra, used with permission).*

the forest floor, the forest canopy, and on or under bark. The most abundant species associated with forest habitats is the ground beetle *Tachys antarcticus*. This species lives in the soils of giant kauri and podocarp forests (Larochelle and Lariviere, 2001). Weevils dominate the tree-feeding fauna. These include *Inosomus rufopiceus* that inhabits giant kauri and the podocarp *Dacrydium cupressinum*, *Heterotyles argentatus* that feeds on various podocarps, *Xenocnema spinipes* that inhabits either giant kauri or *Dacrydium cupressinum*, *Eurynotia hochstetteri* that lives in *Podocarpus totara*, *Macroscytalus parvicornus* that feeds on tree ferns, and *Unas conirostris* which is associated with *Vitex lucens* (the Puriri tree) (Kuschel, 1990). Forest floor habitats are indicated by the presence of several weevils in the tribe Cryptorhynchini that inhabit rotting logs, and several species of Zopheridae including *Pycnomerus perigrinus* that lives in dead or decaying giant kauri trees. A small wetland component in the fauna includes the weevil species *Steriphus acitus* that lives on swamp vegetation (e.g., *Scirpus*) and *Eucossonus setiger* that feeds on the swamp tree *Cordyline* (Kuschel, 1990). The most abundant taxon in the wetland fauna is the pine flower weevil (Nemonychidae) *Rhinorhynchus rufulus*. This species is mostly associated with the swamp forest conifer *Dacrycarpus dacrydioides* (the Kahikatea tree) (Kuschel, 2003).

In spite of the similarities of this fossil record to the modern flora and fauna, both the fossil beetle and pollen assemblages include taxa that are found today at higher elevations in the region, indicating that the temperature regime during late MIS 7 was slightly cooler than today. Marra *et al.* (2006a) estimated that temperatures were less than 1°C cooler than today.

2.3. Banks Peninsula (MIS 6)

A sediment core taken in Gebbies Valley, Banks Peninsula, South Island (Fig. 1, No. 7), yielded a peaty organic unit just below 40 m depth in the core record. This 40-cm thick unit is associated with a TL age of 136,000 ± 10,000 yr BP, placing it toward the end of MIS 6 (Shulmeister *et al.*, 1999). The sample yielded a small beetle fauna of 19 taxa (Marra, 2003b). As with the Auckland region, the modern beetle fauna of Banks Peninsula is relatively well known, because of published regional studies, such as Johns (1986). These inventories provide a basis for comparison with a fossil assemblage from Gebbies Valley. The fossil assemblage is mainly a montane forest fauna, although the site location is close to sea level. Based on the modern ranges of the fossil fauna, Marra (2003b) interpreted the climate as interstadial rather than pleniglacial. The fauna could potentially be found today living at an elevation of 450 m asl. By using a simple environmental lapse rate of 0.55°C per 100 m, Marra estimated a temperature depression of 2–3°C from the current climate.

About 80% of this fauna are forest taxa, including inhabitants of forest floor environments and tree-feeders. The fauna is dominated by weevils (12 taxa), and also contains a pill beetle (Byrrhidae-*Epichorius*), an anobiid (*Sphinditeles*), the rove beetle *Bledius*, and two unidentified ground beetles. Only three of the species on the fossil list are found on Banks Peninsula today. The region's forests have suffered considerable destruction over the past 150 years (Speight *et al.*, 1927), but Marra (2003b) attributes the faunal differences to changes in climate. The site was coastal in MIS 6, but the fauna lived in a climatic regime associated today with montane forests.

2.4. Palliser Bay (MIS 5)

At Ocean Beach, Pallister Bay, on the southwest margin of the Wairarapa Basin, Wharekauhau Stream dissects a marine terrace, exposing about 50 m of marine, fluvial, lacustrine, and estuarine deposits. This site (Fig. 1, No. 8) yielded about 80 kg of organic-rich sediments from two of the laminated silt organic deposits (Marra, 2003b). The lower sample bed, about 12.3 m asl, contains abundant plant macrofossils, including fossil tree fern stems, leaves, and logs and tree stumps in growth position. The upper sample, taken from 21.8 m asl, was excavated from two bands of organic-rich sediment.

A total of 64 species of beetles from 15 families were identified by Marra (2003b). The taxa represent a range of environments including lake margins, swamp forest, and forest. The lower sample yielded 54 taxa, dominated by weevils and rove beetles. The upper sample yielded just 26 identified taxa, dominated by over 600 individuals of the weevil *Macroscytalus parvicornis* that feeds on tree ferns (Kuschel, 1990). The fossil beetle assemblages from both samples indicate ecological succession from swamp forest to a forest environment rich in tree ferns, New Zealand flax (*Phormium*), and swamp forest trees. This swamp forest environment is typical of New Zealand interglacial lowlands. Historical accounts of the study region document the existence of this biological community, prior to European landscape modification.

The lower sample contains at least eight species whose modern range is limited to northern New Zealand, where the climate is significantly warmer. Their presence in the fossil assemblage indicates a southward displacement from their modern ranges of about 700 km during MIS 5e. Modern summer (January) temperatures are 18.8–19.7°C and winter (July) temperatures are 10.4–12.4°C for northern New Zealand locations. At the study site, the modern mean January temperature is 17.2°C and the mean July temperature is 8.1°C. This indicates that the height of the last interglacial was 1.5–2.5°C warmer in the summer (January) and 2–3°C warmer in the winter (July) than it is today. In contrast to this, the upper (younger) faunal assemblage does not contain these more warm-adapted species. This assemblage may represent a decline from the peak of interglacial warming after 125,000 yr BP, but it may also simply reflect local environmental change (Marra, 2003b).

2.5. Westport (MIS 3)

A series of 18 fossil beetle assemblages have been described from a site near Westport on the western coast of South Island (Fig. 1, No. 9) by Burge and Shulmeister (2007). The samples were taken from an organic silt deposit in an in-filled dune hollow. Radiocarbon ages associated with the samples indicate that they accumulated from about 37,000 to 21,300 cal yr BP. The samples yielded 76 taxa in 12 families. Four principal ecological groups are represented in the faunal assemblages (Fig. 3): aquatic and riparian taxa (labeled ''A/R'' in Fig. 3), closed canopy forest taxa (labeled ''Closed'' in Fig. 3), shrubland/forest margin taxa (labeled ''Shrub'' in Fig. 3), and open tussock and herb field taxa (labeled ''Open'' in Fig. 3). As can be seen in the figure, beetles associated with closed canopy forest dominate all assemblages, although they are most dominant in the three oldest assemblages, labeled as beetle zone B1 in the figure. This zone corresponds to the interval 37,000–34,000 cal yr BP. The closed canopy fauna includes the rove beetles *Brachynopus scutellaris*, *Vidamus*, and *Eupines*. Other closed canopy taxa include the scarab *Saphobius edwardsi* and the zopherid *Pycnomerus latitans*, as well as several unidentified species of weevils in the tribe Cryptorhynchini. Another weevil in these assemblages is *Arecocryptus*, whose larvae feed exclusively on palms.

Beetle zone B2a is characterized by faunal assemblages with more open-ground and/or shrubland species, indicating an opening up of local forest stands during the interval 34,000–21,300 cal yr BP. The presence of terrestrial hydrophilids such as *Tormus nitidulus* and *Adolopus hemlsi* indicate that the forest floor remained moist during this interval. Tussock vegetation is indicated by the presence of the ground beetle *Notogonium feredayi*. A stone beetle (Scydmaenidae) *Adrastia nelsoni* was found in these assemblages. This beetle lives today in alpine tundra habitats above 1,200 m elevation. Another highland open-ground beetle found in these assemblages is the rove beetle *Aleochara hammondi*, which lives today in tussock grasslands between 850 and 1,500 m elevation (Klimaszewski and Crosby, 1997). Several weevils associated with open-ground habitats were identified from these assemblages, including *Irenimus* and *Oreocalus latipennis*. The larvae of the latter species feed

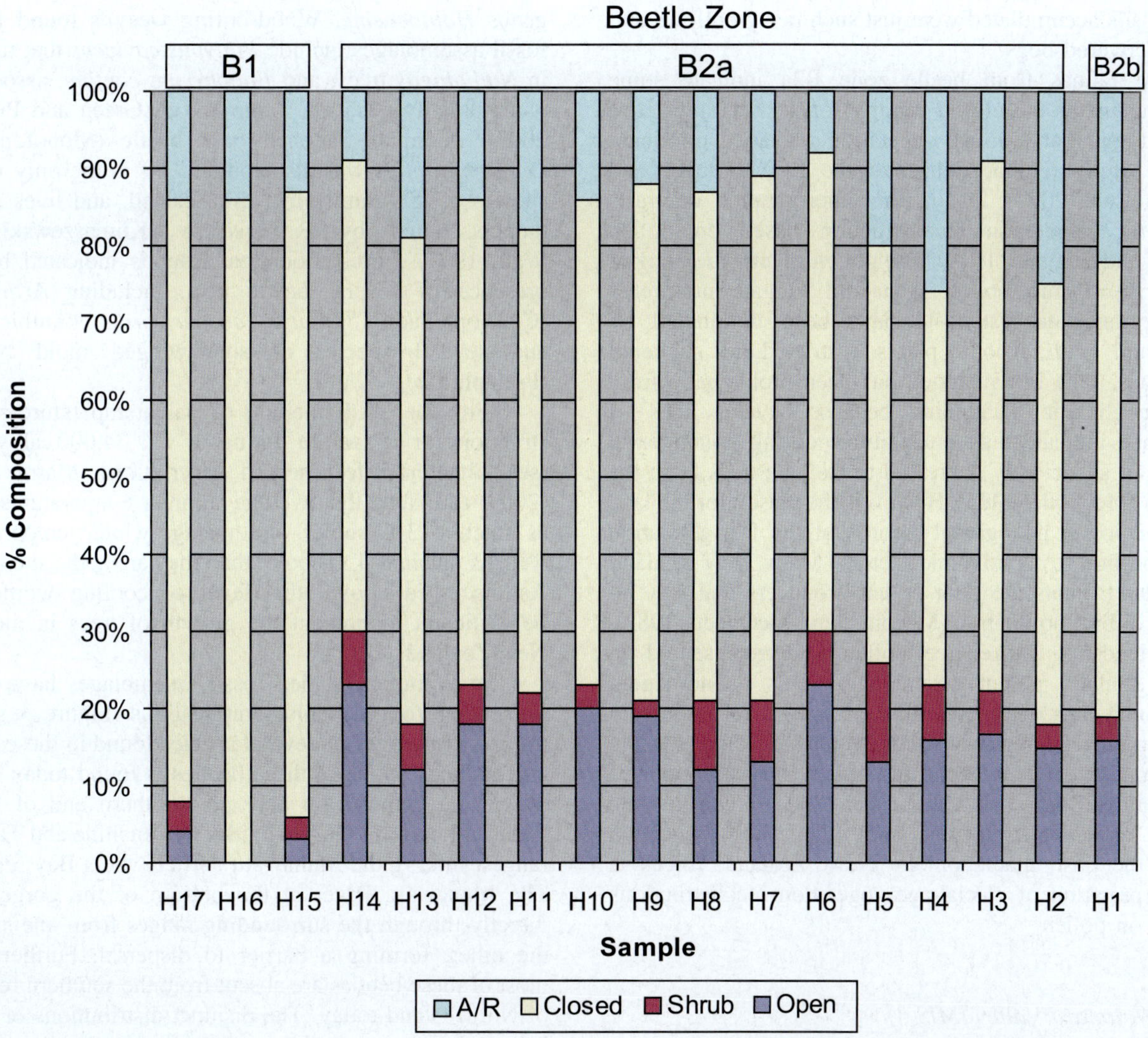

Fig. 3. Percent composition of four ecological groups of beetles in faunal assemblages from MIS 3–2 at the Westport site, New Zealand (after Burge and Shulmeister, 2007). A/R, aquatic and riparian taxa; Closed, closed canopy forest taxa; Shrub, shrubland/forest margin taxa; Open, open tussock and herb field taxa.

exclusively on the plant genus *Hebe*, many species of which are endemic to New Zealand (May, 1993). *Hebe* is typically found along forest margins and in open habitats.

The uppermost assemblage (H1) was separated into its own beetle zone (B2b) based on the presence of the beach beetle (Phycosecidae) *Phycosecis limbata*, an inhabitant of sandy environments (Klimaszewski and Watt, 1997). Otherwise this assemblage is essentially like many of the assemblages in zone B2a (Burge and Shulmeister, 2007).

Interstadial conditions are associated with beetle zone B1, and this climatic interpretation agrees with that of palynologists working on samples from this time interval and other regional sites (Moar and Suggate, 1979). However, the beetle evidence for closed canopy coastal vegetation is at odds with the vegetation cover reconstruction from the pollen, which suggests wet peaty hollows and shallow ponds in an open shrub-grassland setting. Burge and Shulmeister (2007) argue that the pollen rain into small shrub-covered bogs and ponds is dominated by pollen from the immediate vicinity, rather than by the regional pollen signal. Both the pollen and beetle evidence indicate that the environments in which the fossils accumulated were just such peaty hollows and shrub-covered bogs.

The faunas from beetle zone B2a indicate some opening up of the closed canopy forest, but they leave little doubt that arboreal vegetation remained dominant in the study region, right into the LGM (Burge and Shulmeister, 2007). Again, this interpretation contrasts with the vegetation reconstruction based on pollen (Moar and Suggate, 1979). The pollen spectra are marked by a rise in grass pollen. Moar and Suggate interpreted the regional ecosystem to have been dominated by grassland with woody plants restricted to sheltered locations. This interpretation has been problematic from a paleoclimatic standpoint, because several lines of evidence indicate that temperatures during this interval were not sufficiently depressed to exclude trees from the region (McGlone, 1985). However, the reason for the lack of tree pollen in regional records of the last glaciation may be due to taphonomic factors. Many New Zealand woodland plants are poor pollen producers that rely on insect or bird pollination (Macphail and McQueen, 1983). This tends to make tree pollen underrepresented in New Zealand pollen assemblages. This phenomenon, combined with overrepresentation of wind-borne herbaceous pollen in regional records, may account for the discrepancy between beetle and pollen reconstructions of vegetation. Burge and Shulmeister (2007) argue that the beetle evidence from this and other sites associated with the last glaciation in New Zealand forces a reinterpretation of glacial-age vegetation reconstructions based on pollen.

2.6. *Waitotara Valley (MIS 3)*

Marra *et al.* (2007, 2009) have described a series of six insect fossil assemblages from organic-rich silts exposed in a landslide block in the Waitotara Valley in the Taranaki region of southwestern North Island (Fig. 1, No. 11). Radiocarbon ages indicate that these assemblages come from the interval 34,400–33,500 cal yr BP, so these faunas are contemporaneous with some of the older faunal assemblages from the Westport site on South Island, discussed above. The Waitotara faunal assemblages likewise indicate the presence of forest, specifically of Southern Beach (*Nothofagus*) forest in this region. The pollen models (McGlone *et al.*, 1984; McGlone, 1985, 1988; Wardle, 1988) invoke several coastal regions as the most likely refugia. for New Zealand forest trees during glacial intervals. The beetle evidence is beginning to show that lowland forests were much more widespread than has been shown in the pollen record. There are, however, interesting biogeographic gaps in New Zealand, representing dispersal barriers. One of these gaps, called "The Northern Gap," is located in the southern part of North Island. As discussed below, this gap affected insect species' distributions in the Pleistocene.

The beetle assemblages from this site are indicative of a variety of habitats, including forest, shrubland, marsh, and aquatic environments. Beetles associated with forest canopy include the ladybird (Coccinellidae) *Rhyzobius consors*, the leaf beetle *Alema paradoxa*, and the weevil genus *Hoplocneme*. Wood-boring weevils found in the fossil assemblages include *Platypus caviceps* that tunnels in *Nothofagus* trees, and *Euophryum confine*, associated with fungally decayed, damp wood (Green and Pitman, 2003). A small flattened bark beetle (Monotomidae) *Lenax mirandus* was also found. This is the only native species of this family in New Zealand, and lives in the tunnels bored by *P. caviceps* (Klimaszewski and Watt, 1997). Forest floor leaf litter is indicated by the presence of several beetle taxa, including *Arthrolips* (Corylophidae), *Sapintus obscuricornis* (Anthicidae), and several species of short-winged mold beetles (Pselaphidae).

Using the MLE method of paleotemperature reconstruction for a sample dating to ca. 34,000 cal yr BP (see Chapter 5 for method description), Marra *et al.* (2009) estimated that average summer temperatures were as much as 3°C cooler, and average winter temperatures were as much as 4°C cooler than they are at the site today. As discussed above, this level of cooling would not be sufficient to prevent the growth of trees in most of New Zealand.

The Waitotara Valley fossil assemblages have some interesting zoogeographic implications. Figure 4 shows the modern ranges of several species found in these fossil assemblages. None of these beetles is found today in the Manawatu Gap region near the southern end of North Island. Manawatu Gorge divides the Ruahine and Tararua ranges, linking the Manawatu and Hawke's Bay regions. The Manawatu River at the bottom of the gorge runs directly through the surrounding ranges from one side to the other, forming a barrier to dispersal. Furthermore, most of these beetles are absent from the southern regions of North Island today. The disjunct distributions of these insects follows a similar pattern seen in the vegetation, which inspired McGlone (1985) and Wardle (1988) to describe a Northern Gap boundary across southern North Island. Some of the species in the Waitotara fauna

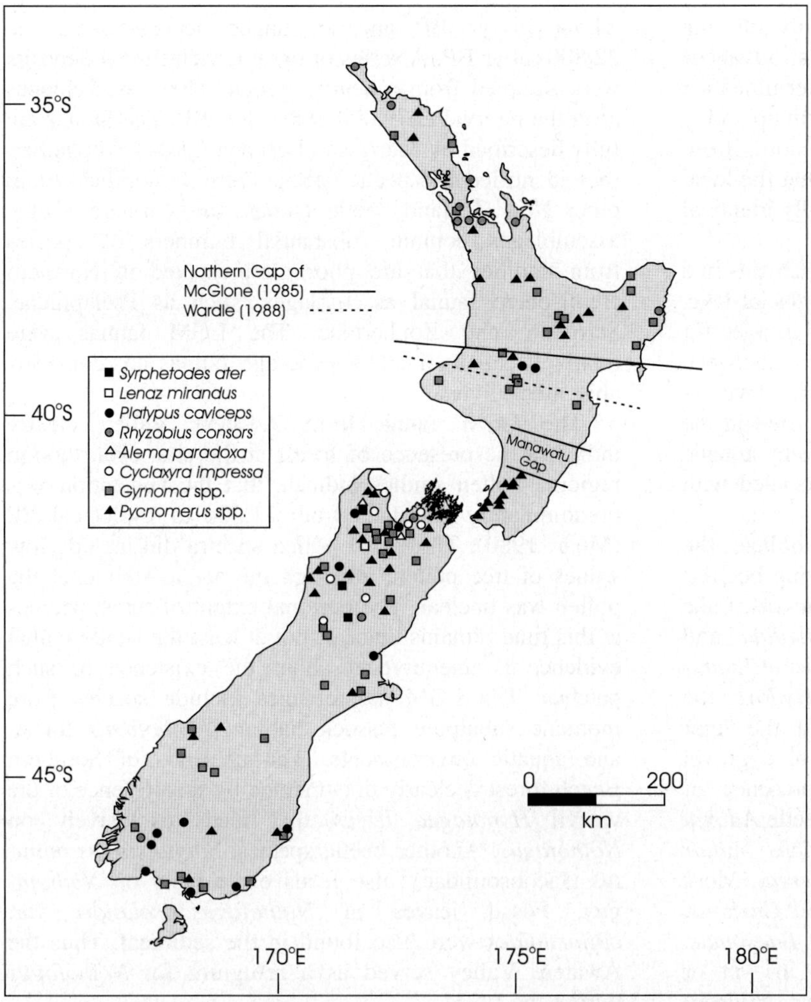

Fig. 4. Map of New Zealand, showing the modern ranges of several species found in the fossil assemblages from the Waitotara Valley. Note that none of these beetles is found today in the Manawatu Gap region near the southern end of North Island. Also shown are proposed Northern Gap boundaries of McGlone (1985) and Wardle (1988) across southern North Island (see text for discussion) (figure after Marra et al., 2009).

live today only on South Island, such as *Syrphetodes ater* and *Cyclaxyra impressa*. McGlone's (1985) explanation for the gaps in species distribution across New Zealand centers on tectonically driven landscape evolution, beginning in the Oligocene. Wardle's (1988) hypothesis focuses on Pleistocene glacial refugia for plants, and he proposed that extreme environmental stress during the last glaciation was the likely cause of plant extinctions in the ''gap.'' The fossil beetle evidence suggests that their disjunctive distribution patterns formed *after* MIS 3, so the beetle evidence tends to nullify McGlone's hypothesis.

It is clear from the fossil record that *Nothofagus* woodland dominated the Waitotara Valley region during MIS 3. Marra *et al.* (2009) hypothesize that this woodland may have been an extension of *Nothofagus* forests that were prevalent in the Nelson region of South Island at that time. Until recently it was thought that a land bridge between North and South islands did not exist during MIS 3, but a recent study by Thompson and Goldstein (2006) indicates that global sea level was sufficiently low (at least 93 m lower than today) at 34,000 cal yr BP to form such a land bridge, or to considerably narrow the gap between the two islands. Marra *et al.* (2009) envision the emergent landscape clothed in Southern Beech forest, acting as a biological passageway for the flora and fauna associated with this kind of forest community.

2.7. Howard Valley (MIS 3)

The Howard Valley site is in the Nelson Lakes district of northern South Island (Fig. 1, No. 10). This site is being studied by Marra and Thackray. Preliminary results are discussed in Marra (2008). Faunal assemblages range in age from 37,000 to 21,000 cal yr BP. As in the Lyndon Stream and Awatere Valley sites discussed below, fossil beetles and fossil leaves from Howard Valley document the presence of *Nothofagus* forest at this site. The presence of forest at this site is doubly surprising, as the site lies 600 m asl, and at the time of fossil deposition the site was situated between two mountain glaciers in the adjacent Rotoiti and Rotora valleys (Marra, 2008).

2.8. Lyndon Stream (LGM)

The Lyndon Stream site is on the eastern side of South Island (Fig. 1, No. 13), at an elevation of about 800 m asl, in the foothills of the central Southern Alps. The stream has down-cut and exposed about 14 m of glacial till, outwash gravels, and lake silts. This is the only site that has yielded a radiometric age on LGM glacial advances east of the Southern Alps. The LGM glacial terminus for the Rakaia Valley glacial system lies about 2 km up valley from the site. Pollen and plant macrofossil records from the site (Soons and Burrows, 1978) indicate that the local vegetation regime during the LGM was virtually identical to the modern one.

Fossil insects were extracted from organic-rich silts in a 6-m sequence. The silts were deposited in a glacial lake. Plant macrofossils have yielded two radiocarbon ages for the deposit, indicating that organic deposits formed between 27,500 and 22,900 cal yr BP (Marra *et al.*, 2006b). Twenty-nine insect taxa and two arachnids were identified in the assemblage. The fossil assemblage is dominated by aquatic and riparian taxa; it also contains beetles associated with tussock, shrub, and forest habitats.

Not surprisingly in a lake sediment assemblage, the most abundant beetle taxa are predaceous diving beetles: *Antiporus strigosulus* and two species of *Liodessus*. Lake margin taxa include the rove beetles *Bledius* and *Stenomalium*, and the minute marsh-loving beetle *Limnichus simplex*. A species of the riffle beetle *Hydora*, the ground beetle *Actenonyx bemibidioides* and the tiger beetle *Cicindela feridayi* are all indicative of a gravel stream with sandy stream margins. The presence of tussock vegetation is inferred from the leaf beetle *Adoxia dilutipes*, the antlike flower beetle *Anthicus minor* (Anthicidae), and the weevil *Niceana cinerea*. Moss feeders in the assemblage include the byrrhid *Curimus*, found in subalpine moss, and the weevil *Baesomus*. The leaf beetle genus *Phyllocharitin* lives in wet or saturated mosses and/or liverworts in forest environments. More specifically, it is found in "cloud forest" habitats, often close to altitudinal tree line. Damp habitat is also indicated by the weevil *Rhopalomerus tenuirostris*, the larvae of which are found in damp, well-decayed wood.

The fossil beetle reconstruction disagrees with the pollen reconstruction of a treeless, dry landscape (Soons and Burrows, 1978). Tree pollen was found in these sediments, but it was interpreted as having come from a distant source. However, the fossil beetle evidence indicates close proximity to damp forest environments. Interestingly, the paleotemperature reconstruction for this fauna indicates that LGM temperatures were only around 1.5–2°C cooler than present day; this is a temperature regime that would certainly be suitable for forest growth. As with several other New Zealand sites discussed here, it appears that the lack of tree pollen in the sediments cannot be taken as an indication of the lack of trees on the landscape. A fundamental scientific philosophy is that the absence of evidence is not evidence of absence.

2.9. *Awatere Valley (LGM through Holocene)*

A series of organic deposits exposed along the banks of Nina Stream in the Awatere Valley yielded fossil insect assemblages from the LGM and the Holocene (Fig. 1, No. 12). The LGM faunas were sampled from two organic-rich alluvial deposits: a lower deposit dated ca. 24,500 cal yr BP and an upper deposit dated ca. 22,800 cal yr BP. A series of organic-rich fluvial deposits were sampled from a nearby trench. These assemblages span the interval ca. 9,600–900 cal yr BP. The faunas are fully described by Marra and Leschen (2004). Altogether, they identified 145 beetle species from 33 families. As in other New Zealand beetle faunas, the Awatere Valley assemblages contain substantial numbers of species from families that are poorly represented in Northern Hemisphere faunal assemblages, such as Pselaphidae, Scirtidae, and Zopheridae. The LGM faunas were relatively small; the Holocene-age faunas are considerably more diverse.

The LGM fauna from Awatere Valley clearly indicates the presence of forest at the site, even though regional pollen studies indicate that the vegetation was predominantly grassland from 31,000 to 14,000 cal BP (Moar, 1980). The LGM pollen spectra did include low values of tree pollen, although the actual source of the pollen was unclear. The regional extent of forest patches at this time remains unclear, but at least the beetle faunal evidence is unequivocal about the existence of such patches. The LGM assemblages include species from montane subalpine tussock habitat, *Nothofagus* forest, and aquatic environments. The presence of Southern Beech forest is clearly documented by the presence of the weevil *Hypotagea lewisi* that feeds exclusively on *Nothofagus*. Another beetle species, *Saphophagus minutus* (Jacobsoniidae), also feeds exclusively on *Nothofagus*. Fossil leaves of *Nothofagus solandri* var. *cliffortioides* were also found in the sediment. Thus the Awatere Valley served as a refugium for *Nothofagus* during the LGM.

In contrast to the results from the Lyndon Stream site, about 300 km to the south, the MLE climate reconstruction for the Awatere Valley shows a much colder climate during the LGM. For the LGM faunas, Marra *et al.* (2004) used the MCR method to estimate that mean February (summer) temperatures were ca. 3.5–4°C cooler, and mean July (winter) daily minimum temperatures were ca. 4–5°C cooler than today. Using the MLE method, they estimated that mean February (summer) temperatures were ca. 2.5–5°C cooler, and that mean July (winter) daily minimum temperatures were 3.5–6°C cooler than today. The apparent conflict between the reconstructions from Lyndon Stream and Awatere Valley needs to be resolved. The two faunal assemblages appear to be of the same age, but tighter age control may reveal that these faunas reflect two distinct episodes within an era of rapid climate change. Marra (2007) also considered another possibility for this discrepancy, that the Lyndon Stream site yielded a warmer temperature reconstruction because the site is sheltered from cold southerly and easterly winds.

The Holocene-age faunas from the Awatere Valley trench site were divided by Marra and Leschen into three principal zones. The oldest of these zones, representing the interval from ca. 9,600 to 6,900 cal yr BP, is characterized by species found today in *Nothofagus* forests, and swamp and pond environments. This fauna includes the weevil *Ectopsis ferugalis* (Fig. 2B) and *Psepholax coronatus* (Fig. 4F). A middle zone, dating from ca. 6,700 to 900 cal yr BP, contains species indicative of podocarp forest and aquatic environments. This fauna includes the ground beetle *Bembidion rotundicolle* (Fig. 2A). The youngest zone represents assemblages from the last 500 years. This is a depauperate fauna, characterized by open-ground species, associated with human clearance of the forest. All of these Holocene faunas yielded paleotemperature estimates close to, or at modern values (Marra *et al.*, 2004).

2.10. *Overview and Prospectus*

As in Australia, Quaternary entomology in New Zealand has gotten off to a very promising start in the last decade. However, another shared trait between the countries is the scarcity of Quaternary entomologists. All of the work in Australia has been done by just one researcher (Nick Porch). There are now two trained paleoentomologists in New Zealand (Philip Burge and Maureen Marra), but both situations are precarious at best. If any of these people were to give up doing this kind of research, it might take decades before another student received the kind of extensive training necessary to replace them. Hopefully the trend will work in the other direction, with the existing researchers training up new students. There is certainly plenty of work to do in both countries.

One of the areas crying out for more research is modern data collection, in concert with the fossil work. As Marra (2008) wrote, ''The greatest challenge for fossil beetle research in New Zealand is the incomplete knowledge of the modern fauna.'' Perhaps only half of the modern New Zealand beetle fauna is known to science. The situation is even direr in Australia, where perhaps only one-quarter of the modern fauna has been described (Lawrence and Britton, 1994). Because of this, many beetle specimens in fossil assemblages from both countries cannot now be identified to the species level. Likewise, incomplete distribution and habitat data for many species hampers their use as paleoenvironmental indicators. It does not help that fundamental entomological research is considered ''19th century science'' by many researchers and funding agency officials. It is therefore increasingly difficult to obtain funding for such activities as collecting expeditions into remote regions, followed by months and years of taxonomic study. As noted by taxonomists Smith and Klopper (2002), ''the funding of taxonomy has been seen to be rather peripheral to the environmental priorities of many countries.'' Yet there has never been a time when we needed this kind of information more than we do now, in the face of global change and vanishing habitats. Governments urgently planning conservation strategies require just this kind of information in order to make sensible plans. Hopefully national and international conservation programs will recognize the importance of building taxonomic, ecological, and distributional databases for all regions, especially the relatively poorly studied Southern Hemisphere.

I believe that the fossil work must continue while the modern taxonomic and ecological research is carried out. It would be a mistake to delay the fossil research, because it will certainly take decades for the modern collecting work to be accomplished, and we cannot afford for the fossil work to fall below the ''radar screen'' of Quaternary science in these regions for any length of time.

3. Postscript: Initial Results from Pacific Islands

Nick Porch has recently begun working on Holocene insect fossil assemblages from Pacific islands, including the Galapagos, Hawaii, and the Austral islands of French Polynesia. He has already made several important discoveries, as discussed below. These islands contain water-lain organic deposits in which insect remains are excellently preserved (Porch, 2007b).

Porch *et al.* (2007) discussed the loss of insect species diversity in Pacific island lowland ecosystems since the advent of human colonization. Oceanic islands are particularly susceptible to large-scale ecological impacts brought on by human colonization. While it is safe to assume that insects are not immune to these effects, there have been no previous analyses of Holocene insect assemblages from Pacific islands that document the history of human impacts on Pacific island biota. Important questions relating to the evolutionary, ecological, biogeographic, and archeological significance of the biota can only be answered through paleontological research. Preliminary results from sites in lowland Hawaii, the Galapagos Islands, and Rimatara (French Polynesia) (Fig. 5) demonstrate the probable extinction of many species of lowland beetles, and the unintended introduction of species from mainland regions, following human arrival.

On the Hawaiian island of Kauai, the Makauwahi Cave site has yielded a sequence of organic deposits that contain insects and plant remains dating from before Polynesian arrival through recent times. Preliminary results of this study were presented by Porch *et al.* (2007). Two genera of beetles, *Blackburnia* (Carabidae) and *Rhyncogonus* (Curculionidae), are represented by at least four and five species respectively. All of these species are probably extinct today. The *Blackburnia* species in the fossil assemblages are morphologically similar to a group known from regions above 650 m on Kauai today. *Rhyncogonus* is only known today above 300–400 m elevation. The fossil *Rhycogonus* remains include one species that is larger than any of the 30+ described species of *Rhyncogonus* found in the highlands of the Hawaiian islands today. Another probable extinction is a species of stag beetle that must have been quite common in the Kauai lowlands before people arrived. It is represented by more than 15 mandibles from a relatively small amount of sediment. It is probably related to the only known Hawaiian lucanid, *Apterocyclus honoluluensis*. However, this name represents an aggregate of species described in the late 19th and early 20th centuries. Several, probably most, of these taxa have become extinct in historic times. Lucanids are found today only in highland regions above 800–900 m. Two water beetle taxa, *Rhantus* (Dytiscidae) and *Limnoxenus* (Hydrophilidae), were also found in the lowland fossil assemblages

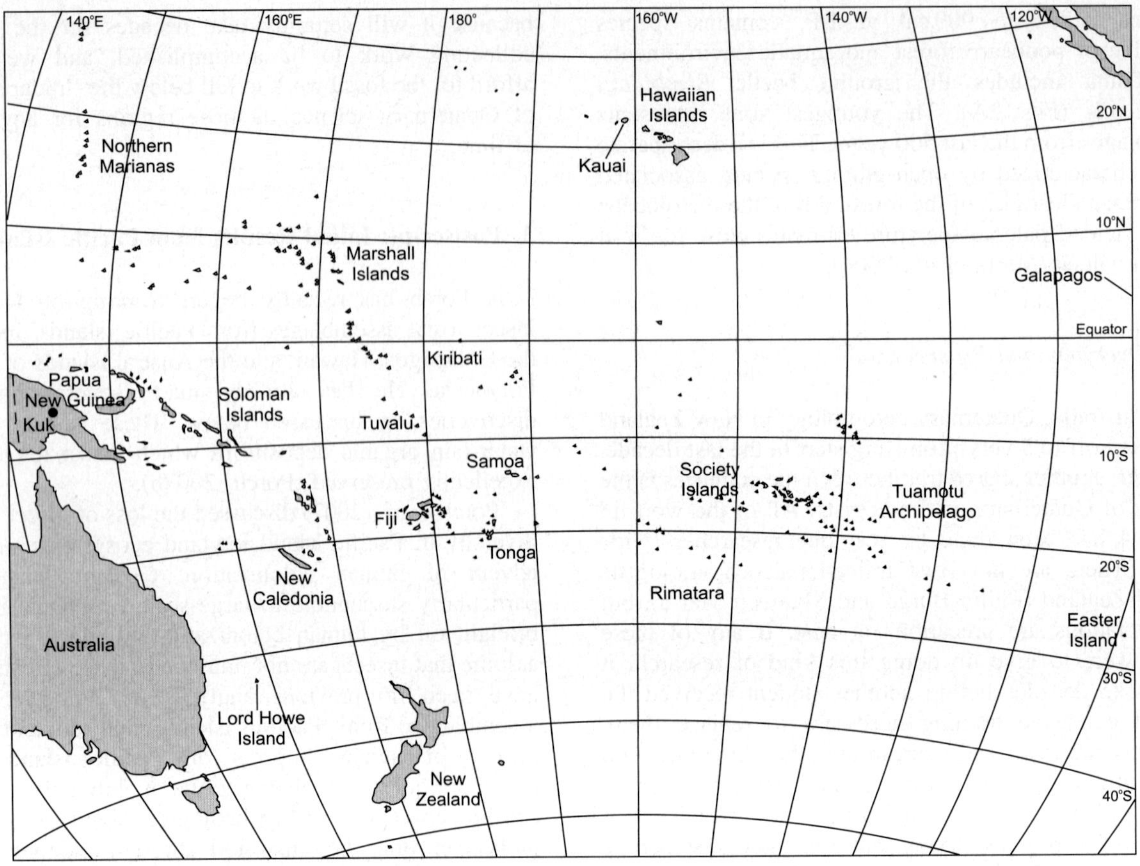

Fig. 5. Map of the Pacific region, showing localities discussed in text (after Porch et al., 2007).

predating human occupation of Kauai. These water beetles are also restricted today to altitudes above 1,000 m elevation.

The Polynesian zone is represented by a layer that dates to about AD 1650. The fauna of this period is markedly different to the indigenous lowland fauna. Many taxa are missing including the lucanid, the large *Rhyncogonus* and *Blackburnia*. However, several other taxa persisted at the site, including *Limnoxenus* and smaller species of *Rhyncogonus*. A range of taxa appear to have been introduced on Kauai by the Polynesians, including cockroaches, some ants, bugs, and a range of beetles. With European colonization most of the remaining indigenous taxa disappeared from the lowland fossil record at Makauwahi Cave, and an increasingly large number of invasive species became established.

The taxa that disappear with the arrival of Polynesians are large, flightless, and probably ground dwelling. These beetles likely fell prey to the Pacific rat (*Rattus exulans*), an accidental introduction that accompanied the Polynesians on their voyages across the Pacific. The extinction of large, flightless beetles on other Pacific islands has also been attributed to the introduction of rats on islands. Reid (2007) recently presented evidence for the extinction of 11 species of beetle previously found only on Lord Howe Island, off the east coast of Australia (Fig. 5). Since 1916 there have been no sightings of these endemic beetle species, despite significant sampling of the island's invertebrate fauna over the years. The culprit responsible for the disappearance of these species is thought to be the black rat (*Rattus rattus*). These rats were accidentally introduced to Lord Howe Island when a supply ship was wrecked off the coast in 1918. Rats are known to eat flightless, ground-dwelling insects on other islands around the world.

Preliminary research on pollen, plant remains, and insect fossils from Maunutu Swamp on the island of Rimatara, French Polynesia, documents large-scale changes in vegetation, accompanied by changes in insect fauna (Porch *et al.*, 2007). Before the arrival of Polynesian colonists, the pollen spectra from Maunutu Swamp sediments were dominated by the mangrove swamp plant, *Achrostichium aureum*, and by screw pine (*Pandanus tectorius*), a coniferous tree that grows throughout the tropical Pacific region (Fig. 6). The Polynesian era pollen spectra show a rapid decline of *Achrostichium* and *Pandanus*, combined with the introduction of taro (*Colocasia esculenta*) and monolete psilate fern spores. The high values of fern spores, combined with a rise in charcoal, and the decline in tree pollen suggests deforestation and other landscape disturbance, with the development of extensive areas of fern land. Horrocks and Weisler (2006) found a very similar pattern of vegetational change with Polynesian occupation and landscape alteration in the Marshall Islands and southeast Polynesia (Pitcairn's Island). In addition to the botanical evidence for Polynesian colonization of the island, these settlers accidentally brought ants with them. There were no ants on Rimatara prior to the arrival of Polynesians. These included *Pheidole fervens*, *Paratrechina*, and *Tetramorium*. The arrival of Europeans is marked by the establishment of *Commelina diffusa*,

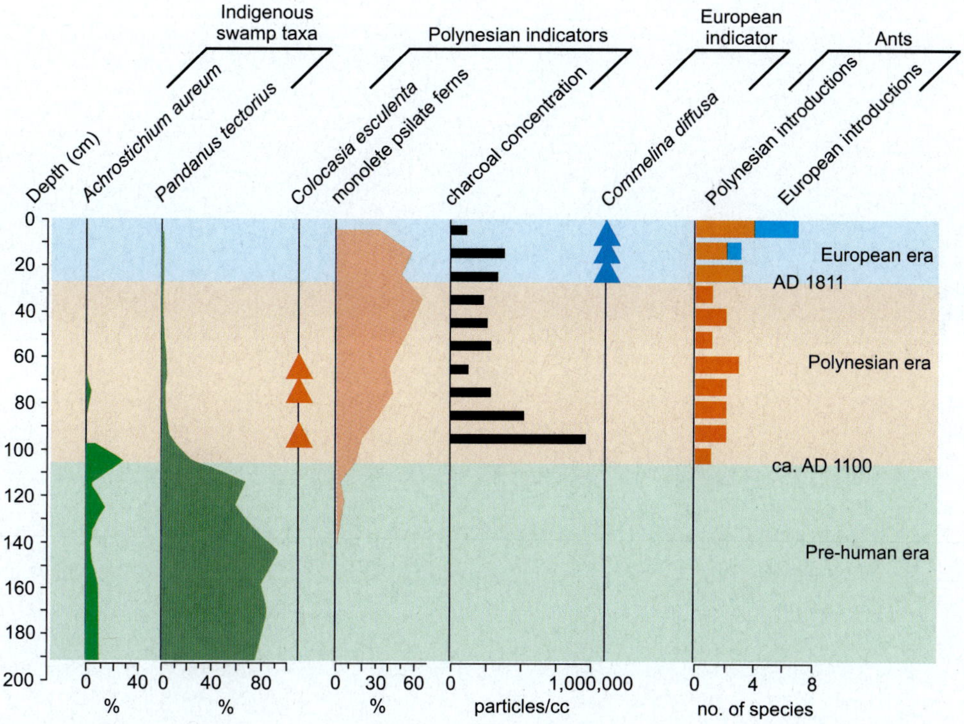

Fig. 6. Summary diagram illustrating changes in percentage pollen spectra, plant macrofossils, and insect fossils at the Maunutu Swamp site, Rimatara, French Polynesia (after Porch et al., 2007).

or spreading dayflower. This plant is indigenous to the Caribbean region, and was introduced by the French in their Polynesian colonies (Welsh, 1998). The arrival of French colonists was also marked by the introduction of European ant species.

Porch has also begun studies of Holocene insect fossil assemblages from other Pacific islands, including the Galapagos Islands and Easter Island; he has also begun work on fossil assemblages from the Island of Mauritius in the western Indian Ocean, where in addition to the Dodo (*Raphus cucullatus*), many species of beetles have also become extinct in historic times. These are exciting, new investigations that will undoubtedly expand our knowledge of the history of tropical island faunas in the near future. Because of the fragility of isolated island ecosystems, many species of insects have become extinct since the arrival of human populations. These extinction events are likely to become more common in continental regions as well, with the increasing destruction and fragmentation of natural habitats and other forms of anthropogenic environmental change. As insect and other wildlife populations are increasingly isolated on habitat ''islands,'' and those ''islands'' grow smaller and further apart, the sad history of tropical island faunas may well be played out again on mainland regions throughout the developed world.

14

Beetle Chitin Isotope Studies

The results of these recent studies show the great potential of beetle chitin isotope geochemistry in the reconstruction of past precipitation and temperature.

– Scott Elias (2006)

In the concluding chapter of my 1994 book on Quaternary entomology, I suggested that the development of isotopic studies of fossil insect cuticle was on the threshold of yielding exciting results. Since then, paleoentomologists have been making considerable headway in the chemical analysis of fossil insect chitin, facilitated by the tandem development of laboratory equipment capable of analyzing very small chemical samples, very rapidly. As discussed in Chapter 1, insect procuticle is made chiefly of chitin, a nitrogenous polysaccharide made up of *N*-acetyl-d-glucosamine units $(C_8H_{13}NO_5)_n$ (Fig. 1). Because chitin is so resistant to decay, and insoluble in water, alcohol, dilute acids, and bases, it has the potential to hold carbon, nitrogen, hydrogen, and oxygen atoms in a stable matrix that can persist for millions of years under the right conditions. The isotopic composition of these elements can tell us a great deal about past environments.

1. Direct AMS Dating of Insect Fossils

The unstable ^{14}C content of fossil insect chitin has been used to obtain radiocarbon ages directly from fossil beetle remains. The advent of AMS radiocarbon dating has meant that it is now possible to obtain a ^{14}C age from just a few fossil insect specimens, such as three to four head capsules or five to six elytra of medium-sized beetles. Samples as small as 100 µg (1×10^{-4} g) of carbon are regularly dated by the AMS method (Jull, 2007).

The first published AMS dates made directly from insect fossil exoskeletons came from my work at Lamb Spring, CO, USA (see Chapter 7). This study allowed the differentiation of age groups between two sets of ecologically incompatible species. Species representing cold-steppe environments yielded an age of more than 3,500 years older than a group of alpine tundra species (Fig. 2) (Elias and Nelson, 1989; Elias and Toolin, 1989). We came to the conclusion that the faunal assemblage in question had been reworked by spring waters. The Lamb Spring story held out the promise of being able to resolve such thorny taphonomic problems through AMS dating of small numbers of fossil specimens.

However, Walker *et al.* (2001) did a much more detailed AMS dating study in which they compared AMS dates in the lateglacial age range, obtained from humic and humin fractions of sediment, plant macrofossils, and a variety of beetle taxa, all from the same sampling interval at the St. Bees site in Cumbria, England. Their beetle-derived AMS dates suffer from small sample sizes in some cases, but the spread of ages obtained from the different beetle taxa (Table 1) is disturbing. In the sample taken from the 22-cm horizon, the aquatic leaf beetle, *Donacia versicolorea*, yielded an age more than 500 years younger than one of the terrestrial species (*Adoxus obscurus*), and almost 900 years younger than the other terrestrial species, the weevil *Barynotus squamus*. All three beetle-derived ages were younger than the plant macrofossil-derived age. In the assemblage from 18 cm, the two terrestrial beetle species yielded ages essentially indistinguishable from those derived from plant macrofossils, but the date obtained from *Donacia versicolorea* was about 200–300 years younger. The aquatic beetles in this study systematically yielded spuriously young radiocarbon ages, but not all of the terrestrial beetles yielded consistently greater ages. In the assemblage from 8 cm, *Donacia versicolorea* yielded an age of $11{,}000 \pm 170$, while fossils of *Barynotus squamus* yielded an age of $10{,}270 \pm 210$. Furthermore, an aquatic predator, *Agabus bipustulatus*, yielded the greatest radiocarbon age in this beetle sample, $11{,}750 \pm 120$. Even this age is almost 500 years younger than the plant macrofossil-derived age for this sample: $12{,}230 \pm 60$.

Why should these different beetles yield significantly different radiocarbon ages? There is no clear-cut answer as yet, but Walker *et al.* (2001) proposed a hypothesis that the age variation is in some way related to the biochemistry of the fossil chitin being dated. They consider that there may have been postmortem incorporation of younger organic residues into the polysaccharide lattice of the chitin. Hodgins *et al.* (2001) examined this possibility, and concluded that significant age differences can exist between the carbon recovered from the amino acids and that recovered from the polysaccharides in a sample of fossil chitin. They conducted experiments in which solvent extractions of fossil chitin were carried out to remove endogenous and exogenous lipid-soluble materials. They found what they described as "surprisingly high levels of contaminating amino acids." Schimmelman and DeNiro (1986) suggested that the most reliable chitin-derived compound for **stable isotope** studies in marine arthropod chitin is glucosamine-HCL. Glucosamine $(C_6H_{13}NO_5)$ is an amino sugar. Stripping the glucosamine out of chitin is a way of eliminating amino acids. These typically make up 41–55% of chitin mass (Miller *et al.*, 1993b), so their removal considerably lowers the mass of

DEVELOPMENTS IN QUATERNARY SCIENCES
VOLUME 12 ISSN 1571-0866

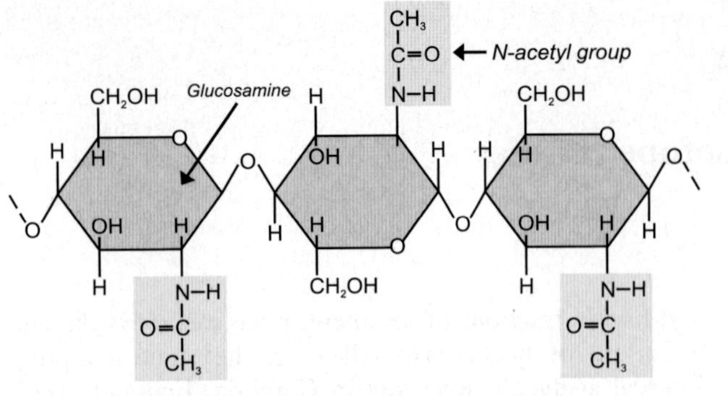

Fig. 1. Molecular configuration of the chitin polymer chain.

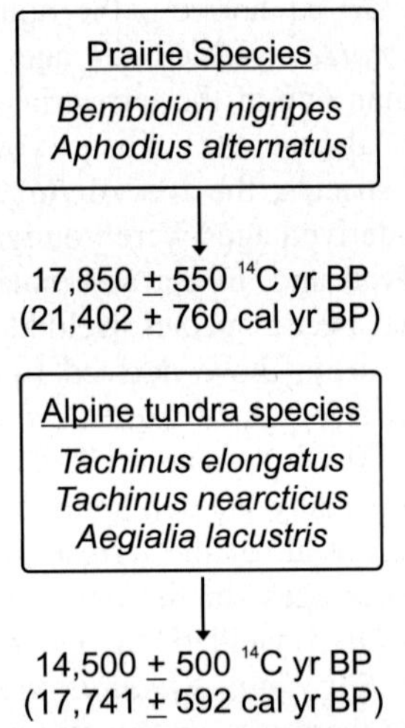

Fig. 2. Accelerator mass spectrometry radiocarbon ages and calibrated ages derived from two ecological groups of beetles from the Lamb Springs fossil site, Colorado, USA (data from Elias and Toolin, 1989).

fossil chitin samples. In the study by Hodgins *et al.* (2001), various methods were used to obtain purified glucosamine samples from fossil chitin samples from the St. Bees beetle faunas. Regardless of the method of deproteinization, only 5% of the starting mass was recovered as glucosamine.

These authors chemically stripped the amino acids from a fossil weevil specimen (*Otiorhynchus nodosus*) from the Redkirk Point site in England. An AMS radiocarbon age obtained from an untreated chitin sample yielded an age of 11,380±70 yr BP; samples of glucosamine obtained through three different deproteinization protocols yielded essentially identical ages (11,670±65, 11,665±65, and 11,645±119 yr BP). The study concluded that while pretreatment of fossil chitin was highly desirable, new methods of pretreatment must be developed in order to overcome the problem of small sample yield.

2. Stable Isotope Studies

The chitinous layers of insect exoskeletons are often hardened, or sclerotized, by chemical cross-linking with non-amino-sugar compounds such as **peptides** and pigments (Muzzarelli, 1977). The carbon–hydrogen bonds in chitin appear to be extremely stable. In cold and temperate regions, hydrogen bound by carbon does not exchange with environmental hydrogen, even over geologic time (Schoell, 1981). Thus the **stable isotopes** of hydrogen and oxygen should remain in the same proportions as they were when they were formed the chitin, during the development of adult insects. Hydrogen exists in two principle stable isotopic forms in nature: normal hydrogen (with one proton and one electron), and deuterium (with one proton, one neutron, and one electron). Deuterium is far less common than normal

Table 1. Radiocarbon dates from fossil beetle specimens from lateglacial assemblages, St. Bees, Cumbria, England.

Depth (cm)	Taxon dated	Ecology	^{14}C age from beetle	^{14}C age from plant macrofossils (seeds)
44	*Otiorhynchus nodosus*	Terrestrial, feeds various plants	10,195±80	11,020±60 (*Carex*)
22	*Donacia versicolorea*	Aquatic, feeds on *Potamogeton*	10,640±120	12,180±60 (*Scirpus* and *Carex*)
	Adoxus obscurus	Terrestrial, feeds on *Epilobium*	11,190±85	
	Barynotus squamosus	Terrestrial, feeds various plants	11,535±95	
18	*D. versicolorea*	Aquatic, feeds on *Potamogeton*	11,850±80	12,230±60 (*Scirpus* and *Carex*); 12,240±40 (*Potamogeton*)
	A. obscurus	Terrestrial, feeds on *Epilobium*	12,160±80	
	B. squamosus	Terrestrial, feeds various plants	12,160±80	
12	*D. versicolorea*	Aquatic, feeds on *Potamogeton*	11,370±180	11,590±60 (*Eleocharis, Scirpus, Carex*)
	A. obscurus	Terrestrial, feeds on *Epilobium*	11,780±85	
	B. squamosus	Terrestrial, feeds various plants	11,860±110	
	Byrrhus sp.	Terrestrial, feeds on mosses	10,940±180	
8	*D. versicolorea*	Aquatic, feeds on *Potamogeton*	11,000±170	12,230±60 (*Eleocharis, Scirpus, Carex*)
	B. squamosus	Terrestrial, feeds various plants	10,270±210	
	Carabus problematicus	Terrestrial predator	11,600±140	
	Plagiodera versicolorea	Terrestrial, feeds on *Salix*	11,340±140	
	Agabus bipustulatus	Aquatic predator	11,750±120	

Source: Data from Walker *et al.* (2001).

hydrogen, accounting for only about 1 out of every 6,500 hydrogen atoms. Deuterium is often abbreviated as ''D'' in the geochemical literature. The term δD is used to denote the ratio of deuterium to normal hydrogen. Oxygen also exists in two principle stable isotopic forms: ''normal'' oxygen, denoted ^{16}O as it has an atomic mass of 16 (8 protons and 8 neutrons), and the much less common ^{18}O, with 8 protons and 10 neutrons. ^{16}O comprises 99.762% of all the oxygen on Earth, while ^{18}O makes up only 0.2%. The ratio of ^{18}O to ^{16}O is abbreviated as $\delta^{18}O$. Because insects incorporate water derived from local precipitation (meteoric water) in their chitin as it forms, the isotopic ratios of oxygen and hydrogen found in fossil chitin should theoretically match the isotopic ratios of those elements that fell as precipitation during the insect's lifetime. The analysis of stable O and H isotope ratios in fossil chitin should therefore be a way of examining the isotope ratios of these elements in ancient precipitation. The hydrogen and oxygen isotope composition of precipitation varies systematically across the globe, and strong and well understood correlation exists between temperature, and the $\delta^{18}O$ and δD values in modern precipitation (Dansgaard, 1964; Rozanski *et al.*, 1993). O and H isotope ratios from fossil chitin may therefore be taken as an independent paleotemperature proxy.

In most terrestrial environments, meteoric precipitation is the ultimate source of the hydrogen atoms incorporated into insect chitin. Insects can either drink water or acquire hydrogen from molecules in food. In the latter case, meteoric water enters the insect's digestive system by way of a food chain, wherein plant organic hydrogen is ingested by herbivorous insects and metabolically converted in contact with digestive enzymes, or predatory insects ingest plant-derived organic hydrogen indirectly as they eat plant-feeding insect prey. As organic hydrogen makes its way up a predatory beetle's food chain, the body fluid of each prey insect is isotopically enriched in deuterium by water-vapor loss through the trachea, thus maintaining or even enhancing deuterium enrichment each time when chitin is synthesized.

Significant D/H isotope fractionation can occur whenever water vapor is lost from liquid water via evapotranspiration (e.g., insects lose water vapor through trachea during breathing in a similar fashion to which plants lose water through stomata; Chikaraishi *et al.*, 2004). It follows that the δD value of meteoric water is an important factor, but not the only factor influencing the δD value of insect chitin. The date of emergence of the adult beetle is an important factor in the final establishment of the isotopic composition of oxygen and hydrogen in its exoskeleton because within a few hours or days of emergence, the exoskeleton becomes hardened (sclerotized) through a series of complex chemical reactions controlled by various enzymes (Hopkins and Kramer, 1992). When the sclerotization process is complete, the beetle retains the hardened exoskeleton for the entire adult stage of its life, and its chitin becomes chemically inert.

Hydrogen atoms bound with carbon make up about 77% of the hydrogen atoms in chitin. When a beetle dies, its chitinous remains are often exposed to ground water in sediments. However, most carbon–hydrogen bonds in chitin are sufficiently strong to render them non-exchangeable with hydrogen derived from water, especially at low temperatures. As a result, the organic hydrogen in fossil chitin is thought to be isotopically conservative, retaining its D/H ratio through geologic time (Schoell, 1981; Sessions *et al.*, 2004). However, Schimmelmann *et al.* (1993) and Schimmelmann and Miller (2002) noted that hydrogen bound to nitrogen and oxygen (in hydroxyl groups) makes up the remaining ca. 23% of hydrogen in chitin, and that these hydrogen atoms may readily exchange with hydrogen in environmental water or even with water vapor. Other minor chemical moieties (functional groups) in chitin, including pigments, lipids, and proteinaceous compounds, contain non-exchangeable and exchangeable organic hydrogen in different proportions. The lipid hydrogen in bulk natural chitin contains hydrogen that has the smallest propensity to exchange isotopically with water hydrogen in a fossil setting (Sessions *et al.*, 2004).

The chemical features of chitin, as described above, needed to be tested in modern and fossil chitin samples before any serious progress could be made in the use of isotopic ratios in fossil chitin as a means of reconstructing past temperatures. It was necessary to determine whether chitin, once formed, exchanges C, H, or O with the environment. The initial evidence, based on results published by Miller (1991a), Miller *et al.* (1985b; 1993b), Gröcke *et al.* (2006) and on the work of others, suggests that chitin is chemically stable. Wooller *et al.* (2003) determined $\delta^{18}O$ in fossil and modern chitin samples from Chironomidae (nonbiting midges) preserved in arctic lake sediments. They found that the chitin in chironomid head capsules remains chemically (and isotopically) stable for many thousands of years.

In theory, the hydrogen atoms bound to carbon in chitin should not exchange with environmental hydrogen. However, Schimmelmann *et al.* (1993) and Schimmelmann and Miller (2002) found that the hydrogen bound to nitrogenous compounds (such as amino acids) and to oxygen (in hydroxyl groups) may readily exchange with environmental hydrogen in the presence of water. These kinds of compounds comprise about 20% of the hydrogen in chitin. Schimmelmann and Miller (2002) developed a pyrolysis method to strip hydrogen atoms bound to either nitrogen or oxygen away from chitin, thereby removing the potential source of contamination by hydrogen isotopic exchange with the environment.

Can chitin isotope ratios serve as a reliable proxy for past temperatures? Miller *et al.* (1988) performed a pilot study in which D/H ratios in modern beetle chitin from North American specimens were found to vary in a similar manner to regional environmental temperatures. The study by Wooller *et al.* (2003) of stable isotopes in arctic fossil chironomids indicates that the isotopic composition found in midge chitin reflects the isotopic composition of regional precipitation. Specifically, Wooller *et al.* (2003) found that $\delta^{18}O$ values in head capsules of midge larvae are equilibrated with the $\delta^{18}O$ of the lake water in which they live. In suitable takes, lake water $\delta^{18}O$ is controlled by the $\delta^{18}O$ of catchment precipitation, which is strongly correlated to mean annual temperature.

Gröcke *et al.* (2006) studied the D/H ratios in several sets of modern beetle chitin samples from throughout Western Europe, testing to see whether the D/H ratios in the chitin reflect the D/H ratio in meteoric water in the region from which beetle specimens are taken. This study was facilitated by the availability of hydrogen isotope data for regional precipitation. The International Atomic Energy Agency (IAEA) developed a global network of international precipitation data (GNIP) in the form of maps and animations (IAEA, 2001), which can be directly related to δD values of precipitation falling during the time of adult beetle metamorphosis. This is a key factor in the analysis, because the date of emergence of adult beetles varies from region to region. For instance, in southern Europe, adult beetles tend to emerge as early as March, whereas in northern Europe they may emerge as late as June. These preliminary results display far less isotopic variability than that found in North America, but Miller *et al.* (1988) suggested that the variability seen in a modern data set relates broadly to the δD of precipitation, and hence should relate to mean annual temperature.

Van Hardenbroek *et al.* (2009) took this testing procedure one step further. They analyzed δD and $\delta^{18}O$ values from chitin sampled from two genera of water beetles (the water-scavenger beetle, *Helophorus*, and the predaceous diving beetle, *Hydroporus*) from a suite of 40 collecting localities for each genus along a latitudinal transect from 25 to 75°N in North America. This transect represents a 25°C range in mean summer temperatures. They found that stable oxygen and hydrogen isotopes in chitin show a strong correlation ($r^2 = 0.88–0.91$) with modeled isotope values for the warmest month (July). This study provides a clear demonstration that isotopic values in modeled July precipitation are strongly correlated with isotopic values in water beetle chitin. As with the European study of Gröcke *et al.* (2006), this North American study also highlighted the importance of the timing of adult beetle emergence. Differences are observed between the trends in isotope ratios in *Helophorus* and *Hydroporus*. The regression lines of *Helophorus* have steeper slopes for both δD and $\delta^{18}O$ in Fig. 3. Van Hardenbroek *et al.* (2009) found differences in the regression of chitin isotopic values on precipitation isotopic values between *Hydroporus* and *Helophorus*. The main differences in slope of the regression lines of the two genera are caused by differences in emergence time. *Helophorus* emerges weeks to months earlier than *Hydroporus* (Smetana, 1985; Larson *et al.*, 2000). Due to seasonal variation in the isotopic composition of precipitation, this results in 24‰ (δD) and 3.7‰ ($\delta^{18}O$) higher modeled values for precipitation at the time

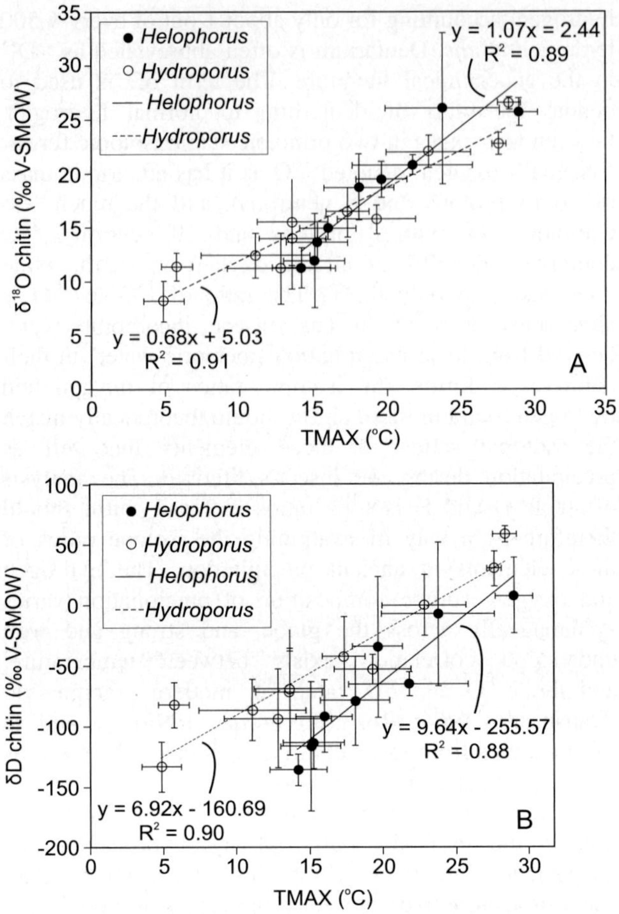

Fig. 3. Correlation between 5° latitude averaged $\delta^{18}O$ values (A) and δD values (B) in chitin and T_{max} (after van Hardenbroek et al., 2009).

Hydroporus emerges. This study is an important reminder that we cannot properly interpret insect chitin data without a good working knowledge of the ecology of the organisms that *made* the chitin.

Work is just now beginning in which stable isotope ratios for O and H are being determined from fossil beetle chitin. Because only minute samples are needed for modern mass spectrometers, it will be possible to obtain δD and $\delta^{18}O$ values from individual beetle sclerites (e.g., heads capsules, pronota, or elytra). We will proceed with fossil water beetle specimens, based on the argument presented by van Hardenbroek *et al.* (2009): in lakes where evaporation is limited and water is in isotopic equilibrium with precipitation, water beetles contain the isotopic signature of precipitation at the time of chitin synthesis.

15

Ancient DNA Studies

> With appropriate attention to details such as selection of specimen type and species, significant advances will be made in the next decade of insect aDNA research.
>
> – Rebecca Reiss (2006)

In recent decades, the study of ancient DNA, or aDNA, has been revolutionizing our view of the genetic history of many species of plants and animals. We now know a great deal about the origins, population genetics, and Late Pleistocene movements of such notable species as Neanderthals (Noonan *et al.*, 2006), woolly mammoth (Poinar *et al.*, 2006; Barnes *et al.*, 2007), Pleistocene bison (Shapiro *et al.*, 2004), and brown bears (Barnes *et al.*, 2002). The study of aDNA from insect fossils got off to a poor start in the 1990s, when it was discovered that DNA thought to have come from insects preserved in amber (Cano *et al.*, 1993) could not be replicated (Austin *et al.*, 1997).

1. Ancient DNA Methods

Ancient DNA is recovered using the polymerase chain reaction (PCR). This is a key technique in molecular genetics that permits the analysis of any short sequence of DNA without having to clone it. PCR is used to reproduce (amplify) selected sections of DNA. PCR makes a large number of copies of a gene. This is necessary to have enough starting template for sequencing. Sequencing is the process of determining the order of **DNA nucleotides** (adenine, thymine, guanine, and cytosine) in a gene.

The study of ancient DNA from Quaternary beetle remains has focused either on Late Holocene fossils, or on specimens preserved either in permanently frozen sediments, or in extremely arid conditions, such as packrat (*Neotoma*) middens from the American Southwest. Both of these extreme environments tend to preserve the integrity of DNA strands over long periods of time. Molecular genetic studies of fossil assemblages have the potential to test the widely held assumption (based on external morphology) that most, if not all, beetle species have remained intact for hundreds of thousands (and in some cases, millions) of years (Coope, 1978; Elias, 1994).

2. Studies of Recent Material

DNA studies of modern- or historical-era insects have paved the way for aDNA studies in recent years, as reviewed by Reiss (2006). mtDNA is most often targeted for these studies, because **eukaryotic** cells contain only one nucleus but thousands of mitochondria, so the survival rate of mtDNA in fossils is much greater. Mitochondrial DNA sequence evolution requires special consideration since it is inherited as a unit from the mother. It thus has a limited recombination rate; but it also has a rapid mutation rate, making it an ideal marker for population-level studies (Reiss, 2006). The mitochondrial genome contains 13 protein-coding regions involved in energy production, 2 regions that code for ribonucleic acids (RNAs) that become part of the ribosome (rRNAs), 22 transfer RNAs (tRNAs) that are also involved in protein synthesis, and the control region, which contains the origin of replication (Avise, 1994). Many studies of insect DNA have targeted portions of the **cytochrome oxidase** subunits I and II (COI and COII), as well as various subunits of the **nicotinamide adenine dinucleotide dehydrogenase** (ND) gene. For instance, Chapco and Litzenberger (2004) were able to recover fragments of the COI, COII, cytochrome *b*, and ND2 genes from the Rocky Mountain grasshopper (*Melanopus spretus*), using both 19th century museum specimens, and specimens recovered from glacial deposits in Wyoming. This species was a major pest on North American prairies in the 19th century, but populations crashed in the 1880s, and it is now thought to be extinct. Some entomologists speculated that *Melanopus spretus* was actually a migratory phase of the extant grasshopper, *Melanopus sanguinipes*. However, the DNA study showed that *Melanopus spretus* was most likely a separate species, and it actually had closer genetic affinities with *Melanopus bruneri*, a grasshopper known today from the prairie regions of Alberta and Saskatchewan.

2.1. Tiger Beetle DNA

In a similar type of study, Goldstein and Desalle (2003) examined DNA extracted from museum specimens of a tiger beetle *Cicindela dorsalis dorsalis* that is threatened with extinction. One of the few surviving populations of this beetle lives on the coast of Martha's Vineyard, an island off the coast of Massachusetts. They removed single hind legs of 92 museum specimens of this beetle that were originally collected along the New England coast between 1885 and 1971. They designed **PCR primers** to amplify a 73-base-pair (bp) fragment of the mitochondrial COIII gene, and their method managed to extract, amplify, and sequence the COIII fragment from

DEVELOPMENTS IN QUATERNARY SCIENCES
VOLUME 12 ISSN 1571-0866

42 individuals. One of the aims of this study was to try to determine which existing populations might best be suited as a source of individuals to be reintroduced into the more northern regions of the beetle's former range. A reintroduction from the Martha's Vineyard population would help maintain the genetic uniqueness of this northern population, but a reintroduction from one of the more southerly extant populations would help restore past genetic diversity of the beetle. Because both the DNA data and ecological/behavioral data indicate the distinctive qualities of the northern populations, it was decided to reinforce this by reintroducing this beetle using specimens from the Martha's Vineyard population.

2.2. Japanese Damaster DNA

Japanese molecular geneticists attempted to determine the origins and diversification patterns of a flightless ground beetle genus in the Japanese islands. They studied the DNA of specimens of the species *Damaster blaptoides* from 78 localities in Japan, using freshly collected specimens (Su *et al.*, 1998). Their study focused on mtDNA from the ND5 gene, extracting samples from thorax muscle tissue. Eight subspecies of this beetle are currently recognized in Japan, based on external morphology. Their study generated a genealogical tree of the mtDNA ND5 gene (Fig. 1) that shows a major split into eastern and

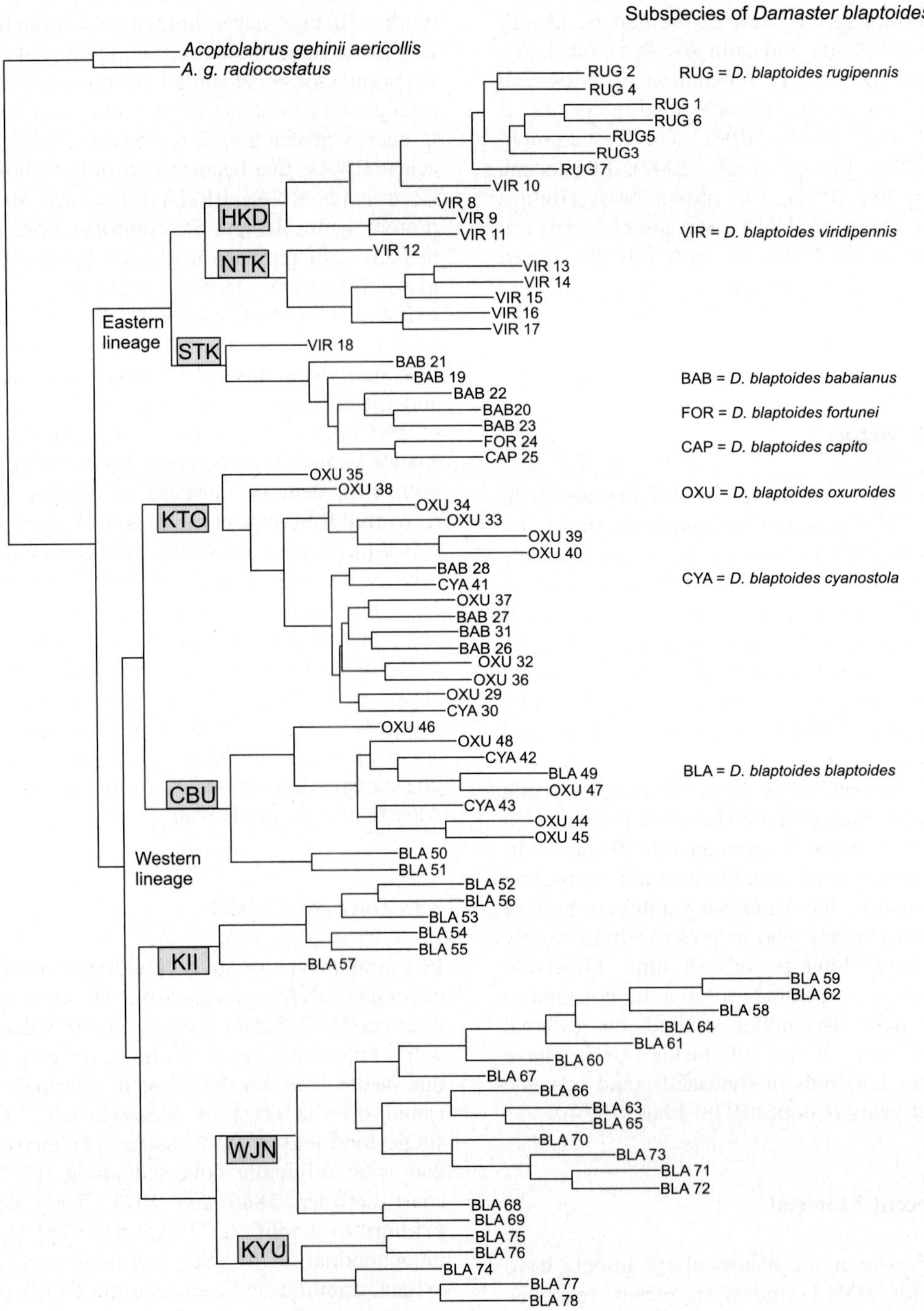

Fig. 1. Genealogical tree of the mitochondrial ND5 gene of the genus Damaster *expressed as a majority-rule bootstrap consensus tree. The sequences of two* Acoptolabrus gehinii *subspecies were taken as an out-group for these phylogenetic analyses. The three letter codes in shaded boxes represent different populations, as shown in Fig. 2B (after Su et al., 1998).*

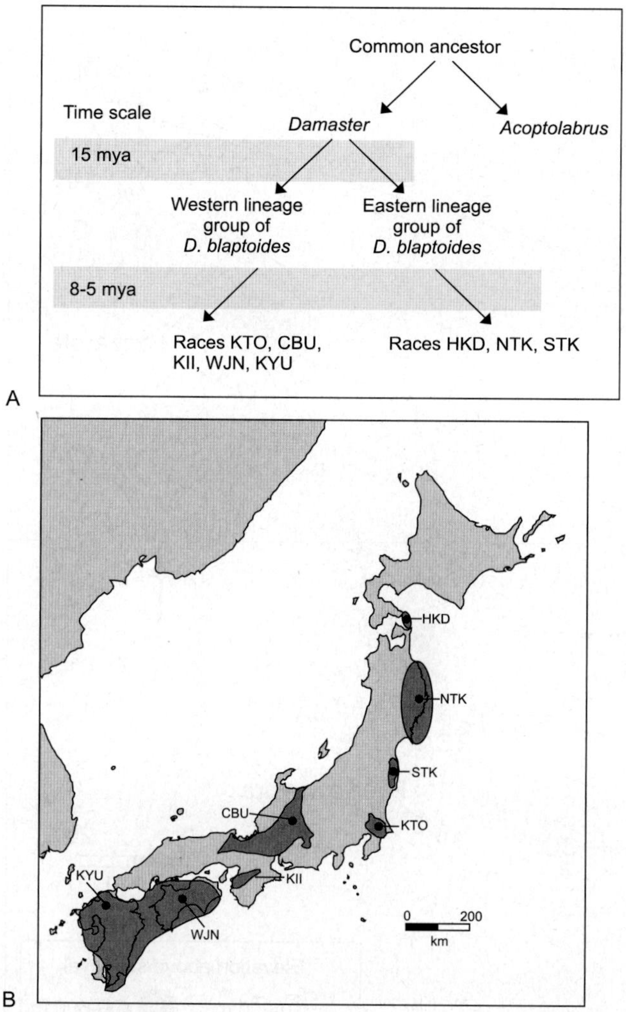

Fig. 2. (A) Reconstructed timescale of the emergence of the western and eastern lineage groups, and phylogenetic races of Damaster blaptoides *in Japan (data from Su et al., 1998). (B) Map of Japan, showing the reconstructed habitat regions for the races of* D. blaptoides *(after Su et al., 1998, permission granted by Oxford University Press).*

western groups, and identifies eight regional races. The study concluded that the eastern and western groups of *Damaster* split about 16.8 ± 1.2 mya, and the races of *D. blaptoides* diverged between 8 and 5 mya (Fig. 2).

2.3. *Canary Islands Hegeter DNA*

Juan *et al.* (1998) reconstructed the phylogenetic history of the darkling beetle *Hegeter politus* from a study of the mtDNA COI gene in various populations from the eastern Canary Islands. This is a flightless species considered endemic to these islands (but one should keep in mind the caveats concerning attribution of endemism based only on modern ranges, as discussed in Chapter 6). Each of the Canary Islands originated from independent underwater volcanic eruptions; the sequence of island building from these eruptions has progressed from south to north, with the oldest island, Fuerteventura, having emerged about 15–20 mya, and the youngest island, Alegranza, having emerged less than 1.5 mya. The authors sampled mtDNA from 53 beetles collected from nine sites on the islands (Fig. 3). A 200-bp region of the COI gene was sequenced, and a sequence from the North African darkling beetle *Pimelia grandis* was used as an out-group in constructing a phylogeny. Based on the genetic evidence shown in Fig. 3, the different **clades** of this species correspond to population groups on the various islands. The most parsimonious colonization history that can be deduced from the phylogenetic tree is a stepwise progression from south to north.

2.4. *Arctic and Alpine Amara DNA*

Reiss *et al.* (1999) reconstructed the pattern of postglacial divergence of populations of the ground beetle *Amara alpina* by studying mtDNA from 22 modern populations collected from arctic and alpine regions of North America. Unlike the other DNA studies discussed here, which involved PCR to facilitate gene sequencing, in this study mitochondrial **restriction site** variation of specimens from the different populations was assayed using radioactively labeled mtDNA to probe southern membranes containing restriction-enzyme-digested total DNA. Restriction sites were then mapped and genetic distances were calculated using pairwise comparisons of presence and absence of restriction sites. Genetic distances were used in a molecular analysis of variance (ANOVA) and to construct a minimal spanning tree (Fig. 4). The investigation detected 15 haplotypes, and modern genetic variation within each of four regions corresponded to the model predicted on the basis of paleontological evidence. Eastern Beringian populations were found to be the most genetically diverse, while populations from the Hudson Bay region showed the least diversity. This agrees with the dispersal model in which postglacial colonization of the Hudson Bay region all came from west-to-east population expansion following deglaciation. The genetic evidence also showed that the two southern refugial population groups in the Rocky Mountains and the Appalachians are genetically distinct from each other. *Amara alpina* lived in lowland regions south of the continental ice sheets during the last glaciation in North America, but its range in the lower 48 state region shrank back to alpine tundra regions of the Rockies and Appalachians at the end of the last glaciation. The mtDNA data suggest that these two refugial populations were derived from genetically diverse populations.

3. Studies from Fossil Material

Reiss (2006) made one of the first attempts to recover aDNA from fossil beetle specimens. She extracted mtDNA from a collection of fossil beetle specimens identified from the Big Bend region of Texas. The fossils came from a packrat midden dated 20,000 ^{14}C yr BP (23,900 cal yr BP), and included elytral specimens of the weevil genus *Ophryastes*, the ground beetle genus *Harpalus*, and the dung beetle genus *Onthophagus*. Reiss targeted portions of the 16s rRNA and ND1 genes, using PCR to obtain sufficient genetic material to allow

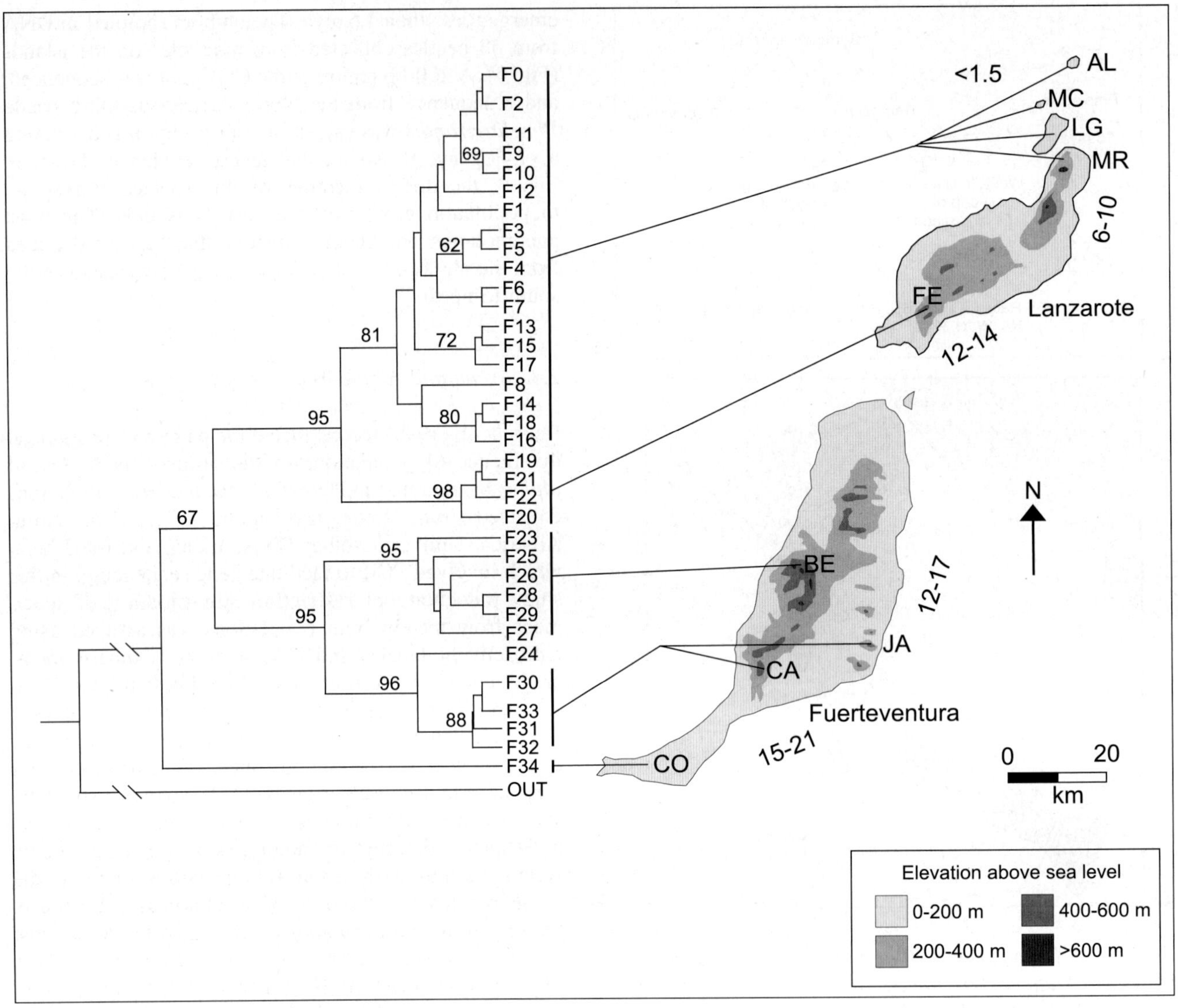

Fig. 3. Phylogenetic tree of Hegeter politus *based on sequencing of the mitochondrial COI gene, using the Kimura two-parameter model-based estimates of genetic distance. Bootstrap values from 500 random replications are shown above the horizontal lines in the tree. The map of the eastern Canary Islands shows the 200, 400, and 600 m contours, and the ages of the islands in millions of years. The various population groups are denoted on the map by two-letter abbreviations of region names: AL, Alegranza; BE, Bentancuria; CA, Montaña Cardón; CO, Degollada de Cofete; FE, Femés; JA, Jacomar; LG, La Graciosa; MC, Montaña Clara; MR, Mirador del Rio (after Juan et al., 1998, permission granted by the Royal Society of London).*

sequencing from the 316-bp fragment being targeted. Figure 5 shows the electrophoresis gel on which two bands of genetic material were separated. The bands marked ''824'' are the product of the reporter template and primers used in conjunction with primers for the mtDNA region of interest (the 316-bp fragment). Bands 1–4 represent mtDNA from modern specimens of *Amara* from New Mexico; bands 5–7 represent mtDNA from museum specimens of *Stereocerus haematopus*; bands 8–10 represent fossil specimens of *Harpalus*; bands 12–14 represent control samples. The photo documents the results of a quality control PCR treatment, based on the first attempt to isolate and amplify mtDNA from fossil and modern beetle specimens. As can be seen, the PCR experiment failed to extract the 316-bp fragment from the fossil specimens. A subsequent trial, using a more refined method, yielded the 316-bp fragment from the fossil ground beetle (*Harpalus*) specimens. Reiss only tried this one group, because she did not have suitable primers for the other families, whereas she was able to find a **GenBank** sequence for the ground beetle genus *Carabus*. The development of species- or genus-specific primers for aDNA research in insect fossils is a critical step that is only now being adequately addressed, as discussed below.

Reiss had success in extracting aDNA from beetle fossils sampled from an arid environment. She noted that evidence from other taxa demonstrates that cold and dry conditions are the least destructive to DNA. Therefore fossil specimens from rodent middens and permafrost environments have the greatest potential to yield aDNA.

Willerslev *et al.* (2007) proved that aDNA does indeed preserve well in permanently frozen conditions. They investigated aDNA from samples of silt-rich ice taken from the Dye 3 and GRIP ice cores from Greenland, as well as from sediments sampled from the Kap København formation in northernmost Greenland (see Chapter 4).

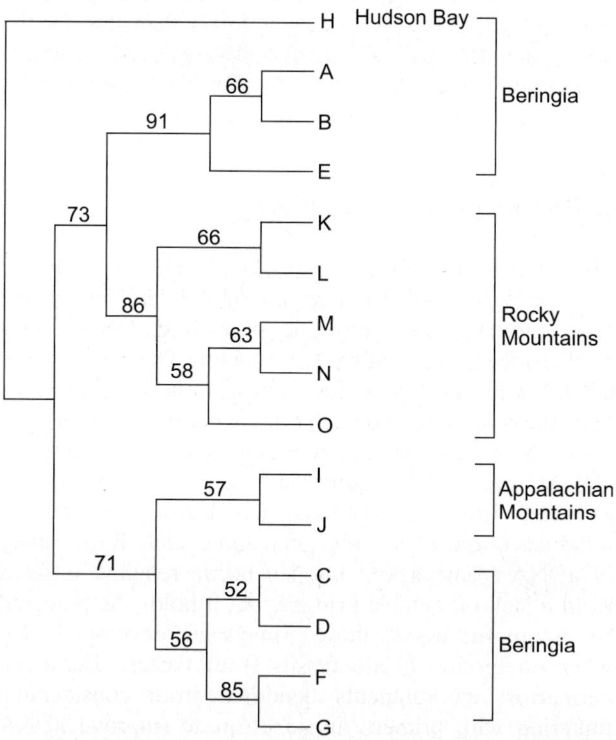

Fig. 4. Phylogenetic tree of North American populations of Amara alpina, based on branch and bound analysis of the restriction site variation data, subjected to bootstrap analysis with 1,000 rounds of sampling. The percentage support for each branch is indicated in numbers shown above each line (after Reiss et al., 1999, Journal of Biogeography; copyright permission granted by Wiley-Blackwell).

They obtained small fragments (97 bp) of mtDNA from the Dye 3 core samples, associated with the COI gene. It is difficult to reliably match such short segments of DNA with known sequences, such as those available in GenBank. Simple comparisons by means of the Basic Local Alignment Search Tool (BLAST) would likely yield misidentifications, so Willerslev *et al.* applied Bayesian statistics to calculate the probability that each sequence belongs to a particular clade by considering its position in a phylogenetic tree, based on similar GenBank sequences. Sequences with >90% posterior probability of membership to a taxonomic group were assigned to that group. The most reliable indication of insect mtDNA in the study was associated with Lepidoptera (moths and butterflies). This identification had 97–99% probability support from the Bayesian statistical treatment. Less substantial evidence was found for the recovery of COI gene sequences associated with beetles, flies, and spiders, but these other identifications got less than 90% support from the probability statistics. Interestingly, no macroscopic remains of any of these arthropods were recovered from the ice core samples. Only their aDNA was found. Based on $^{10}Be/^{36}Cl$ isotope ratios, optically stimulated single-grain luminescence ages, and amino acid racemization, combined with ice sheet modeling, Willerslev *et al.* estimated that their samples are between 800,000 and 450,000 years old, although they could not rule out a much younger age from the last interglacial interval (MIS 5e). The Kap København sediments, thought to date from 2.5 to 2 mya, failed to yield any identifiable fragments of DNA.

How well does beetle aDNA preserve in waterlogged sediments in a temperate environment? An attempt to answer this question was made by King *et al.* (2009) in their study of aDNA from grain weevil (*Sitophilus granarius*) fossils from waterlogged archeological sites in England. They selected six fossil specimens from faunal assemblages analyzed from the sites of Low Petergate, York, and Chester-le-Street, County Durham, for aDNA extraction, as well as modern specimens from

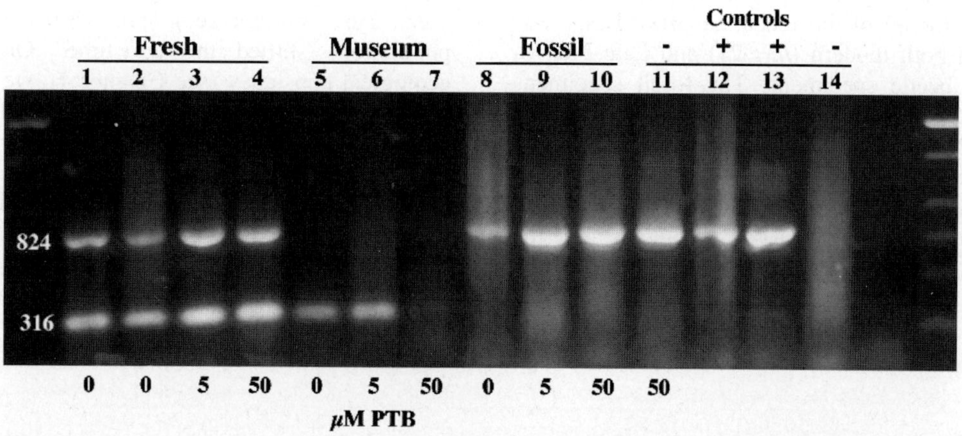

Fig. 5. Photograph of electrophoresis gel showing the results of PCR experiment with aDNA fragments extracted from fossil ground beetles from the Chihuahuan Desert. An 824-bp fragment was sequenced (see text). Bands 1–4 represent mtDNA from modern specimens of Amara *from New Mexico; bands 5–7 represent mtDNA from museum specimens of* Stereocerus haematopus; *bands 8–10 represent fossil specimens of* Harpalus; *bands 12–14 represent control samples (figure reproduced from Reiss, 2006, © Elsevier – QSR).*

a laboratory culture, used as a control. They reported successful extraction of mtDNA from five of the six specimens. The modern and medieval specimens yielded fragments of various lengths from the COI gene. The Roman-age specimens yielded only shorter (98 bp) fragments of this gene; two of the three Roman specimens yielded DNA. Short fragments of nuclear DNA were also extracted and sequenced from the medieval and modern specimens.

Is it wise to focus insect aDNA studies on the COI gene? Developing primers for use in PCR and subsequent genetic sequencing is one of the most difficult and time-consuming aspects of aDNA research. COI data have become available for a number of beetle and other insect taxa, making primer development far simpler. However, King *et al.* (2009) suggest that the COI gene may be too conservative for studies of microevolutionary changes in beetles, and the COII gene may be more conducive to such studies. One of the eventual aims of this research team is to determine whether natural and synanthropic populations of the grain weevil are genetically different. Also, if the genome of modern populations differs significantly from that of ancient populations, this may shed light on shifting habitat preferences of the species. The aDNA may also address the following questions: (1) Were British populations derived from a small number of Roman introductions, or established and maintained by continued trade? (2) Did the population present after the Norman Conquest arise from British survivors of the "Dark Ages," or from a new introduction from Continental Europe?

The author has recently collaborated with Dr. Ian Barnes (School of Biological Sciences, Royal Holloway, University of London), and Laura Mack, a German MSc student, in a pilot study to extract mtDNA from Late Pleistocene beetle fossils from permafrost deposits in Alaska and the Yukon Territory. The work has so far focused on two ground beetle genera, *Amara* and *Pterostichus*. Due to the lack of existing sequence data for the target taxa, the final primer pairs were developed through multiple rounds of primer design, using sequence data from previous experiments to improve fit to the binding sites. Following this optimization, amplification was successful for all of the recent samples. DNA was recovered from both modern ($n = 21$) and Late Pleistocene ($n = 16$) beetle specimens. The fossil specimens come from MIS 3 deposits, dating from about 50,000 to 25,000 cal yr BP. We have been able to obtain a 365-bp fragment of DNA sequence from both taxa for all of the modern samples tested. We had less success with extraction of aDNA from the fossil specimens. For the Late Pleistocene samples, 11 of 16 tested yielded at least one sequence-verified fragment, indicating that DNA is present in around 69% of specimens. Genetic sequences were amplified for at least one of three overlapping fragments within the mitochondrial COI gene. That only four samples (two each from *Pterostichus* and *Amara*) yielded all three fragments can be explained as the result of poor primer sensitivity or sequence variation in the binding site. This success rate is lower than that for the best Late Pleistocene, permafrost-preserved mammal bone (>95%), but it is still considerably better than for non-permafrost Late Pleistocene material.

4. Prospectus for Future Work

The study of aDNA from recent and fossil beetle specimens has only just begun in the last few years, following on the heels of highly publicized aDNA studies of Pleistocene megafauna and humans. The preservation of aDNA in fossil bone does appear to be far better than that found in insect exoskeletons (King *et al.*, 2009), but under the right conditions (extreme cold), even "naked" aDNA (i.e., DNA found in sediments, rather than extracted from exoskeletons) has been preserved for hundreds of thousands of years in Greenland. Reiss' study of aDNA from packrat midden beetle remains suffered from a lack of suitable primers, but it holds the potential for future success as these primers are developed. Our work on ground beetle fossils from Eastern Beringian permafrost environments benefited from considerable tinkering with primers, in an effort to improve aDNA recovery rates. As with aDNA research on other groups of organisms, the insect studies are bound to expand rapidly as more insect genes are sequenced. As the genetic data become available through GenBank, this will facilitate the creation of primers for increasing numbers of beetle taxa. This is painstaking work, but the potential rewards are sufficiently great to stimulate the research. Armed with aDNA evidence, we will be able to address questions on which we could only speculate, based on exoskeletal evidence alone. Where did the modern populations arise? How has the geographic distribution of species' populations shifted through time? On the basis of ecological constancy (see Chapter 4), we *expect* that the genetic evidence will show no indication of cryptic species, or significantly distinct populations that become extinct during the Late Pleistocene, but the genetic evidence will provide the only definitive answers to these questions.

16

Conclusions and Prospectus

> Quaternary paleoentomology is a speciality that requires a lot of time in training to develop identification skills. Skills are concentrated in a few individuals and they may or may not be passed on to the next generation. Paleoentomologists need to develop other ways to provide a legacy of their knowledge.
>
> – Allan Ashworth (2004)

The field of Quaternary entomology has grown from a minor aspect of paleontology in the 19th century to one of the important disciplines within modern Quaternary studies. Enough organic detritus has been examined to show that insect exoskeletons are ubiquitous in many types of organic deposits dating back more than a million years. Studies discussing fossil insect faunas from more than 1,300 sites have been published in this field during the last 40 years, not including chironomid papers. These studies show that insects, in particular beetles, are sensitive, reliable indicators of Quaternary environments. The stability of beetle species over hundreds of thousands or even millions of years has been postulated on the basis of abundant fossil evidence, and studies indicate that some insect species have shifted their distributions over vast regions in response to changing climates.

1. Persistent Obstacles

In spite of the hundreds of publications contributing to the body of knowledge in this field, the body of data accumulated by paleoentomologists remains largely unknown to many neontologists. Some of the blame for this must be placed on the Quaternary entomologists for publishing most of their papers in geological journals and giving most of their public lectures to geological audiences.

Another problem is that Quaternary entomology is perceived by some experimental biologists as "19th century science"; the description of fossil faunas has been equated by some with "stamp collecting." However, as Morgan and Morgan (1987) pointed out, the days of the natural historian did not die at the turn of the 20th century, and gathering "descriptive" data on species is more important now than ever, given the pace of natural habitat destruction around the world (Miller, 1991b).

As Quaternary science grows and expands into many new lines of research, it becomes all too easy to get caught-up in paleoenvironmental reconstructions based on one type of proxy data (pollen, vertebrate remains, oxygen isotopes, etc.). In 1994, I wrote, "Almost every audience I address requires a lengthy introduction to the subject of Quaternary entomology to bring it 'up to speed,' before I can get on with a discussion of my research topic." This remains true, 15 years later. One of the aims of this book has been to provide a broad spectrum of readers with up-to-date background information, in the hope that Quaternary entomology will be more widely understood.

2. Future Directions

To a large extent, Quaternary entomologists need to carry on with their activities of the last few years, that is, continue to find and analyze assemblages of Quaternary insects, especially with the goal of developing knowledge of unstudied regions and intervals of Quaternary time.

2.1. Regional Progress

Some regions are far better studied than others. Research in the United Kingdom, North America, and parts of Russia are well developed, although there are interesting differences in the *kinds* of studies being done in these countries. Some countries are crying out for new kinds of insect fossil studies. Of the 460 sites that have been published from the United Kingdom, 306 are archeological sites. Entomo-archeological research has lagged far behind this level elsewhere. Of the 312 published North American sites, only 27 represent archeological research. This makes some sense, given the relatively short history of European presence in the New World. As discussed in Chapter 7, Native Americans left few traces of their presence on the landscape throughout most of the Holocene. But European Russia has been occupied by people living in the kinds of permanent dwellings that leave abundant archeological records for many centuries; yet of the 147 Russian study sites published thus far, none concerns an archeological site. Nearly all of the Japanese studies (50 out of 54) are paleontological, in spite of that country's rich archeological record. In contrast, almost all of the fossil insect research that has been done on the North Atlantic islands (Iceland and the Faroes) and in the Mediterranean region has been archeological.

2.2. Aids in Training of Students

The quote from Allan Ashworth (2004) at the beginning of this chapter highlights a critical element in the life of

DEVELOPMENTS IN QUATERNARY SCIENCES
VOLUME 12 ISSN 1571-0866

paleoentomology as a branch of Quaternary science. It will not continue to progress without a new generation of scientists being trained in fossil insect identification and interpretation. The amount of time and effort involved in training students is considerably greater than in some other fields. One does not simply acquaint a student with a set of techniques and then set them loose in the laboratory to run a set of samples. It requires many months of hands-on training, working with the student at a microscope, teaching them to recognize the head capsules, pronota, and elytra of hundreds of different species in dozens of families. Ashworth (2004) suggests that we need to incorporate emerging computer technologies in aid of identification training. Of course the development of such electronic tools would, in itself, be a very big task, but it would be one that would reap benefits for many students of the subject for years to come. One of the greatest research tools to come our way in recent years has been the development of the BUGS database (Buckland and Buckland, 2006). BUGS places an incredible wealth of data at our fingertips, especially for European workers, since it catalogues all the species of beetles that have been found in northwest European Quaternary fossil assemblages, provides ecological and modern distribution data for each species, and also shows the localities and time intervals from which each species has previously been found in European fossil assemblages. The BUGS program has also recently added MCR analysis to its toolkit, making it easier for European investigators to perform MCR analyses of their fossil beetle assemblages.

The publishing of good quality photographs of beetles on the internet has grown by leaps and bounds in the last few years. As discussed in Chapter 3, there are now dozens of very useful Web sites where one can find good quality photographs of modern beetle specimens. While nothing can take the place of actual specimens for comparison with fossils, ready access to photographs can save the investigator a lot of time and effort in finding the right family, genus, or species group of beetle to match a fossil specimen. One of the bottlenecks of this line of research in past decades has been the necessity to acquire large numbers of monographs on the various groups of beetles that turn up in fossil assemblages. These taxonomic revisions are costly to buy, often contain less than useful illustrations, and many of them go rapidly out of print, becoming unavailable within just a few years. As more taxonomic papers and illustrations become available via the internet, the job of specimen identification becomes cheaper and quicker. However, it will always be necessary for Quaternary entomologists to have access to well-curated collections of modern specimens in government and university museums. These institutions are constantly under threat of closure from short-sighted administrators and politicians.

3. Vulnerability of Museum Collections

One example of constant upheaval within a uniquely valuable institution has been the history of the Canadian National Collection of Insects (CNC) in Ottawa, Ontario. During its 1st century of operation (1887–1987), this institution had seven directors. It has had another seven directors (now called ''managers'') in the last 22 years, and has gone through five reorganizations since then, each time having to justify its existence to the Canadian government and bureaucrats in Agriculture Canada. The CNC houses what is arguably the most important collection of high latitude North American insects, and has sponsored incredibly valuable work in beetle taxonomy by such entomologists as J. Milton Campbell, Aleš Smetana, Don Bright, the late Ed Becker, Yves Bousquet, and Laurent LeSage.

4. Publication Rates

The publication of Quaternary beetle papers is lead by the UK, followed by the USA, Canada, Russia, and a number of other countries (Fig. 1). Other than fossil midge studies, Quaternary insect research in temperate North America has sadly diminished since the publication of *Quaternary Insects and Their Environments* in 1994. Based on the references cited in Buckland *et al.* (2009b) during the 1980s, 159 papers were published on the topic of North American Quaternary insect fossils (Table 1). During the 1990s this number fell to 135 publications. Since 2000 there have only been 37 papers published on this topic. Many of these have been review articles, rather than papers describing work from new sites. Several researchers, including Alan and Anne Morgan, Don Schwert, Clarke Garry, Scott Miller, Jerry Pilny, and Robert Nelson have not published any new papers on this topic for several years.

5. Potential New Study Regions

As highlighted by Ashworth (2004), we need to find out how the beetle faunas south of the Laurentide ice sheet in eastern and southern United States responded to climate change during the last glaciation. No fossil insect studies have been done along the east coast from New Jersey to Florida. No fossil assemblages have been studied in the central and southern Appalachians, although these areas were likely biological refugia during glacial intervals. Likewise, only a handful of studies have been done from the west coast regions of North America. Pacific Northwest islands such as parts of Queen Charlotte Island are postulated to have remained ice-free during the LGM (Hetherington *et al.*, 2003), yet no one has developed a study of fossil insect faunas from this pivotal location. Ashworth (2004) has also suggested that the important beetle faunas from the Rancho La Brea tar deposits in southern California be reinvestigated, using the AMS radiocarbon technique to directly date beetle exoskeletons. While there are technical obstacles to be overcome in stripping away the petroleum from beetle chitin, we could learn a great deal about the timing of environmental change in this region from directly dated fossil beetle specimens. As discussed in Chapter 1, two species of dung beetles from these deposits, *Onthophagus everestae* and *Copris pristinus*, are thought to have become extinct

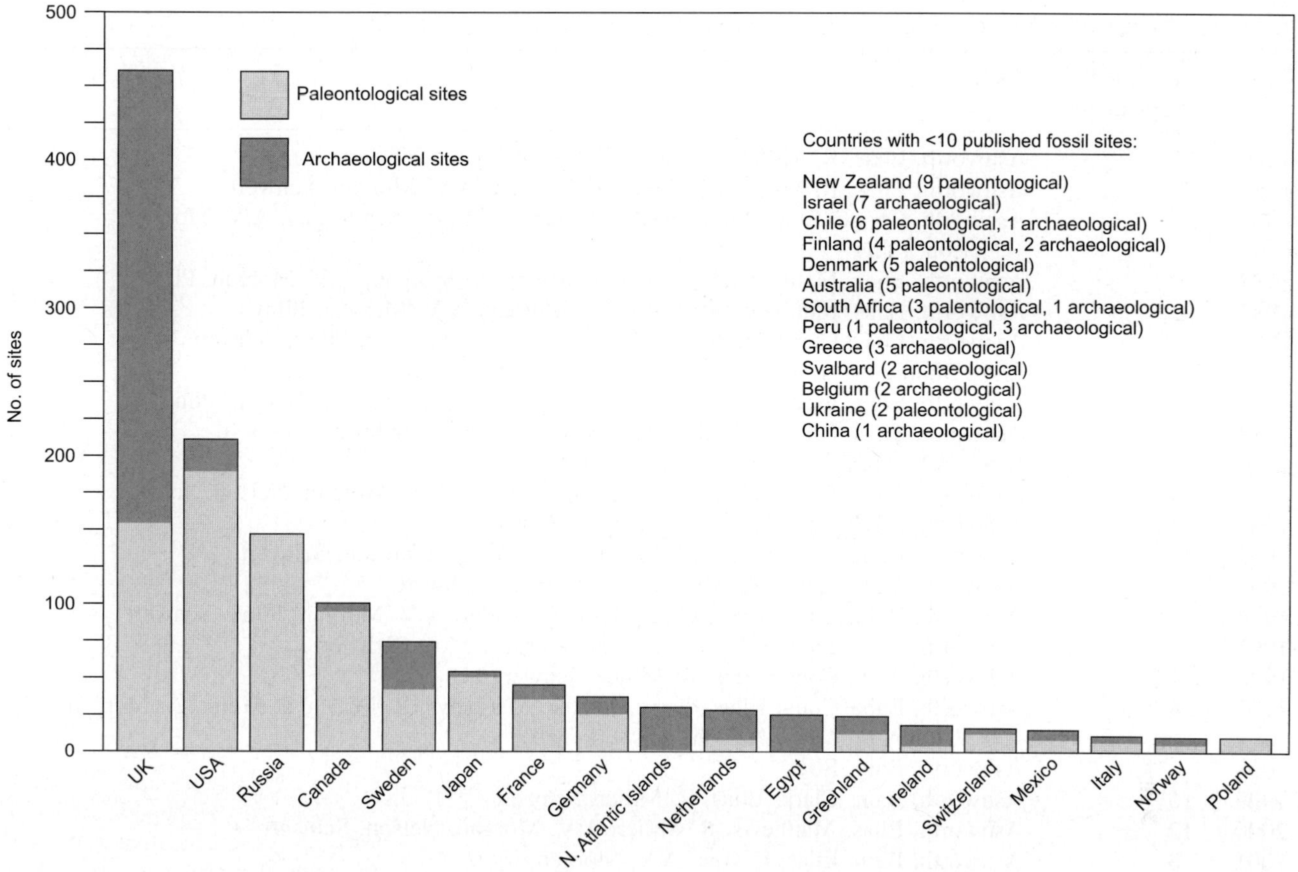

Fig. 1. Summary diagram showing the total number of publications concerning Quaternary beetle fossils, by country. The darker boxes represent archeological publications; the lighter boxes represent paleontological publications. Countries from which less than 10 papers have been published are also listed.

during the last glaciation (Miller *et al.*, 1981). These represent the only two extinctions of Late Pleistocene beetle species in the North American fossil record. It would be quite interesting to know the age of their last appearance in the southern California fossil record.

The situation in Western Europe seems to be healthier, as active research is continuing on new sites in Great Britain, Ireland, Sweden, and France. But the same cannot be said for Russia, where one of the great proponents of this work, Andrei Sher, died in 2008. His frequent collaborator Svetlana Kuzmina has gone to Canada, leaving Evgenij Zinovjev as the only Russian investigator still working there. The number of Quaternary beetle publications produced by European researchers since 2000 is shown in Table 2. Denmark, Germany, the Netherlands, and Norway have each produced less than 10 publications in this decade. France, Ireland, the North Atlantic region (including the Faroes, Iceland, and Greenland), Sweden, and Switzerland have produced between 10 and 20 publications, and the United Kingdom leads all countries in the world, with 167 new publications since 2000. About 90% of the British publications concern fossil insect assemblages from archeological sites. Encouragingly, PhD students have been receiving training in Quaternary beetle identification in recent years in France, Sweden, Northern Ireland, and England.

Ashworth and Hoganson got off to an impressive start with southern South American research in the 1980s, but no one has worked anywhere in South America in recent years. The logistic and taxonomic obstacles are formidable, but the potential rewards of research on this continent are great. Ashworth and Hoganson showed how the work can proceed in a new region. Their first step was to make extensive collections of modern beetles in their proposed fossil research region of southern Chile. They collected in the Parque Nacional Puyehue and other undisturbed habitats in the Chilean Lake Region. They described a diverse fauna of 462 species from habitats collected systematically along an elevational transect from sea level to 1,200 m asl (Ashworth and Hoganson, 1987). Their modern beetle research demonstrated that changes in beetle communities sampled from an altitudinal transect generally correlated with changes in plant communities. In doing this research, they built up a reference collection with which to compare fossil specimens, learned to identify at least the more common species of the regional beetle fauna, and gained an understanding of how regional beetle communities relate to regional environments, including climate and vegetation. These are all vital steps in preparation for paleoentomological research in any region, but in places like the Chilean Lake Region this requires an awful lot of modern background work before meaningful fossil research can commence. It was sensible for Ashworth and Hoganson to attempt their study in one of the most southerly regions of South America, where insect faunal diversity is

Table 1. Number of publications concerning North American Quaternary insect fossils since 1980.

Year	Number of publications	Authors
1980	15	Ashworth, Elias, R. Miller, A. Morgan, A.V. Morgan, Schwert
1981	15	Ashworth, Elias, R. Miller, S. Miller, A. Morgan, A.V. Morgan, Schwert
1982	30	Ashworth, Coope, Elias, Matthews, R. Miller, S. Miller, A. Morgan, A.V. Morgan, Nelson, Pilny
1983	9	Ashworth, Elias, Matthews, R. Miller, S. Miller, A. Morgan, A.V. Morgan, Pilny
1984	12	Ashworth, Elias, Matthews, R. Miller, A. Morgan, A.V. Morgan, Pilny
1985	10	Ashworth, Elias, Garry, R. Miller, A. Morgan, A.V. Morgan, Pilny, Schwert
1986	13	Elias, Matthews, A. Morgan, A.V. Morgan, Mott, Nelson, Pilny, Schwert
1987	16	Ashworth, Elias, Matthews, R. Miller, A. Morgan, A.V. Morgan, Nelson, Pilny
1988	12	Ashworth, Elias, Hall, R. Miller, A.V. Morgan, Nelson, Schwert
1989	9	Elias, Hall, Matthews, A.V. Morgan, Schwert
1990	18	Ashworth, Elias, Garry, Hall, R. Miller, A. Morgan, A.V. Morgan, Nelson, Schwert
1991	9	Ashworth, Elias, R. Miller, A.V. Morgan, Schwert
1992	21	Ashworth, Bain, Elias, Garry, R. Miller, A.V. Morgan, Nelson, Schwert
1993	7	Ashworth, Elias, Garry, R. Miller, A.V. Morgan, Schwert
1994	11	Ashworth, Bain, Cong, Elias, Garry, Hall, R. Miller, A.V. Morgan, Pilny, Schwert
1995	11	Ashworth, Elias, Lavoie, R. Miller, A.V. Morgan, Motz, Schwert
1996	17	Ashworth, Bain, Cong, Elias, R. Miller, Schwert
1997	24	Ashworth, Bain, Cong, Elias, Garry, Lavoie, Matthews, R. Miller, S. Miller, A. Morgan, A.V. Morgan, Motz, Nelson, Schwert, Telka
1998	4	Ashworth, Bain, Elias
1999	10	Ashworth, Bain, Clark, Elias, R. Miller, Schwert
2000	12	Ashworth, Elias, Matthews, R. Miller, A.V. Morgan, Nelson, Schwert
2001	8	Ashworth, Bain, Elias, Lavoie, A.V. Morgan, Motz
2002	2	Elias, Matthews
2003	2	Ashworth, Elias
2004	1	Ashworth
2005	1	Bain
2006	3	Elias, Kuzmina, R. Miller
2007	4	Elias
2008	4	Elias, Kuzmina

Source: Based on references listed in Buckland *et al.* (2009a).

Table 2. Numbers of Quaternary insect publications in European countries since 2000.

Year	Country or region										
	Denmark	France	Germany	Ireland	Netherlands	North Atlantic region (incl. Greenland)	Norway	Russia	Sweden	Switzerland	United Kingdom
2000	1	3	0	0	1	3	3	3	5	5	42
2001	1	2	0	1	0	1	1	5	1	1	29
2002	0	2	1	0	1	2	0	10	3	1	21
2003	0	3	0	2	1	2	0	2	0	1	16
2004	0	4	0	1	0	2	1	2	1	1	23
2005	0	1	2	2	1	4	0	1	0	1	11
2006	0	1	2	2	0	1	1	4	1	0	11
2007	0	0	0	0	0	2	0	1	2	0	9
2008	0	0	0	2	0	1	0	2	1	0	5
Total	2	16	5	10	4	18	6	30	14	10	167

Source: Data from Buckland *et al.* (2009a).

relatively low, compared with the tropics. It seems unlikely that anyone could successfully launch a study of fossil beetles from a rainforest region, as the faunal diversity is simply overwhelming. For instance, Erwin (1983) identified more than 1,200 species of beetles from the canopy of just one species of tree, *Luehea seemannii*, in tropical Panama. He estimated that there could be as many as 11,410 host-specific species of beetles per hectare of tropical forest. This level of faunal diversity is more than twice that of the entire beetle fauna of the British Isles (Duff, 2008).

As in North America, large regions of Europe that lay south of Late Pleistocene ice sheets have not been studied for insect fossils. Just as the Appalachians may have been a refuge for the more cold-adapted flora and fauna of eastern North America, southern European mountain ranges may likewise have played this role. While there were mountain ice caps in the Pyrenees, the Carpathians, the Caucasus, and the higher mountains of Italy, Greece, and the Balkans (Ehlers *et al.*, 2007), there were also unglaciated highland regions in these countries that could yield fossil insect assemblages. Ponel (1997b) studied one site in the Prato Spilla region of northern Italy, but no work has been done in the highland regions of Spain, the Balkans, or any part of southeastern Europe.

Africa is another continent which remains largely unstudied for Quaternary beetles. A few studies have been done from archeological contexts from the dynastic era in Egypt (see Chapter 7), Crossley (1984) has studied fossil termite remains from an archaeological site in Malawi, and Doube (1982) has reported on Late Pleistocene dung beetles from South Africa, but otherwise there has been no work on Quaternary beetle faunas of this great continent. The same obstacles of logistic difficulty and enormous faunal diversity will probably prevent workers from trying to study fossil assemblages from the tropical regions of Africa, but in southern Africa there are many regions with more temperate climates where there would seem to be great potential.

The temperate regions of Asia also remain almost completely unstudied, outside of Japan. Japanese scientists have made great strides in recent decades, but they have not worked outside the Japanese islands. There is tremendous scope for comparative research across the Korea Strait, as well as in temperate regions of China. Only one Chinese study has been published thus far (Chu and Wang, 1975), and this concerned insects found in ancient tombs. But Southeast Asia surely holds fascinating fossil records from earlier times. For instance, during Pleistocene glacial intervals, this region may well have served as refugia for insect species found further north in Asia today. We will not develop this history without the fossil record.

The spread of Quaternary entomology to Australia, New Zealand, and even tropical Pacific islands is a very exciting development in the discipline. Of course this does not happen by accident. Students have to be trained; they have to gain access to well-curated modern regional beetle collections; they have to enlist the aid of beetle taxonomists to identify unknown specimens; and they have to be able to relate their beetle data to environmental parameters through the use of modern beetle distribution data, and climate and vegetation records. It is certainly no accident that these two countries should be receiving attention from paleoentomologists. They are part of the British Commonwealth, and there are strong intellectual and educational links between Australia, New Zealand, and Britain. As such, these two countries are much more like Europe in their educational systems and scientific infrastructure than most countries in Southeast Asia.

In summary, the greatest need in Quaternary entomology is simply to carry on doing the research in new regions and from different intervals of the Pleistocene. A generation of researchers who were trained by Russell Coope in the 1960s and 1970s has spread across Europe and North America, but many of these people either have recently retired, or are about to retire, so the need for training up a new generation of paleoentomologists is paramount.

However, it would be selling the discipline short to infer that the only thing happening in Quaternary entomology these days is the expansion of the discipline to new regions. As discussed in Chapters 5, 14, and 15, there are some new research approaches that promise to take the field new directions in the coming decades. These include the further development of MCR paleoclimate reconstructions, and the extraction of stable isotopes and ancient DNA from fossil beetle chitin.

6. Development of MCR Methods

Following 20 years of MCR reconstructions in Europe, and a decade of MCR reconstructions in North America, the method has shown itself to be robust and reliable, but not without its weaknesses. The linear regression method of calibrating MCR results has proven to be statistically unsound, forcing us to seek other statistical treatments of MCR data, including ubiquity analysis. Such methods can only succeed if a great deal more leg work is done to pin down the exact specifications of climate envelopes for the species found in fossil assemblages. As discussed in Chapter 5, the original species' climate envelopes for beetles found in British Pleistocene assemblages were based chiefly on the modern European distribution of these species. The problem with this is that many of the species of cold-adapted predators and scavengers found in these assemblages have modern ranges that extend well into northern Asia. There were good attempts made to incorporate as much Asian distribution data as possible into those species' climate envelopes, but Coope and his students simply did not have access to all the necessary data. This problem must be rectified through additional compilation of modern records from Asian (chiefly, but not exclusively Russian) beetle collections and literature. Of course the modern distribution data alone is of no use to the development of species' climate envelopes, unless the modern collecting localities can be paired with modern climatic parameters for those sites. Again, this is far easier to do in Europe than it is in northern Asia, where meteorological stations are few and far between. It seems likely that our only recourse will be to rely on modern climate modeling, in which the climatic parameters of remote localities are inferred from a gridded climate model, as used by the author in the development of species' climate envelopes for sites in arctic North America (Elias *et al.*, 1996a). If we are to proceed with ubiquity modeling in order to more closely constrain MCR estimates, then this will require the reworking of all the species' climate envelopes yet again. This is not terribly difficult, but it is very time consuming. I have no doubt that the results could yield very valuable refinements to the method, however.

7. Chitin Isotope Studies

As of this writing, we are just on the cusp of making major advancements in the field of paleothermometry through the analysis of fossil beetle stable isotopes, especially $\delta^{18}O$ studies. Both oxygen and hydrogen isotopes have been studied from modern beetle chitin. The use of hydrogen isotopes has two important drawbacks. Hydrogen is partly exchangeable and fractionates as it passes through the food chain. Oxygen does not have these intrinsic problems, so we intend to use stable oxygen isotopes in future studies. The work of Gröcke *et al.* (2006) with modern ground beetles in Europe, followed by the work of van Hardenbroek *et al.* (2009) on modern water beetles from North America, has demonstrated the strong correlation of the $\delta^{18}O$ signature of modern beetle chitin and the $\delta^{18}O$ ratios of regional precipitation across broad latitudinal transects. Initial tests indicate that the chitin preserved in fossil beetle sclerites should provide accurate, reliable $\delta^{18}O$ signals, allowing us to reconstruct paleotemperatures from beetle remains by a method that is totally independent from MCR.

8. Ancient DNA Studies

Ancient chitin holds another set of useful chemicals, other than oxygen isotopes. Recent work by Reiss (2006), King *et al.* (2009), Thomsen *et al.* (2009), and my own current collaboration with Ian Barnes (unpublished data) has established that ancient DNA is indeed preserved in fossil beetle sclerites, and that identifiable genetic material can be recovered from specimens tens of thousands of years old, especially from permafrost or very arid environments where bacterial decomposition is quite limited. This preliminary research opens up new vistas for future investigations that will reveal things about ancient beetle populations we could learn in no other way.

One of the most fundamental tenets of Quaternary entomology is our deduction, based on the external morphology of specimens, that beetles have not evolved new species in most, if not all, of the Pleistocene. The morphological evidence for this is corroborated by the ecological evidence: the associations of beetle species in Pleistocene (and sometimes even Pliocene) assemblages make ecological "sense" today. That is, the species found together in fossil assemblages are ecologically compatible today, even if they cannot be found together in any single region in the modern world. When I give lectures on Quaternary entomology to audiences including modern biologists, I frequently meet with skepticism on this topic. The typical question is, "How can you be sure of this?" I predict that the ancient DNA evidence is going to answer this question, once and for all. The answers we get concerning the longevity of various species may surprise us, but whether or not they confirm our previous ideas of species' constancy in the Pleistocene, they will tell us a great deal about the population dynamics of these organisms. Unlike some birds and mammals, there are no geographic races of beetles that are discernable from their fossil remains, but ancient DNA has the potential to help unravel the history of these insects through successive climatic oscillations. Did some populations die out, while others flourished? What are the geographic origins of modern populations? These are the kinds of questions we should be able to address in the coming years.

References

Aalto, M.M., Coope, G.R., & Gibbard, P.L., 1984. Late Devensian river deposits beneath the Floodplain Terrace of the River Thames at Abingdon, Berkshire, England. Proceedings of the Geologists' Association 95, 65–79.

Abe, H., 1994. A pictorial guide to the mammals of Japan. Kanagawa, Japan, Tokai University Press.

Addyman, P.V., 1989. The archaeology of public health at York, England. World Archaeology 21, 244–264.

Addyman, P.V., Hood, J.S.R., Kenward, H.K., MacGregor, A., & Williams, D., 1976. Palaeoclimate in urban environmental archaeology at York, England: problems and potential. World Archaeology 8, 220–233.

Alfieri, A., 1931. Les insectes de la tombe de Toutankhamon [Insects from the tomb of Tutankhamun]. Bulletin de la Société Royale Entomologique d'Egypte 3/4, 188–189. [In French].

Alfimov, A.V. & Berman, D.I., 2001. Beringian climate during the Late Pleistocene and Holocene. Quaternary Science Reviews 20, 127–134.

Allison, E., Hall, A., & Kenward, H.K., 1999. Technical report: living conditions and resource exploitation at the Early Christian Rath at Deer Park Farms, Co. Antrim, N. Ireland: evidence from plants and invertebrates. Part 1: text. Reports from the Environmental Archaeology Unit, York, 99/8.

Allison, E.P., Hall, A.R., Kenward, H.K., McKenna, B., Nicholson, C.M., & O'Connor, T.P., 1996. Environmental evidence. In: Foreman, M. (Ed), Further excavations at the Dominican Priory, Beverley, 1986–1989. Sheffield Excavation Reports 4, Sheffield, 195–212.

Allison, E.P. & Kenward, H.K., 1987. Insect remains from Copthall Avenue, London. Ancient Monuments Laboratory Report 194/87, London.

Allison, E.P. & Kenward, H.K., 1996. Insect remains (from a viscera/rubbish pit). In: Emery, N. (Ed), The archaeology and ethnography of St. Kilda, 1. Edinburgh, UK, Her Majesty's Stationery Office, 174–176.

Alluaud, C., 1908. Notes sur les coléoptères trouvés dans les momies d'Égypte [Notes on the beetles found in Egyptian mummies]. Bulletin de la Société Royale Entomologique d'Égypte 1, 29–36. [In French].

Ami, H.M., 1894. Fossil insects from the Leda Clays of Ottawa and vicinity. The Ottawa Naturalist 9, 190–191.

Ammann, B., Chaix, L., Eicher, U., Elias, S.A., Gaillard, M.-J., Hofmann, W., Siegenthaler, U., Tobolski, K., & Wilkinson, B., 1983. Vegetation, insects, mollusks and stable isotopes in Late-Würm deposits at Lobsigensee (Swiss Plateau). Revue de Paléobiologie 2, 184–204.

Ammann, B. & Lotter, A., 1989. Late-Glacial radiocarbon- and palynostratigraphy on the Swiss Plateau. Boreas 18, 109–126.

Amorosi, T., Buckland, P.C., Magnússon, K., McGovern, T.H., & Sadler, J.P., 1994. An archaeozoological investigation of the Midden at Nesstofa, Reykjavík, Iceland. In: Luff, R., Rowley-Conwy, P. (Eds), Whither Environmental Archaeology. Oxbow Monograph 38, 69–80.

Amorosi, T.A., Buckland, P.C., Olafsson, G., Sadler, J., & Skidmore, P., 1992. Site status and the palaeoecological record: a discussion of the results from Bessastaðir, Iceland. In: Morris, C.D., Rackham, D.J. (Eds), Norse and later settlement and subsistence in the North Atlantic. Glasgow, UK, Department of Archaeology, University of Glasgow, 169–192.

Anderson, E., 1984. Who's who in the Pleistocene: a mammalian bestiary. In: Martin, P.S., Klein, R.G. (Eds), Quaternary extinctions: a prehistoric revolution. Tucson, AZ, University of Arizona Press, 40–89.

Anderson, J., 1993. Beetle remains as indicators of the climate in the Quaternary. Journal of Biogeography 20, 557–562.

Anderson, L.L., Hu, F.S., Nelson, D.M., Petit, R.J., & Paige, K.N., 2006. Ice-age endurance: DNA evidence of a white spruce refugium in Alaska. Proceedings of the National Academy of Sciences 103, 12447–12450.

Anderson, P.M. & Lozhkin, A.V. (Eds), 2002. Late Quaternary vegetation and climate of Siberia and Russian Far East (Palynological and Radiocarbon Database). Magadan, Russia, Northeast Scientific Center, Russian Academy of Sciences.

Anderson, R.S., 1997. Overview of beetles of the Yukon. In: Danks, H.V., Downes, J.A. (Eds), Insects of the Yukon. Ottawa, Canada, Biological Survey of Canada (Terrestrial Arthropods), 405–444.

Anderson, R.S., Miller, N.G., Davis, R.B., & Nelson, R.E., 1990a. Terrestrial fossils in the marine Presumpscot Formation: implications for Late Wisconsinan paleoenvironments and isostatic rebound along the coast of Maine. Canadian Journal of Earth Sciences 27, 1241–1246.

Anderson, R.S., Peck, S.B., 1985. The insects and arachnids of Canada, Part 13. The carrion beetles of Canada and Alaska (Coleoptera: Silphidae and Agyrtidae). Agriculture Canada, Research Branch Publication 1778.

Anderson, T.W., Matthews, Jr, J.V., Mott, R.J., & Richard, S.H., 1990b. The Sangamonian Pointe-Fortune site, Ontario-Quebec border. Géographie physique et Quaternaire 44, 271–287.

Andersson, A., 1999. Insektsfynd [Insect remains]. In: Andersson, A. (Ed), Föreningen Gamla Halmstads Årsbok, Föreningen Gamla Halmstads, Old Halmstad Association, Halmstad, 21–24. [In Swedish].

Andersson, G., 1898. Studier öfver Finlands törfmossar och fossila kvartärflora. Fennia 15, 3–30. [In Swedish].

Andersson, G., 1986. Insekter från ostindiefararen Göteborg. Report 1986-12-14. Natural History Museum, Göteborg. [In Swedish].

Andersson, G., 1992a. Funna arter och grupper av insekter och frön I miljöprover. In: Hall, B., Andersson, G. (Eds), Tvävetenskapliga Studien Gården Björsjöås [Faunal species and groups of insects from the environmental samples]. Studier i Nordisk Arkeologi 17, Table 8. [In Swedish].

Andersson, G., 1992b. Arkeo-entomologiska undersökningar i historisk tid i Göteborg [Archaeo-entomological study from historic times in Gothenburg]. Entomologisk Tidskrift 113, 7–14. [In Swedish].

Andreev, A.A., Grosse, G., Schirrmeister, L., Kuzmina, S.A., Novenko, E.Yu., Bobrov, A.A., Tarasov, E., Ilyashuk, B.P., Kuznetsova, T.V., Krbetschek, M., Meyer, H., & Kunitsky, V.V., 2004. Late Saalian and Eemian palaeoenvironmental history of the Bol'shoy Lyakhovsky, Island Laptev Sea region, Arctic Siberia. Boreas 33, 319–348.

DEVELOPMENTS IN QUATERNARY SCIENCES
VOLUME 12 ISSN 1571-0866

Andres, A., 1931. Catalogue of the Egyptian Tenebrionidae. Bulletin Société Entomologique d'Egypte 15, 74–125.

Andrews, J.T. & Dyke, A.S., 2007. Late Quaternary in North America. In: Elias, S.A. (Ed), Encyclopedia of Quaternary science. Amsterdam, Elsevier, 1095–1101.

Angus, R.B., 1973. Pleistocene *Helophorus* (Coleoptera: Hydrophilidae) from Borislav and Starunia in the western Ukraine, with a reinterpretation of M. Lomnicki's species, description of a new Siberian species, and comparison with British Weichselian faunas. Philosophical Transactions of the Royal Society of London, Series B 265, 299–326.

Angus, R.B., 1975. Fossil Coleoptera from Weichselian deposits at Voorthuizen, The Netherlands. Geologie en Mijnbouw 54, 146–147.

Angus, R.B., 1983. Evolutionary stability since the Pleistocene illustrated by reproductive compatibility between Swedish and Spanish *Helophorus lapponicus* Thompson (Coleoptera: Hydrophilidae). Biological Journal of the Linnean Society of London 19, 17–25.

Antipina, E.E., Nazarov, V.I., & Maslov, S.P., 1991. Insects from the well of the Chaika settlement. In: Iron Age monuments in the vicinity of Eupatoria, Moscow University Press, Moscow, 155–161. [In Russian].

Antoine, P., Auguste, P., Bahain, J.-J., Coudret, P., Depaepe, P., Fagnart, J.-P., Falgueres, C., Fontugnes, M., Frechen, M., Hatte, C., Lamotte, A., Laurent, M., Limondin-Lozouet, N., Locht, J.-L., Mercier, N., Moigne, A.-M., Munaut, A.-V., Ponel, P., & Rousseau, D.-D., 2003. Paléoenvironnements pléistocènes et peuplements paléolithiques dans le bassin de la Somme (nord de la France) [Pleistocene paleoenvironments and Paleolithic occupation of the Somme River basin (northern France)]. Bulletin de la Société préhistorique française 100, 5–28. [In French].

Arkhipov, S.A., Isayeva, L.L., Bespaly, V.G., & Glushkova, O., 1986. Glaciation of Siberia and north-east USSR. Quaternary Science Reviews 5, 463–474.

Armelagos, G.J., 1969. Disease in ancient Nubia. Science 163, 255–259.

Armstrong, D.M., 1982. Mammals of the Canyon Country: a handbook of mammals of Canyonlands National Park and vicinity. Moab, UT, Canyonlands Natural History Association.

Arnett, Jr, R.H., 1973. The Beetles of the United States, a Manual for identification. Ann Arbor, MI, The American Entomological Institute.

Arnol'di, L.V., Zherikhin, V.V., Nikritin, L.M., & Ponomarenko, A.G., 1992. Mesozoic Coleoptera. Smithsonian Institution Libraries and National Science Foundation, Washington, DC. [Translation of *Mezozoiskie Zhestkokrylye*, Nauka Publishers, Moscow, 1977].

Aronsson, M., Hedenäs, L., Lagerbäck, R., Lemdahl, G., & Robertsson, A.-M., 1993. Flora och fauna I Norrbotten för 100 000 år sedan [Flora and fauna in the province of Norrbotten, northern Sweden, 100,000 years ago]. Svensk Botaniske Tidskrift 87, 241–253.

Ashworth, A.C., 1972. A Late-Glacial insect fauna from Red Moss, Lancashire, England. Entomologica Scandinavia 3, 211–224.

Ashworth, A.C., 1973. The climatic significance of a Late Quaternary fauna from Rodbaston Hall, Staffordshire, England. Entomologica Scandinavia 4, 191–205.

Ashworth, A.C., 1977. A late Wisconsin Coleopterous assemblage from southern Ontario and its environmental significance. Canadian Journal of Earth Sciences 14, 1625–1634.

Ashworth, A.C., 1979. Quaternary Coleoptera studies in North America: past and present. In: Erwin, T.L., Ball, G.E., Whitehead, D.R. (Eds), Carabid beetles: their evolution, natural history and classification. The Hague, The Netherlands, Dr. W. Junk Publishers, 395–406.

Ashworth, A.C., 1980. Environmental implications of a beetle assemblage from the Gervais Formation (Early Wisconsinan?), Minnesota. Quaternary Research 13, 200–212.

Ashworth, A.C., 2001. Perspectives on Quaternary beetles and climate change. In: Gerhard, L.C., Harrison, W.E., Hanson, B.E. (Eds), Geological perspectives of global climate change, Geologists Studies in Geology 47, American Association of Petroleum, Tulsa, OK, 153–168.

Ashworth, A.C., 2004. Quaternary Coleoptera of the United States and Canada. In: Gillespie, A.R., Porter, S.C., Atwater, B.F. (Eds), The Quaternary period in the United States. Developments in Quaternary science 1. Amsterdam, Elsevier, 505–517.

Ashworth, A.C., 2007. Beetle records: Late Pleistocene of South America. In: Elias, S.A. (Ed), Encyclopedia of Quaternary science. Amsterdam, Elsevier, 212–221.

Ashworth, A.C. & Brophy, J.A., 1972. Late Quaternary fossil beetle assemblage from the Missouri Coteau, North Dakota. Geological Society of America Bulletin 83, 2981–2988.

Ashworth, A.C., Clayton, L., & Bickley, W.C., 1972. The Mosbeck site: a paleoenvironmental interpretation of the Late Quaternary history of Lake Agassiz based on fossil insect and mollusk remains. Quaternary Research 2, 176–186.

Ashworth, A.C. & Hoganson, J.W., 1987. Coleoptera bioassociations along an elevational gradient in the lake region of southern Chile, and comments on the postglacial development of the fauna. Annals of the Entomological Society of America 80, 865–895.

Ashworth, A.C. & Hoganson, J.W., 1993. Magnitude and rapidity of the climate change marking the end of the Pleistocene in the mid-latitudes of South America. Palaeogeography, Palaeoclimatology, Palaeoecology 101, 263–270.

Ashworth, A.C., Hoganson, J., & Gunderson, M., 1989. Fossil beetle analysis. In: Dillehay, T.D. (Ed), Monte Verde: a Late Pleistocene settlement in Chile, Vol. 1, Washington, DC, Smithsonian Institution Press, 211–226.

Ashworth, A.C. & Markgraf, V., 1989. Climate of the Chilean Channels between 11,000 and 10,000 yr B.P. based on fossil beetle and pollen analyses. Revista Chilena de Historia Natural 62, 61–74.

Ashworth, A.C., Markgraf, V., & Villagran, C., 1991. Late Quaternary climatic history of the Chilean Channels based on fossil pollen and beetle analyses, with an analysis of the modern vegetation and pollen rain. Journal of Quaternary Science 6, 279–291.

Ashworth, A.C. & Schwert, D.P., 1992. The Johns Lake site: a late Quaternary fossil beetle (Coleoptera) assemblage from the Missouri Coteau of North Dakota. North Dakota Geological Survey Miscellaneous Series 76, 257–265.

Ashworth, A.C., Schwert, D.P., Watts, W.A., & Wright, Jr, H.E., 1981. Plant and insect fossils at Norwood in south-central Minnesota: a record of late-glacial succession. Quaternary Research 16, 66–79.

Askevold, I.S., 1990. Classification of Tertiary fossil Donaciinae of North America and their implications about evolution of Donaciinae (Coleoptera: Chrysomelidae). Canadian Journal of Zoology 68, 2135–2145.

Atkinson, T.C., Briffa, K.R., & Coope, G.R., 1987. Seasonal temperatures in Britain during the last 22,000 years, reconstructed using beetle remains. Nature 325, 587–592.

Atkinson, T.C., Briffa, K.R., Coope, G.R., Joachim, M., & Perry, D.W., 1986. Climatic calibration of coleopteran data. In: Berglund, B. (Ed), Handbook of Holocene palaeoecology and palaeohydrology. New York, Wiley, 851–858.

Audouin, J.V., 1835. Observation sur la découverte de *Gibbium scotias* dans un ancien tombeau à Thébes [Observations on the discovery of *Gibbium scotias* in an ancient tomb at Thebes]. Bulletin Société Entomologie de France 5, 5–6. [In French].

Austin, J.J., Ross, A.J., Smith, A.B., Fortey, R.A., & Thomas, R.A., 1997. Problems of reproducibility – does geologically ancient DNA survive in amber-preserved insects? Proceedings of the Royal Society of London, Series B 264, 467–474.

Avise, J.C., 1994. Molecular markers, natural history and evolution. New York, Chapman and Hall.

Bain, A., 1998. A seventeenth century beetle fauna from Colonial Boston. Historical Archaeology 32, 38–48.

Bain, A., 2001. Archaeoentomological and archaeoparasitological remains from the Îlot Hunt site, Quebec City (1850–1900). In: Pilon, J.-L., Kirby, M.W., Thériault, C. (Eds), A Collection of Papers Presented at the 33rd Annual Meeting of the Canadian Archaeological Association. Ontario Archaeological Society, Toronto, 42–47.

Bain, A.L., Morgan, A.V., Burns, J.A., & Morgan, A., 1997. The paleoentomology of Rat's Nest Cave, Grotto Mountain, Alberta, Canada. Quaternary Proceedings 5, 23–34.

Bajc, A.F., Schwert, D.P., Warner, B.G., & Williams, N.E., 2000. A reconstruction of Moorhead and Emerson Phase environments along the eastern margin of glacial Lake Agassiz, Rainy River Basin, northwestern Ontario. Canadian Journal of Earth Sciences 37, 1335–1353.

Baker, A.S., 1990. Two new species of *Lardoglyphus* Oudemans (Acari: Lardoglyphidae) found in the gut contents of human mummies. Journal of Stored Products Research 26, 139–147.

Baker, R.G., Rhodes, S.R., Schwert, D.P., Ashworth, A.C., Frest, T.J., Hallberg, G.R., & Janssens, J.A., 1986. A full-glacial biota from southeastern Iowa, USA. Journal of Quaternary Science 1, 91–108.

Baker, R.G., Schwert, D.P., Bettis, III, E.A., & Chumbley, C.A., 1993. Impact of Euro-American settlement on a riparian landscape in northeast Iowa, midwestern USA: an integrated approach based on historical evidence, floodplain sediments, fossil pollen, plant macrofossils and insects. The Holocene 3, 314–323.

Baker, R.G., Schwert, D.P., Bettis, III, E.A., Kemmis, T.J., Horton, D.G., & Semken, H.A., 1991. Mid-Wisconsinan stratigraphy and paleoenvironments at the St. Charles site in south-central Iowa. Geological Society of America Bulletin 103, 210–220.

Ball, G.E., 1966. A revision of the North American species of the subgenus *Cryobius* Chaudoir (*Pterostichus*, Carabidae, Coleoptera). Opuscula Entomologica Supplement 28, 166.

Barnes, I., Matheus, P., Shapiro, B., Jensen, D., & Cooper, A., 2002. Dynamics of Pleistocene population extinctions in Beringian brown bears. Science 295, 2267–2270.

Barnes, I., Shapiro, B., Lister, A., Kuznetsova, T., Sher, A., Guthrie, D., & Thomas, M.G., 2007. Genetic structure and extinction of the woolly mammoth, *Mammuthus primigenius*. Current Biology 17, 1072–1075.

Barnowsky, A., Barnowsky, C.W., Nickmann, R.J., Ashworth, A.C., Schwert, D.P., & Lantz, S.W., 1988. Late Quaternary paleoecology at the Newton site, Bradford Co., northeastern Pennsylvania: *Mammuthus columbi*, palynology, and fossil insects. Bulletin of the Buffalo Society of Natural Sciences 33, 173–184.

Barr, T.C., 1960. The cavernicolous beetles of the subgenus *Rhadine*, genus *Agonum* (Coleoptera: Carabidae). American Midland Naturalist 64, 45–65.

Barry, R.G., 1982. Paleoclimate. In: Hopkins, D.M., Matthews, Jr, J.V., Schweger, C.E., Young, S.B. (Eds), Paleoecology of Beringia. New York, Academic Press, 193–204.

Bartlein, P.J., Lipsitz, B., & Thompson, R.S., 1994. Modern climate data for paleoenvironmental interpretations. American Quaternary Association Thirteenth Biennial Meeting, Program and Abstracts, 197.

Basilewsky, P., 1970. Les Coléopteres Carabidae de L'Ille d'Aldabra (Océan Indien). Bulletin Annuel de la Société Royale Entomologique de Belgique 106, 221–222. [In French].

Bateman, M.D., Buckland, P.C., Carpenter, R., Davies, S., Frederick, C.D., Geary, B., Murton, J.B., & Whitehouse, N.J., 2001. Cove Farm Quarry, Westwoodside (SE 737008). In: Bateman, M.D., Buckland, P.C., Frederick, C.D., Whitehouse, N.J. (Eds), The Quaternary of East Yorkshire and North Lincolnshire field guide. London, UK, Quaternary Research Association, 141–160.

Beets, D.J., Beets, C.J., & Cleveringa, P., 2006. Age and climate of the late Saalian and early Eemian in the type-area, Amsterdam basin, The Netherlands. Quaternary Science Reviews 25, 876–885.

Begét, J., 1996. Tephrochronology and paleoclimatology of the last interglacial-glacial cycle recorded in Alaskan loess deposits. Quaternary International 34–36, 121–126.

Behre, K.-E., Hölzer, A., & Lemdahl, G., 2005. Botanical macro-remains and insects from the Eemian and Weichselian site of Oerel (northwest Germany) and their evidence for the history of climate. Vegetation History and Archaeobotany 14, 31–53.

Bell, C.H., 1991. Activity rhythms linked with foraging behaviour in insecticide-resistant and susceptible strains of *Oryzaephilus surinamensis* (Coleoptera: Silvanidae). Journal of Stored Products Research 27, 171–177.

Bell, F.G., Coope, G.R., Rice, R., & Riley, T., 1972. Mid-Weichselian fossil-bearing deposits at Syston, Leicestershire. Proceedings of the Geological Association 83, 197–211.

Bennett, K.D., 1990. Milankovitch cycles and their effects on species in ecological and evolutionary time. Paleobiology 16, 11–21.

Bennett, K.D., Haberle, S.G., & Lumley, S.H., 2000. The last glacial-Holocene transition in southern Chile. Science 290, 325–328.

Bennike, O., 1992. Paleoecology and paleoclimatology of a Late Holocene Peat Deposit from Brændevinsskær, Central West Greenland. Arctic and Alpine Research 24, 249–252.

Bennike, O., 2000. Notes on the late Cenozoic history of the Washington land area, Western North Greenland. Geology of Greenland Survey Bulletin 186, 29–34.

Bennike, O. & Böcher, J., 1992. Early Weichselian interstadial land biotas at Thule, northwest Greenland. Boreas 21, 111–117.

Bennike, O., Houmark-Nielsen, M., Böcher, J., & Heiberg, E.O., 1994. A multidisciplinary macrofossil study of Middle Weichselian sediments at Kobbelgård, Møn, Denmark. Palaeogeography, Palaeoclimatology, Palaeoecology 111, 1–15.

Bennike, O. & Jensen, J.B., 1995. Near-shore Baltic Ice Lake deposits in Fakse Bugt, southeast Denmark. Boreas 24, 185–195.

Benson, L., 2004. Western lakes. In: Gillespie, A.R., Porter, S.C., Atwater, B.F. (Eds), The Quaternary period in the United States. Amsterdam, Elsevier, 185–204.

Berger, A.L., 1978. Long-term variations in caloric insolation resulting from the earth's orbital elements. Quaternary Research 9, 139–167.

Berglund, B.E. & Digerfeldt, G., 1970. A paleoecological study of the Late-Glacial lake at Torreberga, Scania, South Sweden. Oikos 21, 98–128.

Berglund, B.E., Lemdahl, G., Leidberg-Jonsson, B., & Persson, T., 1984. Biotic response to climatic changes during the time span 13,000–10,000 BP-a case from SW Sweden.

In: Mörner, N.A., Karlén, W. (Eds), Climatic changes on a yearly to Millenial basis. Dordrecht, The Netherlands, Reidel, 25–36.

Berman, D.I., 1990. Ecology of *Morychus viridis* (Coleoptera, Byrrhidae), a moss beetle from Pleistocene deposits in the northeastern USSR. In: Kotlyakov, V.M. Sokolov, V.E. (Eds), Arctic Research: Advances and Prospects. Proceedings of the Conference of Arctic and Nordic Countries on Coordination of Research in the Arctic. Nauka Press, Moscow, 281–288.

Berman, D.I. (Ed), 2001. Cold steppes of the north-east Asia. Russian Academy of Sciences, Far East Branch. Magadan, Russia, Institute of Biological Problems of the North.

Bidashko, F.G. & Proskurin, K.P., 1988. The entomological and carpological reconstruction of the bio-environment of the Singilian (Middle Pleistocene) of the lower Volga. Paleontological Journal 4, 66–72.

Bigelow, N. & Edwards, M.E., 2001. A 14,000 yr paleoenvironmental record from Windmill Lake, central Alaska: Lateglacial and Holocene vegetation in the Alaska Range. Quaternary Science Reviews 20, 203–215.

Binford, M.W., 1982. Ecological history of Lake Valencia, Venezuela: interpretation of animal microfossils and some chemical, physical, and geological features. Ecological Monographs 52, 307–333.

Birks, H.H., Battarbee, R.W., Birks, H.J.B., Bradshaw, E.G., Brooks, S.J., Duigan, C.A., Jones, V.J., Lemdahl, G., Peglar, S.M., Solem, S.M., Solhøy, T., & Stalsberg, M.K., 2000. The development of the aquatic ecosystem at Kråkenes lake, western Norway, during the Late-glacial and Early-Holocene – a synthesis. Journal of Paleolimnology 23, 91–114.

Birks, H.H., Lemdahl, G., Svendsen, J.I., & Landvik, J.Y., 1993. Palaeoecology of a late Allerød peat bed at Godøy, western Norway. Journal of Quaternary Science 8, 147–159.

Bishop, W.W. & Coope, G.R., 1977. Stratigraphical and faunal evidence for Lateglacial and Early Flandrian environments in south-west Scotland. In: Gray, J.M., Lowe, J.J. (Eds), Studies in the Scottish Lateglacial Environment. Oxford, Pergamon Press, 61–88.

Blaisdell, F.E., 1927. A blind beetle excavated from an Egyptian city's ruins dating between 117 and 235 A.D. Proceedings of the Entomological Society of Washington 29, 121–125.

Blake, Jr, W., Matthews, Jr, J.V., 1979. New data on an interglacial peat deposit near Makinson Inlet, Ellesmere Island, District of Franklin, Geological Survey of Canada Papers 79-1A, 157, 164

Böcher, J., 1988a. The Coleoptera of Greenland. Meddelelser om Grønland 26, 1–100.

Böcher, J., 1988b. Insektundersøgelserne. Naturens Verden 11–12, 409–423. [In Danish].

Böcher, J., 1989a. Boreal insects in northernmost Greenland: palaeoentomological evidence from the Kab København Formation (Plio-Pleistocene), Peary Land. Fauna Norvegica, Series B 36, 37–43.

Böcher, J., 1989b. First record of an interstadial insect from Greenland: *Amara alpina* (Paykull, 1790) (Coleoptera: Carabidae). Boreas 18, 1–4.

Böcher, J., 1990. A two-million-year-old insect fauna from north Greenland indicating boreal conditions at the Plio-Pleistocene boundary. Proceedings of the International Conference on the Role of Polar Regions in Global Change 2, 582–584.

Böcher, J., 1995. Paleoentomology of the Kab København Formation, a Plio-Pleistocene sequence in Peary Land, North Greenland. Meddelelser om Grønland, Geoscience 33, 82.

Böcher, J., 1998. Insect remains from Asummiut. In: Arneborg, J., Gulløv, H.C. (Eds), Man, culture and environment in ancient Greenland. Copenhagen, Denmark, Danish National Museum & Danish Polar Centre, 133–134.

Böcher, J. & Bennike, O., 1991. Interglacial land biotas of Jameson Land, East Greenland. LUNQUA Report 33, 129–133.

Böcher, J. & Bennike, O., 1996. Early Holocene insect and plant remains from Jameson land, East Greenland. Boreas 25, 187–193.

Böcher, J. & Fredskild, B., 1993. Plant and arthropod remains from the palaeo-Eskimo site on Qeqertasussuk, West Greenland. Meddelelser om Grønland, Geoscience 30, 1–35.

Böcher, J. & Nachman, G., 2001. Temperature and humidity responses of the arctic-alpine seed bug *Nysius groenlandicus*. Entomologia Experimentalis et Applicata 99, 319–330.

Boessneck, J., 1988. Die Tierwelt des alten Ägypten, untersucht anhand kulturgeschichtlicher und zoologischer Quellen [The animal life of ancient Egypt: studies from cultural material and zoological sources]. Beck, Munich, 73–78. [In German].

Bohncke, S., Vandenberghe, J., Coope, G.R., & Reiling, R., 1987. Geomorphology and palaeoecology of the Mark valley (southern Netherlands): palaeoecology, palaeohydrology and climate during the Weichselian Late Glacial. Boreas 16, 69–85.

Bolton, J., 1862. On a deposit with insects, leaves, etc. near Ulverston. Proceedings of the Geological Society of London 18, 274–277.

Bond, G., Broecker, W., Johnsen, S., McManus, J., Labeyrie, L., Jouzel, J., & Bonani, G., 1993. Correlations between climate records from North Atlantic sediments and Greenland ice. Nature 365, 143–147.

Bondestam, K., Vasari, A., Vasari, Y., Lemdahl, G., & Eskonen, K., 1994. Younger Dryas and Preboreal in Salpausselkä Foreland, Finnish Karelia. Dissertationes Botanicae 234, 161–206.

Borror, D.J., DeLong, D.M., & Triplehorn, C.A., 1981. An introduction to the study of insects (5th ed.). Philadelphia, PA, Saunders College Publishing.

Boswijk, G. & Whitehouse, N.J., 2002. *Pinus* and *Prostomis*: a dendrochronological and palaeoentomological study of a study of a mid-Holocene woodland in Eastern England. The Holocene 12, 585–596.

Bousquet, Y., 1990. Beetles associated with stored products in Canada: an identification guide. Research Branch, Agriculture Canada Publication 1837.

Bowen, G.J., Wassenaar, L.I., & Hobson, K.A., 2005. Global application of stable hydrogen and oxygen isotopes to wildlife forensics. Oecologia 143, 337–348.

Boyer-Lefèvre, N.H., 1971. Influence de la dessiccation sur trois coléoptères de la tribu des Sphodrini vivant dans milieux différents. Bulletin de Societé Histiore Naturel Toulouse 107, 595–605.

Bray, P.J., Blockley, S.P.E., Coope, G.R., Dadswell, L.F., Elias, S.A., Lowe, J.J., & Pollard, A.M., 2006. Refining mutual climatic range (MCR) quantitative estimates of palaeotemperature using ubiquity analysis. Quaternary Science Reviews 25, 1865–1876.

Brayshay, B.A. & Dinnin, M., 1999. Integrated palaeoecological evidence for biodiversity at the floodplain-forest margin. Journal of Biogeography 26, 115–131.

Bresciani, J., Haarlov, N., Nansen, P., & Moller, G., 1983. Head louse (*Pediculus humanus* subsp. *capitis* de Geer) from mummified corpses of Greenlanders, A.D. 1460 (±50). Acta Entomologica Fennica 42, 24–27.

Bresciani, J., Haarlov, N., Nansen, P., & Moller, G., 1989. Head lice in mummified Greenlanders from A.D. 1475. Meddelelser om Grønland, Man & Society 12, 89–92.

Briant, R.M., Coope, G.R., Preece, R.C., Keen, D.H., Boreham, S., Griffiths, H.I., Seddon, M.B., & Gibbard, P.L., 2004. Fluvial system response to Late Devensian (Weichselian) aridity,

Baston, Lincolnshire, England. Journal of Quaternary Science 19, 479–495.

Bridgland, D.R., Allen, T., Coope, G.R., Gibbard, P.L., & Wrayton, R., 1994. Great Totham (Lofts Farm pit; TL 866092). In: Bridgland, D.R. (Ed), Quaternary of the Thames. London, UK, Chapman & Hall, 376–385.

Bridgland, D.R., Preece, R.C., Roe, H.M., Tipping, R.M., Coope, G.R., Field, M.H., Robinson, J.E., Schreve, D.C., & Crowe, K., 2001. Middle Pleistocene interglacial deposits at Barling, Essex, England: evidence for a longer chronology for the Thames terrace sequence. Journal of Quaternary Science 16, 813–840.

Briggs, D.J., Coope, G.R., & Gilbertson, D.D., 1975a. Late Pleistocene terrace deposits at Beckford, Worcestershire, England. Geological Journal 10, 1–16.

Briggs, D.J., Coope, G.R., & Gilbertson, D.D., 1985. The chronology and environmental framework of early man in the upper Thames Valley. British Archaeological Reports 137.

Briggs, D.J., Gilbertson, D.D., Goudie, A.S., Osborne, P.J., Osmaston, H.A., Pettit, M.M.E., Shotton, F.W., & Stuart, A.J., 1975b. New interglacial site at Sugworth. Nature 257, 477–479.

Britton, E.B., 1970. Coleoptera (Beetles). In: Waterhouse, D.F. (Ed), CSIRO Insects of Australia. Melbourne, Melbourne University Press.

Broccoli, A.J. & Manabe, S., 1987. The effects of the Laurentide ice sheet on North American climate during the last glacial maximum. Géographie physique et Quaternaire 41, 291–299.

Brown, A.P. & Robinson, M.A., 1985. A summary of plant and insect remains from the post-medieval pits at Church Street (Site A). In: Hassall, T.G., Halpin, C.E., Mellor, M. (Eds), Excavations in St. Ebbe's, Oxford, 1967–1976: Part II : Post-Medieval Domestic Tenements and the Post-Dissolution Site of the Greyfriars. Oxoniensia 49, 268+Fiche VI, F3.

Brubaker, L.B., 1986. Responses of tree populations to climatic change. Vegetatio 68, 119–130.

Brubaker, L.B., Anderson, P.M., Edwards, M.E., & Lozhkin, A.V., 2005. Beringia as glacial refugium for boreal trees and shrubs: new perspectives from mapped pollen data. Journal of Biogeography 32, 833–848.

Bryant, I.D., Holyoak, D.T., & Moseley, K.A., 1983. Late Pleistocene deposits at Brimpton, Berkshire, England. Proceedings of the Geologists' Association 94, 321–343.

Bryant, V.M., 1974. The role of coprolite analysis in archaeology. Bulletin of the Texas Archaeological Society 45, 1–28.

Bryson, R.A., 1966. Air masses, streamlines and the boreal forest. Geographical Bulletin 8, 228–269.

Büchner, S. & Wolf, G., 1997. Der Kornkäfer-*Sitophilus granarius* (Linné)-aus einer bandkeramischen Grube bei Göttingen [The grain beetle *Sitophilus granarius* from a ribbon-ceramic pit near Göttingen]. Archäologisches Korrespondenzblatt 27, 211–220. [In German].

Buckland, P.C., 1973. Archaeology and environment in the Vale of York. South Yorkshire Doncaster Museum-Studies in Archaeology and Natural History 1, 6–18.

Buckland, P.C., 1974. Archaeology and environment in York. Journal of Archaeological Science 1, 303–316.

Buckland, P.C., 1976a. The use of insect remains in the interpretation of archaeological environments. In: Davidson, D.A., Shackley, M.L. (Eds), Geoarchaeology: earth sciences and the past. London, UK, Duckworth, 360–396.

Buckland, P.C., 1976b. *Niptus hololeucus* Fald. (Col., Ptinidae) from Roman deposits in York. Entomologist's Monthly Magazine 111, 233–234.

Buckland, P.C., 1976c. The environmental evidence from the Church Street Roman Sewer System. Archaeology of York 14/1. Council for British Archaeology for York Archaeological Trust.

Buckland, P.C., 1979. Thorne Moors: a palaeoecological study of a Bronze Age site (a contribution to the history of the British insect fauna). Department of Geography Occasional Publication 8, University of Birmingham, Birmingham, UK.

Buckland, P.C., 1980. Insect remains from the Well. In: Stead, I.M. (Ed), Rudston Roman Villa. Leeds, Yorkshire Archaeological Society, 162–167.

Buckland, P.C., 1981a. The early dispersal of insect pests of stored products as indicated by archaeological records. Journal of Stored Product Research 17, 1–12.

Buckland, P.C., 1981b. An insect fauna from a Roman well at Empingham, Rutland. Transactions of the Leicestershire Archaeological and Historical Society 60, 1–6.

Buckland, P.C., 1981c. Insect remains from beneath the Brigg 'Raft'. In: McGrail, S. (Ed), The Brigg 'Raft' and her prehistoric environment. British Archaeological Reports, Oxford, 89, 155–175.

Buckland, P.C., 1982a. The Malton burnt grain: a cautionary tale. Yorkshire Archaeological Journal 54, 53–61.

Buckland, P.C., 1982b. The Cover Sands of North Lincolnshire and the Vale of York. In: Adlam, B.H., Fenn, C.R., Morris, L. (Eds), Papers in earth studies. Norwich, UK, Geobooks, 143–178.

Buckland, P.C., 1984. North-west Lincolnshire 10,000 years ago. In: Field, N., Norwich, A. (Eds), A prospect of Lincolnshire, Lincoln. Norwich, UK, Geobooks, 11–17.

Buckland, P.C., 1994. A beetle fauna from the organic deposit F23 in the stone-lined pit (F22). In: McCormick, F. (Ed), Excavations at Pluscarden Priory, Moray. Proceedings of the Society of Antiquaries of Scotland, 124, 415–417.

Buckland, P.C., 1996. Insects. In: May, J. (Ed), Dragonby, report on excavations at an Iron Age and Romano-British settlement in North Lincolnshire. Oxbow Monograph 61, Oxford, 165–170.

Buckland, P.C., 2001. West Moor, Armthorpe (SE 637058). In: Bateman, M.D., Buckland, P.C., Frederick, C.D., Whitehouse, N.J. (Eds), The Quaternary of East Yorkshire and North Lincolnshire field guide. London, UK, Quaternary Research Association, 99–102.

Buckland, P.C., 2002. Insect fauna from within the cranium of skeleton 570. Excavations at Glasgow Cathedral 1988–1997. Society for Medieval Archaeology Monograph 18, 157.

Buckland, P.C., Amorosi, T., Barlow, L.K., Dugmore, A.J., Mayewski, P.A., McGovern, T.H., Ogilvie, A.E.J., Sadler, J.P., & Skidmore, P., 1996a. Bioarchaeological and climatological evidence for the fate of Norse farmers in medieval Greenland. Antiquity 70, 88–96.

Buckland, P.C., Ashworth, A.C., & Schwert, D.P., 1996b. By-passing Ellis Island: insect immigration to North America. In: Butlin, R.A., Roberts, N. (Eds), Ecological relations in historical times. London, UK, Royal Geographical Society, 227–244.

Buckland, P.C., Beal, C.J., & Heal, S.V.E., 1990. Recent work on the archaeological and environmental context of the Ferriby boats. In: Ellis, S., Crowther, D.R. (Eds), Humber perspectives: a region through the ages. Hull, UK, Hull University Press, 131–146.

Buckland, P.C., Buckland, P.I., & Skidmore, P., 1998. Insect remains from GUS: an interim report. In: Arneborg, J., Gulløv, H.C. (Eds), Man, culture and environment in ancient Greenland. Copenhagen, Denmark, Danish National Museum & Danish Polar Centre, 74–79.

Buckland, P.C. & Coope, G.R., 1991. A bibliography and literature review of Quaternary entomology. Sheffield, J.R. Collis.

Buckland, P.C., Coope, G.R., & Sadler, J.P., 2009a. Bibliography of Quaternary Entomology. In: Buckland, P.C. and Buckland, P.I. (Eds), Bugs Coleopteran ecology package (Versions: BugsCEP v7.63; Bugsdata v7.11; BugsMCR v2.02; BugStats v1.22) [Downloaded on 10 February 2009 from www.bugscep.com].

Buckland, P.C. & Dinnin, M.H., 1997. The rise and fall of a wetland habitat: recent palaeoecological research on Thorne and Hatfield Moors. Thorne and Hatfield Moors Papers 4, 1–18.

Buckland, P.C. & Dinnin, M.H., 1998. Insect Faunas at Landnám: a palaeoentomological study at Tjørnuvík, Streymoy, Faroe Islands. Fróðskaparrit 46, 277–286.

Buckland, P.C. & Dinnin, M.J., 1993. Holocene woodlands: the fossil insect evidence. In: Kirby, K., Drake, C.M. (Eds), Dead wood matters: the ecology and conservation of saproxylic invertebrates in Britain. English Nature Science Series No. 7, Peterborough, 6–20.

Buckland, P.C., Dugmore, A.J., Perry, D.W., Savory, D., & Sveinbjarndottir, G., 1991. Holt in Eyjafjallasveit, Iceland. A paleoecological study of the impact of Landnám. Acta Archaeologica 61, 252–271.

Buckland, P.C., Edwards, K.J., Blackford, J., Dugmore, A.J., Sadler, J.P., & Sveinbjarnardóttir, G., 1995. A question of Landnám: pollen, charcoal and insect studies on Papey, eastern Iceland. In: Butlin, R., Roberts, N. (Eds), Ecological relations in historical times. Oxford, Institute of British Geographers, Blackwell, 245–264.

Buckland, P.C., Edwards, K.J., Panagiotakopulu, E., & Schofield, E., 2009b. Palaeoecological and historical evidence for manuring and irrigation at Garðar (Igaliku), Norse Eastern Settlement, Greenland. The Holocene 91, 105–116.

Buckland, P.C., Gerrard, A.J., Larsen, G., Perry, D.W., Savory, D.R., & Sveinbjarndóttir, G., 1986a. Late Holocene palaeoecology at Ketilsstadir in Myrdalur, South Iceland. Jökull 36, 41–55.

Buckland, P.C., Greig, J.R.A., Frederick, C., Wagner, P., & Beal, C., 2006a. The Iron Age and Roman landscape. In: Stead, I.M., Flouest, J.-L., Rigby, V. (Eds), Iron Age and Roman burials in Champagne. Oxford, Oxbow Books, 133–156.

Buckland, P.C., Greig, J.R.A., & Kenward, H.K., 1974. York: an early medieval site. Antiquity 48, 25–33.

Buckland, P.C., Holdsworth, P., & Monk, M., 1976. The interpretation of a group of Saxon pits in Southampton. Journal of Archaeological Science 3, 61–69.

Buckland, P.C. & Johnson, C., 1983. *Curimopsis nigrita* (Palm) (Coleoptera: Byrrhidae) from Thorne Moors, South Yorkshire. Naturalist (Hull) 108, 153–154.

Buckland, P.C. & Kenward, H.K., 1973. Thorne Moor: a palaeoecological study of a Bronze Age site. Nature 241, 405–406.

Buckland, P.C., Panagiotakopulu, E., Buckland, P.I., Perdikaris, S., & Skidmore, P., 2006b. Insect faunas from medieval Langenes in Arctic Norway. In: Engelmark, R., Linderholm, J. (Eds), Proceedings from the 8th Nordic conference on the application of scientific methods in archaeology, Umeå 2001. Umeå, Department of Archaeology & Sami Studies, University of Umeå, 17–32.

Buckland, P.C., Parker Pearson, M., Wigley, A., & Girling, M.A., 2001. Is there anybody out there? A reconsideration of the environmental evidence from the Breiddin Hillfort, Powys, Wales. Antiquaries Journal 81, 51–76.

Buckland, P.C. & Perry, D.W., 1989. Ectoparasites of sheep from Storaborg, Iceland and their interpretation. Piss, parasites and people, a palaeoecological perspective. Hikuin 15, 37–46.

Buckland, P.C., Perry, D.W., Gislason, G.M., & Dugmore, A.J., 1986b. The pre-Landnám Fauna of Iceland: a palaeontological contribution. Boreas 15, 173–184.

Buckland, P.C., Perry, D.W., & Sveinbjarnardóttir, G., 1983a. *Hydraena britteni* Joy (Coleoptera, Hydraenidae) fundin í setlögum á Islandi, frá því seint á nútima [Remains of *Hydraena britteni* Joy (Coleoptera, Hydraenidae) found in Icelandic settlements]. Náttúrufræðingurinn 52, 37–44. [In Icelandic].

Buckland, P.C. & Sadler, J., 1985. Late Flandrian alluviation in the Humberhead Levels. East Midland Geographer 8, 239–251.

Buckland, P.C. & Sadler, J., 1989. A biogeography of the human flea, *Pulex irritans* L. (Siphonaptera : Pulicidae). Journal of Biogeography 16, 115–120.

Buckland, P.C. & Sadler, J., 1991. Farm or shieling: an entomological approach. In: Sveinbjarnardóttir, G. (Ed), Shielings in Iceland – an archaeological and historical survey. Acta Archaeologia 61, 93–96.

Buckland, P.C., Sadler, J.P., & Sveinbjarnardóttir, G., 1992. Palaeoecological Investigations at Reykholt, Western Iceland. In: Morris, C.J., Rackham, D.J. (Eds), Norse and later settlement and subsistence in the North Atlantic. Glasgow, UK, Department of Archaeology, University of Glasgow, 149–168.

Buckland, P.C., Sveinbjarnardottir, G., Savory, D., McGovern, T.H., Skidmore, P., & Andreasen, C., 1983b. Norsemen at Nipáitsoq, Greenland: a palaeoecological investigation. Norwegian Archaeological Review 16, 86–98.

Buckland, P.I., 2000. An introduction to palaeoentomology in archaeology and the BUGS database management system. Magister Paper, Archaeology and Sámi Studies, University of Umeå, Sweden.

Buckland, P.I., 2007. The Development and Implementation of Software for Palaeoenvironmental and Palaeoclimatological Research: the bugs Coleopteran ecology package (BugsCEP). PhD Thesis, Umeå University, Sweden. Archaeology and Environment, 23.

Buckland, P.I. & Buckland, P.C., 2006. Bugs Coleopteran ecology package (Versions: BugsCEP v7.63; Bugsdata v7.11; BugsMCR v2.02; BugStats v1.22) [Downloaded on 10 February 2009 from www.bugscep.com].

Burge, P. & Shulmeister, J., 2007. Re-envisioning the structure of last glacial vegetation in New Zealand using beetle fossils. Quaternary Research 68, 121–132.

Calder, A.A., 2002. Adephaga. Australian faunal directory. Australian Biological Resources Study, Canberra. http://www.environment.gov.au/biodiversity/abrs/online-resources/fauna/afd/index.html [Viewed on 29 November 2007].

Callen, E.O., 1967. Analysis of the Tehuacan coprolites. In: Byers, D.S. (Ed), The Prehistory of the Tehuacan Valley, Vol. 1, Environment and Subsistence, Austin, University of Texas Press, 261–289.

Callen, E.O., 1970. Diet as revealed by coprolites. In: Brothwell, D., Higgs, E. (Eds), Science in archaeology. New York, Praeger Publishers, 235–243.

Cameron, A.W., 1958. Mammals of the islands in the Gulf of St. Lawrence. National Museums of Canada Bulletin 154, 1–165.

Campbell, J.M., 1979. A revision of the genus *Tachyporus* Gravenhorst (Coleoptera: Staphylinidae) of North and Central America. Memoirs of the Entomological Society of Canada 109, 95.

Campbell, J.M., 1980. Distribution patterns of Coleoptera in eastern Canada. Canadian Entomologist 112, 1161–1175.

Campbell, J.M., 1982. A revision of the North American Omaliinae (Coleoptera: Staphylinidae) 3. The genus *Acidota* Stephens. Canadian Entomologist 114, 1003–1029.

Campbell, J.M., 1983. A revision of the North American Omaliinae (Coleoptera: Staphylinidae): The genus *Olophrum* Erichson. Canadian Entomologist 115, 577–622.

Cano, R.J., Poinar, H.N., Pienaizek, N.J., Acra, A., & Poinar, G., 1993. Amplification and sequencing of DNA from a 120–135-million-year-old weevil. Nature 363, 536–538.

Capasso, L. & Di Tota, G., 1998. Lice buried under the ashes of Herculaneum. Lancet 351, 992.

Cappers, R.T.J., Bosch, J.H.A., Bottema, S., Coope, G.R., van Geel, B., Mook-Kamps, E., & Woldring, H., 1993. De reconstructie van het landschap [Landscape reconstruction]. In: van der Sanden, W.A.B., Cappers, R.T.J., Beuker, J.R., Mol, D. (Eds), Mens en mammoet. De mammoeten van Orvelte en de vroegste van Noord-Nederland [People and mammoths. The mammoth from Orvelte in northern Netherlands]. Drents, The Netherlands, Drents Museum, 27–41. [In Dutch].

Carrott, J., Fell, V., Large, F., Kenward, H., Kent, P., (2000). Insects and other invertebrate remains. In: Buxton, K., Howard-Davies, C. (Eds), Bremetenacum: excavations at Roman Ribchester 1980, 1989–1990. Lancaster Imprint Series 9, Lancaster, 387–400.

Carrott, J., Hall, A., Kenward, H., Large, F., & Milles, A., 1993. Plant and invertebrate remains from two fill contexts from a Romano-British pit at Glebe Farm, Barton-on-Humber, S. Humberside (site code GFA92). Reports from the Environmental Archaeology Unit, York, 93/13.

Carter, K.D., 1985. Middle and late Wisconsinan (Pleistocene) insect assemblages from Illinois. Unpublished Masters Thesis, University of South Dakota.

Caseldine, A.E., 1987. Charcoal patches from Meare Village East 1982. In: Coles, J.M. (Ed), Meare Village East. The Excavations of A. Bulleid and H. St. George Gray. Somerset Levels Papers 13, 223–226.

Caseldine, C.J., Hatton, J., Gearey, B., Stuijts, I., Reilly, E., & Casparie, W., 2001. From the wet to the dry: palaeoecological studies at Derryville Bog, Co. Tipperary, Ireland. In: Raftery, B., Hickey, J. (Eds), Recent developments in wetland research. Seandálaíocht Monograph 2, Department of Archaeology, UCD and WARP Occasional Paper 14, 99–115.

Chaddick, P.R. & Leek, F.F., 1972. Further specimens of stored products insects found in Ancient Egyptian tombs. Journal of stored Products Research 8, 83–86.

Chapco, W. & Litzenberger, G., 2004. A DNA investigation into the mysterious disappearance of the Rocky Mountain grasshopper, megapest of the 1800s. Molecular Phylogenetics and Evolution 30, 810–814.

Chikaraishi, Y., Naraoka, H., & Poulson, S.R., 2004. Hydrogen and carbon isotopic fractionations of lipid biosynthesis among terrestrial (C3, C4 and CAM) and aquatic plants. Phytochemistry 65, 1369–1381.

Chomko, S.A. & Gilbert, B.M., 1991. Bone refuse and insect remains: their potential for temporal resolution of the archaeological record. American Antiquity 56, 680–686.

Chowne, P., Girling, M., & Greig, J., 1986. Excavations at an Iron Age defended enclosure at Tattershall Thorpe, Lincolnshire. Proceedings of the Prehistoric Society 52, 159–188.

Chu, H.F. & Wang, L.-Y., 1975. Insect carcasses unearthed from Chinese antique tombs. Acta Entomologica Sinica 18, 333–337. [in Chinese].

Churcher, C.S., 1966. The insect fauna from the Talara tar-seeps, Peru. Canadian Journal of Zoology 44, 985–993.

Churchill, L.L. & Nelson, R.E., 1992. Paleoenvironmental analysis of subfossil coleopteran fauna from the Toklat River valley, central Alaska. Maine Geologist 18, 4.

Clark, J.S., Fastie, C., Hurtt, G., Jackson, S.T., Johnson, C., King, G.A., Lewis, M., Lynch, J., Pacala, S., Prentice, C., Schupp, E.W., Webb, III, T., & Wyckoff, P., 1998. Reid's paradox of rapid plant migration. BioScience 48, 13–24.

Clark, W.H. & Sankey, J.T., 1999. Late Holocene Sonoran Desert arthropod remains from a packrat midden, Cataviña, Baja California Norté, México. Pan-Pacific Entomologist 75, 183–199.

Cockerell, T.D.A., 1911. Obituary of Samuel Hubbard Scudder. Science 34, 338–342.

Cole, K.L., 1985. Past rates of change, species richness and a model of vegetational inertia in the Grand Canyon, Arizona. The American Naturalist 125, 289–303.

Coles, J.M., 1987. Tracks across the Wetlands: multi-disciplinary studies of the Somerset Levels of England. In: Coles, J.M., Lawson, A.J. (Eds), European Wetlands in prehistory. Oxford, Clarendon Press, 145–167.

Colhoun, E.A., 1977. A sequence of late Quaternary deposits at Pipe Clay Lagoon, southeastern Tasmania. Papers and Proceedings of the Royal Society of Tasmania 111, 1–12.

Colhoun, E.A., Dickson, J.H., McCabe, A.M., & Shotton, F.W., 1972. A Middle Midlandian freshwater series of Derryvee, Maguiresbridge, County Fermanagh, Northern Ireland. Proceedings of the Royal Society of London B 180, 273–292.

Colhoun, E.A., van de Geer, G., & Mook, W.G., 1982. Stratigraphy, pollen analysis and paleoclimatic interpretation of Pulbeena Swamp, northwestern Tasmania. Quaternary Research 18, 108–126.

Colinvaux, P., 2001. Paradigm lost: Pleistocene environments from the Amazon Basin (continued forest cover in perpetual flux, Part 2). The Review of Archaeology 22, 20–31.

Colledge, S.M., 1981. Environmental Report. In: McNeil-Sale, R. (Ed), Archaeology in Nantwich: Crown Car Park excavations. Interim Report, 36. Bemrose Press, Chester.

Colledge, S.M. & Osborne, P.J., 1980. Plant and insect remains. In: Carver, M., (Ed) The excavation of three medieval craftsmen's tenements in Sidbury, Worcester. Transactions of the Worcestershire Archaeological Society 7, 207–210.

Compte, A. & Perales, J., 1984. Estudios des insectos coleópteros datados en el inicio de la iberización y pertenecientes al poblado de Siriguarach (Alcañiz, Teruel) [Study of Coleoptera dating from the beginning of the Iberization from the village of Siriguarach (Alcañiz, Teruel)]. Kalathos (Revista del Seminario de Arqueologia y Etnologia Turolense), Teruel, Spain 3–4, 121–137. [In Spanish].

Cong, S. & Ashworth, A.C., 1996. Palaeoenvironmental interpretation of Middle and Late Wisconsinan fossil coleopteran assemblages from western Olympic Peninsula, Washington, USA. Journal of Quaternary Science 11, 345–356.

Cong, S., Ashworth, A.C., Schwert, D.P., & Totten, S.M., 1996. Fossil beetle evidence for a short warm interval near 40,000 yr B.P. at Titusville, Pennsylvania. Quaternary Research 45, 216–225.

Coope, G.R., 1959. A Late Pleistocene insect fauna from Chelford, Cheshire. Proceedings of the Royal Society of London, Series B 151, 70–86.

Coope, G.R., 1961. On the study of glacial and interglacial insect faunas. Proceedings of the Linnean Society of London 172, 62–71.

Coope, G.R., 1962a. Coleoptera from a peat interbedded between two boulder clays at Burnhead near Airdrie. Transactions of the Geological Society of Glasgow 24, 279–286.

Coope, G.R., 1962b. A Pleistocene coleopterous fauna with arctic affinities from Fladbury, Worcestershire. Geological Society of London, Quarterly Journal 118, 103–123.

Coope, G.R., 1965. Fossil insect faunas from Late Quaternary deposits in Britain. Advancement of Science, March Issue, 564–575.

Coope, G.R., 1968a. Coleoptera from the "Arctic Bed" at Barnwell Station, Cambridge. Geological Magazine 105, 482–486.

Coope, GR., 1968b. An insect fauna from Mid-Weichselian deposits at Brandon, Warwickshire. Philosophical Transactions of the Royal Society of London, Series B 254, 425–456.

Coope, G.R., 1968c. Fossil beetles collected by James Bennie from Lateglacial silts at Corstorphine, Edinburgh. Scottish Journal of Geology 4, 339–348.

Coope, G.R., 1968d. Insect remains from silts below till at Garfield Heights, Ohio. Geological Society of America Bulletin 79, 753–756.

Coope, G.R., 1969. Insect remains from Mid-Weichselian deposits at Peelo, the Netherlands. Med. Rijks Geol. Dienst NS 20, 79–83.

Coope, G.R., 1970. Interpretations of Quaternary insect fossils. Annual Review of Entomology 15, 97–120.

Coope, G.R., 1971a. The fossil coleoptera from Glen Ballyre and their bearing upon the interpretation of Lateglacial environments. In: Thomas, G.P.S. (Ed), Field guide to the Isle of man. Liverpool, UK, Quaternary Research Association, 13–15.

Coope, G.R., 1971b. Insecta. In: Colhoun, E.A., Mitchell, G.F. (Eds), Interglacial marine formation and lateglacial freshwater formation in Shortalstown Townland, Co. Wexford. Proceedings of the Royal Irish Academy 71, 234–238.

Coope, G.R., 1973. Tibetan species of dung beetle from Late Pleistocene deposits in England. Nature 245, 335–336.

Coope, G.R., 1974a. Interglacial Coleoptera from Bobbitshole, Ipswich, Suffolk. Journal of the Geological Society 130, 333–340.

Coope, G.R., 1974b. Report on the Coleoptera from Wretton. In: West, R.G., Dickson, C.A., Catt, J.A., Weir, A.H., Sparks, B.W., (Eds), Late Pleistocene deposits at Wretton, Norfolk. II. Devensian deposits. Philosophical Transactions of the Royal Society of London, Series B 267, 414–418.

Coope, G.R., 1975. Climatic fluctuations in northwest Europe since the Last Interglacial, indicated by fossil assemblages of Coleoptera. Geological Journal Special Issue 6, 153–168.

Coope, G.R., 1976. Assemblages of fossil Coleoptera from terraces of the Upper Thames near Oxford. Quaternary Research Association, Field Guide to the Oxford Region 20–23,

Coope, G.R., 1977a. Fossil Coleopteran assemblages as sensitive indicators of climatic changes during the Devensian (Last) cold stage. Philosophical Transactions of the Royal Society of London, Series B 280, 313–340.

Coope, G.R., 1977b. Coleoptera as clues to the understanding of climatic changes in North Wales towards the end of the last (Devensian) Glaciation. Cambria 4, 65–72.

Coope, G.R., 1978. Constancy of insect species versus inconstancy of Quaternary environments. In: Mound, L.A., Waloff, N. (Eds), Diversity of Insect Faunas. Royal Entomological Society of London Symposium 9, 176–187.

Coope, G.R., 1979. Late Cenozoic fossil Coleoptera: evolution, biogeography and ecology. Annual Review of Ecology and Systematics 10, 247–267.

Coope, G.R., 1980. Appendix 1. Coleoptera from Lateglacial and Early Flandrian deposits at Folkestone. Philosophical Transactions of the Royal Society of London, Series B 291, 38–39.

Coope, G.R., 1981. Report on the Coleoptera from an eleventh-century house at Christ Church Place, Dublin. In: Bekker-Nielson, H., Foote, P., Olsen, O. (Eds), Proceedings of the Eighth Viking Congress (1977). Odense University Press, Odense, Denmark, 51–56.

Coope, G.R., 1982. Coleoptera from two Late Devensian sites in the lower Colne Valley, West London, England. Quaternary Newsletter 38, 1–6.

Coope, G.R., 1986. Coleoptera analysis. In: Berglund, B.E. (Ed), Handbook of Holocene palaeoecology and palaeohydrology. New York, Wiley, 703–713.

Coope, G.R., 1987a. Fossil beetle assemblages as evidence for sudden and intense climatic changes in the British Isles during the last 45,000 years. In: Berger, W.H., Labeyrie, L.D. (Eds), Abrupt climatic change. Dordrecht, The Netherlands, Reidel, 147–150.

Coope, G.R., 1987b.The response of late Quaternary insect communities to sudden climatic changes. In: Gee, J.H.R., Giller, P.S. (Eds), Organization of Communities. 27th Symposium of the British Ecological Society, Aberystwyth, 421–438.

Coope, G.R., 1988. Lateglacial climates in northwestern Europe interpreted from beetle assemblages. Program and abstracts of the 10th biennial meeting of the American Quaternary Association, Amherst, Massachusetts, 16.

Coope, G.R., 1989. Coleoptera from the Lower Channel, Brandon. In: Keen, D.H. (Ed), West Midlands field guide. Cambridge, Quaternary Research Association, 27–28.

Coope, G.R., 1990. The invasion of northern Europe during the Pleistocene by Mediterranean species of Coleoptera. In: di Castri, F., Hansen, A.J., Debussche, M. (Eds), Biological invasions in Europe and the Mediterranean Basin. Dordrecht, The Netherlands, Kluwer, 203–215.

Coope, G.R., 1994. The response of insect faunas to glacial-interglacial climatic fluctuations. Philosophical Transactions of the Royal Society of London, Series B 344, 19–26.

Coope, G.R., 1995. The effects of Quaternary climatic changes on insect populations: lessons from the past. In: Harrington, R., Stork, N.E. (Eds), Insects in a changing environment. London, UK, Academic Press, 29–48.

Coope, G.R., 1998. Insects. In: Preece, R.C., Bridgeland, D.R. (Eds), Late Quaternary environmental change in North-west Europe: Excavations at Holywell Coombe, South-east England. London, UK, Chapman & Hall, 213–233.

Coope, G.R., 2000a. Coleoptera from Beeston and West Runton, Norfolk. In: Lewis, S.G., Whiteman, C.A., Preece, R.C. (Eds), The Quaternary of Norfolk & Suffolk. London, UK, Quaternary Research Association, 73–76.

Coope, G.R., 2000b. Middle Devensian (Weichselian) coleopteran assemblages from Earith, Cambridgeshire (UK) and their bearing on the interpretation of 'Full glacial' floras and faunas. Journal of Quaternary Science 15, 779–788.

Coope, G.R., 2001. Biostratigraphical distinction of interglacial coleopteran assemblages from southern Britain attributed to Oxygen Isotope Stages 5e and 7. Quaternary Science Reviews 20, 1717–1722.

Coope, G.R., 2006. Insect faunas associated with Palaeolithic industries from five sites of pre-Anglian age in central England. Quaternary Science Reviews 25, 1738–1754.

Coope, G.R., 2007a. Coleoptera from the 2003 excavations of the mammoth skeleton at Niederweningen, Switzerland. Quaternary International 164–165, 130–138.

Coope, G.R., 2007b. Beetle records: Middle Pleistocene of Europe. In: Elias, S.A. (Ed), Encyclopedia of Quaternary science. Amsterdam, Elsevier, 172–179.

Coope, G.R. & Angus, R.B., 1975. An ecological study of a temperate interlude in the middle of the Last glaciation, based on fossil Coleoptera from Isleworth, Middlesex. Journal of Animal Ecology 44, 365–391.

Coope, G.R. & Böcher, J., 2000. Coleoptera from the Late Weichselian deposits at Nørre Lyngby, Denmark and their bearing on palaeoecology, biogeography and palaeoclimate. Boreas 29, 26–34.

Coope, G.R. & Brophy, J., 1972. Lateglacial environmental changes indicated by a coleopteran succession from North Wales. Boreas 1, 97–143.

Coope, G.R., Dickson, J.H., Jones, R.L., & Keen, D.H., 1987. The flora and fauna of late Pleistocene deposits on the Cotentin Peninsula, Normandy. Philosophical Transactions of the Royal Society of London, Series B 315, 231–265.

Coope, G.R., Dickson, J.H., McCutcheon, J.A., & Mitchell, G.A., 1979. The lateglacial and early postglacial deposit at

Drumurcher, Co. Monaghan. Proceedings of the Royal Irish Academy, Series B 79, 63–85.

Coope, G.R. & Elias, S.A., 2000. The environment of Upper Paleolithic (Magdalenian and Azilian) hunters at Hauterive-Champréveyres, Neuchâtel, Switzerland, interpreted from coleopteran remains. Journal of Quaternary Science 15, 157–175.

Coope, G.R., Field, M.H., Gibbard, P.L., Greenwood, M., & Richards, A.E., 2002. Palaeontology and biostratigraphy of Middle Pleistocene river sediment in the Mathon Member, at Mathon, Herefordshire, England. Proceedings of the Geologists' Association 113, 237–258.

Coope, G.R., Gibbard, P.L., Hall, A.R., Preece, R.C., Robinson, J.E., & Sutcliffe, A.J., 1997. Climatic and environmental reconstructions based on fossil assemblages from Middle Devensian (Weichselian) deposits of the river Thames at South Kensington, central London, UK. Quaternary Science Reviews 16, 1163–1195.

Coope, G.R. & Joachim, M.J., 1980. Lateglacial environmental changes interpreted from fossil Coleoptera from St. Bees, Cumbria, NW England. In: Lowe, J.J., Gray, J.M., Robinson, J.E. (Eds), Studies in the Lateglacial of Northwest Europe. Oxford, Pergamon Press, 55–68.

Coope, G.R., Jones, R.L., & Keen, D., 1980. The palaeoecology and age of peat at Fliquet Bay, Jersey, Channel Islands. Journal of Biogeography 7, 187–195.

Coope, G.R., Jones, R.L., Keen, D.H., & Waton, P.V., 1985. The flora and fauna of Late Pleistocene deposits in St. Aubins Bay, Jersey, Channel Islands. Proceedings of the Geologists' Association 96, 315–324.

Coope, G.R. & Lemdahl, G., 1995. Regional differences in the Lateglacial climate of northern Europe based on coleopteran analysis. Journal of Quaternary Science 10, 391–395.

Coope, G.R., Lemdahl, G., Lowe, J.J., & Walkling, A., 1998. Temperature gradients in northern Europe during the last glacial-Holocene transition (14-9 ^{14}C kyr BP) interpreted from coleopteran assemblages. Journal of Quaternary Science 13, 419–433.

Coope, G.R. & Osborne, P.J., 1968. Report on the coleopterous fauna of the Roman well at Barnsley Park, Gloucestershire. Transactions of the Bristol and Gloucestershire Archaeological Society 86, 84–87.

Coope, G.R. & Sands, C.H., 1966. Insect faunas of the last glaciation from the Tame Valley, Warwickshire. Proceedings of the Royal Society of London, Series B 165, 389–412.

Coope, G.R. & Sands, C.H.S., 1963. The discovery, in British late Pleistocene deposits, of the extinct species *Helophorus wandereri* d'Orch. (Col. Hydrophilidae). Opuscula Entomologica 28, 94–96.

Coope, G.R., Shotton, F.W., & Strachan, I., 1961. A Late Pleistocene fauna and flora from Upton Warren, Worcestershire. Philosophical Transactions of the Royal Society of London, Series B 244, 379–417.

Coope, R.C., 2007c. A preliminary investigation of the food and feeding behaviour of Pine Martens (*Martes martes*) in productive forestry from an analysis of the content of their scats collected in Inchnacardoch forest, Fort Agustus. Scottish Forestry 61, 3–14.

Coope, G.R. & Tallon, P.W.J., 1983. A full glacial insect fauna from the Lea Valley, Enfield, North London. Quaternary Newsletter 40, 7–12.

Crossley, R., 1984. Fossil termite mounds associated with stone artifacts in Malawi, Central Africa. Palaeoecology of Africa and of the Surrounding Islands and Antarctica 16, 397–401.

Crowson, R.A., 1958. Some observations on a coleopterological visit to central Italy. Entomologists Monthly Magazine 94, 248–251.

Crowson, R.A., 1975. The evolutionary history of Coleoptera, as documented by fossil and comparative evidence. Proceedings of the Tenth Italian National Congress of Entomology 63–70,

Crowson, R.A., 1981. The Biology of the Coleoptera. New York, Academic Press.

Curry, A., 1979. The insects associated with the Manchester mummies. In: David, R.A. (Ed), The Manchester Mummy Project. Manchester, UK, Manchester University Press, 113–118.

Cymorek, S., Koch, K., 1969. Über Funde von Körperteilen des Messingkäfers *Niptus hololeucus* (Fald.) in Auslagerungen aus dem 15–16. Jahrhundert (Neuss, Neiderrhein) und Folgerungen daraus für die Ausbreitungsgeschichte der Art in Europa [Concerning the discovery of the remains of the spider beetle, *Niptus hololeucus* in deposits from the 15th to 16th centuries at Neuss, Neiderrhein) and inferences therefrom concerning the biogeography of the species in Europe]. Anzeiger für Schädlingskunde und Pflanzenschutz 42, 185–186. [In German].

Dadswell, L., 2003. Developing the Mutual Climatic Range (MCR) method of Coleoptera temperature reconstruction: the application of Ubiquity Analysis. Unpublished MSc Dissertation, Royal Holloway, University of London.

Dal Monte, G., 1956. La presenza di insetti dei granai in frumento trovato negli scavi di Ercolano [The presence of insects in charred grain from the site of Herculaneum]. Redia 41, 23–28. [In Italian].

Damblon, F., Coope, G.R., & Osborne, P.J., 1977. Paleoecological studies of peat bogs in the High Ardenne. Proceedings of the X INQUA Congress, Birmingham. Geobooks, Norwich, 101.

Danks, H.V., 1978. Modes of seasonal adaptation in the insects. 1. Winter survival. Canadian Entomologist 110, 1167–1205.

Danks, H.V., 1979. Terrestrial habitats and distributions of Canadian insects. In: Danks, H.V. (Ed), Canada and Its Insect Fauna. Memoirs of the Entomological Society of Canada, 108, 195–210.

Danks, H.V., 1981. Arctic arthropods. Ottawa, Canada, Entomological Society of Canada.

Dansgaard, W., 1964. Stable isotopes in precipitation. Tellus 16, 436–468.

Dansgaard, W., 1987. Ice core evidence of abrupt climatic changes. In: Berger, W.H., Labeyrie, L.D. (Eds), Abrupt climatic change. Dordrecht, The Netherlands, Reidel, 223–233.

Dansgaard, W., White, J.W.C., & Johnson, S.J., 1989. The abrupt termination of the Younger Dryas climate event. Nature 339, 532–534.

Darlington, P.J., 1938. The American Patrobini. Entomologica Americana 18, 135–183.

Darlington, P.J., 1943. Carabidae of mountains and islands: data on the evolution of isolated faunas and on atrophy of wings. Ecological Monographs 13, 40–61.

Davey, N.D.W., Bridgland, D.R., Keen, D.H., Coope, G.R., Seddon, M.B., & Holyoak, D., 1991. Maxey gravel pit, near Peterborough TF1307. In: Lewis, S.G. (Ed), Central East Anglia and the Fen Basin: field guide. Cambridge, Quaternary Research Association, 185–208.

David, A.R., 1984. The Background. In: David, A.R., Topp, E. (Eds), Evidence embalmed, modern medicine and the mummies of Ancient Egypt. Manchester, UK, Manchester University Press, 35–36.

David, R., 1978. The fauna. In: David, R. (Ed), Mysteries of the mummies. The story of the Manchester University investigation. London, UK, Book Club Associates, 160–167.

Davis, M.B., Shaw, R.G., & Etterson, J.R., 2005. Evolutionary responses to changing climate. Ecology 86, 1704–1714.

Davis, M.B., Woods, K.D., Webb, S.L., & Futyma, R.P., 1986. Dispersal versus climate: expansion of *Fagus* and *Tsuga* into the Upper Great Lakes region. Vegetatio 67, 93–103.

de Bealieu, J.L., Clerc, J., & Reille, M., 1984. Late Weichselian fluctuations in the French Alps and Massif Central from pollen analysis. In: Mörner, N.A., Karlén, W. (Eds), Climatic changes on a yearly to millennial scale. Dordrecht, The Netherlands, Reidel, 75–90.

de Beaulieu, J.L., Pons, A., & Reille, M., 1982. Recherches pollenanalytiques sur l'histoire de la végétation de la bordure nord du massif du Cantal (Massif Central, France) [Palynological research on the history of the vegetation of the northern edge of the massif du Cantal]. Pollen et Spores 34, 251–300. [In French].

Denford, S., 1978. Mites and their potential use in archaeology. In: Brothwell, D.R., Thomas, K.D., Clutton-Brock, J. (Eds), Research problems in zooarchaeology. University of London, Institute of Archaeology, Occasional Paper 3, 77–83.

Denton, G.H., Lowell, T.V., Heusser, C.J., Schlüchter, C., Andersen, B.G., Heusser, L.E., Moreno, P.I., & Marchant, D.R., 1999. Geomorphology, stratigraphy, and radiocarbon chronology of Llanquihue drift in the area of the southern Lake District, Seno Reloncavi, and Isla Grande′ de Chilo., Chile. Geografiska Annaler A81, 167–229.

De Rouffignac, C., Bowen, D.Q., Coope, G.R., Keen, D.H., Lister, A.M., Maddy, D., Robinson, J.E., Sykes, G.A., & Walker, M.J.C., 1995. Late Middle Pleistocene interglacial deposits at Upper Strensham, Worcestershire, England. Journal of Quaternary Science 10, 15–31.

De Veer, G., 1598. The three voyages of William Barents to the Arctic regions (1594, 1595, and 1596). Reprint of the 1609 edition (in English, translation by A. Philip, 1607). London, Hakluyt Society.

Dévai, G. & Moldován, J., 1983. An attempt to trace eutrophication in a shallow lake (Balaton, Hungary) using chironomids. Hydrobiologia 103, 169–175.

Dickson, J.H., Dickson, C.A., & Breeze, D.J., 1979. Flour or bread in a Roman military ditch at Bearsden, Scotland. Antiquity 53, 47–51.

Dillehay, T.D., 1986. The cultural relationships of Monte Verde: a Late Pleistocene settlement site in the sub-antarctic forest of south-central Chile. In: Bryan, A.L. (Ed), New evidence for the Pleistocene peopling of the Americas. Orono, ME, Center for the Study of Early Man, 319–337.

Dillehay, T.D., 1989. An evaluation of the paleoenvironmental reconstruction and the human presence at Monte Verde. In: Dillehay, T.D. (Ed), Monte Verde: a Late Pleistocene settlement in Chile, Vol. 1, Washington, DC, Smithsonian Press, 227–237.

Dinnin, M., 1996. The development of the Outer Hebridean entomofauna: a fossil perspective. In: Gilbertson, D., Kent, M., Grattan, J. (Eds), The Outer Hebrides. The last 14,000 years [Sheffield Environmental and Archaeological Research Campaign in the Hebrides 2]. Sheffield, Sheffield Academic Press, 163–184.

Dinnin, M., 1997. Holocene beetle assemblages from the Lower Trent floodplain at Bole Ings, Nottinghamshire, U.K. In: Ashworth, A.C., Buckland, P.C., Sadler, J.P. (Eds), Studies in Quaternary entomology – an inordinate fondness for insects. Quaternary Proceedings 5, 83–104.

Dinnin, M.H. & Sadler, J.P., 1999. 10,000 years of change: the Holocene entomofauna of the British Isles. In: Edwards, K.J., Sadler, J.P. (Eds), Holocene environments of prehistoric Britain. Quaternary Proceedings 7, 545–562.

Dinnin, M.H. & Skidmore, P., 1995. The insect assemblages associated with Lindow III and their environmental implications. In: Turner, R.C., Scaife, R.G. (Eds), Bog bodies, new discoveries and new perspectives. London, UK, British Museum Press, 31–38.

Dinnin, M.H. & Welsh, M., 2001. Star Carr (SE 744016). In: Bateman, M.D., Buckland, P.C., Frederick, C.D., Whitehouse, N.J. (Eds), The Quaternary of East Yorkshire and North Lincolnshire field guide. London, UK, Quaternary Research Association, 161–164.

Dittmar, K, Mamat, U., Whiting, M., Goldmann, T., Reinhard, T., & Guillen, S., 2003. Techniques of DNA-studies on prehispanic ectoparasites (*Pulex* sp., Pulicidae, Siphonaptera) from animal mummies of the Chiribaya Culture, Southern Peru. Memórias de Instituto Oswaldo Cruz 98, Supplement 1.

Dobney, K., Kenward, H.K., Ottaway, P., & Donel, L., 1998. Down, but not out: biological evidence for complex economic organization in Lincoln in the late 4th century. Antiquity 72, 417–424.

Doube, B.M., 1982. Dung beetle diversity: Pleistocene extinctions and the game park effect. Occasional Bulletin of the Zoological Society of South Africa 2, 22.

Doyen, J.T. & Miller, S.E., 1980. Review of Pleistocene darkling ground beetles of the California asphalt deposits (Coleoptera: Tenebrionidae, Zopheridae). Pan-Pacific Entomologist 56, 1–10.

Dredge, L.A., Morgan, A.V., & Nielsen, E., 1990. Sangamon and pre-Sangamon interglaciations in the Hudson Bay lowlands of Manitoba. Géographie physique et Quaternaire 44, 319–336.

Duff, A.G. (Ed), 2008. Checklist of beetles of the British Isles, 2008 edition. Wells, Somerset, UK, A.G. Duff.

Duplessy, J.C., Labeyrie, L.D., & Paterne, M., 1996. North Atlantic sea surface conditions during the Younger Dryas cold event. In: Andrews, J.T., Austin, W.E.N., Bergsten, H., Jennings, A.E. (Eds), Late Quaternary palaeoceanography of the North Atlantic Margins. Geological Society Special Publication 111, 167–175.

Dyke, A.S. & Matthews, Jr, J.V., 1987. Stratigraphy and paleoecology of Quaternary sediments along Pasley River, Boothia Peninsula, central Canadian arctic. Géographie physique et Quaternaire 41, 323–344.

Dyke, A.S. & Prest, V.K., 1986. Late Wisconsin and Holocene retreat of the Laurentide ice sheet. Geological Survey of Canada Map, Scale = 1:5,000,000, Map 1702A.

Edwards, K.J., Buckland, P.C., Craigie, R., Panagiotakopulu, E., & Stummann Hansen, S., 1998. Landscapes at landnám: palynological and palaeoentomological evidence from Toftanes, Faroe Islands. Fróðskaparrit 46, 299–244.

Edwards, M.E., Bigelow, N., Elias, S.A., Hamilton, T.D., & Schweger, C.E., in press. Last-interglacial climate and environment in interior Alaska. Quaternary Science Reviews (submitted).

Edwards, M.E., Brubaker, L.B., Lozhkin, A.V., & Anderson, P.M., 2005. Structurally novel biomes: a response to past warming in Beringia. Ecology 86, 1696–1703.

Edwards, M.E., Dawe, J.C., & Armbruster, W.S., 1991. Pollen size of *Betula* in northern Alaska and the interpretation of late-Quaternary pollen records. Canadian Journal of Botany 69, 1666–1672.

Edwards, M.E., Hamilton, T.D., Elias, S.A., Bigelow, N.H., & Krumhardt, A.P., 2003. Interglacial extension of the boreal forest limit in the Noatak valley, northwest Alaska: evidence from an exhumed river-cut bluff and debris apron. Arctic, Antarctic, and Alpine Research 35, 49–57.

Egloff, M., 1987. 130 Years of archaeological research in Lake Neuchatel, Switzerland. In: Coles, J.M., Lawson, A.J. (Eds), European Wetlands in Prehistory. Oxford, Clarendon Press, 24–32.

Ehlers, J., Astakhov, V., Gibbard, P.L., Mangerud, J., & Svendsen, J.I., 2007. Late Pleistocene Glaciations in Europe. In: Elias, S.A. (Ed), Encyclopedia of Quaternary science. Amsterdam, Elsevier, 1085–1095.

Elias, S.A., 1982a. Holocene insect fossils from two sites at Ennadai Lake, Keewatin, NWT. Quaternary Research 17, 311–319.

Elias, S.A., 1982b. Bark beetle fossils from two early postglacial age sites at high altitude in the Colorado Rocky Mountains. Journal of Paleontology, Special Issue (Proceedings of the 3rd North American Paleontological Convention) I, 53–57.

Elias, S.A., 1982c. Holocene insect fossil assemblages from northeastern Labrador. Arctic and Alpine Research 14, 311–319.

Elias, S.A., 1983. Paleoenvironmental interpretations of Holocene insect fossil assemblages from the La Poudre Pass site, northern Colorado Front Range. Palaeogeography, Palaeoclimatology, Palaeoecology 41, 87–102.

Elias, S.A., 1985. Paleoenvironmental interpretations of Holocene insect fossil assemblages from four high-altitude sites in the Front Range, Colorado, U.S.A. Arctic and Alpine Research 17, 31–48.

Elias, S.A., 1986. Fossil insect evidence for Late Pleistocene paleoenvironments of the Lamb Spring site, Colorado. Geoarchaeology 1, 381–386.

Elias, S.A., 1987. Paleoenvironmental significance of Late Quaternary insect fossils from packrat middens in south-central New Mexico. Southwestern Naturalist 32, 383–390.

Elias, S.A., 1988a. Climatic significance of late Pleistocene insect fossils from Marias Pass, Montana. Canadian Journal of Earth Sciences 25, 922–926.

Elias, S.A., 1988b. Late Pleistocene paleoenvironmental studies from the Rocky Mountain region: a comparison of pollen and insect fossil records. Geoarchaeology 3, 147–153.

Elias, S.A., 1990a. Observations on the taphonomy of late Quaternary insect fossil remains in packrat middens of the Chihuahuan Desert. Palaios 5, 356–363.

Elias, S.A., 1990b. The timing and intensity of environmental changes during the Paleoindian period in western North America: evidence from the fossil insect record. In: Agenbroad, L.D., Mead, J.I., Nelson, L. (Eds), Megafauna and man. Hot Springs, South Dakota and Flagstaff, AZ, The Mammoth Site of Hot Springs and Northern Arizona University, 11–14.

Elias, S.A., 1991. Insects and climate change: fossil evidence from the Rocky Mountains. BioScience 41, 552–559.

Elias, S.A., 1992a. Late Wisconsin insects and plant macrofossils associated with the Colorado Creek mammoth, southwestern Alaska: taphonomic and paleoenvironmental implications. 22nd Arctic Workshop, Program and Abstracts, Boulder, CO, 45–47.

Elias, S.A., 1992b. Late Quaternary zoogeography of the Chihuahuan Desert insect fauna, based on fossil records from packrat middens. Journal of Biogeography 19, 285–298.

Elias, S.A., 1992c. Late Quaternary beetle fauna of southwestern Alaska: evidence of a refugium for mesic and hygrophilous species. Arctic and Alpine Research 24, 133–144.

Elias, S.A., 1994. Quaternary insects and their environments. Washington, DC, Smithsonian Institution Press.

Elias, S.A., 1995a. A paleoenvironmental setting for early Paleoindians in western North America: evidence from the insect fossil record. In: Johnson, E. (Ed), Worldwide perspectives of man and the Quaternary. Lubbock, TX, Texas Tech University Press, 252–272.

Elias, S.A., 1995b. Insect Fossils. In: Holliday, V.T. (Ed), Stratigraphy and paleoenvironments of Late Quaternary valley fills on the Southern High Plains. Geological Society of America Memoir 186, 50–53.

Elias, S.A., 1996. Late Pinedale and Holocene seasonal temperatures reconstructed from fossil beetle assemblages in the Rocky Mountains. Quaternary Research 46, 311–318.

Elias, S.A., 1997. Paleoenvironmental significance of Lateglacial insect fossil assemblages from the Aubrey Site, North-Central Texas. In: Ashworth, A.C., Buckland, P.C., Sadler, J.P. (Eds), Studies in Quaternary entomology – an inordinate fondness for insects. Quaternary Proceedings 5, 105–112.

Elias, S.A., 1998. The Mutual Climatic Range method of paleoclimate reconstruction based on insect fossils: new applications and interhemispheric comparisons. Quaternary Science Reviews 16, 1217–1225.

Elias, S.A., 1999. Mid-Wisconsin seasonal temperatures reconstructed from fossil beetle assemblages in eastern North America: comparisons with other proxy records from the Northern Hemisphere. Journal of Quaternary Science 14, 255–262.

Elias, S.A., 2000. Late Pleistocene climates of Beringia, based on fossil beetle analysis. Quaternary Research 53, 229–235.

Elias, S.A., 2001. Mutual climatic range reconstructions of seasonal temperatures based on Late Pleistocene fossil assemblages in Eastern Beringia. Quaternary Science Reviews 20, 77–91.

Elias, S.A., 2006. Quaternary beetle research: the state of the art. Quaternary Science Reviews 25, 1731–1737.

Elias, S.A., 2007a. Beetle records: Late Tertiary and Early Quaternary. In: Elias, S.A. (Ed), Encyclopedia of Quaternary science. Amsterdam, Elsevier, 163–172.

Elias, S.A., 2007b. The Bering Land Bridge: a moisture barrier to the dispersal of steppe-tundra biota. Quaternary International 167–168(Supplement), 107.

Elias, S.A., 2007c. Beetle records: Late Pleistocene North America. In: Elias, S.A. (Ed), Encyclopedia of Quaternary science. Amsterdam, Elsevier, 222–236.

Elias, S.A., 2007d. Beetle records: postglacial North America. In: Elias, S.A. (Ed), Encyclopedia of Quaternary science. Amsterdam, Elsevier, 275–286.

Elias, S.A., Anderson, K., & Andrews, J.T., 1996a. Late Wisconsin climate in the northeastern United States and southeastern Canada, reconstructed from fossil beetle assemblages. Journal of Quaternary Science 11, 417–421.

Elias, S.A., Andrews, J.T., & Anderson, K.H., 1999a. New insights on the climatic constraints on the beetle fauna of coastal Alaska derived from the mutual climatic range method of paleoclimate reconstruction. Arctic, Antarctic, and Alpine Research 31, 94–98.

Elias, S.A., Berman, D., & Alfimov, A., 2000. Late Pleistocene beetle faunas of Beringia: where east met west. Journal of Biogeography 27, 1349–1363.

Elias, S.A., Carrara, P.E., Toolin, L.J., & Jull, A.J.T., 1991. New radiocarbon and macrofossil analyses from Lake Emma Colorado: revision of the age of deglaciation. Quaternary Research 36, 307–321.

Elias, S.A. & Crocker, B., 2008. The Bering Land Bridge: a moisture barrier to the dispersal of steppe-tundra biota? Quaternary Science Reviews 27, 2473–2483.

Elias, S.A. & Halfpenny, J.C., 1991. Fox scat evidence of heavy predation on beetles on the alpine tundra, Front Range, Colorado. Coleopterists Bulletin 45, 189–190.

Elias, S.A., Hamilton, T.D., Edwards, M.E., Begét, J.E., Krumhardt, A.P., & Lavoie, C., 1999b. Late Pleistocene environments of the western Noatak Basin, northwestern Alaska. Geological Society of America Bulletin 111, 769–789.

Elias, S.A. & Johnson, E., 1988. Pilot study of fossil beetles at the Lubbock Lake Landmark. Current Research in the Pleistocene 5, 57–59.

Elias, S.A., Kuzmina, S., & Kiselyov, S., 2006. Late Tertiary origins of the Arctic beetle fauna. Palaeogeography, Palaeoclimatology, Palaeoecology 241, 373–392.

Elias, S.A. & Matthews, Jr, J.V., 2002. Arctic North American seasonal temperatures in the Pliocene and Early Pleistocene, based on mutual climatic range analysis of fossil beetle assemblages. Canadian Journal of Earth Sciences 39, 911–920.

Elias, S.A., Mead, J.I., & Agenbroad, L.D., 1992b. Late Quaternary arthropods from the Colorado Plateau, Arizona and Utah. Great Basin Naturalist 52, 59–67.

Elias, S.A. & Nelson, A.R., 1989. Fossil invertebrate evidence for Late Wisconsin environments at the Lamb Spring site, Colorado. Plains Anthropologist 34, 309–326.

Elias, S.A. & Short, S.K., 1992. Paleoecology of an interglacial peat deposit, Nuyakuk, southwestern Alaska. Géographie Physique et Quaternaire 46, 85–96.

Elias, S.A., Short, S.K., & Birks, H.H., 1997. Late Wisconsin environments of the Bering Land Bridge. Palaeogeography, Palaeoclimatology, Palaeoecology 136, 293–308.

Elias, S.A., Short, S.K., & Clark, P.U., 1986. Paleoenvironmental interpretations of the Late Holocene, Rocky Mountain National Park, Colorado, U.S.A. Revue de Paléobiologie 5, 127–142.

Elias, S.A., Short, S.K., Nelson, C.H., & Birks, H.H., 1996b. Life and times of the Bering land bridge. Nature 382, 60–63.

Elias, S.A., Short, S.K., & Phillips, R.L., 1992a. Paleoecology of lateglacial peats from the Bering Land Bridge, Chukchi Sea shelf region, northwestern Alaska. Quaternary Research 38, 371–378.

Elias, S.A., Short, S.K., & Waythomas, C.F., 1996c. Late Quaternary environments, Denali National Park and Preserve, Alaska. Arctic 49, 292–305.

Elias, S.A. & Toolin, L.J., 1989. Accelerator dating of a mixed assemblage of late Pleistocene insect fossils from the Lamb Spring site, Colorado. Quaternary Research 33, 122–126.

Elias, S.A. & Van Devender, T.R., 1990. Fossil insect evidence for late Quaternary climatic change in the Big Bend region, Chihuahuan Desert, Texas. Quaternary Research 34, 249–261.

Elias, S.A. & Van Devender, T.R., 1992. Insect fossil evidence of late Quaternary environments in the northern Chihuahuan Desert of Texas and New Mexico: comparisons with the paleobotanical record. Southwestern Naturalist 37, 101–116.

Elias, S.A., Van Devender, T.R., & DeBaca, R., 1995. Insect fossil evidence of lateglacial and Holocene environments in the Bolson de Mapimi, Chihuahuan Desert, Mexico: comparisons with the paleobotanical record. Palaios 10, 454–464.

Elias, S.A., Webster, L., & Amer, M., 2009. A beetle's eye view of London from the Mesolithic to Late Bronze Age. Geological Journal 44, in press.

Elias, S.A. & Wilkinson, B.J., 1983. Lateglacial insect fossil assemblages from Lobsigensee (Swiss Plateau). Studies in the Late Quaternary of Lobsigensee, 3. Revue de Paléobiologie 2, 184–204.

Eliasson, L. & Kishonti, I., 2003. Rapport över arkeologiska slutundersökning: Öresundsförbindelsen, Lockarp 7B [Report on archaeological post-examination of seaway connections at the Lockarp 7B site]. Rapport Nr 17, Malmö Kulturmiljö. [In Swedish].

Erickson, J.M., 1988. Fossil Oribatid mites as tools for Quaternary paleoecologists: preservation quality, quantities and taphonomy. In: Laub, R.S., Miller, N.G., Steadman, D.W. (Eds.), Late Pleistocene and Early Holocene Paleoecology and Archeology of the Eastern Great Lakes region. Bulletin of the Buffalo Society of Natural Sciences 33, 207–226.

Erickson, J.M. & Platt, Jr, R.B., 2007. Oribatid mites. In: Elias, S.A. (Ed), Encyclopedia of Quaternary science. Amsterdam, Elsevier, 1547–1566.

Erochin, N.G. & Zinovjev, E.V., 1991. The Upper Pleistocene insect faunas from the localities of South and Middle Yamal. In: Ecological Groups of Ground Beetles (Coleoptera, Carabidae) in Natural and Anthropogenic Landscapes of the Ural. Ural Branch, USSR. Academy of Sciences, Sverdlovsk, 18–22. [In Russian].

Erwin, T.L., 1979. Thoughts on the evolutionary history of ground beetles: hypotheses generated from comparative faunal analyses of lowland forest sites in temperate and tropical regions. In: Erwin, T.L., Ball, G.E., Whitehead, D.R. (Eds), Carabid Beetles, their evolution, natural history, and classification. The Hague, The Netherlands, Junk, 539–592.

Erwin, T.L., 1983. Tropical forest canopies: the last biotic frontier. Entomological Society of America Bulletin, Spring 14–19.

Erzinclioglu, Y.Z., 1983. The application of entomology to forensic medicine. Medicine, Science, and the Law 23, 57–63.

Essig, E.O., 1927. Some insects from the adobe walls of the old missions of lower California. Pan-Pacific Entomologist 3, 194–195.

Evans, W.G. & Baldwin, S.J., 1977. Larval exuviae of *Attagenus bicolor* Von Harold (Coleoptera: Dermestidae) from an archeological site at Mesa Verde, Colorado. Quaestiones Entomologicae 13, 309–330.

Extence, C.A., Balbi, D.M., & Chadd, R.P., 1999. River flow indexing using British benthic macroinvertebrates: a framework for setting hydroecological objectives. Regul River 15, 543–574.

Faulkner, C.T., 1991. Prehistoric diet and parasitic infection in Tennessee: evidence from the analysis of desiccated human paleofeces. American Antiquity 56, 687–700.

Field, M.H., De Beaulieu, J.-L., Guiot, J., & Ponel, P., 2000. Middle Pleistocene deposits at La Côte, Val-de-Lans, Isère department, France: plant macrofossil, palynological and fossil insect investigations. Palaeogeography, Palaeoclimatology, Palaeoecology 159, 53–83.

Fiori, G., 1977. La cavita sottoelitrale dei tenebrionidi apomorfi. Redia 60, 1–112. [In Italian].

Fjellberg, A., 1978. Fragments of a Middle Weichselian fauna on Andøya, North Norway. Boreas 7, 39.

Flach, K., 1884. Die Käfer der unterpleistocänen Ablagerungen bei Hösback, unweit Aschaffenburg [Beetles from lower Pleistocene deposits at Hösback, near Aschaffenburg]. Verhandlungen der Physikalisch-medizinischen Gesellschaft zu Würzburg (N.F.) 18, 285–297. [In German].

Flannery, T.F. & Gott, B., 1984. The Spring Creek locality, southwestern Victoria, a late surviving megafaunal assemblage. Australian Zoologist 22, 385–422.

Foddai, D. & Minelli, A., 1994. Fossil arthropods from a full-glacial site in northeastern Italy. Quaternary Research 41, 336–342.

Fletcher, J., 1994. A tale of hair, wigs and lice. Egyptian Archaeology 5, 31–33.

Fletcher, J., 1998. The secrets of the locks unravelled. Nekhen News (Milwaukee) 10, 7–9.

Fliche, P., 1875. Sur les lignites quaternaires de Jarville, prés de Nancy [On the Quaternary lignites of Jarville, near Nancy]. Compte Rendue del'Académie des Sciences, Paris 80, 1233–1236. [In French].

Fliche, P., 1876. Faune et flore des tourbieres de la Champagne [Fauna and flora of the peat bogs of the Champagne region]. Compte Rendu de l'Académie des Sciences, Paris 82, 979–982. [In French].

Floate, K.D. & Gill, B.D., 1998. Seasonal activity of dung beetles (Coleoptera: Scarabaeiidae) associated with cattle dung in southern Alberta and their geographic distribution in Canada. Canadian Entomologist 130, 131–151.

Forman, S.L., 1989. Applications and limitations of thermoluminescence to date Quaternary sediments. Quaternary International 1, 47–60.

Fossil Insect Research Group for Nojiri-ko Excavation (FIRGNE), 1980. Fossil insects from the Nojiri-ko Formation. The Memoirs of the Geological Society of Japan 19, 147–159. [In Japanese].

Fossil Insect Research Group for Nojiri-ko Excavation (FIRGNE), 1984. Fossil insects obtained from the Nojiri-ko Excavations in 1978 to 1982. Monograph of the Association for the Geological Collaboration in Japan 27, 137–156. [In Japanese].

Fossil Insect Research Group for Nojiri-ko Excavation (FIRGNE), 1987. Fossil insects obtained from the Nojiri-ko Excavations in the 9th Nojiri-ko Excavation and the 4th Hill Site Excavation. Monograph of the Association for the Geological Collaboration in Japan 32, 117–136. [In Japanese].

Fossil Insect Research Group for Nojiri-ko Excavation (FIRGNE), 1990. Fossil insects obtained from the Nojiri-ko Formation during the 10th Nojiri-ko and the 5th Hill Site Excavations. Monograph of the Association for the Geological Collaboration in Japan 37, 93–110. [In Japanese].

Fossil Insect Research Group for Nojiri-ko Excavation (FIRGNE), 1993a. Insect fossils obtained from the Nojiri-ko Formation during the 11th Nojiri-ko Excavation. Bulletin of the Nojiri-ko Museum 4, 69–81. [In Japanese].

Fossil Insect Research Group for Nojiri-ko Excavation (FIRGNE), 1993b. Insect fossils obtained from the Nojiri-ko Formation during the 6th Hill Site Excavation. Bulletin of the Nojiri-ko Museum 4, 203–214. [In Japanese].

Fossil Insect Research Group for Nojiri-ko Excavation (FIRGNE), 1996a. Insect fossils obtained from the Nojiri-ko Formation during the 12th Nojiri-ko Excavation. Bulletin of the Nojiri-ko Museum 4, 71–80. [In Japanese].

Fossil Insect Research Group for Nojiri-ko Excavation (FIRGNE), 1996b. Insect fossils obtained from the Nojiri-ko Formation during the 7th Nojiri-ko Hill Site Excavation. Bulletin of the Nojiri-ko Museum 4, 199–206. [In Japanese].

Fossil Insect Research Group for Nojiri-ko Excavation (FIRGNE), 2000a. Insect fossils obtained from the Nojiri-ko Formation during the 13th Nojiri-ko Excavation. Bulletin of the Nojiri-ko Museum 8, 53–58. [In Japanese].

Fossil Insect Research Group for Nojiri-ko Excavation (FIRGNE), 2000b. Insect fossils obtained from the Upper Pleistocene during the 8th Nojiri-ko Hill Site Excavation. Bulletin of the Nojiri-ko Museum 8, 139–147. [In Japanese].

Fossil Insect Research Group for Nojiri-ko Excavation (FIRGNE), 2003. Insect fossils obtained from the Nojiri-ko Formation during the 14th Nojiri-ko Excavation. Bulletin of the Nojiri-ko Museum 11, 83–88. [In Japanese].

Francoeur, A., 1983. The ant fauna near the tree-line in northern Quebec (Formicidae, Hymenoptera). Nordicana 47, 177–180.

Francoeur, A. & Elias, S.A., 1985. *Dolichoderus taschenbergi* Mayr (Hymenoptera: Formicidae) from an early Holocene fossil insect assemblage in the Colorado Front Range. Psyche 92, 303–308.

Franks, J.W., Sutcliffe, A., Kerney, M., & Coope, G.R., 1958. Haunt of elephant and rhinoceros: The Trafalgar Square of 100,000 years ago. Illustrated London News, June 14 issue.

Fredskild, B., 1988. Agriculture in a marginal area-south Greenland from the Norse Landnam (985 A.D.) to the present (1985 A.D.). In: Birks, H.B., Birks, H.J.B., Kaland, P.E., Moe, D. (Eds), The cultural landscape-past, present and future. Cambridge, Cambridge University Press, 381–393.

Fredskild, B., Jacobsen, N., & Roen, U., 1975. Remains of mosses and freshwater animals in some Holocene lake and bog sediments from Greenland. Meddelelser om Grønland 198(5).

Frey, D.G., 1986. Cladocera analysis. In: Berglund, B. (Ed), Handbook of Holocene palaeoecology and palaeohydrology. New York, Wiley, 667–692.

Fritz, P., Morgan, A.V., Eicher, U., & McAndrews, J.H., 1987. Stable isotopes, fossil Coleoptera and pollen stratigraphy in late Quaternary sediments from Ontario and New York State. Palaeogeography, Palaeoclimatology, Palaeoecology 58, 183–202.

Froeschner, R.C., 1960. Cydnidae of the western hemisphere. Proceedings of the United States National Museum 111, 337–680.

Froese, D., Smith, D.G., Westgate, J.A., Ager, T.A., Preece, S.J., Sandhu, A., Enkin, R.J., & Weber, F., 2003. Recurring middle Pleistocene outburst floods in east-central Alaska. Quaternary Research 60, 50–62.

Fry, G.F., 1976. Analysis of prehistoric coprolites from Utah. University of Utah, Anthropological Papers 97, 1–45.

Futuyma, D.J., 1979. Evolutionary biology. Sunderland, MA, Sinauer Associates, Inc.

Fyles, J.G., Hills, L.V., Matthews, Jr, J.V., Barendregt, R.W., Baker, J., Irving, E., & Jetté, H., 1994. Ballast Brook and Beaufort Formations (Late Tertiary) on Northern Banks Island, Arctic Canada. Quaternary International 22/23, 141–172.

Gabus, J.H., Lemdahl, G., & Weidmann, M., 1987. Sur l'age des terraces l'emanniques au SW de Lausanne [On the age of the terraces coming from southwest of Lausanne]. Bulletin de Geologie Lausanne 293, 419–429. [In French].

Gaillard, M.-J. & Lemdahl, G., 1994a. Early-Holocene coastal environments and climate in southeast Sweden: a reconstruction based on macrofossils from submarine deposits. The Holocene 4, 53–68.

Gaillard, M.-J. & Lemdahl, G., 1994b. Lateglacial insect assemblages from Grand-Marais, south-western Switzerland-climatic implications and comparison with pollen and plant macrofossil data. Dissertationes Botanicae 234, 287–308.

Gao, C., Coope, G.R., Keen, D.H., & Petit, M.E., 1998. Middle Devensian deposits of the Ivel Valley at Sandy, Bedfordshire, England. Proceedings of the Geologist's Association 109, 127–137.

Gao, C., Keen, D.H., Boreham, S., Coope, G.R., Pettit, M.E., Stuart, A.J., & Gibbard, P.L., 2000. Last Interglacial and Devensian deposits of the River Great Ouse at Woolpack Farm, Fenstanton, Camridgeshire, UK. Quaternary Science Reviews 19, 787–810.

García Ambrosiani, K. & Robertsson, A.-M., 1992. Early Weichselian interstadial sediments at Härnösand, Sweden. Boreas 21, 305–317.

Garry, C.E., Baker, R.G., Gilbertson, J.P., & Huber, J.K., 1997. Fossil insects and pollen from late Illinoian sediments in midcontinental North America. Quaternary Proceedings 5, 113–124.

Garry, C.E., Baker, R.G, Schwert, D.P., & Schneider, A.F., 1990a. Paleoenvironment of Twocreekan sediments at Kewaunee, Wisconsin as inferred from analyses of fossil beetle (Coleoptera) assemblages. In: Schneider, A.F., Fraser, G.S. (Eds), Late Quaternary history of the Lake Michigan basin. Geological Society of America Special Paper 251, 57–65.

Garry, C.E., Schwert, D.P., Baker, R.G., Kemmis, T.J., Horton, D.G., & Sullivan, A.E., 1990b. Plant and insect remains from the Wisconsinan interstadial/stadial transition at Wedron, north-central Illinois. Quaternary Research 33, 387–399.

Gaunt, G.D., Coope, G.R., & Franks, J.W., 1970. Quaternary deposits at Oxbow opencast coal site in the Aire Valley,

Yorkshire. Proceedings of the Yorkshire Geological Society 38, 175–200.
Gaunt, G.D., Coope, G.R., Osborne, P.J., & Franks, J.W., 1972. An interglacial deposit near Austerfield, southern Yorkshire. Institute of Geological Sciences Report 72/4.
Gaunt, G. & Girling, M., 1996. Southerly-derived fluvio-glacial deposits near Scrooby, Nottinghamshire, U.K., containing a coleopteran fauna. Circaea 12, 191–194.
Gaunt, G. & Girling, M., 1997. Southerly-derived fluvioglacial deposits near Scrooby, Nottinghamshire, U.K., containing a coleopteran assemblage. In: Ashworth, A.C., Buckland, P.C., Sadler, J.P. (Eds), Studies in Quaternary entomology – an inordinate fondness for insects. Quaternary Proceedings 5, 125–128.
Gaunt, G.D., Jarvis, R.A., & Matthews, B., 1971. The Late Weichselian sequence in the Vale of York. Proceedings of the Yorkshire Geological Society 38, 281–284.
Gedda, B., Lemdahl, G., & Gaillard, M.-J., 1999. Lateglacial and Early Holocene environments inferred from a tufa deposit at Fyledalen, S. Sweden. Geologiska Föreningens I Stockholm Förhandlingar (GFF) 121, 33–41.
Geyh, M.A. & Schleicher, H., 1990. Amino acid racemization method (AAR). In: Geyh, M.A., Schleicher, H. (Eds), Absolute age determination. Physical and chemical dating methods and their application. New York, Springer, 346–355.
Gibbard, P.L., Boreham, S., Cohen, K.M., & Moscariello, A., 2005. Global chronostratigraphical correlation table for the last 2.7 million years. Boreas 34 (1) unpaginated, modified/updated 2007.
Gibbard, P.L., Coope, G.R., Hall, A.R., Pierce, R.C., & Robinson, J.E., 1981. Middle Devensian deposits beneath the ''Upper Floodplain'' terrace of the River Thames at Kempton Park, Sunbury, England. Proceedings of the Geological Association 93, 275–289.
Gilbert, M.B. & Bass, W.M., 1967. Seasonal dating of burials from the presence of fly pupae. American Antiquity 32, 534–535.
Gillette, D.D. & Madsen, D.B., 1992. The short-faced bear *Arctodus simus* from the late Quaternary in the Wasatch Mountains of central Utah. Journal of Vertebrate Paleontology 12, 107–112.
Giorgi, J. & Robinson, M., 1985. The Environment. In: Foreman, M., Rahtz, S. (Eds), Excavations at Faccenda Chicken Farm, near Alchester, 1983. Oxoniensia 49, 38–45.
Girling, M., 1974a. Evidence from Lincolnshire of the age and intensity of the mid-Devensian temperate episode. Nature 250, 270.
Girling, M.A., 1974b. Fossil Coleoptera from Crossnacreevy. In: Clayton, B. (Ed), Palaeoecological investigations at Crossnacreevy, Co. Down. Ulster Journal of Archaeology 36–37; 49–51.
Girling, M.A., 1976a. Fossil Coleoptera from the Somerset Levels: the Abbot's Way. Somerset Levels Papers 2, 28–33.
Girling, M.A., 1976b. Changes in the Meare Heath Coleoptera fauna in response to flooding. Proceedings of the Prehistoric Society 42, 297–299.
Girling, M.A., 1977a. Bird pellets from a Somerset Levels Neolithic trackway. Naturalist (Hull) 102, 49–52.
Girling, M.A., 1977b. Fossil insect assemblages from Rowland's track. Somerset Levels Papers 3, 51–60.
Girling, M.A., 1978. The application of fossil insect studies to the Somerset Levels. In: Brothwell, D.R., Thomas, K.D., Clutton-Brock, J. (Eds), Research problems in zooarchaeo logy. University of London, Institute of Archaeology, Occasional Paper 3, 85–90.
Girling, M.A., 1979a. The fossil insect assemblages from the Meare Lake Village. Somerset Levels Papers 5, 25–32.
Girling, M.A., 1979b. Fossil insects from the Sweet track. Somerset Levels Papers 5, 84–93.
Girling, M.A., 1979c. The insects. In: Southwark Excavations 1972–1974. Joint Publication, London and Middlesex Archaeological Society and Surrey Archaeological Society, London 1, 170–172; 414; 466.
Girling, M.A., 1980a. Two late Pleistocene insect faunas from Lincolnshire. Unpublished PhD Thesis, University of Birmingham, Birmingham, UK.
Girling, M.A., 1980b. The fossil insect assemblage from the Baker Site. Somerset Levels Papers 6, 36–42.
Girling, M.A., 1981a. The environmental evidence. In: Mellor, J.E., Pearce, T. (Eds), The Austin Friars, Leicester. Council for British Archaeology, Research Report 35, 169–173.
Girling, M.A., 1981b. Beetle remains. In: Thorn, J.C. (Ed), The burial of John Dygon, Abbot of St. Augustine's. Collectanea historica: Essays in memory of Stuart Rigold. Maidstone, UK, Kent Archaeological Society, 82–84.
Girling, M.A., 1982. The effect of the Meare Heath flooding episodes on the Coleopteran succession. Somerset Levels Papers 8, 46–50.
Girling, M.A., 1983. The environmental implications of the excavations of 1974–76. In: Brown, A.E., Woodfield, C., Mynard, D.C. (Eds), Excavations at Towcester, Northamptonshire: The Alchester Road suburb. Northamptonshire Archaeology 18, 128–130+fiche.
Girling, M.A., 1984a. Aquatic Coleoptera in the fossil insect assemblages from archaeological sites in the Somerset Levels. Balfour-Browne Club Newsletter 30, 1–11.
Girling, M.A., 1984b. The insect fauna. In: Ellis, P. (Ed), Catsgore 1979. Further excavation of the Romano-British village. Oxford, Oxbow Books, 39–40+fiche.
Girling, M.A., 1984c. Investigations of a second insect assemblage from the Sweet Track. Somerset Levels Papers 10, 79–91.
Girling, M.A., 1985a. The insect fauna-City Arms, trench 6, layer 3. In: Shoesmith, R. (Ed), Hereford city excavations. 3, The finds. Council for British Archaeology Research Report 56, 96+fiche M9.B12–14.
Girling, M.A., 1985b. An 'Old-Forest' beetle fauna from a Neolithic and Bronze Age peat deposit at Stileway. Somerset Levels Papers 11, 80–83.
Girling, M.A., 1986a. The insects associated with Lindow man. In: Stead, I.M., Bourke, J.B., Brothwell, D. (Eds), Lindow man. The body in the bog. London, UK, British Museum, 90–91.
Girling, M.A., 1986b. Insect remains: environmental implications of the fauna. In: Booth, P. (Ed), Roman store buildings at Alcester. Transactions of the Birmingham and Warwickshire Archaeological Society 94, 95–96.
Girling, M.A., 1988. The bark beetle *Scolytus scolytus* (Fabricius) and the possible role of elm disease in the early Neolithic. In: Jones, M. (Ed), Archaeology and the Flora of the British Isles. Oxford University Committee for Archaeology Monograph 14, 34–38.
Girling, M.A., 1989a. The insect fauna of the Roman Well at the Cattlemarket. In: Down, A. (Ed), Chichester excavations 6, 234–241. Philimore, Chichester, UK.
Girling, M.A., 1989b. Mesolithic and later landscapes interpreted from the insect assemblages of West Heath Spa Hampstead. In: Collins, D., Lorimer, D. (Eds), Excavations at the Mesolithic Site on West Heath, Hampstead 1976–1981. British Archaeological Reports 217, Oxford, 72–89.
Girling, M.A., 1991. Fossil insect faunas from Buckbean Pond, Breiddin Hillfort. In: Musson, C. (Ed), The Breiddin Hillfort: a later prehistoric settlement in the Welsh Marches. Council for British Archaeology Research Report 76, London, fiche 2; 153–169.

Girling, M.A., 1993. Insect attack in charcoal. In: Ellis, P. (Ed), Beeston Castle, Cheshire. Excavations by Laurence Keen and Peter Hough, 1968–85. Historic Buildings and Monuments Commission for England Archaeological Report 23, London, 79–80.

Girling, M.A. & Greig, J.R.A., 1977. Palaeoecological investigations of a site at Hampstead Heath, London. Nature 268, 45–47.

Girling, M.A. & Greig, J.R.A., 1985. A first fossil record for *Scolytus scolytus* (F.) (Elm Bark Beetle): its occurrence in Elm Decline deposits from London and the implications for Neolithic Elm Disease. Journal of Archaeological Science 12, 347–352.

Girling, M.A., Robinson, M., 1987. The insect fauna. In: Balaam, N.D., Levitan, B., Staker, V. (Eds), Studies in palaeoeconomy and environment in South West England. British Archaeological Reports 181, 239–246.

Girling, M.A., Robinson, M., 1989. Ecological Interpretations from the beetle fauna. In: Hayfield, C., Greig, J, (Eds), Excavation and salvage work on a Moated Site at Cowick, South Humberside, 1976. Yorkshire Archaeological Journal 61, 47–70.

Giterman, R.E., Sher, A.V., & Matthews, Jr, J.V., 1982. Comparison of the development of tundra-steppe environments in West and East Beringia: pollen and macrofossil evidence from key sections. In: Hopkins, D.M., Matthews, Jr, J.V., Schweger, C.E., Young, S.B. (Eds), Paleoecology of Beringia. New York, Academic Press, 43–73.

Goetcheus, V. & Birks, H.H., 2001. Full-glacial upland tundra vegetation preserved under tephra in the Beringia National Park, Seward Peninsula, Alaska. Quaternary Science Reviews 20, 135–147.

Goldstein, P.Z. & Desalle, R., 2003. Calibrating phylogenetic species formation in a threatened insect using DNA from historical specimens. Molecular Ecology 12, 1993–1998.

Golosova, L.D., Druk, A.Ya., Karppinen, E., & Kiselyov, S.V., 1985. Subfossil oribatid mites (Acarina, Oribatei) of northern Siberia. Annales Entomologici Fennici 51, 3–18.

Gould, S.J. & Eldredge, N., 1977. Punctuated equilibria: the tempo and mode of evolution reconsidered. Paleobiology 3, 115–151.

Goulet, H., 1983. The genera of Holarctic Elaphrini and species of *Elaphrus* Fabricius (Coleoptera: Carabidae); classification, phylogeny and zoogeography. Quaestiones Entomologicae 19, 219–482.

Graham, S.A., 1965. Entomology: an aid in archaeological studies. American Antiquity Memoir 19, 167–174.

Green, C.P., Branch, N.P., Coope, G.R., Field, M.H., Keen, D.H., Wells, J.M., Schwenninger, J.-L., Preece, R.C., Schreve, D.C., Canti, M.G., & Gleed-Owen, C.P., 2006. Marine isotope stage 9 environments of fluvial deposits at Hackney, north London, UK. Quaternary Science Reviews 25, 89–113.

Green, C.P., Branch, N., Silva, B., Vaughan-Williams, A., Elias, S., Poole, I., & Valcarcel, I., 2008. Environmental Evidence. In: Mudd, A., Pears, B. (Eds), Bronze Age field system at Tower's Fen, Thorney, Peterborough excavations at 'Thorney Borrow Pit' 2004–2005. British Archaeological Reports 471, 55–67.

Green, C.P., Coope, G.R., Currant, A.P., Holyoak, D.T., Ivanovich, M., Jones, R.L., Keen, D.H., McGregor, D.F., & Robinson, J.E., 1984. Evidence of two temperate episodes in late Pleistocene deposits at Marsworth, UK. Nature 309, 778–781.

Green, C.P., Coope, G.R., Jones, R.L., Keen, D.H., Bowen, D.Q., Currant, A.P., Holyoak, D.T., Ivanovich, M., Robinson, J.E., Rogerson, R.J., & Young, R.C., 1996. Pleistocene deposits at Stoke Goldington, in the valley of the Great Ouse, UK. Journal of Quaternary Science 11, 59–87.

Green, M. & Pitman, A.J., 2003. Effects of temperature and humidity on flight initiation and mortality of *Euophryum confine*. New Zealand Journal of Zoology 30, 11–15.

Greenwood, M. & Smith, D.N., 2005. Changing fluvial conditions and landscapes in the Trent Valley: a review of palaeoentomological evidence. In: Smith, D.N., Brickley, M.B., Smith, W. (Eds), Fertile ground: Papers in honour of Susan Limbrey (Symposia of the Association for Environmental Archaeology No. 22). Oxford, Oxbow Books, 53–67.

Greenwood, M., Wood, P.J., & Monk, W.A., 2006. The use of fossil caddisfly assemblages in the reconstruction of flow environments from floodplain paleochannels of the River Trent, England. Journal of Paleolimnology 35, 747–761.

Gregg, R.E., 1972. The northward distribution of ants in North America. Canadian Entomologist 104, 1073–1091.

Greig, J.R.A., 1985. Agricultural diversity and sub-alpine colonisation; the story from pollen analysis at Fiave. In: Malone, C., Stoddart, S. (Eds), Papers in Italian Archaeology IV, Part 2, Prehistory. British Archaeological Reports, International Series 244, 296–315.

Greig, J.R.A., Girling, M.A., & Skidmore, P., 1982. The plant and insect remains. In: Barker, P., Higham, R. (Eds), Hen Domen, Montgomery: A timber castle on the English-Welsh Border. London, UK, Royal Archaeological Institute, 60–71.

Greig, J.R.A., Pelling, R., Robinson, M., & Stevens, C., 2004. Environmental evidence. In: Hey, G. (Ed), Yarnton: Saxon and medieval settlement and landscape. Results of excavations 1995–96. Oxford Archaeology, Thames Valley Landscapes 20. Oxford, 351–410.

Grinnell, F., 1908. Quaternary myriapods and insects of California. University of California Publications in Geology 5, 207–215.

Gröcke, D.R., Elias, S.A., Miller, R.M., & Schimmelmann, A., 2006. Hydrogen isotopes in chitin. Quaternary Science Reviews 25, 1850–1864.

Grove, K., 1995. A study on insect remains in archaeological deposits. In: Woodward, S. (Ed), The archaeology of Chichester and district 1995. Chichester, Southern Archaeology and Chichester District Council, 39–40.

Grunin, K.Ya., 1973. The first discovery of larvae of the mammoth bot-fly *Cobboldia* (*Mamontia*, subgen. N.) *russanovi* sp. n. (Diptera, Gasterophilidae). Entomological Review 52, 228–230.

Günther, J., 1983. Development of Grossensee (Holstein, Germany): variations in trophic status from the analysis of subfossil microfauna. Hydrobiologia 103, 231–234.

Guthrie, R.D., 1990. Frozen fauna of the mammoth steppe. The story of blue babe. Chicago, University of Chicago Press.

Guyan, W.U., 1981. Puppen der Hausfliege, Puppen der "stallfliege" [Pupae of houseflies; pupae of stable flies]. In: Guyan, W.U. (Ed), Zur Viehhaltung im Steinzeitdorf Thayngen-Weier II. Archäologie der Schweiz 4, 114–116. [In German].

Gvozdetskiy, N.A. & Mikhailov, N.I., 1987. Fizicheskaya geografiya SSSR. Aziatskaya chast' [Physical geography of the USSR. Asian part]. Vysshaya shkola, Moscow. [In Russian].

Haarløv, N., 1967. Arthropoda (Acarina, Diptera) from subfossil layers in West Greenland. Meddelelser om Grønland 184, 1–17.

Hadley, N.F., 1979. Wax secretion and color phases of the desert tenebrionid beetle *Cryptoglossa verrucosa* (LeConte). Science 203, 367–369.

Haeck, J., 1971. The immigration and settlement of carabids in the new Ijselmeer-polders. Miscellaneous Papers, Landbouwhogeschool, Wageningen 8, 33–53.

Hajdas, I., Bonani, G., Furrer, H., Mäder, A., & Schoch, W., 2007. Radiocarbon chronology of the mammoth site at Niederwenigen, Switzerland: results from dating boners, teeth, wood, and peat. Quaternary International 164–165, 98–105.

Hajdas, I., Bonani, G., Moreno, P.I., & Ariztegui, D., 2003. Precise radiocarbon dating of late-glacial cooling in mid-latitude South America. Quaternary Research 59, 70–78.

Hakbijl, T., 1986. Remains of insects. In: Gawronski, J.H.G. (Ed), Amsterdam Project. Annual Report of the VOC Ship ''Amsterdam'' Foundation 1985, Amsterdam, 72–73.

Hakbijl, T., 1987. Insect remains: unadulterated Cantharidum and tobacco from the West Indies. In: Gawronski, J.H.G. (Ed), Amsterdam Project. Annual Report of the VOC Ship ''Amsterdam'' Foundation 1986, Amsterdam, 93–94.

Hakbijl, T., 1988. Insektenresten [Insect remains]. In: Groenman-van Waateringe, W., van Beek, B.L. (Eds), De Romeinse castella te Valkenburg ZH. [The Roman fort of Valkenburg]. Studies in Prae- und Protohistorie 2, 106–107. [In Dutch].

Hakbijl, T., 1989. Insect remains from site Q, an Early Iron Age farmstead of the Assendelver Polders project. Helinium 29, 77–102.

Hakbijl, T. & De Groot, M., 1997. Insect Remains from Willem Barents' 1596 Arctic Exploration preserved in ''Het Behouden Huys'', Novaya Zemlya-with notes on the medicinal use of Cantharides. In: Ashworth, A.C., Buckland, P.C., Sadler, J.P. (Eds), Studies in Quaternary entomology – an inordinate fondness for insects. Quaternary Proceedings 5, 129–134.

Hakbijl, T., Pals, J.P., & Troostheide, C.D., 1989. Plant and insect remains from the Late Neolithic well at Kolhorn. Palaeohistoria 31, 157–163.

Hall, A.R., Jones, A.K.G., Kenward, H.K., & O'Connor, T.P., 1992. The environmental evidence. In: Tweddle, D. (Ed), The Anglian Helmet from Coppergate. Archaeology of York 17/8. Council for British Archaeology for York Archaeological Trust, York, 875–880; 903–907.

Hall, A.R. & Kenward, H.K., 1976. Biological evidence for the usage of Roman riverside warehouses at York. Britannia 7, 274–276.

Hall, A.R. & Kenward, H.K., 1980. An interpretation of biological remains from Highgate, Beverley. Journal of Archaeological Science 7, 33–51.

Hall, A.R. & Kenward, H.K., 1990. Environmental evidence from the Colonia. Archaeology of York, 14/6. Council for British Archaeology for York Archaeological Trust.

Hall, A.R., Kenward, H.K., Hill, M., Large, F., Jaques, D., Dobney, K., Issitt, M., & Lancaster, S., 1994. Technical report: biological remains from excavations on the Leven-Brandesburton by-pass, N. Humberside. Reports from the Environmental Archaeology Unit, York, 94/15.

Hall, A.R., Kenward, H.K., & Robertson, A., 1993. Investigation of medieval and post-medieval plant and invertebrate remains from Area II of the excavations in The Bedern (north-east), York (YAT/Yorkshire Museum sitecode 1973–81.14 II). Ancient Monuments Laboratory Report 58/93.

Hall, A.R., Kenward, H.K., Williams, D., & Greig, J.R.A., 1983. Environment and living conditions at two Anglo-Scandinavian Sites. Archaeology of York, 14/4. Council for British Archaeology for York Archaeological Trust.

Hall, H.J., 1977. A paleoscatalogical study of diet and disease at Dirty Shame Rockshelter, southeast Oregon. Tebiwa. Miscellaneous Papers of the Idaho State University Museum of Natural History 8, 1–14.

Hall, W.E., Olson, C.A., & Van Devender, T.R., 1989. Late Quaternary and modern arthropods from the Ajo Mountains of southwestern Arizona. Pan-Pacific Entomologist 65, 322–347.

Hall, W.E., Van Devender, T.R., & Olson, C.A., 1988. Late Quaternary arthropod remains from Sonoran Desert packrat middens, southwestern Arizona and northwestern Sonora. Quaternary Research 29, 277–293.

Hall, W.E., Van Devender, T.R., & Olson, C.A., 1990. Arthropod history of the Puerto Blanco Mountains, Organ Pipe National Monument, southwestern Arizona. In: Betancourt, J.L., Van Devender, T.R., Martin, P.S. (Eds), Packrat Middens: The last 40,000 years of biotic change. Tucson, AZ, University of Arizona Press, 363–379.

Halstead, D.G.H., 1993. Keys for the identification of beetles associated with stored products-II. Laemophloeidae, Passandridae and Silvanidae. Journal of Stored Product Research 29, 99–197.

Hamilton, T.D., Craig, J.L., & Sellmann, P.V., 1988. The Fox permafrost tunnel: a late Quaternary geologic record in central Alaska. Geological Society of America Bulletin 100, 948–969.

Hammarlund, D. & Lemdahl, G., 1994. A Late Weichselian stable isotope stratigraphy compared with biostratigraphic data. A case study from southern Sweden. Journal of Quaternary Science 9, 13–32.

Hammond, P., Morgan, A., & Morgan, A.V., 1979. On the *gibbulus* group of *Anotylus*, and fossil occurrence of *Anotylus gibbulus* (Staphylinidae). Systematic Entomology 4, 215–221.

Hansen, J.P.H., 1989. The mummies from Qilakitsoq-paleopathological aspects. Meddelelser om Grønland, Man and Society 12, 69–82.

Harde, K.W., 1984. A field guide in colour to beetles. London, UK, Octopus.

Hayashi, M., 1996. Fossils of a Dytiscid beetle, *Ilybius weymarni*, from Late Pleistocene Maebashi Peat in Maebashi City, central Japan. Daiyonki-kenkyu 35, 305–312. [In Japanese].

Hayashi, M., 1998a. Upper Pleistocene fossil beetles from the Ono Peat Bed in Tatsuno-machi, Nagano Prefecture, Japan. Bulletin of the Mizunami Fossil Museum 25, 117–125. [In Japanese].

Hayashi, M., 1998b. Early Pleistocene aquatic beetles from the Ookui Formation in Kitamimaki-mura, Nagano Prefecture, Japan. Bulletin of the Mizunami Fossil Museum 25, 103–115.

Hayashi, M., 1999a. Fossil donaciine beetles (Coleoptera: Chrysomelidae: Donaciinae) from the Pleistocene around the Haruna Volcano, Gunma Prefecture, central Japan. Bulletin of the Mizunami Fossil Museum 26, 149–162. [In Japanese].

Hayashi, M., 1999b. Late Pliocene to early Pleistocene donaciine fossils from the Uonuma Formation in Higashikubiki hills, Niigfata Prefecture, Japan, with description of a new species (Coleoptera: Chrysomelidae: Donaciinae). Bulletin of the Osaka Museum of Natural History 53, 1–22.

Hayashi, M., 2001. Fossil insects and paleoenvironments of the Plio-Pleistocene Uonuma Formation, Niigata Prefecture, Japan. Bulletin of the Mizunami Fossil Museum 28, 223–238. [In Japanese with English abstract].

Hayashi, M., 2002. Late Cenozoic Coleoptera of Japan (Part 4): Donaciinae. Gekkan-Mushi 374, 36–38. [In Japanese].

Hayashi, M., 2004. Faunal changes in Donaciinae during the Quaternary in central Japan (Coleoptera, Chrysomelidae). In: Jolivet, P., Santiago-Blay, J.A., Schmitt, M. (Eds), New developments in the biology of Chrysomelidae. The Hague, The Netherlands, SBP Academic Publishing, 263–274.

Hayashi, M., 2005. Fossil insects and reconstructed paleoenvironments of the Upper Pleistocene Maebashi Peat in Takasaki, Gunma Prefecture, Central Japan. Bulletin of the Gunma Museum of Natural History 9, 93–99. [In Japanese].

Hayashi, M., 2007. Beetle records: Late Pleistocene of Japan. In: Elias, S.A. (Ed), Encyclopedia of Quaternary science. Amsterdam, Elsevier, 196–212.

Hayashi, M., Katoh, S., Yagi, T., Handa, K., Adachi, I., & Nakai, K., 2001a. Age and fossil insects of the Pleistocene Kotari Peat Beds in the eastern part of Hyogo Prefecture, western Japan. Humans and Nature 12, 31–38. [In Japanese with English abstract].

Hayashi, M. & Miyatake, Y., 1996. Last Glacial insect fossils from the Oku Miomote site group, Niigata Prefecture, Japan. Chikyu Kagaku 50, 188–193. [In Japanese].

Hayashi, M. & Shimadu, Y., 2005. Stratigraphy and fossil insects of the Upper Pleistocene Tsukumogawa Peat Bed in Annaka, Gunma Prefecture, Japan. Bulletin of the Gunma Museum of Natural History 9, 101–107. [In Japanese].

Hayashi, M. & Shiyake, S., 2002. Late Pliocene Donaciinae (Coleoptera, Chrysomelidae) from the Koka Formation, Kobiwako Group in Shiga Prefecture, Japan. Elytra (Tokyo) 30, 207–213.

Hayashi, M., Shiyake, S., Miyatake, Y., & Iwai, D., 2003a. Pleistocene fossil insects from the underground of Nagai Park, Osaka City, western Japan, with description of three donaciine leaf beetles (Coleoptera: Chrysomelidae). Bulletin of the Osaka Museum of Natural History 57, 39–45.

Hayashi, M., Yahiro, K., & Kitabayashi, E., 2001b. Middle Pleistocene insects from the Tsumori Formation in Mashiki-machi, Kumamoto Prefecture, Japan. Bulletin of the Mizunami Fossil Museum 28, 239–243. [In Japanese].

Hayashi, M., Yahiro, K., & Kitabayashi, E., 2002. Late Pleistocene insects from the Mizozono Formation in Yoshimatsu-cho, Kagoshima Prefecture, Japan. Bulletin of the Mizunami Fossil Museum 29, 161–168. [In Japanese].

Hayashi, M., Yahiro, K., & Kitabayashi, E., 2003b. Fossil insects and their paleoenvironments of the Pliocene Tsubusagawa Formation in Ajimu Basin, Oita Prefecture, Kyushu, Japan. Bulletin of the Mizunami Fossil Museum 30, 167–173. [In Japanese].

Hayashi, M., Yahiro, K., & Kitabayashi, E., 2004a. Early Pleistocene insect fossils from the Oyama Formation at Oyama-machi, Oita Prefecture, Japan. Bulletin of the Mizunami Fossil Museum 31, 69–72. [In Japanese].

Hayashi, M., Yahiro, K., & Kitabayashi, E., 2005. Middle Pleistocene fossil insects from the Hirabaru Formation of Kitsuki City, Oita Prefecture, Kyushu, Japan. Bulletin of the Mizunami Fossil Museum 32, 227–234. [In Japanese].

Hayashi, M., Yahiro, K., Kitamura, N., & Kitabayashi, E., 2004b. Middle Pleistocene fossil insects from the Tsumori Formation of Mashiki-machi, Kuamoto Prefecture, Japan (Part 2). Bulletin of the Mizunami Fossil Museum 31, 63–67. [In Japanese].

Hayfield, C. & Greig, J., 1989. Excavation and salvage work on a moated site at Cowick, South Humberside, 1976. Yorkshire Archaeological Journal 61, 41–70.

Hebda, R.J., Warner, B.G., & Cannings, R.A., 1990. Pollen, plant macrofossils, and insects from fossil woodrat (*Neotoma cinerea*) middens in British Columbia. Géographie physique et Quaternaire 44, 227–234.

Heer, O., 1865. Die Urwelt der Schweiz (lst ed.) [Primeval Switzerland], Zurich. [In German].

Heizer, R.F., 1970. The anthropology of prehistoric Great Basin human coprolites. In: Brothwell, D., Higgs, E. (Eds), Science in archaeology. New York, Praeger Publishers, 244–250.

Helle, M., Sonstegaard, E., Coope, G.R., & Rye, N., 1981. Early Weichselian peat at Brumunddal, southeastern Norway. Boreas 10, 369–379.

Hellqvist, M., 1996. Insektfynd från Gamla Uppsala [Insect remains from Old Uppsala]. Arkeologi och Miljögeologi i Gamla Uppsala 2, 166–174. [In Swedish].

Hellqvist, M., 1999. Insect assemblages from Iron Age wells in Central Sweden. Norwegian Archaeological Review 32, 1–18.

Hellqvist, M. & Lemdahl, G., 1990. Insektfynd från gamla Faulun [Insects from old Faulun]. Flora och Fauna 85, 234–239. [In Swedish].

Hellqvist, M. & Lemdahl, G., 1996. Insect assemblages and local environment in the mediaeval town of Uppsala, Sweden. Journal of Archaeological Science 23, 873–881.

Henriksen, K.L., 1933. Undersøgelser over Danmark-Skånes kvartaere Insektfauna [Studies of Quaternary insect faunas from Denmark and Skånes]. Videnskabelige Meddelelser fra Danske naturhistorisk Förening i Kjoebhavn 96, 77–355. [In Danish].

Hetherington, R., Barrie, J.V., Reid, R.G.B., MacLeod, R., Smith, D.J., James, T.S., & Kung, R., 2003. Late Pleistocene coastal paleogeography of the Queen Charlotte Islands, British Columbia, Canada, and its implications for terrestrial biogeography and early postglacial human occupation. Canadian Journal of Earth Sciences 40, 1755–1766.

Heusser, C., 1989. Pollen analysis. In: Dillehay, T.D. (Ed), Monte Verde: a Late Pleistocene settlement in Chile, Vol. 1, Palaeoenvironment and context, Washington, DC, Smithsonian Institution Press, 193–199.

Heusser, C.J., Heusser, L.E., & Streeter, S.S., 1980. Quaternary temperatures and precipitation for the north-west coast of North America. Nature 286, 702–704.

Heusser, L.E. & van de Geer, G., 1994. Direct correlation of terrestrial and marine palaeoclimatic records from four glacial-interglacial cycles-DSDP Site 594 Southwest Pacific. Quaternary Science Reviews 13, 273–282.

Hevley, R.H. & Johnson, C.D., 1974. Insect remains from a prehistoric Pueblo in Arizona. Pan-Pacific Entomologist 50, 307–308.

Hewitt, G., 2000. The genetic legacy of the Quaternary ice ages. Nature 405, 907–913.

Heydemann, B., 1967. Über die epigäische Aktivität terrestrischer Arthropoden der Küstenregionen im Tagerhythmus. In: Graff, O., Satchell, J.E. (Eds), Progress in soil biology. Braunschweig, Germany, Vieweg, 249–263.

Hodder, K.H., Bullock, J.M., Buckland, P.C., & Kirby, K.J., 2005. Large herbivores in the wildwood and modern naturalistic grazing systems. English Nature Research Report 648. English Nature, Peterborough.

Hodgins, G.W.L., Thorpe, J.L., Coope, G.R., & Hedges, R.E.M., 2001. Protocol development for purification and characterization of sub-fossil insect chitin for stable isotopic analysis and radiocarbon dating. Radiocarbon 43, 199–208.

Hofmann, W., 1983. Stratigraphy of Cladocera and Chironomidae in a core from a shallow North German lake. Hydrobiologia 103, 235–239.

Hofmann, W., 1986. Chironomid analysis. In: Berglund, B. (Ed), Handbook of Holocene palaeoecology and palaeohydrology. New York, Wiley, 715–727.

Hoganson, J.W. & Ashworth, A.C., 1992. Fossil beetle evidence for climatic change 18,000–10,000 years B.P. in south-central Chile. Quaternary Research 37, 101–116.

Holdhaus, K. & Lindroth, C.H., 1939. Die Europanischen Koleopteren mit boreoalpiner verbreitung [The European beetles with boreo-alpine distributions]. In: Holdhaus, K., Lindroth, C.H. (Eds), Sonder-Abdruck aus dem 50 bande der Annalen des Naturhistorischen Museums in Wien, Tafel VIII, 123–293.

Holdridge, L., 1987. The beetle remains. In: Millett, M., McGrail, S. (Eds), The archaeology of the Hasholme Logboat. Archaeological Journal 144, 88–89.

Holdsworth, P., 1976. An insect fauna from Old Street, St. Helier. Annual Bulletin of the Société Jersiaise 21, 494–496.

Hopkins, D.M., Giterman, R.E., & Matthews, Jr, J.V., 1976. Interstadial mammoth remains and associated pollen and insect fossils, Kotzebue Sound area, northwestern Alaska. Geology 4, 169–172.

Hopkins, D.M., Matthews, Jr, J.V., Wolfe, J.A., & Silberman, M.L., 1971. A Pliocene flora and insect fauna from the Bering Strait region. Palaeogeography, Palaeoclimatology, Palaeoecology 9, 211–231.

Hopkins, T.L. & Kramer, K.J., 1992. Insect cuticle sclerotization. Annual Review of Entomology 37, 273–302.

Horion, A., 1941, Faunistik der deuschen Käfer. Band 1. Adephaga – Caraboidea [Faunal studies of Central European Beetles, 1. Adephaga – Caraboidea]. Hans Goecke, Krefeld. [In German].

Horion, A., 1960. Faunistik der Mitteleuropäischen Käfer, 7. Clavicornia, Sphaeritidae-Phalacridae [Faunal studies of Central European Beetles, 7. Clavicornia, Sphaeritidae-Phalacridae.]. Uberlingen, Bodensee. [In German].

Horn, G.H., 1876. Notes on some Coleopterous remains from the bone cave at Port Kennedy, Penna. Transactions of the American Entomological Society 5, 241–245.

Horne, P., 1979. Head lice from an Aleutian mummy. Paleopathology Newsletter 25, 7–8.

Horrocks, M. & Weisler, M.I., 2006. Analysis of plants in archaeological deposits from two remote archipelagos: the Marshall Islands, Eastern Micronesia, and the Pitcairn Group, Southeast Polynesia. Pacific Science 60, 261–280.

Horton, A., Keen, D.H., Field, M.H., Robinson, J.E., Coope, G.R., Currant, A.P., Graham, D.K., Green, C.P., & Phillips, L.M., 1992. The Hoxnian Interglacial deposits at Woodston, Peterborough. Philosophical Transactions of the Royal Society of London 338, 131–164.

Houlder, D., Hutchinson, M., Nix, H., & McMahon, J., 2001. ANUCLIM 5.1. User's Guide. Centre for Resource and Environmental Studies, Canberra.

Howard, A.J., Bateman, M.D., Garton, D., Green, F.M.L., Wagner, P., & Priest, V., 1999. Evidence of Late Devensian and early Flandrian processes and environments at Tiln, north Nottinghamshire. Proceedings of the Yorkshire Geological Society 52, 383–393.

Howard, L.C., Wood, P.J., Greenwood, M.T., & Rendell, H.M., 2009. Reconstructing riverine paleo-flow regimes using subfossil insects (Coleoptera and Trichoptera): the application of the LIFE methodology to paleochannel sediments. Journal of Paleolimnology (in press).

Howden, H.F., 1969. Effects of the Pleistocene on North American insects. Annual Review of Entomology 14, 39–56.

Howden, H.F. & Cartwright, O.L., 1963. Scarab beetles of the genus *Onthophagus* Latreille north of Mexico (Coleoptera: Scarabaeidae). Proceedings of the United States National Museum No. 3467, 114, 1–135.

Howden, H.F. & Scholtz, C.H., 1986. Changes in a Texas dung beetle community between 1975 and 1985 (Coleoptera: Scarabaeidae, Scarabaeinae). Coleopterists Bulletin 40, 313–316.

Hubbard, R.N.L., 1979. Appendix 2: ancient agriculture and ecology at Servia. In: rescue excavations at Servia 1971–1973: a preliminary report. Annual of the British School of Archaeology at Athens 74, 226–228.

Hughes, O.L., Harrington, C.R., Janssens, J.A., Matthews, Jr, J.V., Morlan, R.E., Rutter, N.W., & Schweger, C.E., 1981. Upper Pleistocene stratigraphy, paleoecology, and archaeology of northern Yukon interior, eastern Beringia 1. Bonnet Plume Basin. Arctic 34, 329–365.

Hughes, P.D.M., Kenward, H.K., Hall, A.R., & Large, F.D., 2000. A high-resolution record of mire development and climatic change spanning the Late-glacial-Holocene boundary at Church Moss, Davenham (Cheshire, England). Journal of Quaternary Science 15, 697–724.

Hulme, M., Conway, D., Jones, P.D., Jiang, T., Barrow, E.M., & Turney, C., 1995. Construction of a 1961–1990 European climatology for climate change modelling and impact applications. International Journal of Climatology 15, 1333–1363.

Hultén, E., 1937. Outline of the history of Arctic and Boreal Biota during the Quaternary period. Lehre, Germany, J. Cramer.

Huntley, B. & Webb, III, T., 1989. Migration: species' response to climatic variations caused by changes in the earth's orbit. Journal of Biogeography 16, 5–19.

Huppert, A. & Solow, A.R., 2004. A method for reconstructing climate from fossil beetle assemblages. Proceedings of the Royal Society of London B271, 1125–1128.

Hutchinson, J.N. & Coope, G.R., 2002. Cambering and valley bulging, periglacial solifluction and Lateglacial Coleoptera at Dowdeswell, near Cheltenham. Proceedings of the Geologists' Association 113, 291–300.

Hyman, P.S., 1992. A review of the scarce and threatened Coleoptera of Great Britain, Part 1. UK Joint Nature Conservation Committee, Peterborough.

IAEA. 2001. GNIP maps and animations. Vienna, International Atomic Energy Agency, http://www-naweb.iaea.org/napc/ih/GNIP/userupdate/Waterloo/index.html (accessed 19 June, 2008).

Iversen, J., 1954. The Late-Glacial flora of Denmark and its relation to climate and soil. Danmarks Geologiska Undersökn, II. Raekke, 80.

Jackson, S.T., Overpeck, J.T., Webb, T., Keattch, S.E., & Anderson, K.H., 1997. Mapped plant-macrofossil and pollen records of Late Quaternary vegetation change in eastern North America. Quaternary Science Reviews 16, 1–70.

Jaques, D., Hall, A., Kenward, H., & Carrott, J., 2004. Technical report: plant, invertebrate and fish remains from deposits at 25 Bridge Street, Chester (site code: CHE/25BS01). Palaeoecology Research Services Report 2004/46.

Jaques, D., Kenward, H., & Johnstone, C., 2000. Assessment of biological remains from excavations at Hayton, East Riding of Yorkshire (site code KINCM1995.1020). Reports from the Environmental Archaeology Unit, York, 2000/35. York.

Jeanne, C., 1965. Carabiques de la Peninsule Iberique (2ª note) [Ground Beetles of the Iberian Peninsula]. Actes de la Société linnéenne de Bordeaux 102(10). serie A: 1–34. [In French].

Jeanne, C., 1966. Carabiques de la Peninsule Iberique (3ª note). Actes de la Société linnéenne de Bordeaux 103(7). serie A: 3–18

Jeanne, C., 1967a. Carabiques de la Peninsule Iberique (4ª note). Actes de la Société linnéenne de Bordeaux 104(3). serie A: 1–24. [In French].

Jeanne, C., 1967b. Carabiques de la Peninsule Iberique (5ª note). Actes de la Société linnéenne de Bordeaux 104(10). serie A: 1–22. [In French].

Jeanne, C., 1967c. Carabiques de la Peninsule Iberique (6ª note). Actes de la Société linnéenne de Bordeaux 104(13). serie A: 1–19. [In French].

Jeanne, C., 1968a. Carabiques de la Peninsule Iberique (7ª note). Actes de la Société linnéenne de Bordeaux 105(1). serie A: 1–25. [In French].

Jeanne, C., 1968b. Carabiques de la Peninsule Iberique (8ª note). Actes de la Société linnéenne de Bordeaux 105(6). serie A: 1–40. [In French].

Jeanne, C., 1968c. Carabiques de la Peninsule Iberique (9ª note). Actes de la Société linnéenne de Bordeaux 105(8). serie A: 1–22. [In French].

Jeanne, C., 1969. Carabiques de la Peninsule Iberique (lére note). Archives del instituto de Aclimatacion Almeria 14, 101–124.

Jeanne, C., 1971a. Carabiques de la Peninsule Iberique (10ª note). Bulleten de la societe linneenne de Bordeaux 1(2), 5–18.

Jeanne, C., 1971b. Carabiques de la Peninsule Iberique (11ª note). Bulleten de la societe linneenne de Bordeaux 1(4), 87–96.

Jeanne, C., 1971c. Carabiques de la Peninsule Iberique (12[a] note). Bulleten de la societe linneenne de Bordeaux 1(9), 203–220.

Jeanne, C., 1972a. Carabiques de la Peninsule Iberique (13[a] note). Bulleten de la societe linneenne de Bordeaux 2(2), 27–42.

Jeanne, C., 1972b. Carabiques de la Peninsule Iberique (14[a] note). Bulleten de la societe linneenne de Bordeaux 2(5), 99–116.

Jeanne, C., 1973. Carabiques de la Péninsule Iberique (ler suplément). Bulleten de la societe linneenne de Bordeaux 3(l), 3–20.

Johns, P.M., 1986. Arthropods of Banks Peninsula reserves. Report to the Commissioner of Crown Lands, Christchurch. Zoology Department, University of Canterbury, Christchurch, New Zealand.

Johnstone, J.F. & Chapin, F.S., 2003. Non-equilibrium succession dynamics indicate continued northern migration of lodgepole pine. Global Change Biology 9, 1401–1409.

Jones, A.K.G., 1976. The insect remains. In: Rogerson, A. (Ed), Excavations on Fuller's Hill, Great Yarmouth. East Anglian Archaeology 2, 225.

Jones, F.G.W. & Jones, M.G., 1974. Pests of field crops (2nd ed.). London, UK, Arnold.

Jones, G., 1983. The LM 11 Plant Remains. In: Popham, M.R. (Ed), The minoan unexplored mansion at Knossos. London, UK, British School at Athens, Thames & Hudson, 303–306.

Jones, J., 1998. Plant and insect remains. In: Price, R., Ponsford, M. (Eds), St. Bartholomew's Hospital, Bristol. The excavation of a medieval hospital: 1976–8. Council for British Archaeology Research Report 110, 193–197.

Jónsson, J., 1892–1894. Faestebondens Kaar paa Island í det 18 Aarhundrede [Inhabitants of Iceland in the 18th Century]. Dansk Historisk Tidsskrift 6, 563–604. [In Danish].

Jørgensen, G., 1986. Medieval plant remains from the settlements in Mollergade 6. In: Jansen, H.M. (Ed), The archaeology of Svendborg, Denmark. 4. The analysis of medieval plant remains, textiles and wood from Svendborg. Odense, Denmark, Odense University Press, 45–83.

Jost-Stauffer, M., Coope, G.R., & Schlüchter, C., 2005. Environmental and climatic reconstructions during Marine Oxygen Isotope Stage 3 from Gossau, Swiss Midlands, based on coleopteran assemblages. Boreas 34, 53–60.

Joy, N.H., 1932. A practical handbook of British beetles. Edinburgh, UK, Riverside Press.

Juan, C., Ibrahim, K.M., Oromi, P., & Hewitt, G., 1998. The phylogeography of the darkling beetle, *Hegeter politus*, in the eastern Canary Islands. Proceedings of the Royal Society of London B 265, 135–140.

Jull, A.J.T., 2007. Radiocarbon dating: AMS Method. In: Elias, S.A. (Ed), Encyclopedia of Quaternary science. Amsterdam, Elsevier, 2911–2918.

Karppinen, E. & Koponen, M., 1974. Further observations on subfossil remains of oribatids (Acar., Oribatei) and insects in Piilonsuo, a bog in southern Finland. Annales Entomologici Fennici 40, 172–175.

Karrow, P.F., McAndrews, J.H., Miller, B.B., Morgan, A.V., Seymour, K.L., & White, O.L., 2001. Illinoian to Late Wisconsinan stratigraphy at Woodbridge, Ontario. Canadian Journal of Earth Sciences 38, 921–942.

Karrow, P.F., Morris, T.F., McAndrews, J.H., Morgan, AV., Smith, A.J., & Walker, I.R., 2007. A diverse late-glacial (Mackinaw Phase) biota from Leamington, Ontario. Canadian Journal of Earth Sciences 44, 287–296.

Kashiwazaki Naumann's Elephant Research Group., 1991. *Palaeoloxodon naumanni* (Makiyama) found in the upper Pleistocene, Yasuda Formation in the Kashiwazaki City, Niigata Prefecture, Central Japan, and its palaeoenvironment. Chikyu-Kagaku 45, 161–176. [In Japanese].

Kasse, C., Huitzer, A.S., Krzyszkowski, D., Bohncke, S.J.P., & Coope, G.R., 1998. Weichselian Late-Pleniglacial and Late-Glacial depositional environment, Coleoptera and periglacial climate records from central Poland (Belchatów). Journal of Quaternary Science 13, 455–469.

Kavanaugh, D.H., 1979. Rates of taxonomically significant differentiation in relation to geographical isolation and habitat: examples from a study of the nearctic *Nebria* fauna. In: Erwin, T.L., Ball, G.E., Whitehead, D.R. (Eds), Carabid Beetles, their evolution, natural history, and classification. The Hague, The Netherlands, Junk, 37–57.

Kawamura, Y., 1991. Late Pleistocene to Holocene mammalian faunal succession in the Japanese islands, with comments on the Late Quaternary extinctions. 13th INQUA Congress Abstracts, p. 160.

Keen, D.H., Bateman, M.D., Coope, G.R., Field, M.H., Langford, H.E., Merry, J.S., & Mighall, T.M., 1999. Sedimentology, palaeoecology and geochronology of Last Interglacial deposits from Deeping St James, Lincolnshire, England. Journal of Quaternary Science 14, 411–436.

Keen, D.H., Coope, G.R., Jones, R.L., Field, M.H., Griffiths, H.I., Lewis, S.G., & Bowen, D.Q., 1997. Middle Pleistocene deposits at Frog Hall Pit, Stetton-on-Dunsmore, Warwickshire, Englidh Midlands, and their implications for the age of the type Wolstonian. Journal of Quaternary Science 12, 183–208.

Keen, D.H., Hardaker, T., & Lang, A.T.O., 2006. A Lower Palaeolithic industry from the Cromerian (MIS 13) Baginton Formation of Waverley Wood and Wood Farm Pits, Bubbenhall, Warwickshire, UK. Journal of Quaternary Science 21, 457–470.

Keepax, C.A., Girling, M.A., Jones, R.T., & Arthur, J.R.B., 1978. The environmental analysis. In: Williams, J. (Ed), St. Peter's Square, Northampton, excavations, 1973–1978. Northampton, UK, Northampton Development Corporation, 337

Keepax, C.A., Girling, M.A., Jones, R.T., Arthur, J.R.B., Paradine, P.J., & Keeley, H., 1979. The environmental analysis. In: Williams, J. (Ed), St. Peter's Square, Northampton, excavations, 1973–1976. Northampton, UK, Northampton Development Corporation, 337

Keigwin, L.D. & Gorbarenko, S.A., 1992. Sea level, surface salinity of the Japan Sea, and the Younger Dryas event in the Northwestern Pacific Ocean. Quaternary Research 37, 346–360.

Kelly, M. & Osborne, P.J., 1965. Two faunas and floras from the alluvium at Shustoke, Warwickshire. Proceedings of the Linnean Society of London 176, 37–65.

Kenward, H.K., 1975a. Pitfalls in the environmental interpretation of insect death assemblages. Journal of Archaeological Science 2, 85–94.

Kenward, H.K., 1975b. The biological and archaeological implications of the beetle *Aglenus brunneus* (Gyllenhall) in ancient faunas. Journal of Archaeological Science 2, 63–69.

Kenward, H.K., 1976. Reconstructing ancient ecological conditions from insects remains: some problems and an experimental approach. Ecological Entomology 1, 7–17.

Kenward, H.K., 1977. A note on the insect remains from Column Sample IV. In: Armstrong, P. (Ed), Excavations in Sewer Lane, Hull, UK, 1974. East Riding Archaeologist 3, 31–32.

Kenward, H.K., 1978a. The value of insect remains as evidence of ecological conditions on archaeological sites. In: Brothwell, D.R., Thomas, K.D., Clutton-Brock, J. (Eds), Research problems in zooarchaeology. Institute of Archaeology, University of London, Occasional Publication 3, 25–38.

Kenward, H.K., 1978b. The Analysis of Archaeological Insect Assemblages: a New Approach. Archaeology of York, 19/1. Council for British Archaeology for York Archaeological Trust.

Kenward, H.K., 1979a. Five insect assemblages. In: Carver, M.O.H., Three Saxo-Norman tenements in Durham City. Medieval Archaeology 23, 60–67.

Kenward, H.K., 1979b. The insect death assemblages. In: Ayres, B.S., Excavations at Chapel Lane Staithe, 1978. Hull Old Town Reports Series, 3. East Riding Archaeologist 5, 65–72.

Kenward, H.K., 1980. Insect remains. In: Schia, E. (Ed), Feltene "Oslogate 3 og 7". De Arkeologiske Utgravninger i Gamlebyen, Oslo 2, 134–137.

Kenward, H.K., 1982. Insect communities and death assemblages, past and present. In: Hall, A.R., Kenward, H.K. (Eds), Environmental archaeology in the urban context. Council for British Archaeology Research Report 43, 71–78.

Kenward, H.K., 1984. *Helophorus tuberculatus* Gyll. (Col., Hydrophilidae) from Roman Carlisle. Entomologist's Monthly Magazine 120, 236.

Kenward, H.K., 1985a. The insect fauna. Berrington Street 4, period 6, pit 651. In: Shoesmith, R. (Ed), Hereford city excavations. Council for British Archaeology Research Report, 56, 96+fiche M9.B4–11.

Kenward, H.K., 1985b. Outdoors-indoors? The outdoor component of archaeological insect assemblages. In: Fieller, N.R.J, Gilbertson, D.D., Ralph, N.G.A. (Eds), Palaeobiological investigations. Research Design, Methods and Data Analysis. Association for Environmental Archaeology Symposium 5B, 97–104.

Kenward, H.K., 1987. Insect remains from putlog holes. In: Wenham, L.P., Hall, R.A., Briden, C.M., Stoker, D.A. (Eds), St. Mary Bishophill Junior and St. Mary Castlegate. Archaeology of York, 8/2. York, Council for British Archaeology for York Archaeological Trust, 128–130.

Kenward, H.K., 1988a. *Helophorus tuberculatus* Gyllenhall (Col., Hydrophilidae) from Roman York. Entomologist's Monthly Magazine 124, 90.

Kenward, H.K., 1988b. Insect remains. In: Griffin, K., Okland, R.H., Jones, A.K., Kenward, H.K., Lie, R.W., Schia, E. (Eds), Animal bones, moss, plant, insect and parasite remains. De arkeologiske utgravninger i Gamlebyen, Oslo 5, 115–140. Alveheim & Eide, Ovre Ervik.

Kenward, H.K., 1990. The insect remains (1981). In: McCarthy, M.R. (Ed), A Roman, Anglian and Medieval Site at Blackfriars Street, Carlisle: Excavations 1977–9. Cumberland & Westmoreland Antiquarian & Archaeological Society Research Series 4, Carlisle, 317–318 & fiche 3/1–3.

Kenward, H.K., 1997. Synanthropic decomposer insects and the size, remoteness and longevity of archaeological occupation sites: applying concepts from biogeography to past 'islands' of human occupation. In: Ashworth, A.C., Buckland, P.C., Sadler, J.P. (Eds), Studies in Quaternary entomology – an inordinate fondness for insects. Quaternary Proceedings 5, 135–152.

Kenward, H.K., 1998. Sheep parasites: keds and lice. In: Walton Rogers, P. (Ed), Textile Production at 16–22 Coppergate. Archaeology of York, 17/11. York, Council for British Archaeology for York Archaeological Trust, 1715–1717.

Kenward, H.K., 1999. Pubic lice (*Pthirus pubis*) were present in Roman and medieval Britain. Antiquity 73, 911–915.

Kenward, H.K., 2004. Do insect remains from historic-period archaeological occupation sites track climate change in northern England? Environmental Archaeology 9, 47–59.

Kenward, H.K., 2006. The visibility of past trees and woodland: testing the value of insect remains. Journal of Archaeological Science 33, 1368–1380.

Kenward, H.K. & Allison, E., 1987. Waterlogged insect remains from Southampton. Ancient Monuments Laboratory Report, 124/87. London. (Microfiche).

Kenward, H.K. & Allison, E., 1994a. A preliminary view of insect assemblages from the early Christian rath site at Deer Park Farms, Northern Ireland. In: Rackham, D. J. (Ed), Environment and economy in Anglo-Saxon England. Council for British Archaeology Research Report 89, 89–107.

Kenward, H.K. & Allison, E., 1994b. Insects. In: Ayers, B.S. (Ed), Excavations at Fishergate, Norwich, 1985. East Anglian Archaeology 68, 45–48.

Kenward, H.K. & Carrott, J., 2006. Insect species associations characterise past occupation sites. Journal of Archaeological Science 33, 1452–1473.

Kenward, H.K., Engleman, C., Robertson, A., & Large, F., 1988. Insect remains from the Roman fort at Papcastle, Cumbria. Ancient Monuments Laboratory Report 145/88. London. (Microfiche)

Kenward, H.K. & Hall, A.R., 1995. Biological evidence from Anglo-Scandinavian deposits at 16–22 Coppergate. Archaeology of York 14/7. York, Council for British Archaeology for York Archaeological Trust.

Kenward, H.K. & Hall, A.R., 2001. Plants, intestinal parasites and insects. In: Cameron, A.S., Stones, J.A. (Eds), Aberdeen: an in-depth view of the city's past. (Society of Antiquaries of Scotland Monograph Series 19) Society of Antiquaries of Scotland, Edinburgh, UK, 280–297.

Kenward, H.K., Hall, A.R., & Jones, A.K.G., 1986. Environmental evidence from a Roman well and Anglian pits in the Legionary fortress. Archaeology of York, 14/5. London, UK, Council for British Archaeology for York Archaeological Trust, 241–288

Kenward, H.K., Hill, M., Jaques, D., Kroupa, A., & Large, F., 2000. Evidence from beetles and other insects: evidence for living conditions on the crannog: Coleoptera analysis. In: Crone, A. (Ed), The history of a Scottish lowland crannog: excavations at Buiston, Ayrshire, 1989–90. Loanhead, Scotland, Scottish Trust for Archaeological Research 76–78; 99–101; 230–247; 300–320.

Kenward, H.K. & Large, F., 2001. The invertebrate remains. In: Roberts, I., Burgess, A., Berg, D. (Eds), A new link to the past. The archaeological landscape of the M1-A1 link road. Yorkshire Archaeology 7. Wakefield, 227–231; 240–246.

Kenward, H.K., Large, F., & Issitt, M., 1997. The invertebrate assemblage. In: Crone, A., O'Sullivan, J. (Eds), Excavation of a cropmark enclosure at Brixwold (Dalhousie Mains), Bonnyrigg, Midlothian. Proceedings of the Society of Antiquaries of Scotland 127, 396–400.

Kenward, H.K. & Tomlinson, P.R., 1991. Environmental Analysis. In: Higgins, D.A. (Ed), Speke Hall: excavations in the west range. Journal of the Merseyside Archaeological Society 8, 75–78.

Kenward, H.K. & Williams, D., 1979. Biological evidence from the Roman warehouses in Coney Street. Archaeology of York, 14/2. Council for British Archaeology for York Archaeological Trust.

Kenward, H.K., Williams, D., Spencer, P. J., Greig, J.R.A., Rackham, D.J., & Brinklow, D.A., 1978. The environment of Anglo-Scandinavian York. In: Hall, R.A. (Ed), Viking Age York and the North. Council for British Archaeology Research Report 27, 58–70.

Kershaw, A.P., 1995. Environmental change in Greater Australia. Antiquity 69, 656–675.

Kershaw, A.P., Tibby, J., Penny, D., Yezdani, H., Walkley, R., & Cook, E., 2004. Latest Pleistocene and Holocene vegetation and environmental history of the Western Plains of Victoria, Australia. Proceedings of the Royal Society of Victoria 116, 139–161.

Kershaw, P. & van der Kaars, S., 2007. Pollen records: Late Pleistocene Australia and New Zealand. In: Elias, S.A. (Ed), Encyclopedia of Quaternary science. Amsterdam, Elsevier, 2613–2623.

King, G.A., Gilbert, T.P., Willerslev, E., Collins, M.J., & Kenward, H., 2009. Recovery of DNA from archaeological insect remains: first results, problems and potential. Journal of Archaeological Science 36, 1179–1183.

Kirkpatrick, J.B. & Fowler, M., 1998. Locating likely glacial forest refugia in Tasmania using palynological and ecological information to test alternative climatic models. Biological Conservation 85, 171–182.

Kiselyov, S.V., 1973. Late Pleistocene Coleoptera of Transuralia. Paleontological Journal 4, 507–510.

Kiselyov, S.V., 1981. Late Cenozoic coleoptera of North-East Siberia. Moscow, Nauka Press, [In Russian].

Kiselyov, S.V., 1988. Pleistocene and Holocene Coleoptera of western Siberia. In: Modern condition and history of the living world of the western Siberian lowlands. USSR Academy of Sciences, Ural Section, 97–118. [In Russian].

Kiselyov, S.V., 1994. Northern Eurasia environment in the Pleistocene and Holocene, based on Coleoptera research. PhD dissertation, Moscow State University, Moscow [In Russian].

Kiselyov, S.V., Druk, A.R., & Krivolotski, D.A., 1982. On the fauna of beetles and Acari from buried mammoths. In: Sokolov, V.E. (Ed), Yuribeisku Mamont. Moscow, Akademiya Nauk SSSR, 44–51. [In Russian].

Kiselyov, S.V., Kolesnikov, S., & Rybakova, N.O., 1987. Climate of the vegetation period during the 'Ice Complex' deposits forming from the Omolon River Site. Byulleten Moskovskogo obshchestva ispytatelei prirody 62, 113–119. [In Russian].

Kiselyov, S.V. & Nazarov, V.I., 1984. Late Pleistocene insects. In: Wright, Jr, H.E., Barnowsky, C.W. (Eds), Late Quaternary environments of the Soviet Union. Minneapolis, MN, University of Minnesota Press, 223–226.

Kislev, M.E. & Melamed, Y., 2000. Ancient infested wheat and horsebean from Horbat Rosh Zayit. In: Gal, Z., Alexandre, Y. (Eds), Horbat Rosh Zayit. An Iron Age storage fort and village. Jerusalem, Israel, Israel Antiquities Authority, 206–220.

Kislev, M.E. & Simchoni, O., 2001. Recent advances in storage archaeoentomology in Israel. Abstracts, 12th Symposium of the International Workgroup for Palaeoethnobotany, Sheffield, UK, 17–23 June 2001.

Klicka, J. & Zink, R.M., 1997. The importance of recent ice ages in speciation: a failed paradigm. Science 277, 1666–1669.

Klimaszewski, J. & Crosby, T.K., 1997. A revision of the New Zealand species of the parasitoid genus *Aleochara*, with descriptions of four new species (Coleoptera: Staphylinidae). Journal of the Royal Society of New Zealand 27, 243–269.

Klimaszewski, J. & Watt, J.C., 1997. Coleoptera: Family group review and keys to identification. Fauna of New Zealand 37. New Zealand, Manaaki Whenua Press.

Klinger, L.E., Elias, S.A., Behan-Pelletier, V.M., & Williams, N.E., 1990. The bog climax hypothesis: fossil arthropod and stratigraphic evidence in peat sections from southeast Alaska, USA. Holarctic Ecology 13, 72–80.

Klink, A., 1989. The Lower Rhine: paleoecological analysis. In: Petts, G.E. (Ed), Historical change of large Alluvial Rivers: Western Europe. London, UK, Wiley, 183–201.

Koch, C., 1962. The Tenebrionidae of southern Africa XXXI. Comprehensive notes on the tenebrionid fauna of the Namib Desert. Annals of the Transvaal Museum 24, 61–103.

Koch, K., 1970. Subfossile Käferreste aus romerzeitlichen und mittelalterlichen Ausgrabungen im Rheinland [Subfossil beetle remains from Roman and Medieval excavations in the Rhein Land]. Entomologische Blatter für Biologie und Systematik der Käfer 66, 41–56. [In German].

Koch, K., 1971. Zur Untersuchung subfossiler Käferreste aus römerzeitlichen und mittelalterlichen Ausgrabungen im Rheinland [On the study of subfossil beetle remains from Roman-age and Medieval excavations in the Rhein Land]. Rheinische Ausgrabungen 10, 378–448. [In German].

Kolbe, H., 1894. Über fossile Reste von Coleopteren aus einen alten Torflager (Schmierkohle) bei Gr. Raschen in der nieder Lausitz [On fossil beetle remains from a peat deposit near Gr. Raschen in lower Lausitz]. Sitzberichte naturforschender Freunde zu Berlin 236–238. [In German].

Koponen, M. & Nuorteva, M., 1973. Über subfossile Waldinsekten aus dem Moor Piilonsuo in Sudfinnland [On subfossil forest insects from Piilonsuo bog in southern Finland]. Acta Entomologica Fennica 29, 4–81. [In German].

Korotyayev, B.A., 1977. An ecological-faunistic survey of snout beetles (Coleoptera, Curculionidae) of the Northeast of the USSR. Entomologicheskoe Obozrenie 56, 60–70. [In Russian].

Krell, F.T., 2006. Fossil record and evolution of Scarabaeoidea (Coleoptera: Polyphaga). Coleopterists Society Monograph 5, 120–143.

Krivolutsky, D.A. & Druk, A.Ya., 1986. Fossil Oribatid mites. Annual Review of Entomology 31, 533–545.

Krivolutsky, D.A., Druk, A.Ya., Eitminaviciute, I.S., Laskova, L.M., & Karppinen, E., 1990. Fossil Oribatid Mites. Vilnius, Mosklas, [in Russian].

Kuschel, G., 1990. Beetles in a suburban environment: a New Zealand case study. DSIR Plant Protection Report 3, 118.

Kuschel, G., 2003. Nemonychidae, Belidae, Brentidae (Insecta: Coleoptera: Curculionoidea). Fauna of New Zealand 45. Lincoln, New Zealand, Manaaki Whenua Press/Landcare Research.

Kuzmina, S., 1989. Late Pleistocene insects from the Alazea River (Kolyma lowland). Bulletin of the Moscow Society of Naturalists, Geology Series 64, 42–55. [in Russian].

Kuzmina, S., Elias, S.A., Matheus, P., Storer, J.E., & Sher, A., 2008. Paleoenvironmental reconstruction of the Last Glacial Maximum, inferred from insect fossils from a tephra buried soil at Tempest Lake, Seward Peninsula, Alaska. Palaeoclimatology, Palaeogeography, Palaeoecology 267, 245–255.

Kuzmina, S. & Kolesnikov, S., 2000. [Insects from the Late Pleistocene and Holocene deposits of the Medvezhi Islands (East Siberian Sea)]. Bulletin of the Moscow Society of Naturalists (MOIP) (ser. geol.) 75, 68–71. [In Russian].

Kuzmina, S. & Sher, A., 2006. Some features of the Holocene insect faunas of northeastern Siberia. Quaternary Science Reviews 25, 1790–1820.

Larochelle, A. & Lariviere, M.-C., 2001. Carabidae (Insecta: Coleoptera): Catalogue. Fauna of New Zealand 43. Lincoln, New Zealand, Manaaki Whenua Press.

Larson, D.J., Alarie, Y., & Roughley, R.E., 2000. Predaceous Diving Beetles (Coleoptera: Dytiscidae) of the Nearctic Region, with emphasis on the fauna of Canada and Alaska. Ottawa, Canada, NRC Research Press.

Lavoie, C., 2001. Reconstructing the late-Holocene history of a subalpine environment (Charlevoix, Québec) using insect fossils. The Holocene 11, 89–99.

Lavoie, C. & Arseneault, D., 2001. Late Holocene climate of the James Bay area, Québec, Canada, reconstructed using fossil beetles. Arctic, Antarctic, and Alpine Research 33, 13–18.

Lavoie, C., Elias, S.A., & Filion, L., 1997a. A 7000-year record of insect communities from peatland environments, southern Québec. Ecoscience 4, 394–403.

Lavoie, C., Elias, S.A., & Payette, S., 1997b. Holocene fossil beetles from a treeline peatland in subarctic Québec. Canadian Journal of Zoology 75, 227–236.

Lawrence, J.F. & Britton, E.B., 1994. Australian beetles. Melbourne, Melbourne University Press.

Lea, P.D., 1989. Quaternary environments and depositional systems of the Nushagak lowlands, southwestern Alaska. Ph.D. Dissertation, University of Colorado, Boulder, CO, 328.

Lea, P.D., Elias, S.A., & Short, S.K., 1991. Stratigraphy and paleoenvironments of Pleistocene nonglacial units in the Nushagak Lowland, Southwestern Alaska. Arctic and Alpine Research 23, 375–391.

LeConte, J.L., 1859. The Coleoptera of Kansas and eastern New Mexico. Smithsonian contributions to knowledge 1–6, 58.

Leech, R. & Matthews, Jr, J.V., 1971. *Xysticus archaeopalpus* (Arachnidae: Thomisidae): a new species of crab spider from Pliocene sediments in western Alaska. Canadian Entomologist 103, 1337–1340.

Lemdahl, G., 1983. Beetle remains from the refuse layer of the bog site Ageröd V, Appendix V. In: Larsson, L. (Ed), Ageröd V, an Atlantic bog site in Central Scania. Acta Archaeologica Lundensia 12, 169–172.

Lemdahl, G., 1985. Fossil insect faunas from Late-Glacial deposits in Scania (South-Sweden). Ecologia Mediterranea 11, 185–191.

Lemdahl, G., 1988a. Late Weichselian insect assemblages from the Kullen Peninsula, south Sweden-palaeoenvironmental interpretations. Lunqua Report 30.

Lemdahl, G., 1988b. Insect remains from an Allerød peat at Lake Bysjon, south Sweden. Boreas 17, 265–266.

Lemdahl, G., 1990. Insect assemblages from an Iron Age settlement in the clay district of Butjadingen, NW Germany. International Quaternary Union Subcommission for the Study of the Holocene, Cultural Landscapes Meetings Abstracts, 12–22.

Lemdahl, G., 1991a. A rapid climatic change at the end of the Younger Dryas in south Sweden-paleoclimatic and paleoenvironmental reconstructions based on fossil insect assemblages. Palaeogeography, Palaeoclimatology, Palaeoecology 83, 313–331.

Lemdahl, G., 1991b. Late Vistulian insect assemblages from Zabinko, western Poland. Boreas 20, 71–77.

Lemdahl, G., 1991c. Tidigmedeltida insektfynd från Lund [Early medieval insect remains from Lund]. FaZett 4, 34–39. [In Swedish].

Lemdahl, G., 1991d. Insetker. In: Carlsson, R., Elfwendahl, M., Perming, A. (Eds), Bryggaren ett kvarter i centrum-En medeltidsarkeologisk undersökning i Uppsala 1990 [Medieval studies from Uppsala, 1990]. Riksantikvarieämbetet och Statens Historika Museer Rapport UV (1991) 1, 222–223. [In Swedish].

Lemdahl, G., 1991e. Insekter från Oskarshamkoggen-pilotundersökning av sedimentprover från vraket [Insects from the boat, 'Oskarshamm,' pilot study on sediment samples from the shipwreck]. Uppsats i Påbyggnadskurs i arkeologi vid Stockholms Universitet, 45–47. [In Swedish].

Lemdahl, G., 1991f. Paleoekologia kopalnych owadów-przyklad z poznoglacjalnego stanowiska w Lednickim Parku Krajobrazowym [Fossil insects and palaeoecology-an example from a Lateglacial site in the Lednica Region]. In: Tobolski, K. (Ed), Wstep do paleoekologii Lednickiego Parku Krajobrazowego [Introduction to palaeoecology of the Lednica Landscape Park]. Biblioteka Studiów Lednickich, Poznan, 129–141.

Lemdahl, G., 1997. Early Weichselian insect faunas from Northern Sweden: climatic and environmental implications. Arctic and Alpine Research 29, 63–74.

Lemdahl, G., 2000a. Late-glacial and early Holocene insect assemblages from sites at different altitudes in the Swiss Alps-implications on climate and environment. Palaeogeography, Palaeoclimatology, Palaeoecology 159, 293–312.

Lemdahl, G., 2000b. Late-glacial and early-Holocene Coleoptera assemblages as indicators of local environment and climate at Kråkenes Lake, western Norway. Journal of Paleolimnology 23, 57–66.

Lemdahl, G., 2003. Insektfynd Från brunnar på jarlens gård [Insect finds from wells on Earl's Farm]. In: Söderberg, B. (Ed), Järrestad – Huvudgård i centralbygd. Riksantikvarieämbetet Arkeologiska undersökningar Skrifter 51, 271–286. [In Swedish with English summary].

Lemdahl, G., 2004. Insect remains from Arbon Bleiche. In: Jacomet, S., Leuzinger, U., Schibler, J. (Eds), Die jungsteinzeitlische Seefersiedlung Arbon Bleiche 3. Umwelt and Wirtschaft [The Neolithic Lake Settlement of Arbon Bleiche 3. Environment and Economy]. Archaeologie im Thurgau 12, 365–371.

Lemdahl, G. & Coope, G.R., 2007. Beetle records: Late Pleistocene of Europe. In: Elias, S.A. (Ed), Encyclopedia of Quaternary science. Amsterdam, Elsevier, 190–196.

Lemdahl, G. & Gustavson, G., 1997. Lateglacial and middle Holocene Coleoptera assemblages from coastal environments in south-western Sweden. Entomologisk Tidskrift 118, 117–187.

Lemdahl, G. & Nilsson, A., 1982. Ett postglacialt subfossilfynd av dykarbaggen *Cybister lateralimarginalis* från Skane [A postglacial subfossil discovery of the predaceous diving beetle, *Cybister lateralimarginalis* from Scania (Sweden)]. Entomologisk Tidskrift 103, 121–122. [In Swedish].

Lemdahl, G. & Persson, T., 1989. Late Weichselian biostratigraphy at Mickelsmossen, Skåne, southern Sweden. Geologiska Föreningens i Stockholm Förhandligar 111, 251–259.

Lemdahl, G., Sronsson, M., & Hedenäs, L., 1995. Insekter från ett medeltida handelsfartyg [Insects from a medieval commerce ship]. Entomologiske Tidskrift 116, 169–174. [In Swedish].

Lemdahl, G. & Thelaus, M., 1989. Subfossila insektfynd från det medeltida Halmstad [Subfossil insect fossils from medieval Halmstad]. Entomologiska Tidskrift 110, 39–41. [In Swedish].

Lentacker, A., Bakels, C.C., Verbeeck, M., & Desender, K., 1992. The archaeology, fauna and flora of a Roman Well at Erps-Kwerps (Brabant, Belgium). Helinium 32, 110–131.

Leonard, E., 2007. Modeled patterns of Late Pleistocene glacier inception and growth in the Southern and Central Rocky Mountains, USA: sensitivity to climate change and paleoclimatic implications. Quaternary Science Reviews 26, 2152–2166.

Levinson, H. & Levinson, A., 1994. Origin of grain storage and insect species consuming desiccated food. Anzieger Schädlingskunde, Pflanzenschutz, Umweltschutz 67, 47–59.

Lewis, S.G., Maddy, D., Buckingham, C., Coope, G.R., Field, M.H., Keen, D.H., Roe, D.A., Scaife, R.G., & Scott, K., 2006. Pleistocene fluvial sediments, palaeontology and archaeology of the upper River Thames at Latton, Wiltshire, England. Journal of Quaternary Science 21, 181–205.

Lindroth, C.H., 1948. Interglacial insect fossils from Sweden. Arsbok Sveriges Geologiska Undersökn, Series C 42, 1–29.

Lindroth, C.H., 1949. Die Fennoskandischen Carabidae. Kungliga Vetenskaps Vitterhets Samfundets Handlingar, Series B4 1, 1–709.

Lindroth, C.H., 1957. The Faunal Connections Between North America and Europe. New York, Wiley.

Lindroth, C.H. (Ed), 1960. Catalogus Coleopterorum Fennoscandiae et Daniae. Lund, Sweden, Entmologiska Sällskapet.

Lindroth, C.H., 1961. The ground beetles of Canada and Alaska, part 2. Opuscula Entomologica Supplement 20, 1–200.

Lindroth, C.H., 1963. The ground beetles of Canada and Alaska, part 3. Opuscula Entomologica Supplement 24, 201–408.

Lindroth, C.H., 1966. The ground beetles of Canada and Alaska, part 4. Opuscula Entomologica Supplement 29, 409–648.

Lindroth, C.H., 1968. The ground beetles of Canada and Alaska, part 5. Opuscula Entomologica Supplement 33, 649–944.

Lindroth, C.H., 1969. The ground beetles of Canada and Alaska, part 1. Opuscula Entomologica Supplement 35, I-XLVIII.

Lindroth, C.H., 1971. Biological investigations on the new volcanic island Surtsey, Iceland. In den Boer, P.J. (Ed),

Dispersal and dispersal power of Carabid Beetles. Miscellaneous Papers, Landbouwhogeschool, Wageningen 8, 65–69.

Lindroth, C.H., 1974. On elytral microsculpture of Carabid beetles. Entomologica Scandinavia 5, 251–264.

Lindroth, C.H., 1985. The Carabidae (Coleoptera) of Fennoscandia and Denmark. Fauna Entomologica Scandinavica 15, 1–227.

Lindroth, C.H., 1986. The Carabidae (Coleoptera) of Fennoscandia and Denmark. Fauna Entomologica Scandinavica 15, 228–499.

Lindroth, C.H., 1992. Ground Beetles (Carabidae) of Fennoscandia: A Zoogeographical Study. Parts 1–3 (English Translation of 1945 original German text), Intercept, Andover, UK.

Lindroth, C.H. & Coope, G.R., 1971. The insects from the interglacial deposits at Leveaniemi. Sveriges Geologiska Undersokn Series C, 658, Appendix 2.

Lomnicki, A.M., 1894. Pleistocenskie owady z Borislawia. (Fauna Pleistocenica insectorum Boryslaviensium). Wydawnictwa Muzeum imienia Dzieduszyckich 4, 1–116. [In Polish].

Lowe, J.J., Coope, G.R., Sheldrick, C., Harkness, D.D., & Walker, M.J.C., 1995. Direct comparison of UK temperatures and Greenland snow accumulation rates, 15000–12000 yr ago. Journal of Quaternary Science 10, 175–180.

Lowell, T., Savage, K., Dell, A., Morgan, A.V., Pilny, J., Miller, N., Shane, L., & Stuckenrath, R., 1990. Late Wisconsin glacial maximum biota, Cincinnati, Ohio. Canadian Quaternary Association and American Quaternary Association, First Joint Meeting, Program and Abstracts, 23.

Lucy, D., Aykroyd, R.G., & Pollard, A.M., 2002. Nonparametric calibration for age estimation. Journal of The Royal Statistical Society Series C (Applied Statistics) 51, 183–196.

Luff, M.L., 1998. Provisional Atlas of the Ground Beetles (Coleoptera, Carabidae) of Britain. Centre for Ecology and Hydrology, National Environmental Research Council, Swindon.

Lundqvist, J., 1967. Submoräna sediment I Jämtlands län. Sverigen Geologiska Undersökning C618, 267 pp. [In Swedish].

Lundqvist, J., 1986. Stratigraphy of the central area of the Scandinavian glaciation. Quaternary Science Reviews 5, 251–268.

MacDonald, G.M. & Cwynar, L.C., 1986. A fossil pollen based reconstruction of the late Quaternary history of lodgepole pine (*Pinus contorta* ssp. *latifolia*) in the western interior of Canada. Canadian Journal of Forest Research 15, 1039–1044.

MacDonald, G.M. & Cwynar, L.C., 1991. Post-glacial population growth rates of *Pinus contorta* ssp. *latifolia* in western Canada. Journal of Ecology 79, 417–429.

MacDonald, G.M., Velichko, A.A., Kremenetski, C.V., Borisova, C.K., Goleva, A.A., Andreev, A.A., Cwynar, L.C., Riding, R.T., Forman, S.L., Edvards, T.W.D., Aravena, R., Hammarlund, D., Szeicz, J., & Gattaulin, V.N., 2000. Holocene treeline history and climate change across northern Eurasia. Quaternary Research 53, 302–311.

Mackay, W.P. & Elias, S.A., 1992. Late Quaternary ant fossils from Chihuahuan Desert packrat middens (Hymenoptera: Formicidae). Psyche 99, 169–184.

Macphail, M.K. & McQueen, D.R., 1983. The value of New Zealand pollen and spores as indicators of Cenozoic vegetation and climates. Tuatara 26, 37–59.

Maddy, D., Coope, G.R., Gibbard, P.L., Green, C.P., & Lewis, S.G., 1994. Reappraisal of Middle Pleistocene fluvial deposits near Brandon, Warwickshire and their significance for the Wolston glacial sequence. Journal of the Geological Society 151, 221–233.

Maddy, D., Lewis, S.G., Scaife, R.G., Bowen, D.Q., Coope, G.R., Green, C.P., Hardaker, T., Keen, D.H., Rees-Jones, J., Parfitt, S., & Scott, K., 1998. The Upper Pleistocene deposits at Cassington, near Oxford, UK. Journal of Quaternary Science 13, 205–231.

Madsen, D.B. & Kirkman, J.E., 1988. Hunting hoppers. American Antiquity 53, 593–604.

Mani, M.S., 1968. Ecology and biogeography of high altitude insects. The Hague, The Netherlands, Junk.

Marggi W.A., 1992. Faunistik der Sandlaufkäfer und Laufkäfer der Schweiz (Cicindellidae & Carabidae, Coleoptera) [Faunal studies of tiger beetles and ground beetles of Switzerland]. Documenta Faunistica Helvitica 13, Parts 1& 2, 1–477; 1–243. [In German].

Markgraf, V., 1986. Plant inertia reassessed. The American Naturalist 127, 725–726.

Markgraf, V., 1993. Younger Dryas in southernmost South America – an update. Quaternary Science Reviews 12, 351–355.

Marra, M.J., 2003a. A last interglacial beetle fauna from New Zealand. Quaternary Research 59, 122–131.

Marra, M.J., 2003b. A description and interpretation of a fossil beetle assemblage from marine isotope stage 6 from Banks Peninsula, New Zealand. New Zealand Journal of Geology and Geophysics 46, 523–528.

Marra, M.J., 2007. Beetle records: Late Pleistocene of New Zealand. In: Elias, S.A. (Ed), Encyclopedia of Quaternary science. Amsterdam, Elsevier, 236–246.

Marra, M.J., 2008. Quaternary fossil beetles from New Zealand. New Zealand Entomologist 31, 3–14.

Marra, M.J., Alloway, B.V., & Newnham, R.M., 2006a. Paleoenvironmental reconstruction of a well-preserved Stage 7 forest sequence catastrophically buried by basaltic eruptive deposits, northern New Zealand. Quaternary Science Reviews 25, 2143–2161.

Marra, M.J., Crozier, M., & Goff, J., 2007. Paleoenvironment and biogeography of a late MIS 3 fossil beetle fauna from Southern Taranaki, New Zealand. Quaternary International 167–168(Supplement), 270.

Marra, M.J., Crozier, M., & Goff, J., 2009. Paleoenvironment and biogeography of late MIS 3 fossil beetles from New Zealand. Journal of Quaternary Science 24, 97–107.

Marra, M.J. & Leschen, R.A.B., 2004. Late Quaternary paleoecology from fossil beetle communities in the Awatere Valley, South Island, New Zealand. Journal of Biogeography 31, 571–586.

Marra, M.J., Shulmeister, J., & Smith, E.G.C., 2006b. Reconstructing temperature during the Last Glacial Maximum from Lyndon Stream, South Island, New Zealand using beetle fossils and maximum likelihood envelopes. Quaternary Science Reviews 25, 1841–1849.

Marra, M.J., Smith, E.G.C., Shulmeister, J., & Leschen, R., 2004. Late Quaternary climate change in the Awatere Valley, South Island, New Zealand using a sine model with a maximum likelihood envelope on fossil beetle data. Quaternary Science Reviews 23, 1637–1650.

Masefield, R., Branch, N., Couldrey, P., Goodburn, D., & Tyers, I., 2003. Insects from selected samples. In: Masefield, R., Branch, N., Couldrey, P., Goodburn, D., Tyers, I. (Eds), A Later Bronze Age well complex at Swalecliffe, Kent. Antiquaries Journal 83, 87–88.

Masuda, H., Amaoka, K., Araga, C., Uyeno, T., & Yoshino, T., 1984. The fishes of the Japanese Archipelago. Kanagawa, Japan, Tokai University Press.

Matheus, P., Begét, J., Mason, O., & Gelvin-Reymiller, C., 2003. Late Pliocene to late Pleistocene environments preserved at the Palisades site, central Yukon River, Alaska. Quaternary Research 60, 33–43.

Matthews, Jr, J.V., 1968. A paleoenvironmental analysis of three Late Pleistocene coleopterous assemblages from Fairbanks, Alaska. Quaestiones Entomologicae 4, 202–224.

Matthews, Jr, J.V., 1970. Two new species of *Micropeplus* from the Pliocene of western Alaska with remarks on the evolution of Micropeplinae (Coleoptera: Staphylinidae). Canadian Journal of Zoology 48, 779–788.

Matthews, Jr, J.V., 1974a. Insect fossils from the early Pleistocene Olyor Suite (Chukochya River: Kolymian lowland, USSR). Geological Survey of Canada Papers 74-1A, 207–211.

Matthews, Jr, J.V., 1974b. Quaternary environments at Cape Deceit (Seward Peninsula, Alaska): evolution of a tundra ecosystem. Geological Society of America Bulletin 85, 1353–1384.

Matthews, Jr, J.V., 1974c. Wisconsin environment of interior Alaska: pollen and macrofossil analysis of a 26 meter core from the Isabella Basin (Fairbanks, Alaska). Canadian Journal of Earth Sciences 11, 828–841.

Matthews, Jr, J.V., 1974d. A preliminary list of insect fossils from the Beaufort Formation, Meighen Island, District of Franklin, Geological Survey of Canada Papers 74-1A, 203–206.

Matthews, Jr, J.V., 1975a. Incongruence of macrofossil and pollen evidence: a case from the Late Pleistocene of the northern Yukon coast. Geological Survey of Canada Paper 75-1B, 139–146.

Matthews, Jr, J.V., 1975b. Insects and plant macrofossils from two Quaternary exposures in the Old Crow-Porcupine region, Yukon Territory, Canada. Arctic and Alpine Research 7, 249–259.

Matthews, Jr, J.V., 1976a. Evolution of the subgenus *Cyphelophorus* (Genus *Helophorus*, Hydrophilidae): description of two new fossil species and discussion of *Helophorus tuberculatus* Gyll. Canadian Journal of Zoology 54, 653–673.

Matthews, Jr, J.V., 1976b. Insect fossils from the Beaufort Formation: geological and biological significance. Geological Survey of Canada Papers 76-1B, 217–227.

Matthews, Jr, J.V., 1977a. Tertiary Coleoptera fossils from the North American arctic. Coleopterists Bulletin 31, 297–308.

Matthews, Jr, J.V., 1977b. Coleoptera fossils: their potential value for dating and correlation of late Cenozoic sediments. Canadian Journal of Earth Sciences 14, 2339–2347.

Matthews, Jr, J.V., 1979a. Fossil beetles and the Late Cenozoic history of the tundra environment. In: Gray, J., Boucot, A.J. (Eds), Historical biogeography, plate tectonics, and the changing environment. Corvallis, OR, Oregon State University Press, 371–378.

Matthews, Jr, J.V., 1979b. Tertiary and Quaternary environments: historical background for an analysis of the Canadian insect fauna. In: Danks, H.V. (Ed), Canada and its insect fauna. Memoirs of the Entomological Society of Canada 108, 31–86.

Matthews, Jr, J.V., 1980a. Tertiary land bridges and their climate: backdrop for development of the present Canadian insect fauna. Canadian Entomologist 112, 1089–1103.

Matthews, Jr, J.V., 1980b. Paleoecology of John Klondike Bog, Fisherman Lake region, Southwest District of Mackenzie. Geological Survey of Canada Paper 80–22, 12.

Matthews, Jr, J.V., 1981. Tertiary land bridges and their climate: backdrop for the development of the present Canadian insect fauna. Canadian Entomologist 112, 1089–1103.

Matthews, Jr, J.V., 1982. East Beringia during Late Wisconsin time: a review of the biotic evidence. In: Hopkins, D.M., Matthews, Jr, J.V., Schweger, C.E., Young, S.B. (Eds), Paleoecology of Beringia. New York, Academic Press, 127–150.

Matthews, Jr, J.V., 1983. A method for comparison of northern fossil insect assemblages. Géographie physique et Quaternaire 37, 297–306.

Matthews, Jr, J.V., 1989. Plants and other fossils from a late Tertiary beaver pond, Ellesmere Island, N.W.T. In: Paleobotany in Canada: Status and prospects. Canadian Museum of Nature, Ottawa, 2.

Matthews, Jr., J.V. & Fyles, J.G., 2000. Late Tertiary plant and arthropod fossils from the high-terrace sediments on Fosheim Peninsula, Ellesmere Island, Nunavut. Geological Survey of Canada Bulletin 529, 295–317.

Matthews, Jr, J.V., Mott, R.J., & Vincent, J.-S., 1986. Preglacial and Interglacial Environments of Banks Island: pollen and macrofossils from Duck Hawk Bluffs and related sites. Géographie physique et Quaternaire 40, 279–298.

Matthews, Jr, J.V. & Ovenden, L.E., 1990. Late Tertiary plant macrofossils from localities in arctic/subarctic North America: a review of the data. Arctic 43, 364–392.

Matthews, Jr, J.V., Ovenden, L.E., & Fyles, J.G., 1990a. Plant and insect fossils from the late Tertiary Beaufort Formation on Prince Patrick Island, N. W. T. In: Harrington, C.R. (Ed), Canada's missing dimension: science and history in the Canadian Arctic Islands. Ottawa, Canada, National Museums of Canada, 105–139.

Matthews, Jr, J.V., Schweger, C.E., & Hughes, O.L., 1990b. Plant and insect fossils from the Mayo Indian Village section (Central Yukon): new data on middle Wisconsinan environments and glaciation. Géographie physique et Quaternaire 44, 15–26.

Matthews, Jr, J.V., Schweger, C.E., & Janssens, J.A., 1990c. The Last (Koy-Yukon) interglaciation in the northern Yukon: evidence from Unit 4 at Ch'ijee's Bluff, Bluefish Basin. Géographie physique et Quaternaire 44, 341–362.

Matthews, Jr, J.V., Smith, S.L., & Mott, R.J., 1987. Plant macrofossils, pollen, and insects of arctic affinity from Wisconsinan sediments in Chaudière Valley, southern Quebec. Geological Survey of Canada Papers, Current Research A 87–1A, 165–175.

Matthews, Jr, J.V. & Telka, A., 1997. Insect fossils from the Yukon. In: Danks, H.V., Downes, J.A. (Eds), Insects of the Yukon. Ottawa, Canada, Biological Survey of Canada (Terrestrial Arthropods), 911–962.

May, B.M., 1993. Larvae of Curculionoidea (Insecta: Coleoptera): a systematic overview. Fauna of New Zealand, Vol. 28, Lincoln, New Zealand, Manaaki Whenua Press.

Mayr, E., 1970. Populations, species, and evolution. Cambridge, MA, Harvard University Press.

McCabe, A.M., Coope, G.R., Gennard, D.E., & Doughty, P., 1987. Freshwater organic deposits and stratified sediments between Early and Late Midlandian (Devensian) till sheets, at Aghnadarragh, County Antrim, Northern Ireland. Journal of Quaternary Science 2, 11–33.

McDougall, K., 1995. Age of the Fishcreekian transgression. Palaios 10, 199–220.

McGlone, M.S., 1985. Plant biogeography and the late Cenozoic history of New Zealand. New Zealand Journal of Botany 23, 723–749.

McGlone, M.S., 1988. New Zealand. In: Huntley, B.J., Webb, T., III (Eds), Vegetation history. The Hague, The Netherlands, Kluwer, 557–599.

McGlone, M.S., Neall, V.E., & Pillans, B.J., 1984. Inaha Terrace deposits: a Late Quaternary terrestrial record in South Taranaki, New Zealand. New Zealand Journal of Geology and Geophysics 27, 35–49.

McGovern, T.H., Buckland, P.C., Savory, D., Sveinbjarnardottir, G., Andreasen, C., & Skidmore, P., 1983. A study of the faunal and floral remains from two Norse farms in the western settlement, Greenland. Arctic Anthropology 20, 93–120.

Mead, J.I., Agenbroad, L.D., Davis, O.K., & Martin, P.S., 1986. Dung of *Mammuthus* in the arid southwest, North America. Quaternary Research 25, 121–127.

Merritt, J.W., Coope, G.R., Taylor, B.J., & Walker, M.J.C., 1990. Late Devensian organic deposits beneath till in the Teith Valley, Perthshire. Scottish Geological Magazine 26, 15–24.

Merritt, J.W., Coope, G.R., & Walker, M.J.C., 2003. The Torrie Late-Glacial organic site and Auchenlaich Pit, Callander. In: Evans, D.A. (Ed), The Quaternary of the Western Highland boundary, Field Guide. London, UK, Quaternary Research Association, 126–133.

Meyer, F.G., 1980. Carbonised food plants of Pompeii, Herculaneum, and the villa at Torre Annunziata. Economic Botany 34, 401–437.

Mighall, T.M., Timberlake, S., Clark, S.H.E., & Caseldine, A.E., 2002. A palaeoenvironmental investigation of sediments from the prehistoric mine of Copa Hill, Cwmystwyth, mid-Wales. Journal of Archaeological Science 29, 1161–1188.

Milankovitch, M., 1941. Kanon der Erdbestrahlung une seine Anwendung auf das Eiszeitproblem [Canon of Solar Radiation of the Earth and its Application to the Problem of Glaciations]. Belgrade, Académie Royale Serbe, 633. [In German].

Mildenhall, D.C., 2003. Deep-sea record of Pliocene and Pleistocene terrestrial palynomorphs from offshore eastern New Zealand (ODP Site 1123, Leg 181). New Zealand Journal of Geology and Geophysics 46, 343–361.

Miller, G.H. & Brigham-Grette, J., 1989. Amino acid geochronology: resolution and precision in carbonate fossils. Quaternary International 1, 111–128.

Miller, G.H., Magee, J.W., & Jull, A.J.T., 1997. Low-latitude glacial cooling in the Southern Hemisphere from amino-acid racemization in emu eggshells. Nature 385, 241–244.

Miller, N., Williams, D., & Kenward, H.K., 1993a. Plant macrofossils and insect remains from Mytongate. In: Evans, D.H. (Ed), Excavations in Hull 1975–76. East Riding Archaeologist 4, 195–198+fiche.

Miller, R.F., 1990. Coleoptera from the Sangamon interglacial (?) in northern Ontario. Canadian Quaternary Association and American Quaternary Association, First Joint Meeting, Program and Abstracts, 26.

Miller, R.F., 1991a. Chitin paleoecology. Biochemical Systematics and Ecology 19, 401–411.

Miller, R.F. & Elias, S.A., 2000. Late-glacial climate in Maritime Canada, reconstructed from Mutual Climatic Range analysis of fossil Coleoptera. Boreas 29, 79–88.

Miller, R.F., Fitzgerald, W.D., & Buhay, D.N., 1987. Fossil Coleoptera from the postglacial spruce-pine transition period near Minesing Swamp, Ontario. Canadian Journal of Earth Sciences 24, 2099–2103.

Miller, R.F., Fritz, P., & Morgan, A.V., 1988. Climatic implications of D/H ratios in beetle chitin. Palaeogeography, Palaeoclimatology, Palaeoecology 66, 277–288.

Miller, R.F. & Morgan, A.V., 1982. A postglacial coleopterous assemblage from Lockport Gulf, New York. Quaternary Research 17, 258–274.

Miller, R.F., Morgan, A.V., & Hicock, S.R., 1985a. Pre-Vashon fossil Coleoptera of Fraser age from the Fraser Lowland, British Columbia. Canadian Journal of Earth Sciences 22, 498–505.

Miller, R.F., Orr, G.L., Fritz, P., Downer, R.G.H., & Morgan, A.V., 1985b. Stable carbon isotope ratios in *Periplanata americana* L., the American cockroach. Canadian Journal of Zoology 63, 584–589.

Miller, R.F., Voss-Foucart, M.-F., Toussaint, C., & Jeuniaux, C., 1993b. Chitin preservation in Quaternary Coleoptera: preliminary results. Palaeogeography, Palaeoclimatology, Palaeoecology 103, 133–140.

Miller, S.E., 1983. Late Quaternary insects of Rancho La Brea and McKittrick, California. Quaternary Research 20, 90–104.

Miller, S.E., 1991b. Biological diversity and the need to nurture systematics collections. American Entomologist 37, 76.

Miller, S.E., Gordon, R.D., & Howden, H.F., 1981. Reevaluation of Pleistocene scarab beetles from Rancho La Brea, California (Coleoptera: Scarabaeidae). Proceedings of the Entomological Society of Washington 83, 625–630.

Miller, S.E. & Peck, S.B., 1979. Fossil carrion beetles of Pleistocene California asphalt deposits, with a synopsis of Holocene California Silphidae (Insecta: Coleoptera: Silphidae). Transactions of the San Diego Society of Natural History 19, 85–106.

Millien-Parra, V. & Jaeger, J.-J., 1999. Island biogeography of the Japanese terrestrial mammal assemblages: an example of a relict fauna. Journal of Biogeography 26, 959–972.

Mithen, S., 2003. After the ice: A global human history, 20,000–5000 BC. London, UK, Weidenfeld and Nicolson.

Mjöberg, E., 1904. Über eine schwedische interglaciale Coleopteren-species [On a Swedish interglacial beetle species]. Geologiska Föreningens i Stockholm Förhandlingar 26, 493–497. [In German].

Mjöberg, E., 1905. Über eine schwedische interglaciale *Gyrinus*-Species [On a Swedish interglacial *Gyrinus* species]. Geologiska Föreningens i Stockholm Förhandlingar 27, 233–236. [In German].

Mjöberg, E., 1915. Über die Insektenreste der sogennanten ''Härnögyttja'' im nordlichen Schweden [On insect remains from mineralized 'Härnögyttja' in northern Sweden]. Sveriges Geologiska Undersokning C268, 1–14. [In German].

Mjöberg, E., 1916. Über die lnsektenreste der sog. ''Härnögyttja'' im nördlichen Schweden. Geologiska föreningens i Stockholm Fördhandlingar 38, 222–223. [In German].

Moar, N.T., 1980. Late Otiran and early Aranuian grassland in central South Island. New Zealand Journal of Ecology 3, 4–12.

Moar, N.T. & Suggate, R.P., 1979. Contributions to the Quaternary history of the New Zealand flora 8. Interglacial and glacial vegetation in the Westport District, South Island. New Zealand Journal of Botany 17, 361–387.

Moffett, L. & Smith, D., 1997. Insects and plants from a late medieval and early post-medieval tenement in Stone, Staffordshire, U.K. Circaea 12, 157–175.

Moore, P.D., 1981. Life seen from a medieval latrine. Nature 294, 614.

Moore, P.D., 1986. Bears versus beetles. Nature 320, 385–386.

Moreno, P.I., 1997. Vegetation and climate near Lago Llanquihue in the Chilean Lake District between 20,200 and 9500 14C yr BP. Journal of Quaternary Science 12, 485–500.

Moreno, P.I., Lowell, T.V., Jacobson, Jr, G.L., & Denton, G.H., 1999. Abrupt vegetation and climate changes during the last glacial maximum and last termination in the Chilean Lake District: A case study from Canal de la Puntilla (41^{a}S). Geografiska Annaler 81A, 285–311.

Moret, P. & Martín-Cantarino, C., 1996. L'utilisation des Coléoptères subfossiles dans la reconstruction des paléoenvironnements: l'exemple du port antique de Santa Pola (Espagne) [Use of subfossil Coleoptera in paleoenvironmental reconstruction: an example from the ancient port of Santa Pola, Spain]. Bulletin de la Société entomologique de France 101, 225–229. [In French].

Morgan, A., 1969. A Pleistocene fauna and flora from Great Billing, Northamptonshire, England. Opuscula Entomologica 34, 109–129.

Morgan, A., 1970. Weichselian insect faunas of the English Midlands. Unpublished PhD Thesis, University of Birmingham, Birmingham, UK.

Morgan, A., 1972. The fossil occurrence of *Helophorus arcticus* Brown (Coleoptera Hydrophilidae) in Pleistocene deposits of the Scarborough Bluffs, Ontario. Canadian Journal of Zoology 50, 555–558.

Morgan, A., 1973. Late Pleistocene environmental changes indicated by fossil insect faunas of the English Midlands. Boreas 2, 173–212.

Morgan, A., 1975. Fossil beetle assemblages from the Early Wisconsinan Scarborough Formation. Quaternary Non-marine Paleoecology Conference, Abstracts, 10.

Morgan, A., Morgan, A.V., & Elias, S.A., 1985. Holocene insects and palaeoecology of the Au Sable River, Michigan. Ecology 66, 1817–1828.

Morgan, A., Morgan, A.V., & French, H.M., 1982a. A Late Quaternary insect fauna from Belchatow, central Poland. XI INQUA Crongress, Moscow, Abstracts, 2, 197.

Morgan, A.V., 1987. Late Wisconsin and early Holocene paleoenvironments of east-central North America based on assemblages of fossil Coleoptera. In: Ruddiman, W.F., Wright, H.E., Jr (Eds), North America and adjacent oceans during the last deglaciation. The Geology of North America, K-3. Boulder, CO, Geological Society of America, 353–370.

Morgan, A.V., 1988. Late Pleistocene and early Holocene Coleoptera in the lower great lakes region. In: Laub, R.S., Miller, N.G., Steadman, D.W. (Eds), Late Pleistocene and Early Holocene paleoecology and archeology of the eastern great lakes region. Bulletin of the Buffalo Society of Natural Sciences 33, 195–206.

Morgan, A.V., 1989. Late Pleistocene zoogeographic shifts and new collecting records for *Helophorus arcticus* Brown (Coleoptera: Hydrophilidae) in North America. Canadian Journal of Zoology 67, 1171–1179.

Morgan, AV., 1997. *Aphodius holdereri* Reitter (Coleoptera: Scarabaeidae): ecological notes and new localities. Coleopterists Bulletin 51, 154.

Morgan, A.V., Elias, S.A., & Morgan, A., 1981. Paleoenvironmental implications of a lateglacial insect assemblage from southeast Michigan. Geological Society of Canada Annual Meeting, Calgary, Alberta, Abstracts, A-41.

Morgan, A.V., Elias, S.A., & Morgan, A., 1982b. Insect fossils from a lateglacial site at Longswamp, Pennsylvania. Geological Association of Canada Annual Meeting, Calgary, Alberta, Abstracts, A-41.

Morgan, A.V. & Freitag, R., 1982. The occurrence of *Cicindela limbalis* Klug (Coleoptera: Cicindelidae) in a lateglacial site at Brampton, Ontario. Coleopterists Bulletin 36, 105–108.

Morgan, A.V., McAndrews, J.M., Pilny, J.J., Goodwin, A.G., & Morgan, A., 1983a. Paleoecological reconstruction at the Rostock mammoth site, Ontario. Geological Association of Canada, Program with Abstracts, 8A, 48.

Morgan, A.V. & Morgan, A., 1976. Climatic interpretations from the fossil faunas of the Don and Scarborough Formations. Geological Society of America, Annual Meeting, Abstracts, 8, 1020.

Morgan, A.V. & Morgan, A., 1979. The fossil Coleoptera of the Two Creeks Forest Bed, Wisconsin. Quaternary Research 12, 226–240.

Morgan, A.V. & Morgan, A., 1980. Faunal assemblages and distributional shifts of Coleoptera during the Late Pleistocene in Canada and the northern United States. Canadian Entomologist 112, 1105–1128.

Morgan, A.V. & Morgan, A., 1986. A preliminary note on fossil insect faunas from central Illinois. Quaternary Records of Central and North Illinois 1, 81–87.

Morgan, A.V. & Morgan, A., 1987. Paleoentomology-towards the next decade. Episodes 10, 38–40.

Morgan, A.V., Morgan, A., Ashworth, A.C., & Matthews, Jr, J.V., 1984a. Late Wisconsin fossil beetles in North America. In: Porter, S.C. (Ed), Late Quaternary environments of the United States, Vol. 1, The Late Pleistocene, Boulder, CO, Geological Society of America, 354–363.

Morgan, A.V., Morgan, A., & Carter, L.D., 1979. Paleoenvironmental interpretation of a fossil insect fauna from bluffs along the lower Colville River, Alaska. In: The United States Geological Survey in Alaska: accomplishments during 1978. U.S. Geological Survey Circular 804-B, 41–44.

Morgan, A.V., Morgan, A., & Miller, R.F., 1983b. A preliminary report on two fossil insect assemblages from west-central Indiana. In: Bleuer, N.K., Melhorn, W.N., Pavey, R.R. (Eds), Interlobate stratigraphy of the Wabash Valley, Indiana. Lafayette, IN, Purdue University Press, 133–136.

Morgan, A.V., Morgan, A., & Miller, R.F., 1984b. Range extension and fossil occurrences of *Holoboreaphilus nordenskioeldi* (Mäklin) (Coleoptera: Staphylinidae) in North America. Canadian Journal of Zoology 62, 463–467.

Mori, Y., 1995. Fossil insects and the paleoenviromnets. In: Reports on Owatari II Site Excavation. Iwate Archaeological Research Center, Iwate, Japan, 351–370. [In Japanese].

Mori, Y. & Itoh, T., 1992. Paleoenvironments of the Tomizawa Site reconstructed by fossil insects and diatoms. In: Reports on Tomizawa Site (Paleolithic). Board of Education, Sendai City, Japan, 126–141. [In Japanese].

Mori Y., Nakamura, T., & Yoshii, R., 1997. Insect and macroplant fossils found from the Miyanomae Site. Abstract of the Japan Association for Quaternary Research 27, 130–131. [In Japanese].

Morlan, R.E., & Matthews, Jr, J.V., 1978. New dates for early man. Geos, Winter issue, 2–5.

Morlan, R.E. & Matthews, Jr, J.V., 1983. Taphonomy and paleoecology of fossil insect assemblages of Old Crow River Locality (CRH-15), northern Yukon Territory. Géographie physique et Quaternaire 37, 147–158.

Morris, D.P., 1986. Archaeological investigations at antelope house. Washington DC, National Park Service.

Moseley, K.A., 1983. Fossil fauna. In: Bryant, I.D., Gibbard, P.L., Holyoak, D.T., Switsur, V.R., & Wintle, A.G. (Eds), Stratigraphy and palaeontology of Pleistocene cold-stage deposits at Alton Road Quarry, Farnham, Surrey, England. Geological Magazine 120, 597

Mott, R.J., 1990. Sangamonian forest history and climate in Atlantic Canada. Géographie physique et Quaternaire 44, 257–270.

Mott, R.J., Anderson, T.W., & Matthews, Jr, J.V., 1981. Lateglacial paleoenvironments of sites bordering the Champlain Sea based on pollen and macrofossil evidence. In: Mahany, W.C. (Ed), Quaternary Paleoclimate. Geoabstracts, Norwich, England, 129–172.

Mott, R.J., Anderson, T.W., & Matthews, Jr, J.V., 1982. Pollen and macrofossil study of an interglacial deposit in Nova Scotia. Géographie physique et Quaternaire 36, 197–208.

Mott, R.J. & DiLabio, R.N.W., 1990. Paleoecology of organic deposits of probable last interglacial age in northern Ontario. Géographie physique et Quaternaire 44, 309–318.

Mott, R.J. & Matthews, Jr., J.V., 1990. The Last (?) interglaciation in Canada. Géographie physique et Quaternaire 44, 245–248.

Mott, R.J., Matthews, Jr, J.V., Grant, D.R., & Beke, G.J., 1986. A lateglacial buried organic profile near Brookside, Nova Scotia. Current Research, Part B, Geological Survey of Canada Paper 86-1B, 289–294.

Mumcuoglu, K.Y., Zias, J., Tarshis, M., Lavi, M., & Stiebel, G.D., 2003. Body louse remains found in textiles excavated at Masada, Israel. Journal of Medical Entomology 40, 585–587.

Musson, C.R., Smith, A.G., & Girling, M.A., 1977. Environmental evidence from the Breiddin Hillfort. Antiquity 51, 147–151.

Muzzarelli, R.A., 1977. Chitin. Oxford, Pergamon Press.

Nakagawa, T., Tarasov, P.E., Nishida, K., Gotanda, K., & Yasuda, Y., 2002. Quantitative pollen-based climate reconstruction in central Japan: application to surface and Late Quaternary spectra. Quaternary Science Reviews 21, 2099–2113.

National Institute of Water and Atmospheric Research (New Zealand) (NIWA), 2008. Overview of New Zealand climate. Web site: http://www.niwa.cri.nz/edu/resources/climate/overview [Accessed on 15 January 2008].

Nazarov, V.I., 1979. Coleoptera from the Rubezhnitsa locality and their environment. Paleontological Journal 13, 462–470.

Nazarov, V.I., 1984. A reconstruction of the Anthropogene landscapes in the northeastern part of Belorussia according to palaeoentomological data. Trudy Paleontological Institute AN SSR Publication, 205, 95. Nauka Press, Moscow. [In Russian].

Nazarov, V.I., 1991. Written communication. Institute of Geochemistry and Geophysics, Byelorrussian Academy of Sciences, Minsk, Byelorussia.

Nelson, R.E., 1982. Late Quaternary environments of the western Arctic Slope, Alaska. PhD Thesis, University of Washington, 90.

Nelson, R.E., 1986. Mid-Wisconsin aridity in northern Alaska: incongruity of a mesic pollen record and xeric insect indicators. Current Research in the Pleistocene 3, 57–59.

Nelson, R.E., 1987. A postglacial flora and subfossil insect fauna from central Maine. Twelfth International Congress, International Union for Quaternary Research (INQUA), Ottawa, Ontario, Programme with Abstracts, 231.

Nelson, R.E. & Carter, L.D., 1987. Paleoenvironmental analysis of insects and extralimital Populus from an early Holocene site on the Arctic Slope of Alaska. Arctic and Alpine Research 19, 230–241.

Nelson, R.E., & Coope, G.R., 1982. A Late-Pleistocene insect fauna from Seattle, Washington. American Quaternary Association, Seventh Biennial Meeting, Seattle, Abstracts and Program, 146.

Nichols, H., 1975. Palynological and paleoclimatic study of the Late Quaternary displacement of the boreal forest-tundra ecotone in Keewatin and Mackenzie, N. W. T., Canada. Institute of Arctic and Alpine Research, Occasional Paper 15, 87.

Nielsen, B.O., 1989. House fly puparia (*Musca domestica* L) from a Neolithic field in Thayngen-Weier, Switzerland. Mitteilungen der Schweizerischen Entomologischen Gesellschaft 62, 5–8.

Nielsen, B.O., Mahler, V., & Rasmussen, P., 2000. An arthropod assemblage and the ecological conditions in a byre at the Neolithic settlement of Weier, Switzerland. Journal of Archaeological Science 27, 209–218.

Nielsen, E., Morgan, A.V., Morgan, A., Mott, R.J., Rutter, N.W., & Causse, C., 1986. Stratigraphy, paleoecology, and glacial history of the Gillam area, Manitoba. Canadian Journal of Earth Sciences 23, 1641–1661.

Nilsson, A.N., Petterson, R.B., & Lemdahl, G., 1993. Macroptery in altitudinal specialists versus brachyptery in generalists – a paradox of alpine Scandinavian Carabid beetles (Coleoptera: Carabidae). Journal of Biogeography 20, 227–234.

Nirei, T. & Hayashi, M., 1998. Fossil pollen and insect assemblage from Late Pleistocene peat deposits at Taguro, Tamagawa village, western margin of Kanto Plain, Central Japan. Bulletin of the Saitama Museum of Natural History 16, 15–21.

Nirei, T. & Hayashi, M., 2004. Fossil pollen assemblages of the Late Pleistocene Maebashi Peat Bed and palaeoclimatic oscillation at Takasaki, Gunma, central Japan. The Natural Environmental Science Research 17, 43–49.

Nissen, K., 1973. Analysis of human coprolites from Bamert Cave, Amador County, California. In: Heizer, R.F., Hester, T.R. (Eds), The archaeology of Bamert Cave, Amadore County, California. Berkeley, CA, University of California Archaeological Facility, 65–71.

Noe-Nygaard, B., 1982. The applicability of insect finds to the interpretation of archaeological sites. In: Second Nordic Conference on the application of scientific methods in archaeology, Helsingfors, PACT, Journal of the European Study Group on Physical, Chemical and Mathematical Techniques applied to Archaeology 7, 239–245.

Noonan, G.R., 1985. The influences of dispersal, vicariance, and refugia on patterns of biogeographical distributions of the beetle family Carabidae. In: Ball, G.E. (Ed), Taxonomy, Phylogeny and Zoogeography of Beetles and Ants. Junk, Dordrecht, The Netherlands, 322–350.

Noonan, G.R., 1988. Biogeography of North American and Mexican insects, and a critique of vicariance biogeography. Systematic Zoology 37, 366–384.

Noonan, G.R., 1990. Biogeographical patterns of North American *Harpalus* Latreille (Insecta: Coleoptera: Carabidae). Journal of Biogeography 17, 583–614.

Noonan, J.P., Coop, G., Kudaravalli, S., Smith, D., Krause, J., Alessi, J., Chen, F., Platt, D., Pääbo, S., Pritchard, J.K., & Rubin, E.M., 2006. Sequencing and analysis of Neanderthal genomic DNA. Science 314, 1113–1118.

Numata, M., 1974. The flora and vegetation of Japan. Tokyo, Japan, Kodansha, Ltd.

O'Brien, C., Selby, K., Ruiz, Z., Brown, A., Dinnin, M., Caseldine, C., Langdon, P., & Stuijts, I., 2005. A sediment-based multi-proxy palaeoecological approach to the environmental archaeology of lake dwellings (crannogs), central Ireland. The Holocene 15, 707–719.

O'Connell, J.F. & Allen, J., 2004. Dating the colonisation of Sahul (Pleistocene Australia–New Guinea): A review of recent research. Journal of Archaeological Science 31, 835–853.

O'Connor, J.P., 1979. *Blaps lethifera* Marsham (Coleoptera: Tenebrionidae), a beetle new to Ireland from Viking Dublin. Entomologist's Gazette 30, 295–297.

O'Connor, J.P., 1987. Appendix V. Insecta. In: Mitchell, G.F. (Ed), Archaeology and environment in Early Dublin. Medieval Dublin Excavations, C1, 39–40. Royal Irish Academy.

O'Connor. J.P., Hall. A.R., Jones, A.K.G., & Kenward, H.K., 1984. Ten years of environmental archaeology at York. In: Addyman, P.V., Black, V.E. (Eds), Archaeological papers from York presented to M. W. Barley. York Archaeological Trust, 166–172.

O'Neill, J., Plunkett, G., & Whitehouse, N., 2007. The archaeological and palaeoecological investigation of a Middle Bronze Age settlement at Ballyarnet Lake, County Derry. Ulster Journal of Archaeology 66, 39–49.

Osborne, P.J., 1969. An insect fauna of Late Bronze Age date from Wilsford, Wiltshire. Journal of Animal Ecology 38, 555–566.

Osborne, P.J., 1971a. An insect fauna from the Roman site at Alcester, Warwickshire. Britannia 2, 156–165.

Osborne, P.J., 1971b. The insect fauna from the Roman harbour. In: Cuncliffe, B. W., Excavations at Fishbourne, 1961–1969. Research Reports of the Society of Antiquaries of London 27, 393–396.

Osborne, P.J., 1972. Insect faunas of Late Devensian and Flandrian age from Church Stretton, Shropshire. Philosophical Transactions of the Royal Society of London B 263, 327–367.

Osborne, P.J., 1973. Insects in archaeological deposits. Science and Archaeology 10, 4–6.

Osborne, P.J., 1974. An insect assemblage of Early Flandrian age from Lea Marston, Warwickshire, and its bearing on contemporary climate and ecology. Quaternary Research 4, 471–486.

Osborne, P.J., 1975. The Coleoptera from the Roman Well (on the intervallum road, east of the gyrus). In: Hobley, B., The Lunt Roman Fort and Training School for Cavalry,

Baginton, Warwickshire. Transactions of the Birmingham and Warwickshire Archaeological Society, 87, 44–46.

Osborne, P.J., 1976. The insect fauna of the peat within the lake deposits. In: Clarke, B. B. (Ed), Quaternary features of Porth Meare Cove, North Cornwall. Transactions of the Royal Geological Society of Cornwall 20, 283–285.

Osborne, P.J., 1977. Stored product beetles from a Roman site at Droitwich, England. Journal of Stored Products Research 13, 203–204.

Osborne, P.J., 1979. Insect remains. In: Smith, C. (Ed), Fisherwick: The Reconstruction of an Iron Age Landscape. British Archaeological Reports 61, 85–87, 189–193.

Osborne, P.J., 1980a. The insect fauna of the organic deposits at Sugworth and its environmental and stratigraphic implications. Philosophical Transactions of the Royal Society of London, Series B 289, 119–133.

Osborne, P.J., 1980b. The Late Devensian-Flandrian transition depicted by serial insect faunas from West Bromwich, Staffordshire, England. Boreas 9, 139–147.

Osborne, P.J., 1980c. Report of the insects from two organic deposits on the Smithfield Market site, Birmingham. In: Watts, L. (Ed), Birmingham Moat: its history, topography, and destruction. Transactions of the Birmingham and Warwickshire Archaeological Society 89, 63–66.

Osborne, P.J., 1980d. Insect remains from the Sidbury site, Worcester, Pit F41. In: Carver, M.O.H. (Ed), Medieval Worcester. An archaeological framework. Transactions of the Worcestershire Archaeological Society, 3rd Series 7, 211–212.

Osborne, P.J., 1981a. The insect fauna. In: Stanford, S.C. (Ed), Midsummer Hill, an Iron Age Hillfort on the Malverns. Published Privately in Hereford, 156–157.

Osborne, P.J., 1981b. The insect fauna. In: Jarrett, M.G., Wrathmell, S. (Eds), Whitton: an Iron Age and Roman Farmstead in South Glamorgan. Cardiff, UK, University of Wales Press, 245–248.

Osborne, P.J., 1981c. Coleopterous fauna from Layer 1. In: Greig, J.R.A. (Ed), The investigation of a Medieval barrel latrine from Worcester. Journal of Archaeological Science 8, 268–271.

Osborne, P.J., 1982. Report on the insect remains from Cirencester. In: Wacher, J., McWhirr, A. (Eds), Early Roman occupation at Cirencester. Cirencester Excavations 1, 232. Cirencester Excavation Committee.

Osborne, P.J., 1983. An insect fauna from a modern cesspit and its comparison with probable cesspit assemblages from archaeological sites. Journal of Archaeological Science 10, 453–463.

Osborne, P.J., 1984. Insect Remains. In: Leach, P. (Ed), The archaeology of taunton excavations and fieldwork to 1980. Western Archaeological Trust, Excavation Monograph 8, 165–166.

Osborne, P.J., 1985. The insect remains from Fiave F4, 170 cm. In: Malone, C., Stoddart, S. (Eds), Papers in Italian archaeology IV, part ii, prehistory. British Archaeological Reports S244, 308 & 311–312.

Osborne, P.J., 1986. The Wilsford Shaft: Report on the insect fauna. Ancient Monuments Laboratory Report, 8/86. London. (Microfiche)

Osborne, P.J., 1988. A late Bronze Age insect fauna from the River Avon, Warwickshire, England: its implications for the terrestrial and fluvial environment and for climate. Journal of Archaeological Science 15, 715–727.

Osborne, P.J., 1989. Insects. In: Ashbee, P., Bell, M., Proudfoot, E. (Eds), Wilsford Shaft: excavations 1960–62. London, UK, English Heritage, 96–99 and fiche C1–7.

Osborne, P.J., 1994. The insects. In: Bidwell, P., Speak, S. (Eds), Excavations at South Shields Roman Fort I. Society of Antiquaries of Newcastle upon Tyne & Tyne & Wear Museums Monograph 4. Newcastle upon Tyne, 266.

Osborne, P.J., 1996. An insect fauna of Roman date from Stourport, Worcestershire, U.K., and its environmental implications. Circaea, The Journal of the Association for Environmental Archaeology 12, 183–189.

Osborne, P.J., 1997. Insects, man and climate in the British Holocene. In: Ashworth, A.C., Buckland, P.C., Sadler, J.P. (Eds), Studies in Quaternary entomology – an inordinate fondness for insects. Quaternary Proceedings 5, 193–198.

Osborne, P.J., 2003. The insects. In: Field, N., Parker Pearson, M. (Eds), Fiskerton. An Iron Age timber causeway with Iron Age and Roman votive offerings: the 1981 excavations. Oxford, Oxbow Books, 22–23.

Osborne, P.J., 2006. Insects. In: Greig, J., Osborne, P.J., Rackham, D.J. (Eds), Environmental evidence from Well V 18. Roman Droitwich. Dodderhill fort, Bays Meadow villa, and roadside settlement. Council for British Archaeology Research Report 146, York, 227–234.

Osborne, P.J. & Shotton, F.W., 1968. The fauna of the channel deposit of Early Saalian age at Brandon, Warwickshire. Philosophical Transactions of the Royal Society of London, Series B 254, 417–424.

Palmer, C., & Whitehouse, N., 2006. Plant macrofossil and insect evidence from the water hole. In: Millett, M. (Ed), Shiptonsthorpe, East Yorkshire: archaeological studies of a Romano-British settlement. Yorkshire Archaeological Report 5. Yorkshire Archaeological Society, Roman Antiquities Section & East Riding Archaeological Society, Leeds, 280–291.

Pals, J.P. & Hakbijl, T., 1992. Weed and insect infestation of a grain cargo in a ship at the Roman fort of Laurium im Woerden (Province of Zuid-Holland). Review of Palaeobotany and Palynology 73, 287–300.

Panagiotakopulu, E., 1998. An insect study from Egyptian stored products in the Liverpool Museum. Journal of Egyptian Archaeology 84, 231–234.

Panagiotakopulu, E., 1999. An examination of biological materials from coprolites from XVIII Dynasty Amarna, Egypt. Journal of Archaeological Science 26, 547–551.

Panagiotakopulu, E., 2000. Archaeology and entomology in the eastern Mediterranean. Research into the history of insect synanthropy in Greece and Egypt. British Archaeological Reports International Series 836. Oxford.

Panagiotakopulu, E., 2001a. New records for ancient pests: archaeoentomology in Egypt. Journal of Archaeological Science 28, 1235–1246.

Panagiotakopulu, E., 2001b. Fleas from pharaonic Amarna. Antiquity 75, 499–500.

Panagiotakopulu, E., 2004a. The insect remains. In: Caseldine, A., Earwood, C. (Eds), A Neolithic wetland site at Abercynafon, Talybont, South Wales. Journal of Wetland Archaeology 4, 27–29.

Panagiotakopulu, E., 2004b. Pharaonic Egypt and the origins of plague. Journal of Biogeography 31, 269–275.

Panagiotakopulu, E. & Buckland, P.C., 1991. Insect pests of stored products from Late Bronze Age Santorini, Greece. Journal of Stored Products Research 27, 179–184.

Panagiotakopulu, E. & Buckland, P.C., 1999. The bed bug, *Cimex lectularius* L. from Pharaonic Egypt. Antiquity 73, 908–911.

Panagiotakopulu, E., Buckland, P.C., & Day, P.M., 1995. Natural insecticides and insect repellents in antiquity: a review of the evidence. Journal of Archaeological Science 22, 705–710.

Panagiotakopulu, E., Buckland, P.C., Day, P.M., Doumas, C., Sarpaki, A., & Skidmore, P., 1997. A lepidopterous cocoon from Thera and evidence for silk in the Aegean Bronze Age. Antiquity 71, 420–429.

Panagiotakopulu, E., Skidmore, P., & Buckland, P., 2007. Fossil insect evidence for the end of the Western Settlement in Norse Greenland. Naturwissenschaften 94, 300–306.

Panagiotakopulu, E. & Van der Veen, M., 1997. Synanthropic insect faunas from Mons Claudianus, a Roman quarry site in the Eastern Desert, Egypt. In: Ashworth, A.C., Buckland, P.C., Sadler, J.P. (Eds), Studies in Quaternary entomology – an inordinate fondness for insects. Quaternary Proceedings 5, 199–206.

Parfitt, S.A., Barendregt, R.W., Breda, M., Candy, I., Collins, M.J., Coope, G.R., Durbridge, P., Field, M.H., Lee, J.R., Lister, A.M., Mutch, R., Penkman, K.E.H., Preece, R.C., Rose, J., Stringer, C.B., Symmonds, R., Whitaker, J.E.P., Wymer, J.J., & Stuart, A.J., 2005. The earliest humans in Northern Europe: artefacts from the Cromer Forest Formation at Pakefield. Nature 438, 1008–1012.

Parker-Pearson, M., 1995. Bronze Age Britain. London, UK, Batsford.

Parkin, E.A., 1956. Stored product entomology: the assessment and reduction of losses caused by insects to stored foodstuffs. Annual Review of Entomology 1, 223–240.

Pauketat, T.R., Kelly, L.S., Fritz, G.J., Lopinot, N.H., Elias, S.A., & Hargrave, E., 2002. The residues of feasting and public ritual at early Cahokia. American Antiquity 67, 52–65.

Peake, D.S. & Osborne, P.J., 1971. The Wandle Gravels in the vicinity of Croydon. Proceedings and Transactions of the Croydon Natural History and Scientific Society 14, 145–175.

Pearson, R.G., 1962. The Coleoptera from a Lateglacial Deposit at St. Bees, West Cumberland. Journal of Animal Ecology 31, 129–150.

Pearson, R.G., 1967. The significance of the Colney Heath beetle fauna. Entomologist's Monthly Magazine 103, 169–170.

Pearson, S., 1999. Late Holocene biological records from the middens of stick-nest rats in the central Australian arid zone. Quaternary International 59, 39–46.

Pennington, W., 1986. Lags in adjustment of vegetation to climate caused by the pace of soil development: evidence from Britain. Vegetatio 67, 105–118.

Penny, L.F., Coope, G.R., & Catt, J.A., 1969. The age and insect fauna of the Dimlington silts, East Yorkshire. Nature 224, 65–67.

Perry, D.W., Buckland, P.C., & Snaesdottir, M., 1985. The application of numerical techniques to insect assemblages from the site of Stóraborg, Iceland. Journal of Archaeological Science 12, 335–345.

Peteet, D.M. & Mann, D.H., 1994. Late-glacial vegetational, tephra, and climatic history of southwestern Kodiak Island, Alaska. Ecoscience 1, 255–267.

Petit, R.J., Hu, F.S., & Dick, C.W., 2008. Forests of the past: a window to future changes. Science 320, 1450–1452.

Pettigrew, T.J., 1834. Insects. In: Pettigrew, T.J. (Ed), A history of Egyptian mummies and an account of the worship and embalming of the sacred animals. London, UK, Longman, Rees, Orme, Brown, Green & Longman, 221–225 +pl.V.

Pierce, K.L., 2004. Pleistocene glaciations in the Rocky Mountains. In: Gillespie, A.R., Porter, S.C. (Eds), The Quaternary period in the United States. Amsterdam, Elsevier, 63–76.

Pierce, W.D., 1946. Fossil arthropods of California. 11. Description of the dung beetles (Scarabaeidae) of the tar pits. Bulletin of the Southern California Academy of Sciences 45, 119–131.

Pierce, W.D., 1948. Fossil arthropods of California. 16. The carabid genus *Elaphrus* in the asphalt deposits. Bulletin of the Southern California Academy of Sciences 47, 53–54.

Pierce, W.D., 1949. Fossil arthropods of California. 17. The silphid burying beetles in the asphalt deposits. Bulletin of the Southern California Academy of Sciences 48, 55–70.

Pierce, W.D., 1954a. Fossil arthropods of California. 19. The Tenebrionidae-Scaurinae of the asphalt deposits. Bulletin of the Southern California Academy of Sciences 53, 93–98.

Pierce, W.D., 1954b. Fossil arthropods of California. 20. The Tenebrionidae-Coniontinae of the asphalt deposits. Bulletin of the Southern California Academy of Sciences 53, 142–156.

Pierce, W.D., 1957. Insects. Geological Society of America Memoirs 67, 943–952.

Pilny, J.J. & Morgan, A.V., 1987. Paleoentomology and paleoecology of a possible Sangamonian site near Innerkip, Ontario. Quaternary Research 28, 157–174.

Pilny, J.J., Morgan, A.V., & Morgan, A., 1987. Paleoclimatic implications of late Wisconsinan insect assemblages from Rostock, southwestern Ontario. Canadian Journal of Earth Sciences 24, 617–630.

Plunkett, G., Whitehouse, N.J., Hall, V.A., Charman, D.J., Blaauw, M., Kelly, E., & Mulhall, I., 2009. A multi-proxy palaeoenvironmental investigation of the findspot of an Iron Age bog body from Oldcroghan, Co. Offaly, Ireland. Journal of Archaeological Sciences 36, 265–277.

Poinar, H.N., Schwarz, C., Qi, J., Shapiro, B., MacPhee, R.D.E., Buigues, B., Tikhonov, A., Huson, D.H., Tomsho, L.P., Auch, A., Rampp, M., Miller, W., & Schuster, S.C., 2006. Metagenomics to paleogenomics: large-scale sequencing of mammoth DNA. Science 311, 392–394.

Ponel, P., 1994. Les fluctuations climatiques au Pléniglacaire würmien déduites des assemblages d'Arthropodes fossiles à La Grande Pile (Haute Saône, France) [Climatic fluctuations of the Würm Glaciation inferred from fossil arthropod assemblages from La Grand Pile (Haute Saône, France)]. Comptes-Rendus de l'Académie des Sciences, Paris 319(2), 845–852. [In French].

Ponel, P., 1995. Rissian, Eemian and Würmian Coleoptera assemblages from La Grand Pile (Vosges, France). Palaeogeograhy, Palaeoclimatology, Palaeoecology 114, 1–41.

Ponel, P., 1997a. Sucession des assemblages de Coléoptères à Chalain 3. Monographies de Chalain et Clairvaux, 3 [Beetle faunal succession at the Chalain 3 site]. Que Chalain 3, le Néolithique final à la fin du Ive millénaire, 113–118. [In French].

Ponel, P., 1997b. Late Pleistocene Coleopteran fossil assemblages in high altitude sites: a case study from Prato Spilla (Northern Italy). In: Ashworth, A.C., Buckland, P.C., Sadler, J.P., (Eds), Studies in Quaternary entomology – an inordinate fondness for insects. Quaternary Proceedings 5, 207–218.

Ponel, P., Andrieu-Ponel, V., Parchoux, F., Juhasz, I., & de Beaulieu, J.-L., 2001. Late-glacial and Holocene high altitude environmental changes in Vallée des Merveilles (Alpes-Maritimes, France): insect evidence. Journal of Quaternary Science 16, 795–812.

Ponel, P. & Coope, G.R., 1990. Lateglacial and early Flandrian Coleoptera from La Taphanel, Massif Central, France: climatic and ecological implications. Journal of Quaternary Science 5, 235–249.

Ponel, P., Coope, G.R., Andrieu-Ponel, V., & Reille, M., 1999. Coleopteran evidence for a mosaic of environments at high altitude in the Eastern Pyrénées, France during the climatic transition between the Allerød and Younger Dryas. Journal of Quaternary Science 14, 169–174.

Ponel, P., Coope, G.R., Antoine, P., Limondin-Lozouet, N., Leroyer, C., Munaut, A.V., Pastre, J.F., & Guiter, F., 2005. Lateglacial palaeoenvironments and palaeoclimates from Conty and Houdancourt, northern France, reconstructed from beetle remains. Quaternary Science Reviews 24, 2449–2465.

Ponel, P., deBeaulieu, J.L., & Tobolski, K., 1992. Holocene paleoenvironment at the timberline in the Taillefer Massif: pollen analysis, study of plant and insect macrofossils. The Holocene 2, 117–130.

Ponel, P. & Gadbin, C., 1988. Contribution of palaeoentomology to an episode in the vegetational history of the Lac d'Issarles region. Compte-Rendu de l'Academie des Sciences Paris 307, 755–758. [In French].

Ponel, P. & Lowe, J.J., 1992. Coleopteran, pollen and radiocarbon evidence from the Prato Spilla 'D' succession, N. Italy. Comptes Rendus de l'Académie des Sciences, Paris 315(II), 1425–1431.

Ponel, P., Matterne, V., Coulthard, N., & Yvenic, J.-H., 2000. La Tène and Gallo-Roman natural environments and human impact at the Touffréville rural settlement, reconstructed from Coleoptera and plant macroremains (Calvados, France). Journal of Archaeological Science 27, 1055–1072.

Ponel, P., Orgeas, J., Samways, M.J., Andrieu-Ponel, V., de Beaulieu, J.-L., Reille, M., Roche, P., & Tatoni, T., 2003. 110000 years of Quaternary beetle diversity change. Biodiversity and Conservation 12, 2077–2089.

Ponel, P. & Yvinec, J.-H., 1997. L'archéoentomologie en France [Archaeo-entomology in France] Les nouvelles de l'archéologie, 68, 31–44. [In French].

Porch, N., 2007a. Beetle records: Late Pleistocene of Australia. In: Elias, S.A. (Ed), Encyclopedia of Quaternary science. Amsterdam, Elsevier, 179–190.

Porch, N., 2007b. Insect fossils from Pacific Island prehuman and archaeological contexts: reconstructing lost 'Edens' and human impacts. XVII INQUA Congress Abstracts, Cairns, Australia, Quaternary International 167–168; 328.

Porch, N., Burney, D., & Prebble, M., 2007. Here yesterday, gone yesterday: using the subfossil insect record to rediscover the lost diversity of heavily modified Pacific lowland ecosystems. Combined 8th Invertebrate Biodiversity and Conservation/Society of Australian Systematic Biologists Conference, University of Queensland, 2–7 December 2007, Abstracts.

Porch, N. & Elias, S.A., 2000. Quaternary beetles: a review and issues for Australian studies. Australian Journal of Entomology 39, 1–9.

Preece, R.C. & Penkman, K.E.H., 2005. Molluscan assemblages and amino acid dating of the Lower Palaeolithic site at East Farm, Barnham, Suffolk. Proceedings of the Geologists' Association 116, 363–377.

Prentice, I.C., 1986. Vegetation responses to past climatic variation. Vegetatio 67, 131–141.

Prest, V.K., Terasme, J., & Matthews, Jr, J.V., 1976. Late-Quaternary history of Magdalen Islands, Quebec. Maritime Sediments 12, 39–59.

Price, P.W., 1984. Insect Ecology. New York, Wiley.

Pringle, H., 1997. Death in Norse Greenland. Science 275, 924–926.

Rasmussen, S.O., Andersen, K.K., Svensson, A.M., Steffansen, J.P., Vinther, B.M., Clausen, H.B., Siggard-Andersen, M.-L., Johnsen, S.J., Larsen, L.B., Dahl-Jensen, D., Bigler, M., Rothlisberger, R., Fischer, H., Goto-Azuma, K., Hanssen, M.E., & Ruth, U., 2006. A new Greenland ice core chronology for the last glacial termination. Journal of Geophysical Research 111, D06102.

Ratcliffe, B.C. & Fagerstrom, J.A., 1980. Invertebrate *lebensspuren* of Holocene floodplains: their morphology, origin and paleoecological significance. Journal of Paleontology 54, 614–630.

Reid, C., 2007. Beetle Extinction on Lord Howe Island. Combined 8th Invertebrate Biodiversity and Conservation/ Society of Australian Systematic Biologists Conference, University of Queensland, 2–7 December 2007, Abstracts.

Reilly, E., 1996. The insect fauna (Coleoptera) from the Neolithic trackways Corlea 9 and 10: the environmental implications. In: Raferty, B. (Ed), Trackway excavations in the Mountdillon Bogs, Co. Longford, 1985–1991. Irish Archaeological Wetlands Unit Transactions 3. Dublin, Crannog Publications.

Reilly, E., 2003. The contribution of insect remains to an understanding of the environment of Viking age and medieval Dublin. In: Duffy, S. (Ed), Medieval Dublin IV. Proceedings of the friends of medieval Dublin symposium 2002. Four Courts Press, Dublin, 40–62.

Reiss, R.A., 2006. Ancient DNA from ice age insects: proceed with caution. Quaternary Science Reviews 25, 1877–1893.

Reiss, R.A., Ashworth, A.C., & Schwert, D.P., 1999. Molecular genetic evidence of post-Pleistocene population divergence of the arctic/alpine ground beetle *Amara alpina* (Paykull) (Copeoptera: Carabideae). Journal of Biogeography 26, 785–794.

Rhode, D., 2000. Middle and Late Wisconsin vegetation in the Bonneville Basin. In: Madsen, D. (Ed), Late Quaternary paleoecology in the Bonneville Basin. Salt Lake City, UT, Utah Geological Survey, 149–164.

Roback, S. S., 1970. The Chironomidae. In: Hutchinson, G. E. (Ed), An account of the history and development of the Lago di Monterosi, Latium, Italy. Transactions of the American Philosophical Society 60, 150–162.

Robinson, M.A., 1975. The environment of the Roman defenses at Alchester and its implications. Oxoniensia 40, 161–170.

Robinson, M. A., 1978. Insect and seed remains. In: Brodribb, A.C.C., Hands, A.R., Walker, D.R., Excavations at Shakenoak Farm, near Wilcote, Oxfordshire. Part V: Sites K and E. British Archaeological Reports, Oxford, 161–164.

Robinson, M.A., 1979a. The biological evidence. In: Lambrick, G., Robinson, M.A. (Eds), Iron Age and Roman riverside settlements at Farmoor, Oxfordshire. Council for British Archaeology Research Report 32, 77–133.

Robinson, M.A., 1979b. Calcified seeds and arthopods. In: Parrington, M. (Ed), Excavations at Stert Street, Abingdon, Oxon. Oxoniensia 44, 23–24.

Robinson, M.A., 1980. Appendix 10: Insect Remains. In: Christie, P.M., Coad, J.G. (Eds), Excavations at Denny Abbey. Archaeological Journal 137, 267.

Robinson, M.A., 1981a. Roman waterlogged plant and invertebrate evidence. In: Hinchliffe, J., and Thomas, R. (Eds), Archaeological investigations at Appleford. Oxoniensia 45, 90–106.

Robinson, M.A., 1981b. Waterlogged plant and invertebrate evidence. In: Palmer, N. (Ed), A beaker burial and medieval tenements in the Hamel, Oxford. Oxoniensia 45, 199–206+fiche.

Robinson, M.A., 1983. The insect remains. In: Williams, J.H., Farwell, D. (Eds), Excavations of a Saxon site in St. James' square, Northampton, 1981. Northamptonshire Archaeology 18, Fiche M33 & M38.

Robinson, M.A., 1984. Plant and Invertebrate Remains. In: Halpin, C. (Ed), Late Saxon evidence and excavation of Hinxey Hall, Queen Street, Oxford. Oxoniensia 48, 69+fiche.

Robinson, M.A., 1986. Plant and invertebrate remains from the Priory drains. In: Lambrick, G. (Ed), Further excavations on the second site of the Dominican Priory, Oxford. Oxoniensia 50, 196–201+fiche E10–F1.

Robinson, M.A., 1991. The Neolithic and Late Bronze Age insect assemblages. In: Needham, S. (Ed), Excavation and salvage at Runnymede Bridge, 1978: the Late Bronze Age waterfront site. London, UK, British Museum Press, 277–326.

Robinson, M.A., 1992a. Insect assemblages. In: Butterworth, C.A., Lobb, S.J. (Eds), Archaeological investigations at Anslow's Cottages, Burghfield. Excavations in the Burghfield Area, Berkshire; developments in the Bronze Age and Saxon landscapes. Wessex Archaeological Report 1, 143–148.

Robinson, M.A., 1992b. The Coleoptera from Flag Fen. Antiquity 66, 467–469.

Robinson, M.A., 1992c. Plant and invertebrate remains from 13th-century silt beneath the hospital buildings. In: Durham, B. (Ed), The infirmary and hall of the medieval Hospital of St. John the Baptist, Oxford. Oxoniensia 56, 60.

Robinson, M.A., 1993. The scientific evidence. In: Allen, T.G., Robinson, M.A. (Eds), The prehistoric landscape and Iron Age enclosed settlement at Mingies Ditch, Hardwich-with-Yelford Oxon. Oxford Archaeological Unit, Thames Valley Landscapes. The Windrush Valley 2, 101–141. Council for British Archaeology Research Report, London.

Robinson, M.A., 1996. Plant and invertebrate remains. In: Mudd, A. (Ed), The excavation of a Late Bronze Age/Early Iron Age site at Eight Acre Field, Radley. Oxoniensia 60, 41–49.

Robinson, M.A., 1997. The Insects. In: Whittle, A. (Ed), Sacred mound holy rings. Silbury Hill and the West Kennet palisade enclosures: a Later Neolithic complex in north Wiltshire, Oxbow Monograph 74, Oxford, 36–46.

Robinson, M.A., 1999. Insect remains. In: Fitzpatrick, A.P., Butterworth, C.A., Grove, J. (Eds), Prehistoric and Roman sites in East Devon: the A30 Honiton to Exeter Improvement DBFO, 1996–1999. Wessex Archaeology Report 16, 329–332; 342–345.

Robinson, M.A., 2000a. Coleopteran evidence for the Elm Decline, Neolithic activity in woodland, clearance and the use of the landscape. In: Fairbairn, A.S. (Ed), Plants in Neolithic Britain and beyond. Neolithic Studies Group Seminar Papers 5. Oxbow Books, Oxford, 27–36.

Robinson, M.A., 2000b. Middle Mesolithic to Late Bronze Age insect assemblages and an Early Neolithic assemblage of waterlogged macroscopic plant remains. In: Needham, S.P. (Ed), The passage of the Thames. Holocene environment and settlement at Runnymede. London, UK, British Museum, 146–167.

Robinson, M.A., 2001a. Insects as palaeoenvironmental indicators. In: Brothwell, D.R., Pollard, A.M. (Eds), Handbook of archaeological sciences. Chichester, Wiley, 121–133.

Robinson, M.A., 2001b. Macroscopic plant and insect remains. In: Walker, G., King, R. (Eds), Early medieval and later tenements at 113–119 High Street, Oxford: excavations in 1993–5. Oxoniensia 65, 433–437.

Robinson, M.A., 2001c. Waterlogged macroscopic plant and invertebrate remains. In: Booth, M.J., Evans, P., Hiller, J. (Eds), Excavations in the extramural settlement of Roman Alchester. Oxford Archaeology Monograph 1, Oxford, 417–418.

Robinson, M.A., 2001d. Late Bronze Age Coleoptera from Flag Fen. In: Pryor, F. (Ed), The Flag Fen Basin. Archaeology and environment of a fenland landscape. English Heritage Archaeological Reports, London, 384–389.

Robinson, M.A., 2003a. Palaeoenvironmental studies. In: Barclay, A., Lambrick, G., Moore, J., Robinson, M. (Eds), Lines in the landscape. Cursus monuments in the Upper Thames Valley. Oxford Archaeology, Thames Valley Landscapes Monograph 15, Oxford, 163–178.

Robinson, M.A., 2003b. Environmental reports. In: Dodd, A. (Ed), Oxford before the University. The late Saxon and Norman archaeology of the defences and the town. Thames Valley Landscapes Monograph 17, Oxford, 365–398.

Rogerson, R.J., Keen, D.H., Coope, G.R., Robinson, E., Dickson, J.H., & Dickson, C.A., 1992. The fauna, flora and palaeoenvironmental significance of deposits beneath the low terrace of the River Great Ouse at Radwell, Bedfordshire, England. Proceedings of the Geologists' Association 103, 1–13.

Roper, T., 1996. Fossil insect evidence for the development of raised mire at Thorne Moors, near Doncaster. Biodiversity and Conservation 5, 503–521.

Roper, T., 1999. The beetles. In: Parker Pearson, M., Sharples, N. (Eds), Between land and sea. Excavations at Dun Vulan, South Uist. Sheffield, Sheffield Academic Press, 55–56; 192–194; 290–297.

Roper, T. & Whitehouse, N., 1997. Beetle remains. In: Parker Pearson, M., Sydes, R.E. (Eds), Iron Age enclosures and prehistoric landscape of Sutton Common, South Yorkshire. Proceedings of the Prehistoric Society 63, 238–245.

Rose, J., Turner, C., Coope, G.R., & Bryan, M.D., 1980. Channel changes in a lowland river catchment over the last 13,000 years. In: Cullingford, R.A., Davidson, D.A., Lewin, J. (Eds), Timescales in geomorphology. London, UK, Wiley, 159–175.

Rowland, S., Kenward, H.K., Jaques, D., Hall, A., & Carrott, J., 2000. Evaluation of biological remains from excavations at Skellgarths, Ripon, North Yorkshire (site code: HARGM 10426). Reports from the Environmental Archaeology Unit 2001/24.

Rozanski, K., Araguas-Araguas, L., & Gonfiantini, R., 1993. Isotopic patterns in modern global precipitation. In: Swart, P.K., Lohmann, K.C., McKenzie, J., Savin, S. (Eds), Climate change in continental isotopic records. American Geophysical Union, Geophysical Monograph 78, 1–36.

Ruddiman, W.F. & McIntyre, A., 1981. The North Atlantic Ocean during the Last deglaciation. Palaeogeography, Palaeoclimatology, Palaeoecology 35, 145–214.

Ruffer, M.A., 1914. Pathological notes on the royal mummies of the Cairo Museum. Mitteilungen Geschichte Med. Naturwiss. 13, 239–240

Ryder, M.L., 1983. Sheep & man. London, UK, Duckworth.

Sadler, J.P, 1987. The analysis of insect remains from Norse sites in the former Western Settlement of Greenland. Unpublished. MSc Dissertation, University of Birmingham, Birmingham, UK.

Sadler, J.P., 1990. Records of ectoparasites on humans and sheep from Viking Age deposits in the former Western Settlement on Greenland. Journal of Medical Entomology 27, 628–631.

Sadler, J. P., & Buckland, P.C., 1998. The biogeography of the Coleoptera of the Orkney Isles of Britain. Proceedings of the 2nd Symposium, Fauna and Flora of the Atlantic Islands. Boletim do Museu Municipal do Funchal (História Natural) Supplement 5, 363–398.

Samuels, J. & Buckland, P.C., 1978. A Romano-British settlement at Sandtoft, South Humberside. Yorkshire Archaeological Journal 50, 65–75.

Sanders, C.J., 1964. The biology of carpenter ants in New Brunswick. Canadian Entomologist 96, 894–909.

Schaff, E., 1892. Über Insektenreste aus dem Torflager von Klinge [On insect remains from peat beds at Klinge]. Sitzungsberichte der Gesellschaft naturforschender Freunde zu Berlin (1892). [In German].

Schelvis, J., 1987. Some aspects of research on mites (Acari) in archaeological samples. Palaeohistoria 29, 211–218.

Schelvis, J., 1989. Mites (Acari) from the Late Neolithic well at Kolhorn (The Netherlands). Palaeohistoria 31, 165–171.

Schelvis, J., 1990a. Mites from medieval Oldeboorn. Experimental and Applied Entomology, Proceedings of the Netherlands Entomological Society 1, 96–97.

Schelvis, J., 1990b. The reconstruction of local environments on the basis of remains of oribatid mites (acari; Oribatida). Journal of Archaeological Science 17, 559–571.

Schelvis, J., 1997. Mites in the background: use and origin of remains of mites (Acari) in Quaternary deposits. In: Ashworth, A.C., Buckland, P.C., Sadler, J.P. (Eds), Studies in Quaternary entomology – an inordinate fondness for insects. Quaternary Proceedings 5, 233–236.

Schelvis, J., 1999. Remains of sheep ectoparasites as indicators of wool processing in the past. In: Dewilde, M., Ervynck, A.,

Wielemans, A. (Eds), Ieper en de middeleeuwse lakennijverheid in Vlaanderen. Archeologische en historische bijdragen [Ypres and the medieval cloth industry in Flanders. Archaeological and historical contributions]. Archeologie in Vlaanderen Monograph 2, 89–100. [In Dutch].
Schelvis, J. & Ervynck, A., 1992. Mijten (Acari) als ecologische indicatoren in de archeologie Onderzoek op de Romeinse vindplaats Oudenburg (prov. West-Vlaanderen) [Mites as ecological indicators in archaeology. A case-study in Roman Oudenberg (West Flanders)]. Archeologie in Vlaanderen 2, 175–189. [In Dutch].
Schelvis, J. & Koot, C., 1995. Sheep or goat? *Damalinia* deals with the dilemma. Experimental and Applied Entomology, Proceedings of the Netherlands Entomological Society 6, 161–162.
Schimmelman, A. & DeNiro, M.J., 1986. Stable isotopic studies on chitin II. The $^{13}C/^{12}C$ and $^{15}N/^{14}N$ ratios in arthropod chitin. Contributions in Marine Science 29, 113–130.
Schimmelmann, A. & Miller, R.F., 2002. Review of methods for the determination of deuterium/hydrogen stable isotope ratios in chitin. In: Muzzarelli, R.A., Muzzarelli, C. (Eds), Chitosan in pharmacy and chemistry. Italy, Atec Edizioni, 469–474.
Schimmelmann A., Miller, R.F., & Leavitt, S.W., 1993. Hydrogen isotopic exchange and stable isotope ratios in cellulose, wood, chitin, and amino compounds. In: Swart, P.K., Lohmann, K.C., McKenzie, J., Savin S. (Eds), Climate Change in Continental Isotopic Records, American Geophysical Union, Geophysical Monograph 78, 367–374.
Schlüter, T. & Dreyer, G., 1984. Subfossil darkling beetle remains from the Prehistoric site of the Presumable King Horus Aha at Abydos, Upper Egypt (Coleoptera: Tenebrionidae). Entomologia Generalis 10, 143–148. [In German with English Abstract].
Schoell, M., 1981. D/H-Isotopenverhältnisse in organischen Substanzen, Erdölen und Erdgasen. Forschungsbericht T 81–204 [D/H isotope ratios in organic substances, petroleum and natural gas]. Bundesanstalt für Geowissenschaften und Rohstoffe, Hannover, Germany, 1–79.
Schweger, C.E., Matthews, Jr, J.V., Hopkins, D.M., & Young, S.B., 1982. Paleoecology of Beringia: a synthesis. In: Hopkins, D.M., Matthews, Jr, J.V., Schweger, C.E., Young, S.B. (Eds), Paleoecology of Beringia. New York, Academic Press, 425–444.
Schweiger, H., 1967. Gebirgssysteme als Zentren der Artbildung [Mountain chains as centers of speciation]. Deutsch Entomologisches Zeitschrift 16, 159–174. [In German].
Schweigert, G., 2003. Alpenbock und Hirschkäfer im Pliozän von Willershausen [Mountain goats and stag beetles from the Pliocene of Willershausen]. Fossilien 20, 178–182. [In German].
Schwert, D.P., 1992. Faunal transitions in response to an ice age: the late Wisconsinan record of Coleoptera in the north-central United States. Coleopterists Bulletin 46, 68–94.
Schwert, D.P., 1996. Effect of Euro-American settlement on an insect fauna: Paleontological analysis of the recent chitin record of beetles (Coleoptera) from Northeastern Iowa. Annals of the Entomological Society of America 89, 53–63.
Schwert, D.P., Anderson, T.W., Morgan, A., Morgan, A.V., & Karrow, P., 1985. Changes in Late Quaternary vegetation and insect communities in southern Ontario. Quaternary Research 23, 205–226.
Schwert, D.P. & Ashworth, A.C., 1985. Fossil evidence for late Holocene faunal stability in southern Minnesota (Coleoptera). Coleopterists Bulletin 39, 67–79.
Schwert, D.P. & Ashworth, A.C., 1988. Late Quaternary history of the northern beetle fauna of North America: a synthesis of fossil distributional evidence. Memoirs of the Entomological Society of Canada 144, 93–107.
Schwert, D.P. & Bajc, A.F., 1989. Fossil Coleoptera associated with the Moorhead Low-Water Phase of Lake Agassiz in northwestern Ontario. Geological Society of America Annual Meeting, Program and Abstracts, 345.
Schwert, D.P. & Morgan, A.V., 1980. Paleoenvironmental implications of a lateglacial insect assemblage from northwestern New York. Quaternary Research 13, 93–110.
Schwert, D.P., Torpen-Kreft, H.J., & Hajic, E.R., 1997. Characterisation of the late-Wisconsinan tundra/forest transition in midcontinental North America using assemblages of beetle fossils. Quaternary Proceedings 5, 237–244.
Scudder, S.H., 1877. Description of two new species of Carabidae found in the interglacial deposits at Scarborough Hts., in Toronto, Canada. United States Geological Survey of the Territories (Hayden) B3, 763–764.
Scudder, S.H., 1890. The Tertiary insects of North America. Report to the United States Geological Survey of the Territories (Hayden) B3, 763–764.
Scudder, S.H., 1898. The Pleistocene beetles of Fort River, Massachusetts. Monographs of the United States Geological Survey 24, 740–746.
Scudder, S.H., 1900. Canadian fossil insects. Geological Survey of Canada, Publication No. 710, Contributions to Canadian Palaeontology 2 (Part 2), 67–92.
Seidenkrantz, M.-S., Bornmalm, L., Johnsen, S.J., Knudsen, K.L., Kuijpers, A., Lauritzen, S.-E., Leroy, S.A.G., Mergeai, I., Schweger, C., & van Vliet-Lanoe, B., 1996. Two-step deglaciation at the oxygen isotope stage 6/5E transition: the Zeifen-Kattegat climate oscillation. Quaternary Science Reviews 15, 63–75.
Selby, K.A., O'Brien, C.E., Brown, A.G., & Stuijs, I., 2005. A multi-proxy study of Holocene lake development, lake settlement and vegetation history in Central Ireland. Journal of Quaternary Science 20, 147–168.
Sessions, A.L., Sylva, S.P., Summons, R.E., & Hayes, J.M., 2004. Isotopic exchange of carbon-bound hydrogen over geologic timescales. Geochimica et Cosmochimica Acta 68, 1545–1559.
Shapiro, B., Drummond, A.J., Rambaut, A., Wilson, M.C., Matheus, P.E., Sher, A.V., Pybus, O.G., Gilbert, M.T., Barnes, I., Binladen, J., Willerslev, E., Hansen, A.J., Baryshnikov, G.F., Burns, J.A., Davydov, S., Driver, J.C., Froese, D.G., Harington, C.R., Keddie, G., Kosintsev, P., Kunz, M.L., Martin, L.D., Stephenson, R.O., Storer, J., Tedford, R., Zimov, S., & Cooper, A., 2004. Rise and fall of the Beringian steppe bison. Science 306, 1561–1565.
Shaw, J.W. & Shaw, M.C., 1995. Insects. In: Shaw, J.W., Shaw, M.C. (Eds), Kommos, an excavation on the south coast of Crete. I, part 1, Princeton, New Jersey, Princeton University Press, 277–278.
Sher, A. & Kuzmina, S., 2007. Beetle records: Late Pleistocene of Northern Asia. In: Elias, S.A. (Ed), Encyclopedia of quaternary science. Amsterdam, Elsevier, 246–267.
Sher, A.V., 1991. Problems of the last interglacial in Arctic Siberia. Quaternary International 10–12; 215–222.
Sher, A.V., Giterman, R.Ye., Zazhigan, V.S., & Kiselyov, S.V., 1977. New data on the Late Cenozoic sediments of the Kolyma lowland. Proceedings of the USSR Academy of Sciences, Geology Series 5, 69–83. [In Russian].
Sher, A.V., Kaplina, T.N., Giterman, R.E., Lozhkin, A., Arkhangelov, A.A., Kiselyov, S.V., Kouznetsov, Yu.V., Virina, E.I., & Zazhigin, V.S., 1979. Late Cenozoic of the Kolyma Lowland: XIV Pacific Science Congress, Tour Guide XI, Khabarovsk August 1979. Academy of Sciences of the USSR, Moscow, 1–116.
Sher, A.V., Kuzmina, S.A., Kuznetsova, T.V., & Sulerzhitsky, L.D., 2005. New insights into the Weichselian environment

and climate of the Eastern-Siberian Artic, derived from fossil insects, plants, and mammals. Quaternary Science Reviews 24, 533–569.

Short, S.K., 1985. Palynology of Holocene sediments, Colorado Front Range: vegetation and treeline changes in the subalpine forest. American Association of Stratigraphic Palynologists Contribution Series 16, 7–30.

Short, S.K. & Elias, S.A., 1987. New pollen and beetle analysis at the Mary Jane site, Colorado: evidence for Late-Glacial tundra conditions. Geological Society of America Bulletin 98, 540–548.

Short, S.K., Elias, S.A., Waythomas, C.F., Williams, N.E., & Lea, P.D., 1992. Holocene environments, Nushagak and Holitna lowlands, southwestern Alaska. Arctic 45, 381–392.

Shotton, F.W. & Coope, G.R., 1983. Exposures in the Power House Terrace of the River Stour at Wilden, Worcestershire, England. Proceedings of the Geological Association 94, 33–44.

Shotton, F.W., Keen, D.H., Coope, G.R., Currant, A.P., Gibbard, P.L., Pegler, S.M., & Robinson, T.E., 1993. The Middle Pleistocene deposits of Waverley Wood Pit, Warwickshire, England. Journal of Quaternary Science 8, 293–325.

Shotton, F.W. & Osborne, P.J., 1965. The fauna of the Hoxnian Interglacial deposits of Nechells, Birmingham. Philosophical Transactions of the Royal Society of London, Series B 248, 353–378.

Shotton, F.W., Osborne, P.J., & Greig, J.R.A., 1977. The fossil content of a Flandrian deposit at Alcester. Proceedings of the Coventry and District Natural History and Scientific Society 5, 19–32.

Shotton, F.W., Sutcliffe, A.J., & West, R.G., 1962. The fauna and flora from the brick pit at Lexden, Essex. Essex Naturalist 31, 15–22.

Shotton, F.W., Williams, R.E.G., & Johnson, A.S., 1974. Birmingham University radiocarbon dates VIII. Radiocarbon 16, 285–303.

Shulmeister, J., Barrett, P., & Marra, M., 2000. Modern and ancient sedimentary environments of Southern Wairarapa. Field trip guide GSNZ conference. Geological Society of New Zealand Miscellaneous Publication 108B, 31–47.

Shulmeister, J., Soon, J.M., Berger, G.W., Harper, M., Holt, S., Moar, N., & Carter, J., 1999. Environmental and sea-level changes on Banks Peninsula (Canterbury New Zealand) through three glaciation-interglaciation cycles. Palaeogeography, Palaeoclimatology, Palaeoecology 152, 101–127.

Simmons, I.G., 1996. The environmental impact of later Mesolithic cultures. Edinburgh, UK, Edinburgh University Press.

Simpson, T., 2001. The Roman well at Piddington, Northamptonshire, England: an investigation of the Coleopterous fauna. Environmental Archaeology 6, 91–96.

Sinka, K.J., 1993. Developing the mutual climatic range method of palaeoclimate reconstruction. Unpublished PhD Thesis, University of East Anglia, Norwich, UK.

Skidmore, P., 1986. The dipterous remains. In: Stead, I.M., Bourke, J.B., Brothwell, D. (Eds), Lindow man. The body in the bog. London, UK, British Museum, 92

Skidmore, P., 1994. Analysis of fly remains from F23 in the stone-lined pit F22. In: McCormick, F. (Ed), Excavations at Pluscarden Priory, Moray. Proceedings of the Society of Antiquaries of Scotland 124, 418–420.

Skidmore, P., 1996. A dipterological perspective on the Holocene history of the North Atlantic area. Unpublished PhD Dissertation, University of Sheffield, Sheffield, UK.

Smetana, A., 1985. Revision of the subfamily Helophorinae of the Nearctic region (Coleoptera: Hydrophilidae). Memoirs of the Entomological Society of Canada 131, 153.

Smith, B., 1981. Insect remains. In: Neil, N.R.J. (Eds), A Bronze Age burial at Holland, St. Ola, Orkney. Glasgow Archaeological Journal 8, 45.

Smith, C., 1978. The landscape and natural history of Iron Age settlement on the Trent Gravels. In: Cunliffe, B.W., Rowley, T. (Eds), Lowland Iron Age Communities in Europe. British Archaeological Reports S48, 91–101.

Smith, D., 1989. Insect remains from site DSR. In: Buckland, P.C., Magilton, J.R., Hayfield, C. (Eds), The archaeology of Doncaster 2: the medieval and later town. British Archaeological Reports 202, 447–450.

Smith, D., Osborne, P., & Barratt, J., 2000. Beetles and evidence of past environments at Goldcliff. In: Bell, M., Caseldine, C., Neumann, H. (Eds), Prehistoric intertidal archaeology in the Welsh Severn Estuary. Council for British Archaeology Report 120, London, 245–260.

Smith, D.N., 1996. Thatch, turves and floor deposits: a survey of Coleoptera in materials from abandoned Hebridean blackhouses and the implications for their visibility in the archaeological record. Journal of Archaeological Science 23, 161–174.

Smith, D.N., 1999. Analysis of beetles in smoke blackened thatch. In Letts, J. B. (Ed), Smoke blackened thatch; a unique source of late medieval plant remains from southern England. English Heritage and Reading University, 42–43.

Smith, D.N., 2000a. Beetles. In: Rippon, S. (Ed), Romano-British exploitation of coastal wetlands: North Somerset Levels. Britannia 31, 112–115.

Smith, D.N., 2000b. Disappearance of elmid ''riffle beetles'' from lowland river systems in Britain-the impact of alluviation. In: Nicholson, R.A., O'Connor, T.P. (Eds), People as an agent of environmental change. Oxford, Oxbow Books, 75–80.

Smith, D.N., 2000c. Detecting the nature of materials on farms from Coleoptera: a number of taphonomic problems. In: Huntley, J.P., Stallibrass, S. (Eds), Taphonomy and interpretation. Oxford, Oxbow Books, 71–83.

Smith, D.N., Gaffney, V., Grossman, D., Howard, A.J., Milosevic, A., Ostir, K., Todobnikar, T., Smith, W., Tetlow, E., Tingle, M., & Tinsley, H., 2006. Assessing the later prehistoric environmental archaeology and landscape of the Cetina valley, Croatia. Environmental Archaeology 11, 171–187.

Smith, D.N. & Howard, A.J., 2004. Identifying changing fluvial conditions in low gradient alluvial archaeological landscapes: can Coleoptera provide insights into changing discharge rates and floodplain evolution? Journal of Archaeological Science 31, 109–120.

Smith, D.N., Letts, J., & Jones, M., 2005a. The insects from non-cereal stalk smoked blackened thatch. Environmental Archaeology 10, 171–178.

Smith, D. N., Osborne, P. J., & Barrett, J., 1997. Preliminary palaeoentomological research at the Iron Age sites at Goldcliff, Gwent, Wales, 1991–1993. In: Ashworth, A.C., Buckland, P.C., Sadler, J.P. (Eds), Studies in Quaternary entomology – an inordinate fondness for insects. Quaternary Proceedings 5, 255–268.

Smith, D.N., Roseff, R., Bevan, L., Brown, A.G., Butler, S., Hughes, G., & Monckton, A., 2005b. Archaeological and environmental investigations of a Lateglacial and Holocene River Valley sequence on the River Soar, at Croft, Leicestershire. The Holocene 15, 353–377.

Smith, D.N., Roseff, R., & Butler, S., 2001. The sediments, pollen, plant macro-fossils and insects from a Bronze age channel fill at Yoxall Bridge, Staffordshire. Environmental Archaeology 6, 1–12.

Smith, D. N. & Whitehouse, N. J., 2005. Not seeing the woods for the trees; a palaeoenvironmental perspective on woodland decline. In: Brickley, M., Smith, W., Smith, D. N. (Eds), The Fertile Ground. Papers in Honour of Susan Limbrey. Oxbow Press, Oxford, 136–161.

Smith, G.F. & Klopper, R.R., 2002. A southern perspective on the All Species Project and the Global Taxonomic Imperative. Taxon 51, 359–361.

Smith, K.G.V., 1986. A manual of forensic entomology. Trustees of the British Museum (Natural History), London.

Solomon, M.E., 1965. Archaeological records of storage pests: *Sitophilus granarius* (L.) (Coleoptera, Curculionidae) from an Egyptian Pyramid Tomb. Journal of Stored Product Research 1, 105–107.

Soons, J.M. & Burrows, C.J., 1978. Dates for Otiran deposits, including plant microfossils and macrofossils, from Rakaia Valley. New Zealand Journal of Geology and Geophysics 21, 607–615.

Sota, T. & Hayashi, M., 2007. Comparative historical biogeography of *Plateumaris* leaf beetles (Coleoptera: Chrysomelidae) in Japan: interplay between fossil and molecular data. Journal of Biogeography 34, 977–993.

Sota, T., Hayashi, M., & Yagi, T., 2007. Geographic variation in body and ovipositor sizes in the leaf beetle *Plateumaris constricticollis* (Coleoptera: Chrysomelidae) and its association with climatic conditions and host plants. European Journal of Entomology 104, 165–172.

Speight, R., Wall, A., & Laing, R.M., 1927. The natural history of Canterbury. Christchurch, New Zealand, Simpson and Williams.

Stanford, D., Wedel, W.R., & Scott, G.R., 1982. Archaeological investigations of the Lamb Spring site. Southwestern Lore 47, 14–27.

Stafford, F., 1971. Insects of a medieval burial. Science and Archaeology 7, 6–10.

Stanley, S.M., 1979. Macroevolution: pattern and process. San Francisco, Freeman.

Stanley, S.M., 1985. Rates of evolution. Paleobiology 11, 13–26.

Stewart, J.H., 1938. Basin-plateau aboriginal sociopolitical groups. Bureau of American Ethnology Bulletin 33, 3.

Stiger, M.A., 1977. Anasazi diet: the coprolite evidence. Unpublished Masters Thesis, Anthropology Department, University of Colorado, Boulder, CO, 88.

Straker, V., Robinson, M., & Robinson, E., 1984. Biological investigations of waterlogged deposits in the Roman fortress ditch at Exeter. Devon Archaeological Society Proceedings 42, 59–69.

Stringer, C., 2005. The Ancient Human Occupation of Britain (AHOB) Project. Transactions of the Leicester Literary and Philosophical Society 99, 29–32.

Strobel, P. & Pigorni, L., 1864. Le terremare e le palafitte del Parmense, seconda relazione [The 'terremare' and pile-dwellings in the Parma region, second report]. Atti della Societa italiana di Scienze Naturali, Milan 7, 36–37. [In Italian].

Strong, L., 1981. Dermestids-an embalmer's dilemma. Antenna 5, 136–139.

Su, Z.-H., Tominaga, O., Okamoto, M., & Osawa, S., 1998. Origin and diversification of hindwingless *Damaster* ground beetles within the Japanese islands as deduced from mitochondrial ND5 gene sequences. (Coleoptera, Carabidae). Molecular Biology and Evolution 18, 1026–1039.

Sudell, T. L., 1990. The insects from well 1. In: Wrathmell, S., Nicholson, A. (Eds), Dalton Parlours: Iron Age settlement and Roman villa site. Yorkshire Archaeology 3. West Yorkshire Archaeological Service, Wakefield, 267–271.

Suggate, R.P., Stevens, G.R., & Te Punga, M.T., 1978. The Geology of New Zealand. Wellington, New Zealand, Government Printing Office.

Sveinbjarnardottir, G., 1983. Palaeoekologiske undersogelser pa Holt i Eyjafjallasveit, Sydisland [Paleoecological investigations from Holt in Eyjafjallasveit, southern Iceland]. In: Olafsson, G. (Ed), Hus, Gard och Bebyggelse. Föredrag fran det XVI nordiska arkeologmötet, Island, 1982. 4, 241–250. [In Icelandic].

Sveinbjarnardóttir, G. & Buckland, P.C., 1983. An uninvited guest. Antiquity 57, 127–130.

Sveinbjarnardóttir, G., Buckland, P.C., Gerrard, A. J., Greig, J. R. A., Perry, D., Savory, D., & Snæsdóttir, M., 1981. Excavations at Stóraborg: a palaeoecological approach. Arbók hins Islenzka Fornleifarfélags (1980), 113–129.

Swetnam, T.W. & Betancourt, J.L., 1990. Fire-southern oscillation relations in the southwestern United States. Science 249, 1017–1020.

Tado Organizations Research Group., 1998. Paleoenvironments of the Plio-Pleistocene Tokai Group in the northern part of Mie Prefecture, Central Japan. Earth Science (Japan) 52, 115–135. [in Japanese].

Taylor, B.J. & Coope, G.R., 1985. Arthropods in the Quaternary of East Anglia-their role as indices of local paleoenvironments and regional paleoclimates. Modern Geology 9, 159–185.

Ters, M., Azema, C., Brebion, P., Churcher, C.S., Delibrias, G., Denefle, M., Guyader, J., Lauriet, A., Mathieu, R., Michel, J.P., Osborne, P.J., Rouvillois, A., & Shotton, F.W., 1971. Sur le remblaiement holocène dans l'estuaire de la Seine, au Havre (Seine Maritime), France [On the Holocene backfilling of the Seine estuary at Le Havre, France]. Quaternaria 14, 151–174. [In French].

Thiele, H.U., 1977. Carabid beetles in their environments. New York, Springer.

Thomsen, P.F., Elias, S., Gilbert, M.T.P., Haile, J., Munch, K., Kuzmina, S., Froese, D.G., Sher, A., Holdaway, R.N., & Willerslev, E., 2009. Non-destructive sampling of ancient insect DNA. PLoS ONE 4(4): e5048. doi:10.1371/journal.pone.0005048.

Thompson, R.S., Shafer, S.L., Strickland, L.E., Van de Water, P.K., & Anderson, K.H., 2004. Quaternary vegetation and climate change in the western United States: developments, perspectives, and prospects. In: Gillespie, A.R., Porter, S.C., Atwater, B.F. (Eds), The Quaternary period in the United States. New York, Elsevier, 403–426.

Thompson, W.G. & Goldstein, S.L., 2006. A radiometric calibration of the SPECMAP timescale. Quaternary Science Reviews 25, 3207–3215.

Tobolski, K., 1988. Palaeobotanical study of Bølling sediments of Zabinko in the vicinity of Poznan, Poland. Quaestiones Geographicae 10, 119–124.

Tominaga, O., Su, Z.-H., Kim, C.-G., Okamoto, M., Imura, Y., & Osawa, S., 2000. Formation of the Japanese Carabina fauna inferred from a phylogenetic tree of mitochondrial ND5 gene sequences (Coleoptera, Carabidae). Journal of Molecular Evolution 50, 541–549.

Tomlinson, P., Allison, E.P., Innes, J.B., & Kenward, H.K., 2002. Environmental evidence from the prehistoric period. In: Freke, D. (Ed), Excavations on St Patrick's Isle Peel, Isle of Man, 1982–88. Prehistoric, Viking, medieval and later. Centre for Manx Studies Monograph 3, Liverpool University Press, 231–247.

Totten, S.M., 1971. The occurrence of beetle remains in Pleistocene deposits, east-central United States. Geological Society of America, Annual Meeting, Washington, DC, Abstracts with Program 3, 733–734.

Tyers, I., 1988. Environmental evidence from Southwark and Lambeth. In: Hinton, P. (Ed), Excavations in Southwark 1973–76, Lambeth 1973–79. Joint Publication 3, London & Middlesex Archaeological Society and Surrey Archaeological Society, 443–477.

Tyers, I., 2003. Insects from selected samples. In: Masefield, R., Branch, N., Couldrey, P., Goodburn, D., Tyers, I. (Eds),

A Later Bronze Age well complex at Swalecliffe, Kent. Antiquaries Journal 83, 87–88.

Ullrich, W.G. & Coope, G.R., 1974. Occurrence of the East Palaearctic beetle *Tachinus jacuticus* Poppius (Col. Staphylinidae) in deposits of the Last glacial period in England. Journal of Entomology B42, 207–212.

Uutala, A.J., 1990. *Chaoborus* (Diptera: Chaoboridae) mandibles-paleolimnological indicators of the historical status of fish populations in acid-sensitive lakes. Journal of Paleolimnology 4, 139–151.

Valamoti, S.M. & Buckland, P.C., 1995. An early find of *Oryzaephilus surinamensis* L. (Coleoptera: Silvanidae) from Final Neolithic Mandalo, Macedonia, Greece. Journal of Stored Product Research 31, 307–309.

Van Devender, T.R., 1990a. Late Quaternary vegetation and climate of the Chihuahuan Desert, United States and Mexico. In: Betancourt, J.L., Van Devender, T.R., Martin, P.S. (Eds), Packrat Middens, the last 40,000 years of biotic change. Tucson, AZ, University of Arizona Press, 104–133.

Van Devender, T.R., 1990b. Late Quaternary vegetation and climate of the Sonoran Desert, United States and Mexico. In: Betancourt, J.L., Van Devender, T.R., Martin, P.S. (Eds), Packrat Middens, the last 40,000 years of biotic change. Tucson, AZ, University of Arizona Press, 134–165.

Van Dyke, E.C., 1939. The origin and distribution of the coleopterous insect fauna of North America. Proceedings of the Sixth Pacific Science Congress 4, 255–268.

van Geel, B., Buurman, J., Brinkkemper, O., Schelvis, J., Aptroot, A., Van Reenen, G., & Hakbijl, T., 2003. Environmental reconstruction of a Roman period settlement site in Uitgeest (The Netherlands), with special reference to coprophilous fungi. Journal of Archaeological Science 30, 873–883.

van Geel, B., Coope, G.R., & Van der Hammen, T., 1989. Palaeoecology and stratigraphy of the lateglacial type section at Usselo (The Netherlands). Review of Palaeobotany and Palynology 60, 25–129.

van Hardenbroek, M.R., Gröcke, D.R., & Elias, S.A., 2009. Isotopes in water beetles: $\delta^{18}O$ and δD through North America. Earth and Planetary Science Letters (submitted).

van Wijngaarden-Bakker, L.H. & Pals, J.P., 1981. Life and work in Smeerenburg: the bio-archaeological aspects. Proceedings of the International Symposium, Early Exploitation of the North Atlantic, 800–1700. University of Groningen, The Netherlands, 133–151.

Varley, G.C., 1980. Insect remains. In: Hassall, T.G. (Ed), Excavations at Oxford Castle 1965–1973. Oxoniensia 41, 308.

Veevers, J.J., 2006. Updated Gondwana (Permian–Cretaceous) earth history of Australia. Gondwana Research 9, 231–260.

Velichko, A.A. & Faustova, M.A., 1986. Glaciations in the east European region of the USSR. Quaternary Science Reviews 5, 447–461.

Vickers, K., Bending, J., Buckland, P.C., Edwards, K.J., Stummann Hansen, S., & Cook, G., 2005. Toftanes: The paleoecology of a Faroese Landnám farm. Human Ecology 33, 685–710.

Von Knorre, D., 1978. Fossile Insektenreste aus letzt interglazialen Travertinen Thüringens [Fossil insect remains from last interglacial travertine deposits in Thuringia]. In: Kahlke, H.D. (Ed), Das Pleistozän von Burgtonna in Thüringen [The Pleistocene of Burgtonna in Thüringen]. Quartärpalaeontologie 3, 93–95; 340–343. [In German].

Vuilleumier, B.S., 1971. Pleistocene changes in the fauna and flora of South America. Science 173, 771–780.

Waage, J.K., 1976. Insect remains from ground sloth dung. Journal of Paleontology 50, 991–994.

Walker, I.R., 1987. Chironomidae (Diptera) in paleoecology. Quaternary Science Reviews 6, 29–40.

Walker, I.R., 1995. Chironomids as indicators of past environmental change. In: Armitage, P.D., Cranston, P.S., Pinder, L.C.V. (Eds), The Chironomidae: biology and ecology of nonbiting midges. London, UK, Chapman & Hall, 405–422.

Walker, I.R., 2001. Midges: Chironomidae and related Diptera. In: Smol, J.P., Birks, H.J.B., Last, W.M. (Eds), Tracking environmental change using lake sediments, Vol. 4, Zoological indicators, Dordrecht, The Netherlands, Kluwer, 43–66.

Walker, I.R., 2007. Chironomid overview. In: Elias, S.A. (Ed), Encyclopedia of Quaternary science. Amsterdam, Elsevier, 360–366.

Walker, I.R. & Mathewes, R.W., 1987. Chironomidae (Diptera) and postglacial climate at Marion Lake, British Columbia, Canada. Quaternary Research 27, 89–102.

Walker, I.R., Mott, R.J., & Smol, J.P., 1991a. Alleröd-Younger Dryas Lake Temperatures from Midge Fossils in Atlantic Canada. Science 253, 1010–1012.

Walker, I.R., Smol, J.P., Engstrom, D.R., & Birks, H.J.B., 1991b. An assessment of Chironomidae as quantitative indicators of past climatic change. Canadian Journal of Fisheries and Aquatic Science 48, 975.

Walker, M.J.C., Bryant, C., Coope, G.R., Harkness, D.D., Lowe, J.J., & Scott, E.M., 2001. Towards a radiocarbon chronology of the Late-Glacial: sample selection strategies. Radiocarbon 43, 1007–1019.

Walker, M.J.C., Coope, G.R., & Lowe, J.J., 1993. The Devensian (Weichselian) Lateglacial palaeoenvironmental record from Gransmoor, East Yorkshire, England. Quaternary Science Reviews 12, 659–680.

Walkling, A., 1997. Käferkundliche Untersuchungen an wechselzeitlichen Ablagerungen der Borrung Groß Todtshorn (Kr. Harburg; Niedersachsen) [Beetle investigations of sampling residues from the bore hole at Groß Todtshorn]. Schriftenreihe der Deutschen Geologischen Gesellschaft 4, 145–159. [In German].

Walkling, A.P. & Coope, G.R., 1996. Climatic reconstructions from the Eemian/Early Weichselian transition in Central Europe based on the coleopteran record from Gröbern, Germany. Boreas 25, 145–159.

Walter, R.C., 1989. Application and limitation of fission-track geochronology to Quaternary tephras. Quaternary International 1, 35–46.

Wardle, P., 1988. Effects of glacial climates on floristic distribution in New Zealand 1. A review of the evidence. New Zealand Journal of Botany 26, 541–555.

Warner, B.G., Karrow, P.F., Morgan, A.V., & Morgan, A., 1987. Plant and insect fossils from Nipissing sediments along the Goulais River, southeastern Lake Superior. Canadian Journal of Earth Sciences 24, 1526–1536.

Warner, B.G., Morgan, A.V., & Karrow, P.F., 1988. A Wisconsinan interstadial arctic flora and insect fauna from Clarksburg, southwestern Ontario, Canada. Palaeogeography, Palaeoclimatology, Palaeoecology 68, 27–47.

Warsop, C. & Skidmore, P., 1998. Insect remains from the rock-cut pit. In: Ewart, G., Baker, F. (Eds), Carrick Castle: symbol and source of Campbell power in south Argyll from the 14th to the 17th century. Proceedings of the Society of Antiquaries of Scotland 128, 987–992.

Wastegård, S., Aronsson, M., Hedenas, L., Johansson, K., Lemdahl, G., & Ransed, G., 1995. Lateglacial and Early Holocene stratigraphy at Ranstad, Västergötland, S Sweden. In: Robertsson, A.M., Hackens, T., Hicks, S., Risberg, J., Äkerlund, A. (Eds), Landscape and Life. Studies in honour of Urve Miller. PACT 50, 97–108.

Watts, W.A., 1979. Late Quaternary vegetation of central Appalachia and the New Jersey Coastal Plain. Ecological Monographs 49, 427–469.

Waythomas, C.F. 1990. Quaternary geology and late-Quaternary environments of the Holitna lowland, and Chuilnuk-Kioluk Mountains region, interior southwestern Alaska. PhD Dissertation, University of Colorado, Boulder, CO.

Waythomas, C.F., Lead, P.D., & Walter, R.C., 1993. Stratigraphic context of Old Crow tephra, Holitna lowland, interior southwest Alaska. Quaternary Research 40, 20–29.

Waythomas, C.F., Short, S.K., & Elias, S.A., 1989. Deposits of interglacial character, Nenana Valley, Alaska. Current Research in the Pleistocene 6, 95–98.

Welsh, S.L., 1998. *Flora societensis*: a summary revision of the flowering plants of the Society Islands. Orem, UT, E.P.S. Inc., 420.

Werner, R.A., Raffa, K.F., & Illman, B.L., 2006. Dynamics of phytophagous insects and their pathogens in Alaskan boreal forests. In: Chapin, F.S., Oswood, M.W., Van Cleve, K., Viereck, L.A., Verbyla, D.L. (Eds), Alaska's changing boreal forest. Oxford, Oxford University Press, 133–146.

Wesenberg-Lund, C., 1896. Om Ferskvandsfaunaens Kitin- og Kisellemninger i Torvelagene [Concerning chitinous- and silicon remains of the freshwater fauna in peat layers]. Meddelelser fra Dansk geologisk Forening 3, 51–84. [In Danish].

Westgate, J.A., Preece, S.J., & Péwé, J.A., 2003. The Dawson Cut Forest Bed in the Fairbanks area, Alaska, is about two million years old. Quaternary Research 60, 2–8.

White, J.P. & Flannery, T., 1995. Late Pleistocene fauna at Spring Creek, Victoria: a re-evaluation. Australian Archaeology 40, 13–17.

White, J.M., Ager, T.A., Adam, D.P., Leopold, E.B., Liu, G., Jetté, H., & Schweger, C.E., 1997. An 18 million year record of vegetation and climate change in northwestern Canada and Alaska: Tectonic and global climatic correlates. Palaeogeography, Palaeoclimatology, Palaeoecology 130, 293–306.

White, J.M., Ager, T.A., Adam, D.P., Leopold, E.B., Liu, G., Jetté, H., & Schweger, C.E., 1999. Neogene and Quaternary quantitative palynostratigraphy and paleoclimatology from sections in Yukon and adjacent northwest territories and Alaska. Geological Survey of Canada Bulletin 543, 1–30.

White, R.E., 1983. A field guide to the beetles of North America. Boston, MA, Houghton Mifflin.

Whitehead, P.F., 1989. Changing environments and Coleoptera faunas from Aston Mill, Worcestershire, England. Entomologist's Monthly Magazine 125, 187–197.

Whitehead, P.F., 1992. Terraces of the River Avon at Twyning, Gloucestershire: their stratigraphy, climate and biota. Quaternary Newsletter 67, 3–24.

Whitehouse, N.J., 1997. Insect faunas associated with *Pinus sylvestris* L. From the mid-Holocene of the Humberhead Levels, Yorkshire, U.K. Quaternary Proceedings 5, 293–303.

Whitehouse, N.J., 2000. Forest fires and insects: palaeoentomological research from a sub-fossil burnt forest. Palaeogeography, Palaeoclimatology, Palaeoecology 164, 247–262.

Whitehouse, N.J., 2004. Mire ontogeny and climate change inferred from fossil beetle assemblages at Hatfield Moors, eastern England. The Holocene 14, 79–93.

Whitehouse, N.J., 2006. The Holocene British and Irish ancient forest fossil beetle fauna: implications for forest history, biodiversity and faunal colonisation. Quaternary Science Reviews 25, 1755–1789.

Whitehouse, N.J., 2007. The study of fossil insect remains in environmental and archaeological investigations: an Irish perspective. In: Murphy, E.M., Whitehouse, N.J. (Eds), Environmental archaeology in Ireland. Oxford, Oxbow Books, 136–163.

Whitehouse, N.J. & Smith, D.N., 2004. "Islands" in Holocene forests: implications for forest openness, landscape clearance and "culture-steppe" species. Environmental Archaeology 9, 203–212.

Whittington, G., Buckland, P.C., Edwards, K.J., Greenwood, M.T., Hall, A.M., & Robinson, M., 2003. Multiproxy Devensian Late-glacial and Holocene environmental records at an Atlantic coastal site in Shetland. Journal of Quaternary Science 18, 151–168.

Wickham, H.F., 1917. Some fossil beetles from the Sangamon peat, Champaign Co., Ill. American Journal of Science (4th. Ser.) 44, 137–145.

Wickham, H.F., 1919. Fossil beetles from Vero, Florida. American Journal of Science (4th Ser.) 47, 355–357.

Wiggins, G.B., 1977. Larvae of the North American caddisfly genera (Trichoptera). Toronto, Canada, University of Toronto Press.

Willerslev, E., Cappellini, E., Boomsma, W., Nielsen, R., Hebsgaard, M.B., Brand, T.B., Hofreiter, M., Bunce, M., Poinar, H.N., Dahl-Jensen, D., Johnsen, S., Steffensen, J.P., Bennike, O., Schwenninger, J.L., Nathan, R., Armitage, S., de Hoog, C.J., Alfimov, V., Christl, M., Beer, J., Muscheler, R., Barker, J., Sharp, M., Penkman, K.E.H., Haile, J., Taberlet, P., Gilbert, M.T.P., Casoli, A., Campani, E., & Collins, M.J., 2007. Ancient biomolecules from deep ice cores reveal a forested southern Greenland. Science 317, 111–114.

Williams, C.T., 1985. Mesolithic exploitation patterns in the central Pennines: a palynological study of Soyland Moor. British Archaeological Reports, 139.

Williams, N.E., 1988. The use of caddisflies (Trichoptera) in paleoecology. Palaeogeography, Palaeoclimatology, Palaeoecology 62, 493–500.

Williams, N.E. & Morgan, A.V., 1977. Fossil caddisflies (Insecta: Trichoptera) from the Don Formation, Toronto, Ontario, and their use in paleoecology. Canadian Journal of Zoology 55, 519–527.

Williams, N.E., Westgate, J.A., Williams, D.D., & Morgan, A.V., 1981. Invertebrate fossils (Insecta: Trichoptera, Diptera, Coleoptera) from the Pleistocene Scarborough Formation at Toronto, Ontario and their paleoenvironmental significance. Quaternary Research 16, 146–166.

Wilmshurst, J.M., Higham, T.F.G., Allen, H., Johns, D., & Phillips, C., 2004. Early Maori settlement impacts in northern coastal Taranaki, New Zealand. New Zealand Journal of Ecology 28, 167–179.

Wilson, D.E. & Reeder, D.M. (Eds), 1993. Mammal species of the World. Washington, DC, Smithsonian Institution Press.

Wilson, E.C., 1986. Type specimens of fossil Invertebrata in the Natural History Museum of Los Angeles County: Supplement II. Natural History Museum of Los Angeles County, Technical Report 1, 1–150.

Wilson, M.J. & Elias, S.A., 1986. Paleoecological significance of Holocene insect fossil assemblages from the north coast of alaska. Arctic 39, 150–157.

Wilson, M.V.H., 1983. Is there a characteristic rate of radiation for the insects? Paleobiology 9, 79–85.

Wollaston, T.V., 1863. Note on the remains of Coleoptera from the peat of Lexden Brick-pit. Quarterly Journal of the Geological Society 19, 400–401.

Woodroffe, G.E., 1953. An ecological study of the insects and mites in the nests of certain birds in Britain. Bulletin of Entomological Research 44, 739–772.

Wooller, M.J., Francis, D., Fogel, M.L., Miller, G.H., Walker, I.R., & Wolfe, A.P., 2003. Quantitative paleotemperature estimates from $\delta^{18}O$ in chironomid head capsules from arctic lake sediment. Sixteenth International Quaternary Union Congress, Programs with Abstracts, 158.

Worsley, P., 1991. Possible early Devensian glacial deposits in the British Isles. In: Ehlers, J., Gibbard, P.L., Rose, J. (Eds), Glacial deposits in Great Britain and Ireland. Rotterdam, The Netherlands, Balkema, 47–52.

Worsley, P., Coope, G.R., Good, T.R., Holyoak, D.T., & Robinson, J.E., 1983. A Pleistocene succession from beneath Chelford Sands at Oakwood Quarry, Chelford, Cheshire. Geological Journal 18, 307–324.

Yansa, C.H. & Ashworth, A.C., 2005. Late Pleistocene palaeoenvironments of the Southern Lake Agassiz Basin, USA. Journal of Quaternary Science 20, 255–267.

Young, F.N., 1959. Fossil beetles from the Vero Pleistocene. Coleopterists Bulletin 13, 103–106.

Yvinec, J.-H., 1997. Infestation par les insectes d'un grenier carbonisé de la fin du IIe siècle A.D. à Amiens (Somme) [Insect infestation of charred grain from the end of the second century AD at Amiens, Somme]. L'Entomologiste (Paris) 53, 113–128. [In French].

Zazula, G.D., Froese, D.G., Elias, S.A., Kuzmina, S., La Farge, C., Reyes, A.V., Sanborn, P.T., Schweger, C.E., Smith, C.A.S., & Mathewes, R.W., 2006. Vegetation buried under Dawson tephra (25,000 yr ^{14}C BP) and locally diverse late Pleistocene paleoenvironments of Goldbottom Creek, Yukon, Canada. Palaeogeography, Palaeoclimatology, Palaeoecology 242; 253–286.

Zazula, G.D., Froese, D.G., Elias, S.A., Kuzmina, S., & Mathewes, R.W., 2007. Arctic ground squirrels of the mammoth-steppe: paleoecology of late Pleistocene middens ($\sim$24 000–29 450 ^{14}C yr BP), Yukon Territory, Canada. Quaternary Science Reviews 26, 979–1003.

Zazula, G.D., Froese, D.G., Telka, A.M., Mathewes, R.W., & Westgate, J.A., 2002. Plants, bugs, and a giant mammoth tusk: paleoecology of Last Chance Creek, Yukon Territory. Yukon Exploration and Geology 2002, 251–258.

Zias, J. & Mumcuoglu, Y., 1991. Pre-Pottery Neolithic B head lice from Nahal Hemar Cave. Atiqot 20, 167–168.

Zinovjev, E.V., 2006. Problems of ecological interpretation of Quaternary insect faunas from the central part of Northern Eurasia. Quaternary Science Reviews 25, 1821–1840.

Zinovjev, E.V., 2008. A history of ground-beetle faunas of West Siberia and the Urals during the Late Pleistocene to Holocene. In: Penev, L., Erwin, T., Assmann, T. (Eds), Back to the roots of the future. Towards a new synthesis amongst taxonomic, ecological and biogeographical approaches in carabidology. Proceedings of the XIII European Carabidologists Meeting, Blagoevgrad, August 20–24, 2007. Pensoft Publishers, Sofia, 241–254.

Taxonomic Index

Topic Index